Lecture Notes in Applied and Computational Mechanics

Volume 85

Series editors

Peter Wriggers, Leibniz Universität Hannover, Hannover, Germany
e-mail: wriggers@ikm.uni-hannover.de
Peter Eberhard, University of Stuttgart, Stuttgart, Germany
e-mail: peter.eberhard@itm.uni-stuttgart.de

About this Series

This series aims to report new developments in applied and computational mechanics—quickly, informally and at a high level. This includes the fields of fluid, solid and structural mechanics, dynamics and control, and related disciplines. The applied methods can be of analytical, numerical and computational nature.

More information about this series at http://www.springer.com/series/4623

Paolo Vannucci

Anisotropic Elasticity

Springer

Paolo Vannucci
Laboratoire de Mathématiques de Versailles
Université de Versailles Saint-Quentin-en-Yvelines
Versailles
France

ISSN 1613-7736 ISSN 1860-0816 (electronic)
Lecture Notes in Applied and Computational Mechanics
ISBN 978-981-10-5438-9 ISBN 978-981-10-5439-6 (eBook)
DOI 10.1007/978-981-10-5439-6

Library of Congress Control Number: 2017944298

Printed on acid-free paper

This Springer imprint is published by Springer Nature
The registered company is Springer Nature Singapore Pte Ltd.
The registered company address is: 152 Beach Road, #21-01/04 Gateway East, Singapore 189721, Singapore

To the beloved memory of my mother,
Maria Pia

Preface

This book was born as the class notes of the course *Anisotropic structures: theory and design* that I hold at Ecole Polytechnique for the students of Master *M4S—Multiphysics and Multiscale Modeling of Materials and Structures*, a specialization curriculum of the second year of *Master in Mechanics* of the University Paris-Saclay, a federation of Universities in the Paris region.

As such, this book, intended as an introduction to anisotropic elasticity, is specially addressed to graduate students of courses in theoretical mechanics and engineering, as well as to Ph.D. students and researchers in mechanics. The book is composed of some classical parts and some more advanced topics.

Chapter 1 is a rapid introduction to the matter: the mathematical consequences of anisotropy, the true topic of this book, are briefly introduced along with some typical anisotropic phenomena. Also, a quick nod to anisotropy classifications is given. Some basic elements of elasticity of crystals are proposed too and the chapter ends with a recall of some fundamental equations of elasticity, needed in the following.

Chapter 2 concerns general anisotropic elasticity, that is, the three-dimensional elasticity of anisotropic bodies. Starting from the classical Hooke's law for anisotropic materials, the subject is treated in depth: the classical Voigt's and Kelvin's notations are introduced and the choice of Kelvin's one is argued. After having explained the mechanical meaning of the anisotropic elastic constants, the elastic symmetries are treated extensively. Then, the technical constants are introduced and the problem of determining the bounds on the elastic and technical constants is tackled. This chapter ends with the determination of the symmetry planes and with the notion of curvilinear anisotropy. Finally, a list of anisotropic materials is given, with an original way of plotting the technical constants.

Chapter 3 is devoted to a classical subject: plane anisotropic elasticity. The classical cases of plane strain, plane stress and generalized plane stress are developed and compared. Then, the Lekhnitskii theory is treated in detail and the chapter ends with an introduction to the Stroh's problem.

Chapter 4 is an introduction to the polar formalism. This mathematical technique for plane anisotropy was introduced by Prof. G. Verchery in 1979 and I worked extensively on it during a long period, first with Prof. Verchery, then with different collaborators. The polar method is an effective tool for analyzing plane anisotropic phenomena and it is especially useful in the resolution of design problems. This part of the book is hence a complete presentation of the polar formalism; classical results are revisited and presented in a new way, along with some more recent results concerning special cases of elastic bodies as well as other properties, like piezo-electricity, anisotropy induced by damage, strength. Some examples of plane anisotropic materials end the chapter. The polar formalism is applied extensively in the last two chapters of the book.

Chapter 5 is an application of the previous results and concepts to the mechanics of anisotropic laminates, the main anisotropic bodies used in modern structural applications. The classical theory of laminated plates is first introduced and developed also for the polar formalism. Some general results, easily obtained applying this last, are highlighted. Then, the most used types of laminates are described and some sufficient rules for obtaining general elastic properties are given. Also, a study on the sensitivity of uncoupling and quasi-homogeneity to ply angles errors is detailed. This chapter ends with an extension of the classical theory to thermal and hygral properties and with a brief recall of higher order theories for laminated plates, given here also in the framework of the polar formalism.

Chapter 6 is an excursion into engineering design problems. Based upon the use of the polar formalism, that in design problems gives its best results, the chapter proposes first the theoretical problem of the optimal material orientation and then an original classification of anisotropic structures design problems and a presentation of classical and modern methods of design. More recent and modern problems of design are also considered: the optimization of anisotropy fields, that is, of structures with anisotropic properties depending upon the position, the case of modular systems and some multiphysics problems, concerning the thermal and piezoelectric response of an anisotropic structure. Several examples and details on modern numerical approaches complete the chapter.

The book makes use of modern mathematical tools: matrix and tensorial algebra are widely used throughout all the text, and the reader is considered to be familiar with standard tensor algebra.

The topic being very wide, this text is necessarily incomplete, as an introductory text usually is. Nevertheless, the author hopes sincerely that it will help the reader to progress in the field of anisotropic elasticity, a so particular world, unusual to mechanicians normally used to deal with isotropic materials.

I am indebted to many people for this book, namely Prof. Verchery, who introduced me to the polar method and to the research in this field when we were together at the University of Burgundy. Collaborating with him has been a privilege and a true opportunity for me. I wish to thank also Prof. M. Potier-Ferry, of the University of Lorraine, who has been a key encounter in my professional life, and to remember Prof. P. Villaggio, who has been much more than my Ph.D. supervisor at the University of Pisa: he remains for me an invaluable example of scientist, scholar, intellectual and the person who more than anyone else made me passionate about science and research.

I have had the opportunity to work with different colleagues on several topics presented in this book, especially A. Vincenti, E. Valot, M. Montemurro, A. Catapano, B. Desmorat, R. Barsotti, J. Pouget. I wish sincerely to thank them all: without their help, this book would not exist.

Finally, I need to thank my family: my wife Carla and my children Bianca and Alessandro. Their patience and support has been important during these last years.

Versailles, France
April 2017

Paolo Vannucci

Contents

About the Author

Paolo Vannucci is Professor of Mechanics at the LMV—Laboratory of Mathematics of Versailles, University of Versailles and Saint-Quentin. His main research activities concern plane anisotropic elasticity and multiphysics problems, optimization methods for anisotropic structures, metaheuristics for structural optimization, mechanics of no-tension materials applied to the study of monumental structures.

Chapter 1
Basic Concepts on Anisotropy

Abstract In this chapter the general concepts of anisotropy are introduced, starting from the basic question: what is anisotropy? The mathematical consequences of anisotropy on algebraic operators are introduced along with the concept of geometrical symmetries. Different examples of anisotropic physical phenomena are given, and a brief account of the anisotropy of crystals is proposed. The Chapter ends recalling some fundamental laws of elasticity, useful in the following.

1.1 Introduction: What is Anisotropy?

In ancient Greek, the word ἄνισος means *different*, *unequal*, and τρόπος, *direction*: *anisotropy*[1] indicates hence the concept of *not equivalent directions*. In physics, an *anisotropic phenomenon* is a property changing with the direction. As a consequence, any two directions are not equivalent with regards to such a phenomenon, but if a couple or more directions can be found whereon the phenomenon has the same behavior or value, then these directions are called *equivalent*.

Several physical properties can be anisotropic, namely because the medium where they take place or propagate is anisotropic. This is essentially due to the structure of the solid matter itself, the most part of times organized, at a given scale, according to a geometrical scheme. The typical example is that of crystals, where the molecules of the constituents are distributed on a regular lattice. At a greater scale, several stones are anisotropic due to their formation process (sedimentation or/and tectonic pressure), like sandstone or marble. Also structures and materials created by Nature can be anisotropic, like pack ice, bones, leaves or wood, where the matter is organized along preferential directions (e.g. along the vertical one) for biological reasons. Examples of artificial anisotropic materials are laminated steel, whose anisotropy is induced by the lamination process, and composite materials, where the union of an isotropic matrix and aligned fibers or tissues gives a final material that is macroscopically anisotropic. At a larger scale, some structures can be modeled as anisotropic shells or membranes, though constituted by materials that are isotropic, see Fig. 1.1.

---

[1]In old texts, anisotropy is sometimes called *æolotropy*, like in the classical book of Love (1944); this term comes from ancient Greek αἰόλος, that means *changeful*.

P. Vannucci, *Anisotropic Elasticity*, Lecture Notes in Applied and Computational Mechanics 85, DOI 10.1007/978-981-10-5439-6_1

(a) Giant crystals.

(b) A quartz crystal.

(c) An expensive carbon crystal: a diamond.

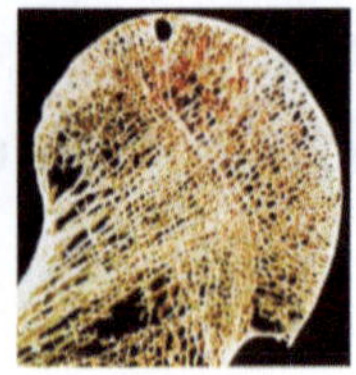

(d) A human bone (section).

(e) A natural anisotropic material: wood

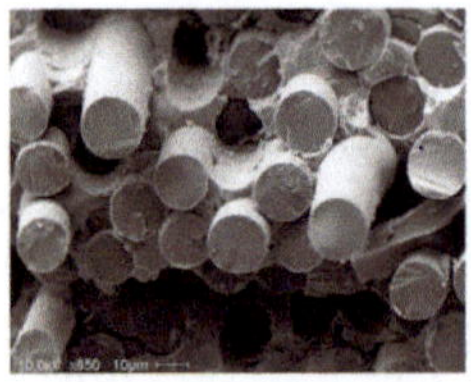

(f) An anisotropic composite: carbon fibres in an epoxy matrix

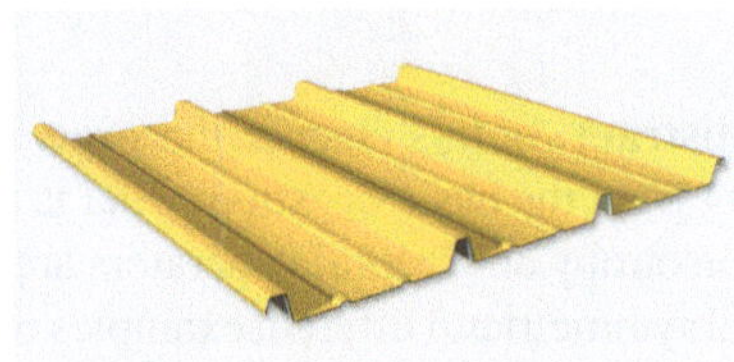

(g) An anisotropic plate of isotropic material

(h) An anisotropic concrete slab

(i) An isotropic membranes: ship sails

(j) An anisotropic structure: a grid shell (courtesy of ENPC, Paris)

Fig. 1.1 Some examples of anisotropic materials or structures

We are interested here in the anisotropic behavior of elastic materials and structures; the few examples in Fig. 1.1 show that a large part of them can be considered or modeled as anisotropic. Several structural materials and biological tissues, not only wood or bones, are actually anisotropic, while in different cases a structure can be modeled, at a sufficiently large scale, as composed by a fictitious, equivalent anisotropic material. In modern industrial applications, like aircrafts and sport devices, the composite materials reinforced by oriented fibers find an increasing use, thanks to their excellent mechanical performances in terms of stiffness, strength and lightness. In civil engineering, structural solutions making use of organized schemes that induce a macroscopic anisotropic behavior (e.g. stiffened plates, spatial trusses) are more and more used, because of their capacity of bearing loads and covering large spans.

There is hence a real need of understanding how to model and design anisotropic structures and materials: on one side, for understanding the mechanical behavior of natural structures, in botanics, bio-engineering, geophysics and so on. On the other side, for designing practical applications for different fields of the human activity.

1.2 Mathematical Consequences of Anisotropy

The anisotropy of a physical property has some consequences on its mathematical description. On one side, the effect of the dependence upon the direction must be described, and this usually leads to an increase of the number of parameters to be used for the description of the phenomenon. On the other side, the possible geometrical symmetries must be taken into account, because they normally give some relations about the parameters describing the property. The following sections give a short account of the algebraic and geometrical consequences of anisotropy, introducing some notions, definitions and concepts useful in the following.

1.2.1 Effects on the Algebraic Operators

A physical property or phenomenon is mathematically described by an *operator*; this can be simply a scalar, or a vector, or more frequently a tensor of a given rank. Often, when the property is isotropic, some scalar quantities are sufficient to completely describe it, but when the same property is anisotropic, this is no more sufficient, because the dependence upon the direction must be taken into account. Let us just consider some examples to show this.

As a first example, we consider thermal expansion: the strain $\boldsymbol{\varepsilon}$ produced in an elastic body by a change of temperature Δt can be described by a law of the type[2]

$$\boldsymbol{\varepsilon} = \Delta t \boldsymbol{\alpha}; \tag{1.1}$$

if the body is isotropic, the thermal expansion is the same for all the directions and a single coefficient α completely describes the phenomenon: the second rank symmetric tensor $\boldsymbol{\alpha}$ reduces to $\alpha\mathbf{I}$. But when the body is anisotropic, $\boldsymbol{\alpha}$ cannot be proportional to the identity, because the difference of behavior with the direction must be described by $\boldsymbol{\alpha}$: it remains a symmetric second-rank tensor, but in this case defined by six independent quantities $\alpha_{ij} = \alpha_{ji}$.

[2]Following common standard rules, we will indicate by $\mathscr{E}$ the natural Euclidean space, by $\mathscr{V}$ the vector space of translations on $\mathscr{E}$, or generally speaking any other vector space, when not differently indicated, by Lin the manifold of second-order tensors and by $\mathbb{L}$in that of fourth-rank tensors. Normally, elements of $\mathscr{E}$, points, will be denoted by italic lowercase letters, e.g. p, elements of $\mathscr{V}$, vectors, by bold lowercase letters, e.g. $\mathbf{v}$, elements of Lin, second-rank tensors, by bold uppercase letters, e.g. $\mathbf{L}$, and elements of $\mathbb{L}in$, fourth-rank tensors, by special characters, for instance like $\mathbb{E}$.

As a second example, let us consider the Ohm's law, relating the vectors of electric field $\mathbf{E}$ and current density $\mathbf{j}$. The law can be written as

$$\mathbf{j} = \boldsymbol{\mu}\mathbf{E}; \tag{1.2}$$

for an isotropic conductor, $\boldsymbol{\mu} = \mu\mathbf{I}$: the property depends uniquely on the conductivity μ, $\mathbf{j}$ is parallel to $\mathbf{E}$ and its magnitude is proportional to the magnitude of $\mathbf{E}$. Now, if the conductor is anisotropic, $\boldsymbol{\mu}$ is not proportional to the identity and the Ohm's law depends hence upon a second-rank symmetric tensor, the *conductivity tensor*; $\mathbf{j}$ is different in the different directions of the space and it is parallel to $\mathbf{E}$ only when this is aligned with one of the principal directions of $\boldsymbol{\mu}$.

A final example is that of elasticity, the topic of this text. Anticipating what will be shown in the following, the mechanical behavior of an isotropic elastic body is completely known when two quantities, usually the Young's modulus E and the Poisson's coefficient ν, are given. These two independent parameters describe the only two mechanical effects present in an isotropic elastic body: the direct stretching effect, i.e. the rigidity of the material with respect to a traction applied to it, and the lateral contraction effect, known as the *Poisson's effect*. The Hooke's law, the linear relation between the stress, $\boldsymbol{\sigma}$, and strain, $\boldsymbol{\varepsilon}$, tensors is, for all the materials,

$$\boldsymbol{\sigma} = \mathbb{E}\boldsymbol{\varepsilon} \tag{1.3}$$

where $\mathbb{E}$ is a fourth-rank tensor. In the isotropic case $\mathbb{E}$ is completely described by two parameters, e.g. E and ν, but if the material is anisotropic, 21 independent parameters are needed to determine $\mathbb{E}$. In addition, some of these quantities describe mechanical effects unknown in the isotropic case. The number of the independent components of $\mathbb{E}$ can be progressively reduced, until only two for the isotropic case, if some symmetry conditions exist for the material. All this will be the topic of Chap. 2, but for the while the last consideration about the existence of possible symmetry conditions introduces us to the following Section.

1.2.2 Geometrical Symmetries

A body is said to *have a symmetry* when it can be brought to coincidence with itself by a transformation, called a *covering operation*, that moves any of his points. In such a case, we said that the body *allows* the transformation. Possible covering operations are:

- *rotation* about an angle through a definite, or even indefinite, angle; in this case the body possesses an *axis of symmetry*;
- *reflexion* in a plane; the body has then a *plane of symmetry*;
- a combination of rotations and reflexions.

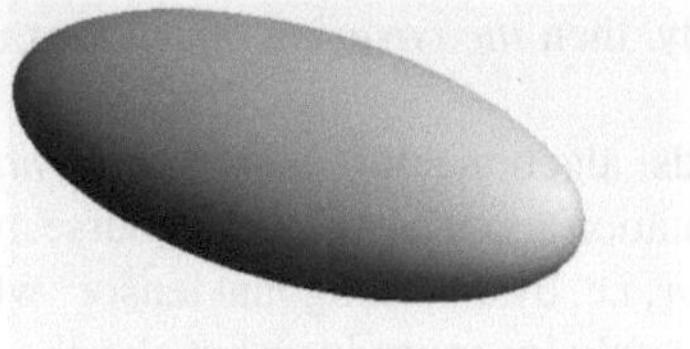

(a) Ellipsoid: the three principal planes are planes of symmetry and the three principal axes are axes of symmetry.

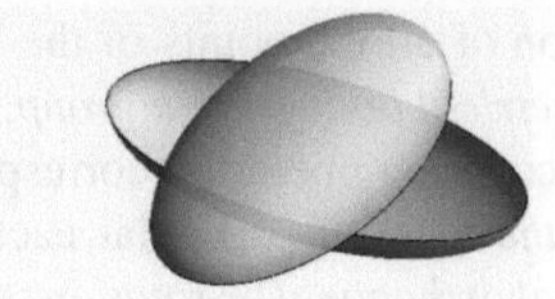

(b) The splitted ellipsoid: the upper part is rotated through $\pi/2$ about the vertical axis.

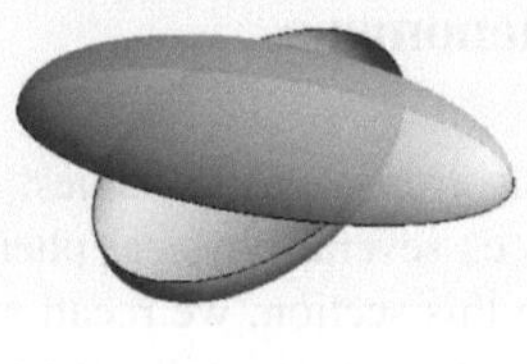

(c) Rotation through $\pi/2$ about the vertical axis (or reflexion in the horizontal plane).

(d) Reflexion in the horizontal plane (or rotation through $\pi/2$ about the vertical axis): the original figure (b) is recovered.

Fig. 1.2 An example of a figure with plane symmetries or with an axis of alternating symmetry

A particularly important one of such combinations is a rotation combined with a reflexion in the plane perpendicular to the symmetry axis. A body that allows such a combined covering is said to possess an *axis of alternating symmetry*.

A special case of alternating symmetry is that with an angle of rotation about the symmetry axis of π: each ray going out from a point is replaced by its opposite and the direction of the axis of alternating symmetry is arbitrary. This operation is called a *central perversion* and a body that allows this operation is said to have a *centre of symmetry*.

Figure 1.2, inspired by Love (1944, p. 150), shows two simple examples of bodies having planes of symmetry or an axis of alternating symmetry. The first, Fig. 1.2a is an ellipsoid: it allows a reflexion in each one of the three principal planes and a rotation of amount π about its three principal axes. The second body is obtained from the first one, cutting it in half along the horizontal plane of symmetry and rotating the upper part through $\pi/2$. The body so obtained, Fig. 1.2b, does not allow a rotation of $\pi/2$ about the vertical axis, Fig. 1.2c, nor a reflexion in the horizontal plane, still Fig. 1.2c, but it allows their combination, Fig. 1.2d: the vertical axis is an axis of alternating symmetry.

It can be shown that the effect of covering operations performed successively, in any order, is the same of a single covering operation. If we include among the covering operations the *identical covering*, i.e. the operation that leaves unchanged the position of all the points of the body, then *the covering operations allowed by any symmetrical body form a group.*

Every covering operation corresponds, algebraically, to an *orthogonal transformation of the space*; in particular, each rotation about an axis is, of course, represented by a *special orthogonal* or *rotation tensor*, i.e. by an orthogonal tensor[3] whose determinant is $+1$; such a tensor preserves not only inner products but also the orientation of the space.[4] Any reflexion in a plane, as well as any combination of rotations and reflexions, can be always obtained as the product of $-\mathbf{I}$, the opposite of the identity tensor, and a rotation, i.e. by an orthogonal tensor whose determinant is -1.

1.3 Some Anisotropic Physical Phenomena

The subject of this text is the theory of anisotropic elastic bodies. Nevertheless, the anisotropic structure of the matter induces several physical phenomena to be anisotropic, not only the elastic response. In this section, we recall some of them, besides those already mentioned in the previous Section, without any ambition of

---

[3] An *orthogonal tensor* $\mathbf{U}$ is a tensor that preserves the inner product (and hence angles):

$$\mathbf{U}\mathbf{v}_1 \cdot \mathbf{U}\mathbf{v}_2 = \mathbf{v}_1 \cdot \mathbf{v}_2 \ \forall \mathbf{v}_1, \mathbf{v}_2 \in \mathscr{V};$$

choosing $\mathbf{v}_1 = \mathbf{v}_2$ we see that an orthogonal tensor preserves not only angles but also lengths: that is why any rigid transformation of the space can be represented uniquely by an orthogonal tensor. From the above definition, it follows immediately that

$$\mathbf{U}\mathbf{U}^\top = \mathbf{U}^\top\mathbf{U} = \mathbf{I},$$

and finally, by the uniqueness of the inverse tensor,

$$\mathbf{U}^{-1} = \mathbf{U}^\top,$$

while by the Binet's theorem on the determinant of a product of tensors we get

$$\det \mathbf{U} = \pm 1.$$

[4] An orthonormal basis $\mathscr{B} = \{\mathbf{e}_1, \mathbf{e}_2, \mathbf{e}_3\}$ is said to be *positively oriented* if

$$\mathbf{e}_1 \times \mathbf{e}_2 \cdot \mathbf{e}_3 = +1.$$

Because of the identity, see e.g. (Gurtin 1981, p. 8),

$$\mathbf{L}\mathbf{v}_1 \times \mathbf{L}\mathbf{v}_2 \cdot \mathbf{L}\mathbf{v}_3 = \det \mathbf{L}\ \mathbf{v}_1 \times \mathbf{v}_2 \cdot \mathbf{v}_3 \ \ \forall \mathbf{v}_1, \mathbf{v}_2, \mathbf{v}_3 \in \mathscr{V} \text{ and } \mathbf{L} \in \mathrm{Lin},$$

we see that only orthogonal tensors with positive determinant preserve the space orientation and hence can represent rigid rotations. Because an inversion of the space orientation corresponds to change the orientation of one of the vectors of the triad, reflexions are represented by orthogonal tensors with negative determinant.

completeness. A rather exhaustive description of these phenomena can be found in the classical book of Nye (1957).

1.3.1 Paramagnetism and Diamagnetism

Be **H** the *magnetic field intensity*, **L** the *intensity of magnetization* and **B** the *magnetic induction*; they are related by the law

$$\mathbf{B} = \mu_0 \mathbf{H} + \mathbf{L}, \tag{1.4}$$

with $\mu_0 = 4\pi \times 10^{-7}$ H/m the *magnetic permeability of vacuum.* For isotropic bodies,

$$\mathbf{L} = \mu_0 \Psi \mathbf{H}, \tag{1.5}$$

where Ψ is the *magnetic susceptibility* of the medium, per unit of volume. We can then write

$$\mathbf{B} = \mu \mathbf{H}, \tag{1.6}$$

with $\mu = \mu_0(1 + \Psi)$ the *magnetic permeability of the substance.* **H**, **L** and **B** are all parallel for the isotropic case. When the magnetic field is applied to an anisotropic medium, like a crystal, then

$$\begin{aligned} \mathbf{L} &= \mu_0 \boldsymbol{\Psi} \mathbf{H}, \\ \mathbf{B} &= \boldsymbol{\mu} \mathbf{H}. \end{aligned} \tag{1.7}$$

Now, $\boldsymbol{\Psi}$ is the *magnetic susceptibility tensor* and $\boldsymbol{\mu} = \mu_0(\mathbf{I} + \boldsymbol{\Psi})$ is the *permeability tensor* of the substance. It can be proved, using energetic arguments, that $\boldsymbol{\Psi}$ and $\boldsymbol{\mu}$ are symmetric second order tensors; **H**, **L** and **B** are all *parallel* if and only if **H** is parallel to one of the eigenvectors (principal axes) of $\boldsymbol{\Psi}$. A crystal is said to be *paramagnetic* along the principal directions $\mathbf{e}_i$ of $\boldsymbol{\Psi}$ if the corresponding eigenvalue $\Psi_i > 0$, *diamagnetic* if $\Psi_i < 0$.

1.3.2 Dielectric Susceptibility

A property similar to the previous one, at least formally, is the *electric polarization*: be **E** the *electric field intensity*, **P** the *electric polarization* and **D** the *electric displacement* or *electric flux density*; they are related by the law

$$\mathbf{D} = \kappa_0 \mathbf{E} + \mathbf{P}, \tag{1.8}$$

with $\kappa_0 = 8.854 \times 10^{-12}$ F/m the *dielectric permittivity of vacuum.* For isotropic bodies,

$$\mathbf{P} = \kappa_0 \chi \mathbf{E}, \tag{1.9}$$

where χ is the *dielectric susceptibility* of the medium, per unit of volume. We can then write

$$\mathbf{D} = \kappa \mathbf{E}, \tag{1.10}$$

with $\kappa = \kappa_0(1+\chi)$ the *dielectric permittivity of the substance*. Of course, also **D**, **E** and **P** are all parallel for the isotropic case. When the medium is anisotropic, then

$$\begin{aligned} \mathbf{P} &= \kappa_0 \boldsymbol{\chi} \mathbf{E}, \\ \mathbf{D} &= \boldsymbol{\kappa} \mathbf{E}. \end{aligned} \tag{1.11}$$

Now, $\boldsymbol{\chi}$ is the *dielectric susceptibility tensor* and $\boldsymbol{\kappa} = \kappa_0(\mathbf{I} + \boldsymbol{\chi})$ is the *permittivity tensor* of the substance. Also in this case, still using energetic arguments it can be proved that $\boldsymbol{\chi}$ and $\boldsymbol{\kappa}$ are symmetric second order tensors; hence, once again **E**, **P** and **D** are all parallel if and only if **E** is parallel to one of the eigenvectors (principal axes) of $\boldsymbol{\chi}$.

1.3.3 Thermal Conductivity

The heat flux vector in an anisotropic solid **h** (i.e. the vector whose components represent the quantity of heat passing across unit areas perpendicular to the axes, in unit time) is related to the temperature gradient by the law

$$\mathbf{h} = -\mathbf{k}\nabla t, \tag{1.12}$$

with **k** the *thermal conductivity tensor*. Hence, the heat flux is not necessarily parallel to ∇t, while if the medium is isotropic, $\mathbf{k} = k\mathbf{I}$, so **h** is parallel to ∇t. The symmetry of **k** can be derived as a consequence of the Onsager's principle, Nye (1957, p. 205).

1.3.4 Piezoelectricity

Some crystals, thanks to their particular type of anisotropy, develop an electric polarization if stressed by applied forces: it is the *direct piezoelectric effect*, from ancient Greek πιέζειν that means to press, described by the law

$$\mathbf{P} = \mathscr{D}\boldsymbol{\sigma}, \tag{1.13}$$

where $\mathscr{D}$ is the *tensor of piezoelectric moduli*. Relating linearly a vector and a second-rank symmetric tensor, $\mathscr{D}$ is a third-rank tensor; in components, the previous equation reads like

$$P_i = D_{ijk}\sigma_{jk}. \tag{1.14}$$

The *converse piezoelectric effect* or *Lippmann's effect*, is the one connecting the strain to the electric field,

$$\boldsymbol{\varepsilon} = \mathscr{D}\mathbf{E}: \tag{1.15}$$

the electric current strains a piezoelectric crystal. A little bit surprisingly, tensor $\mathscr{D}$ describes also the converse effect; this can be proved using thermodynamical considerations. Now, $\mathscr{D}$ relates a second-rank symmetric tensor to a vector; hence, explicitly, the last equation is written as[5]

$$\varepsilon_{jk} = D_{ijk}E_i. \tag{1.16}$$

The 27 components of $\mathscr{D}$ are not all independent, because the following nine relations exist:

$$D_{ijk} = D_{ikj}, \tag{1.17}$$

leaving only 18 independent piezoelectric moduli, in the most general case; their number can be further reduced taking into account the specific symmetries of a crystal.

1.3.5 Photoelastic and Electro-Optical Effects

The change of refractive index of a medium caused by an electric field is called *electro-optical effet*, while the *photoelastic effect* is its dependence upon the state of strain.

The refractive index is detailed by the *indicatrix*, an ellipsoid whose equation can be written explicitly as

$$B_{ij}x_ix_j = 1, \tag{1.18}$$

with the B_{ij} the components of the *relative dielectric impermeability tensor at optical frequencies*, **B**, a second-rank symmetric tensor. The change of the B_{ij}s produced by an electric field **E** and a strain field $\boldsymbol{\varepsilon}$ is given by the law

$$\Delta B_{ij} = z_{ijk}E_k + p_{ijrs}\varepsilon_{rs}; \tag{1.19}$$

the z_{ijk} are the *electro-optical coefficients*, components of a third-rank tensor, while the p_{ijrs} are the *elasto-optical coefficients*, components of a fourth-rank tensor. The symmetries of **B** and of $\boldsymbol{\varepsilon}$ give the following relations

$$z_{ijk} = z_{jik}, \quad p_{ijrs} = p_{jirs} = p_{ijsr} = p_{jisr}, \tag{1.20}$$

which reduce the number of independent coefficients to 18 and 36 respectively.

[5]Unless explicitly indicated, we will use the Einstein's convention on dummy indexes for denoting summation.

1.3.6 A General Consideration About Anisotropic Phenomena

Before going on, we need now to precise an important point. From what said in the previous Sections, it appears that anisotropy is an intrinsic quality of a continuum, originated by the existence, or absence, of some symmetries in the geometric distribution of the matter, see in particular the next Section.

This is indeed a classical point of view, but it is important to understand that the anisotropy of the continuum is just a necessary, but not sufficient, condition for a given physical property to be anisotropic on such a continuum.

In fact, they can exist, and really they do, some media that are intrinsically anisotropic, i.e. they have some kind of particular geometric disposition of the matter, but that with regards to some physical properties they behave just like an isotropic continuum.

The fact of a property being or not anisotropic is actually determined by two factors: the intrinsic anisotropy of the continuum, i.e. its true material symmetries, and the type of algebraic operator describing a given physical property.

Roughly speaking, increasing the tensorial order, which corresponds to have a richer operator, described by a larger number of parameters, allows for taking into account for more and more complex situations, hence also of anisotropic properties.

So, for the same continuum, depending upon its material symmetries, it is possible that some physical properties described by high order tensor operators are anisotropic, while some other ones, described by low order tensors, are perfectly isotropic. A classical example is that of the so-called square symmetry in planar media.

Such a material symmetry corresponds to a complete covering operation of rotation of $\pi/2$ about a vertical axis. Such a material can be easily obtained reinforcing an isotropic matrix by a balanced fabric, i.e. by a tissue having the same amount of fibres in warp and weft, a technique currently used in the fabrication of composite laminated plates and shells.

For such a material, the elastic behavior, described by a fourth-rank tensor, is not isotropic, contrarily to what is often believed, but the thermoelastic behavior, described by the second-rank tensor of thermal expansion coefficients, is isotropic, as well as any other physical property depending upon a second-rank tensor.

This fact has been condensed, in some way, in the empirical Principle of Neumann, see Sect. 1.4.4, where an example concerning cubic syngony, the 3D corresponding of square symmetry, is given. Anyway, the Principle of Neumann does not give a precise correspondence between the anisotropy of the continuum and that of the property; actually, it is just a naturalistic principle concerning the properties of crystals.

As a conclusion, it is apparent that it is much more correct to talk about the anisotropy of a physical property on a continuum rather than that of anisotropy of the continuum itself. More complicate situations can occur in special cases, like that of laminates, where the mechanical behaviors, e.g. the response in extension or in bending, can have different types of anisotropy and the material symmetries can also be completely absent, see Chap. 5.

1.4 Some Basic Elements About Crystals

1.4.1 Lattices and Cells

A crystal is a typical anisotropic body: the geometrical disposition of the elementary constituents, atoms or molecules, on a *lattice* induces the anisotropy of the physical phenomena that take place in the crystal.

We give here a brief account about crystals, far from any ambition of completeness. Much more concerned, in this text, with the algebraic structure of the operators describing an anisotropic property, the matter is not of the primary importance here, so the reader is addressed to classical texts of crystallography and physical properties of crystals, e.g. (Nye 1957), for a deeper insight in the matter.

A lattice is an array of *equally spaced and similarly situated points*; in the physical reality of the crystals, these points are occupied by atoms or molecules. The points are said to be similarly situated when the rest of the lattice appears the same, and in the same orientation, when it is viewed from them.

One can consider three sets of evenly spaced parallel planes that pass through the lattice points and divide up the crystal into identical elementary parallelepipeds, the *primitive unit cells*, that have lattice points only at its corners. Nevertheless, it is often better to chose a larger unit cell, having lattice points not only in the corners, but for instance at the centre of the faces; such cells are called *multiply primitive unit cells*. In all the cases, a cell is defined by the lengths and directions of three vectors, denoted in the following by **a**, **b** and **c**, corresponding to three non parallel of its edges, see Fig. 1.3.

The possible dispositions of points in a regular lattice in a two- or three-dimensional space have been studied by A. Bravais (1850). A *Bravais lattice* is an infinite lattice of points generated by a set of discretized translation operations ruled by a law of the type

$$\mathbf{u} = p\mathbf{a} + q\mathbf{b} + r\mathbf{c}, \tag{1.21}$$

with p, q and r integers. The lattice is independent from the choice of the position of the vector **u**, so it appears the same when viewed from any lattice point. It can be shown that there are only 14 possible Bravais lattices in the three-dimensional space.

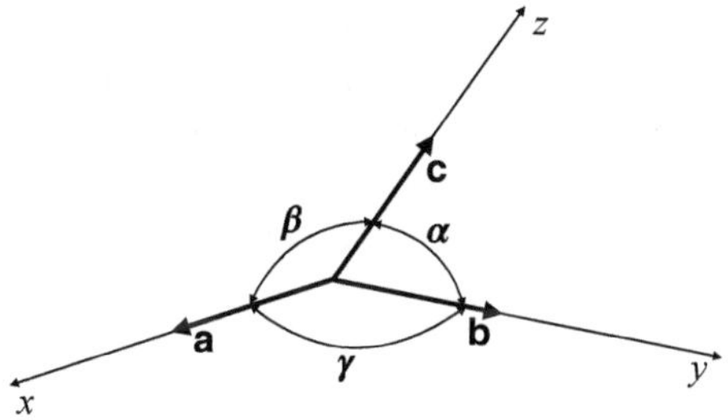

Fig. 1.3 Geometrical elements of a unit cell

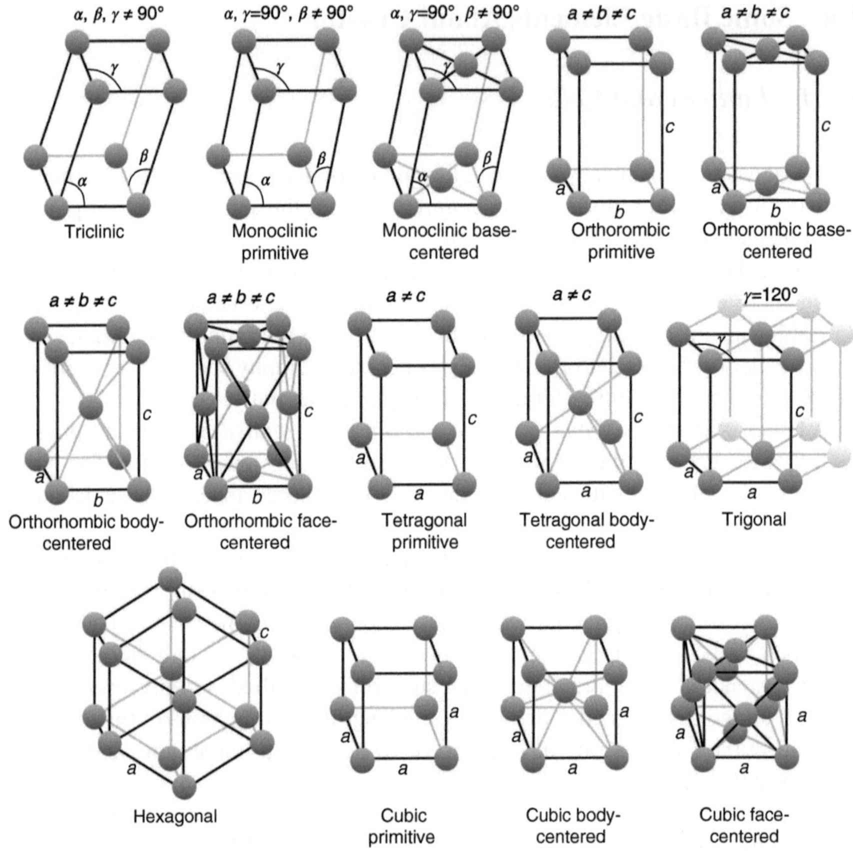

Fig. 1.4 The 14 Bravais cells

They constitute the only possibilities for filling the space through the repetition of an elementary *motive*, the unit cell of the Bravais lattice, called also the *Bravais cell.* The 14 Bravais cells are represented in Fig. 1.4. They are usually grouped into seven distinct *crystal systems*: *Triclinic*, *Monoclinic*, *Orthorhombic*, *Tetragonal*, *Trigonal*, *Hexagonal* and *Cubic*.

A crystal system is sometimes called a *syngony* and authors do not agree completely on their number (some authors group together the hexagonal and trigonal systems, so listing only 6 crystal syngonies, like Love (1944, p.158) who follows the classification given by Lewis (1899)), nor on their names (e.g. Lewis names *Rhombic* or *Prismatic* the Orthorombic syngony, and *Rhombohedral* the Trigonal system, included in the Hexagonal syngony).

1.4.2 The Symmetries of Crystals

The physical properties of crystals, namely, the anisotropy of such properties, depend upon the *symmetry elements* of the crystals and on their combinations. It can be shown that the only possible symmetry elements for a crystal are:

- a centre of symmetry;
- a plane of symmetry;
- a 2-, 3-, 4- or 6-fold axis of symmetry;
- a 2-, 3-, 4- or 6-fold axis of alternating symmetry.

Of course, a crystal can have more than one element of symmetry. A $n-$fold axis of symmetry allows a covering that is a rotation through an angle of $2\pi/n$. Hence, the fact that a crystal can have only a 2-, 3-, 4- or 6-fold axis of symmetry reflects the circumstance that the angles $\pi, 2\pi/3, \pi/2$ and $\pi/3$ are the only ones allowed for rotation coverings in a crystal; the same is true also for an axis of alternating symmetry.

It can be shown that the number of groups of covering operations for crystals are 32, defining the 32 *classes of crystals*. Once the class of a crystal known, i.e. all its symmetries, one can determine the form of the strain energy function, see Sect. 2.2, and hence the number and type of independent elastic constants. It turns out that there are only 10 possible cases, that we will call *elastic syngonies* (*classes* for Lekhnitskii, (1950, p. 28–32)), to differentiate them from the crystal syngonies, whose number is limited to 7 (or to 6 for some authors). Though the elastic syngonies are 10, the number of independent anisotropic elastic constants can get only 8 distinct values: 21, 13, 9, 7, 6, 5, 3 and 2, see Sect. 2.2.11; the last case is that of isotropy, impossible for crystals.

1.4.3 Classifications of the Crystals

There exist different possible classifications for crystals, depending upon the elements of symmetry; the different classifications use different nomenclatures and make a different use of the symmetry elements. The most well known are the *Hermann–Mauguin classification* Nye (1957), with a notation based upon the essential symmetry elements of a class, adopted as the standard one since 1935 by the *International Tables for Crystallography* and the *classification of Schoenflies*, used in spectroscopy, Schoenflies (1891). On its turn, Lekhnitskii gives a different nomenclature for the crystal classes, based upon a complete listing of the symmetry elements, Lekhnitskii (1950, p.27). Voigt (1910) has given a number to each one of the 32 classes, while Lewis (1899) and Miers (1902) have named them, of course with different names, Love (1944, p. 158). In Table 1.1 the 32 crystal classes are listed, subdivided in seven syngonies. The number N of independent elastic constants is also

Table 1.1 Crystal syngonies

Syngony	Symmetry elements	Examples	Classification			
			Voigt	Sch	H-M	N
Triclinic	none	H_3BO_3, $K_2Cr_2O_7$, $CuSO_4 \cdot 5H_2O$	1	S_2	$\bar{1}$	21
			2	C_1	1	21
Monoclinic	a 2-fold a-s or a-a-s	$Na_2SO_4 \cdot 10H_2O$	3	C_2^h	$2/m$	13
			4	S	m	13
			5	C_2	2	13
Orthorhombic	3 mutually orthogonal 2-fold a-s or a-a-s	KNO_3, $BaSO_4$	6	V_h	mmm	9
			7	V	222	9
			8	C_2^v	$mm2$	9
Trigonal	a 3-fold a-s or a-a-s	HgS, $CaCO_3$	9	S_6^u	$\bar{3}m$	6
			10	D_3	32	6
			11	C_3^v	$3m$	6
			12	S_6	$\bar{3}$	7
			13	C_3	3	7
Tetragonal	a 4-fold a-s or a-a-s	$CaSO_4$, TiO_2	14	D_4^h	$4/mmm$	6
			15	D_4	4222	6
			16	C_4^v	$4mm$	6
			17	C_4^h	$4/m$	7
			18	C_4	4	7
			19	S_4^u	$\bar{4}2m$	6
			20	S_4	$\bar{4}$	7
Hexagonal	a 6-fold a-s or a-a-s	CdS, ZnO, graphite	21	D_6^h	$6/mmm$	5
			22	C_6^v	$6mm$	5
			23	D_6	622	5
			24	C_6^h	$6/m$	5
			25	C_6	6	5
			26	D_3^h	$\bar{6}m2$	5
			27	C_3^h	$\bar{6}$	5
Cubic	four 3-fold a-s arranged as the cubic diagonal	$NaCl$, Cu, Zn, Au, diamond	28	O^h	$m\bar{3}m$	3
			29	O	432	3
			30	T^d	$\bar{4}3m$	3
			31	T^h	$m\bar{3}$	3
			32	T	332	3

Note Sch = Schoenflies, H-M = Hermann–Mauguin, a-s = axis of symmetry, a-a-s = axis of alternate symmetry. N = number of independent elastic constants given, along with the classifications of Voigt, Schoenflies and Hermann–Mauguin. Finally, the symmetry elements of the syngony are also indicated and some examples given.

1.4.4 The Neumann's Principle

The Neumann's Principle (Neumann 1885) is an empirical principle, i.e. not derived from a mathematical principle or demonstration, that relates the symmetry class of a crystal and the symmetries of its anisotropic properties. It may be stated as Love (1944, p.155):

> *Any kind of symmetry, possessed by the crystallographic form of a material, is possessed also by the material in respect of every physical property.*

This principle does not establish a one to one correspondence between the crystal symmetries and the physical symmetries, but merely the *inclusion of all the symmetries of the crystal in the set of physical symmetries*. In other words, some physical properties can have, and generally have, some symmetries that are not possessed by the crystal. For instance, cubic crystals are optically isotropic; isotropy includes all the cubic symmetries, so the physical property has a symmetry that includes those of the crystal, as the principle demands.

1.5 Some Fundamental Equations of the Mechanics of Elastic Bodies

We recall here some fundamental equations, that will occasionally be used in the following of this text.

- *Strain tensors*: for a displacement vector field $\mathbf{u}(x_1, x_2, x_3)$ a general strain measure is given by the *Green-Lagrange strain tensor* $\mathbf{L}$:

$$\mathbf{L} = \frac{1}{2}(\nabla\mathbf{u} + \nabla\mathbf{u}^\top) + \frac{1}{2}\nabla\mathbf{u}^\top\nabla\mathbf{u}, \tag{1.22}$$

where[6,7]

$$\nabla\mathbf{u} = u_{i,j}\mathbf{e}_i \otimes \mathbf{e}_j,\ i, j = 1, 2, 3, \tag{1.23}$$

[6]Generally speaking, the comma as subscript will be used to denote differentiation: $u_{i,j} = \frac{\partial u_i}{\partial x_j}$.

[7]$\forall \mathbf{a}, \mathbf{b}, \mathbf{c} \in \mathscr{V}$, the *dyad* $(\mathbf{a} \otimes \mathbf{b})$ is the second-rank tensor defined by the operation $(\mathbf{a} \otimes \mathbf{b})\mathbf{c} := \mathbf{b} \cdot \mathbf{c}\, \mathbf{a}$. Given an orthonormal basis $\mathscr{B} = \{\mathbf{e}_1, \mathbf{e}_2, \mathbf{e}_3\}$ of $\mathscr{V}$, any second-rank tensor $\mathbf{L}$ can be expressed as

$$\mathbf{L} = L_{ij}\mathbf{e}_i \otimes \mathbf{e}_j,$$

where the Cartesian components L_{ij} of $\mathbf{L}$ are

is the *displacement gradient*. In the limit

$$|u_{i,j}| \ll 1 \ \forall i, j = 1, 2, 3, \tag{1.24}$$

we get the *infinitesimal strain tensor* $\boldsymbol{\varepsilon}$

$$\boldsymbol{\varepsilon} = \frac{1}{2}(\nabla\mathbf{u} + \nabla\mathbf{u}^\top), \tag{1.25}$$

already introduced in Eq. (1.3).

- *Cauchy's stress tensor*: we consider bodies for which the stress is a local quantity described by the *Cauchy's stress tensor* $\boldsymbol{\sigma}$, with $\boldsymbol{\sigma} = \boldsymbol{\sigma}^\top$. The *stress vector* $\mathbf{t}$ upon a surface in p is a linear function of the outward normal $\mathbf{n}$ (*Cauchy's theorem on stress*):

$$\mathbf{t} = \boldsymbol{\sigma}\mathbf{n}. \tag{1.26}$$

- *Equation of motion*:

$$\mathrm{div}\boldsymbol{\sigma} = \mathbf{b} \ \rightarrow \ \sigma_{ij,j} = b_i \ \forall i, j = 1, 2, 3, \tag{1.27}$$

where $\mathbf{b}$ is the body vector, that includes the mass acceleration and the reversed body force $\mathbf{f}$:

$$\mathbf{b} = \rho\frac{\partial^2\mathbf{u}}{\partial t^2} - \mathbf{f}. \tag{1.28}$$

When the body has a null acceleration everywhere, then the above equation becomes the *equilibrium equation*.

- *Compatibility equations*: they give the integrability necessary conditions for the components of $\boldsymbol{\varepsilon}$:

$$\begin{aligned}
2\varepsilon_{12,12} &= \varepsilon_{11,22} + \varepsilon_{22,11},\\
2\varepsilon_{23,23} &= \varepsilon_{22,33} + \varepsilon_{33,22},\\
2\varepsilon_{31,31} &= \varepsilon_{33,11} + \varepsilon_{11,33},\\
\varepsilon_{11,23} + \varepsilon_{23,11} &= \varepsilon_{13,12} + \varepsilon_{12,13},\\
\varepsilon_{22,13} + \varepsilon_{13,22} &= \varepsilon_{12,23} + \varepsilon_{23,12},\\
\varepsilon_{33,12} + \varepsilon_{12,33} &= \varepsilon_{23,13} + \varepsilon_{13,23}.
\end{aligned} \tag{1.29}$$

The compatibility equations can be derived also for other strain measures, namely for the Green-Lagrange tensor, see for instance Rand and Rovenski (2005, p. 10).

(Footnote 7 continued)

$$L_{ij} = \mathbf{e}_i \cdot \mathbf{L}\mathbf{e}_j.$$

References

M.E. Gurtin, *An Introduction to Continuum Mechanics* (Academic Press Inc., New York, 1981)
S.G. Lekhnitskii, *Theory of elasticity of an anisotropic elastic body*, 1950. (English translation by P. Fern, Holden-Day, San Francisco, 1963)
W.J. Lewis, *Treatise on Cystallography* (Cambridge University Press, Cambridge, 1899)
A.E.H. Love, *A Treatise on the Mathematical Theory of Elasticity* (Dover, New York, 1944)
H.A. Miers, *Mineralogy* (Oxford University Press, Oxford, 1902)
F. Neumann, *Vorlesungen über die Theorie der Elasticität* (B. G. Teubner, Leipzig, 1885)
J.F. Nye, *Physical Properties of Crystals* (Clarendon Press, Oxford, 1957)
O. Rand, V. Rovenski, *Analytical Methods in Anisotropic Elasticity* (Birkhäuser, Boston, 2005)
A.M. Schoenflies, *Krystallsysteme und Krystallstructur* (B. G. Teubner, Leipzig, 1891)
W. Voigt, *Lehrbuch der Kristallphysik* (B. G. Teubner, Leipzig, 1910)

Chapter 2
General Anisotropic Elasticity

Abstract This Chapter is an introduction to general anisotropic elasticity, i.e. to the elasticity of 3D anisotropic bodies. The main classical topics of the matter are treated in detail: starting from the Hooke's law for anisotropic bodies, the two principal notations of Voigt and Kelvin are introduced and the reasons for the use of the last one are argued. Then, after an explanation of the mechanical meaning of the anisotropic elastic constants, the key topic of elastic symmetries is treated in detail. The technical elastic constants are then introduced as well as the elastic bounds for the components of the elastic tensor and for the technical constants. After an observation on the decomposition of the strain energy for anisotropic bodies, the Chapter ends with the determination of the symmetry planes, the curvilinear anisotropy and some examples of typical anisotropic materials.

2.1 The Hooke's Law for Anisotropic Bodies

Be Ω a body acted upon by body forces $\mathbf{f}$ and by surface tractions $\mathbf{t}$ on its frontier $\partial\Omega$ whose outward unit normal is $\mathbf{n}$. We consider a small arbitrary variation $\delta\mathbf{u}$ of the displacement field on Ω, compatible with the given boundary conditions and satisfying the kinematical conditions (1.25). The total mechanical work dW done by the applied forces can be transformed as follows

$$\begin{aligned} dW = \int_\Omega \mathbf{f}\cdot\delta\mathbf{u}\,d\omega + \int_{\partial\Omega}\mathbf{t}\cdot\delta\mathbf{u}\,ds = \int_\Omega \mathbf{f}\cdot\delta\mathbf{u}\,d\omega + \int_{\partial\Omega}\boldsymbol{\sigma}\mathbf{n}\cdot\delta\mathbf{u}\,ds = \\ \int_\Omega \mathbf{f}\cdot\delta\mathbf{u}\,d\omega + \int_{\partial\Omega}\boldsymbol{\sigma}\delta\mathbf{u}\cdot\mathbf{n}\,ds = \int_\Omega[\mathbf{f}\cdot\delta\mathbf{u} + \operatorname{div}(\boldsymbol{\sigma}\delta\mathbf{u})]\,d\omega = \\ \int_\Omega[(\mathbf{f}+\operatorname{div}\boldsymbol{\sigma})\cdot\delta\mathbf{u} + \boldsymbol{\sigma}\cdot\nabla\delta\mathbf{u}]\,d\omega = \int_\Omega \boldsymbol{\sigma}\cdot\delta\boldsymbol{\varepsilon}\,d\omega \end{aligned} \tag{2.1}$$

P. Vannucci, *Anisotropic Elasticity*, Lecture Notes in Applied and Computational Mechanics 85, DOI 10.1007/978-981-10-5439-6_2

In establishing Eq. (2.1) we have used successively the Cauchy's stress theorem (1.26), the symmetry of $\boldsymbol{\sigma}$, the Gauss theorem, two standard results for tensor and vector fields[1] and the motion equation (1.27).

The quantity

$$\delta V = \boldsymbol{\sigma} \cdot \delta \boldsymbol{\varepsilon} = \sigma_{ij} \delta \varepsilon_{ij}, \tag{2.2}$$

represents the variation of the internal energy of the body per unit of volume produced by a small variation of the strain state.

Following the energetic approach of Green (1839), we define as *elastic* a body for which the total variation ΔV of the internal energy due to a finite transformation from a state A to a state B is independent from the integration path. In particular, ΔV must then be null for any transformation where A=B:

$$\Delta V = \int_A^B \delta V = V_B - V_A. \tag{2.3}$$

Hence, for an elastic body δV must be the exact differential dV of a scalar function $V(\boldsymbol{\varepsilon})$, the *strain energy density* or *elastic potential*[2]:

$$V = V(\boldsymbol{\varepsilon}) : \ dV = \frac{\partial V}{\partial \varepsilon_{ij}} d\varepsilon_{ij}. \tag{2.4}$$

In such a case, Eq. (2.2) must be rewritten as

$$dV = \boldsymbol{\sigma} \cdot d\boldsymbol{\varepsilon} = \sigma_{ij} d\varepsilon_{ij}, \tag{2.5}$$

so that we get the *Green's formula*

$$\sigma_{ij} = \frac{\partial V}{\partial \varepsilon_{ij}}. \tag{2.6}$$

---

[1]Namely, we have used the identity

$$\text{div}(\mathbf{L}^\top \mathbf{v}) = \text{div}\mathbf{L} \cdot \mathbf{v} + \mathbf{L} \cdot \nabla \mathbf{v},$$

with $\mathbf{L}$ a second-rank tensor field and $\mathbf{v}$ a vector field, see (Gurtin 1981, p. 30), and the fact that, $\forall \mathbf{L} : \mathbf{L} = \mathbf{L}^\top$,

$$\mathbf{L} \cdot \frac{\nabla \mathbf{v} + \nabla^\top \mathbf{v}}{2} = \frac{1}{2}(\mathbf{L} \cdot \nabla \mathbf{v} + \mathbf{L} \cdot \nabla \mathbf{v}^\top) = \frac{1}{2}(\mathbf{L} \cdot \nabla \mathbf{v} + \mathbf{L}^\top \cdot \nabla \mathbf{v}^\top) = \mathbf{L} \cdot \nabla \mathbf{v},$$

because of the property of tensor scalar product:

$$\mathbf{A} \cdot \mathbf{B} = A_{ij} B_{ij} = \mathbf{A}^\top \cdot \mathbf{B}^\top.$$

[2]The existence of such a function can be established upon physical arguments, using the first law of thermodynamics for adiabatic transformations or the second law for the isothermal ones, (Love 1944, p. 94).

We now postulate that in the initial state, i.e. when the body is not acted upon by forces, $\boldsymbol{\varepsilon} = \mathbf{O}$ and $\boldsymbol{\sigma} = \mathbf{O}$, i.e. the body is unstrained and unstressed in its initial state. Then, developing $V(\boldsymbol{\varepsilon})$ in a Taylor series about $\boldsymbol{\varepsilon} = \mathbf{O}$ we get

$$V(\boldsymbol{\varepsilon}) = V(\boldsymbol{\varepsilon} = \mathbf{O}) + \left.\frac{\partial V}{\partial \varepsilon_{ij}}\right|_{\boldsymbol{\varepsilon}=\mathbf{O}} \varepsilon_{ij} + \frac{1}{2} \left.\frac{\partial^2 V}{\partial \varepsilon_{ij} \partial \varepsilon_{kl}}\right|_{\boldsymbol{\varepsilon}=\mathbf{O}} \varepsilon_{ij}\varepsilon_{kl} + \ldots \tag{2.7}$$

Choosing arbitrarily $V(\boldsymbol{\varepsilon} = \mathbf{O}) = 0$, always possible for a potential, and limiting the development to the first non null term, which is correct for small strains, gives

$$V = \frac{1}{2} \left.\frac{\partial^2 V}{\partial \varepsilon_{ij} \partial \varepsilon_{kl}}\right|_{\boldsymbol{\varepsilon}=\mathbf{O}} \varepsilon_{ij}\varepsilon_{kl}; \tag{2.8}$$

the second derivatives in the above equation are linear operators depending upon four indexes; they are the components of a fourth-rank tensor $\mathbb{E}$,

$$E_{ijkl} := \left.\frac{\partial^2 V}{\partial \varepsilon_{ij} \partial \varepsilon_{kl}}\right|_{\boldsymbol{\varepsilon}=\mathbf{O}}, \tag{2.9}$$

the *(stiffness) elasticity tensor*, so that

$$V = \frac{1}{2} E_{ijkl}\varepsilon_{ij}\varepsilon_{kl} = \frac{1}{2} \boldsymbol{\varepsilon} \cdot \mathbb{E}\boldsymbol{\varepsilon}. \tag{2.10}$$

Collecting all the parameters describing the elastic response of the material, $\mathbb{E}$ is the operator that describes the elastic response of the continuum. It has 81 Cartesian components, the *elastic moduli* E_{ijkl}. Nevertheless, the number of *independent* elastic moduli is far less than 81. In fact, first of all, by the Schwarz theorem we get

$$E_{ijkl} = \frac{\partial^2 V}{\partial \varepsilon_{kl} \partial \varepsilon_{ij}} = \frac{\partial^2 V}{\partial \varepsilon_{ij} \partial \varepsilon_{kl}} = E_{klij}; \tag{2.11}$$

the above 15 relations are known as *major symmetries* and reduce the number of independent Cartesian components of $\mathbb{E}$ from 81 to 66.

Now, if we apply the Green's formula (2.6) to Eq. (2.10) we get

$$\sigma_{ij} = E_{ijkl}\varepsilon_{kl} \rightarrow \boldsymbol{\sigma} = \mathbb{E}\boldsymbol{\varepsilon}. \tag{2.12}$$

This is the *Hooke's law* (1660), establishing a linear relation between stress and strain.[3] This linearity is a direct consequence of the quadratic structure of V and of the Green's formula; though initially formulated for isotropic bodies, it is the basic

[3] An alternative, classical, approach to elasticity is to postulate the Hooke's law and the existence of V; once obtained the Green's formula, using the Schwarz theorem gives again the major symmetries, while the minor ones are still given by the symmetry of $\boldsymbol{\varepsilon}$ and $\boldsymbol{\sigma}$. Then, the expression of V is obtained integrating dV:

law of elasticity also for the more general case of anisotropic continua. Using now the Hooke's law, we obtain another reduction of the number of independent elastic moduli due to the symmetry of $\boldsymbol{\sigma}$ and $\boldsymbol{\varepsilon}$:

$$\sigma_{ij}=\sigma_{ji} \text{ and } \varepsilon_{ij}=\varepsilon_{ji} \Rightarrow E_{ijkl}=E_{jikl}=E_{ijlk}=E_{jilk} \ \forall i,j,k,l\in\{1,2,3\}. \tag{2.13}$$

The above 45 relations among the components of $\mathbb{E}$ are called the *minor symmetries* and reduce the number of independent elastic moduli to only 21. This is the highest number of independent moduli that an elastic material can have. In such a case, the material is completely anisotropic or *triclinic*. Further reductions of the number of independent moduli can be obtained only if special conditions, not universal but depending upon the material type, are introduced. Such conditions are called *elastic symmetries*, and indicate the invariance of some elastic moduli under some geometric transformations. Injecting the Hooke's law into the expression (2.10) of V we get also

$$V=\frac{1}{2}\boldsymbol{\sigma}\cdot\boldsymbol{\varepsilon}. \tag{2.14}$$

Let us now consider the inverse of the Hooke's law:

$$\varepsilon=\mathbb{Z}\sigma,\ \ \mathbb{Z}=\mathbb{E}^{-1}, \tag{2.15}$$

with $\mathbb{Z}$ the *compliance elasticity tensor*; introducing this last equation for $\boldsymbol{\varepsilon}$ into (2.10) gives

$$V=\frac{1}{2}\boldsymbol{\sigma}\cdot\mathbb{Z}\boldsymbol{\sigma}, \tag{2.16}$$

an expression called *stress energy density* in the literature.

A last remark: in this section, the word *symmetry* has been used for denoting the equivalence of the positions of an index for two or more components of the elasticity tensor; to make the distinction with the concept of *elastic symmetry*, the expression *tensor* or *index symmetry* could be used. Anyway, the reader should be aware of the fact that the same word, *symmetry*, can have two rather different meanings in our context.

2.1.1 The Voigt's Notation

The general, tensorial, expression of the Hooke's law needs the use of quantities with four indexes,

---

(Footnote 3 continued)

$$dV=\boldsymbol{\sigma}\cdot d\boldsymbol{\varepsilon}=\mathbb{E}\boldsymbol{\varepsilon}\cdot d\boldsymbol{\varepsilon}=E_{ijkl}\varepsilon_{kl}d\varepsilon_{ij}\ \rightarrow\ V=\frac{1}{2}E_{ijkl}\varepsilon_{ij}\varepsilon_{kl}=\frac{1}{2}\boldsymbol{\varepsilon}\cdot\mathbb{E}\boldsymbol{\varepsilon}.$$

$$\sigma_{ij} = E_{ijkl}\varepsilon_{kl}, \tag{2.17}$$

which can be somewhat cumbersome and heavy. That is why some simplified notations have been proposed. In particular, they allow for a matrix representation of (2.17); these formalisms switch the algebra from that of a fourth-rank tensor to that of a 6×6 square symmetric matrix.

The most well known of the matrix formalisms for anisotropic elasticity is that of Voigt (1910): the stress and strain tensors are written as follows:

$$\{\sigma\} = \begin{Bmatrix} \sigma_1 = \sigma_{11} \\ \sigma_2 = \sigma_{22} \\ \sigma_3 = \sigma_{33} \\ \sigma_4 = \sigma_{23} \\ \sigma_5 = \sigma_{31} \\ \sigma_6 = \sigma_{12} \end{Bmatrix}, \quad \{\varepsilon\} = \begin{Bmatrix} \varepsilon_1 = \varepsilon_{11} \\ \varepsilon_2 = \varepsilon_{22} \\ \varepsilon_3 = \varepsilon_{33} \\ \varepsilon_4 = 2\varepsilon_{23} \\ \varepsilon_5 = 2\varepsilon_{31} \\ \varepsilon_6 = 2\varepsilon_{12} \end{Bmatrix}. \tag{2.18}$$

Equation (2.18) shows the relations and order established for the indexes by the Voigt's notation:

$$11 \to 1, \;\; 22 \to 2, \;\; 33 \to 3, \;\; 23 \to 4, \;\; 31 \to 5, \;\; 12 \to 6. \tag{2.19}$$

The introduction of the coefficient 2 for the terms ε_4, ε_5 and ε_6 is needed for taking into account for the symmetry of $\boldsymbol{\sigma}$ and $\boldsymbol{\varepsilon}$ in the Hooke's law. This fact imposes some accuracy in the use of the Voigt's notation, because the algebras for tensors $\boldsymbol{\sigma}$ and $\boldsymbol{\varepsilon}$ are not completely the same, namely for their transformation upon axes rotation and tensor inversion.

The Voigt's notation transforms hence second rank symmetric tensors into column vectors; correspondingly, the fourth-rank elasticity tensor is transformed into a 6×6 symmetric square matrix, the symmetry of such a matrix corresponding to the major symmetries of $\mathbb{E}$. According to the index transformation rule (2.19), the matrix form of the Hooke's law with the Voigt's notation is

$$\{\sigma\} = [C]\{\varepsilon\} \;\; \to \;\; \begin{Bmatrix} \sigma_1 \\ \sigma_2 \\ \sigma_3 \\ \sigma_4 \\ \sigma_5 \\ \sigma_6 \end{Bmatrix} = \begin{bmatrix} C_{11} & C_{12} & C_{13} & C_{14} & C_{15} & C_{16} \\ C_{12} & C_{22} & C_{23} & C_{24} & C_{25} & C_{26} \\ C_{13} & C_{23} & C_{33} & C_{34} & C_{35} & C_{36} \\ C_{14} & C_{24} & C_{34} & C_{44} & C_{45} & C_{46} \\ C_{15} & C_{25} & C_{35} & C_{45} & C_{55} & C_{56} \\ C_{16} & C_{26} & C_{36} & C_{46} & C_{56} & C_{66} \end{bmatrix} \begin{Bmatrix} \varepsilon_1 \\ \varepsilon_2 \\ \varepsilon_3 \\ \varepsilon_4 \\ \varepsilon_5 \\ \varepsilon_6 \end{Bmatrix}. \tag{2.20}$$

The name $[C]$ is usually preferred to $\mathbb{E}$ to make a clear distinction between the tensor and matrix representation (for the same reason, we will name differently the compliance tensor and matrix). Thanks to the introduction of coefficients 2 in $(2.18)_2$, there is a perfect coincidence between the E_{ijkl} and the C_{pq}; it is sufficient to remind rule (2.19) to make correspond to each pair of indexes ij and kl in E_{ijkl} the correct p and q in C_{pq}; for instance, $E_{2312} = C_{46}$, $E_{1322} = C_{52}$ and so on.

Let us now consider the inverse of the Hooke's law (2.15), that we will write in the Voigt's notation as

$$\{\varepsilon\} = [S]\{\sigma\} \quad \rightarrow \quad \begin{Bmatrix} \varepsilon_1 \\ \varepsilon_2 \\ \varepsilon_3 \\ \varepsilon_4 \\ \varepsilon_5 \\ \varepsilon_6 \end{Bmatrix} = \begin{bmatrix} S_{11} & S_{12} & S_{13} & S_{14} & S_{15} & S_{16} \\ S_{12} & S_{22} & S_{23} & S_{24} & S_{25} & S_{26} \\ S_{13} & S_{23} & S_{33} & S_{34} & S_{35} & S_{36} \\ S_{14} & S_{24} & S_{34} & S_{44} & S_{45} & S_{46} \\ S_{15} & S_{25} & S_{35} & S_{45} & S_{55} & S_{56} \\ S_{16} & S_{26} & S_{36} & S_{46} & S_{56} & S_{66} \end{bmatrix} \begin{Bmatrix} \sigma_1 \\ \sigma_2 \\ \sigma_3 \\ \sigma_4 \\ \sigma_5 \\ \sigma_6 \end{Bmatrix}. \tag{2.21}$$

Unlike the case of the stiffness matrix $[C]$, not all the components of $[S]$ are equal to the corresponding ones of $\mathbb{Z}$. This is a consequence of the introduction of the factors 2 in $(2.18)_2$ and the correct transformation is

$$[S_{ij}] = \left[\begin{array}{c|c} Z_{ppqq} & 2Z_{pprs} \\ \hline 2Z_{pprs} & 4Z_{pqrs} \end{array}\right] \rightarrow$$

$$\left[\begin{array}{ccc|ccc} S_{11} = Z_{1111} & S_{12} = Z_{1122} & S_{13} = Z_{1133} & S_{14} = 2Z_{1123} & S_{15} = 2Z_{1131} & S_{16} = 2Z_{1112} \\ & S_{22} = Z_{2222} & S_{23} = Z_{2233} & S_{24} = 2Z_{2223} & S_{25} = 2Z_{2231} & S_{26} = 2Z_{2212} \\ & & S_{33} = Z_{3333} & S_{34} = 2Z_{3323} & S_{35} = 2Z_{3331} & S_{36} = 2Z_{3312} \\ \hline & & & S_{44} = 4Z_{2323} & S_{45} = 4Z_{2331} & S_{46} = 4Z_{2312} \\ & sym & & & S_{55} = 4Z_{3131} & S_{56} = 4Z_{3112} \\ & & & & & S_{66} = 4Z_{1212} \end{array}\right]. \tag{2.22}$$

The above equations show that passing to the Voigt's notation implies a different algebra for stiffness and compliance, as anticipated above. Mathematically, matrices $[C]$ and $[S]$ do not represent some second order tensors in $\mathbb{R}^6$. Practically, the use of the Voigt's notation needs some carefulness, not only in the differences between $[C]$ and $[S]$, but also in the transformation of these matrices produced by axes rotation.

2.1.2 The Kelvin's Notation

The Kelvin's notation (by somebody named *Mandel's notation*) was proposed by W. Thomson, baron of Kelvin, as early as 1856 (Kelvin 1856, 1878), but, probably because making use of irrational quantities, it has not known an as large success as the Voigt's notation. Nevertheless, rather recently a new attention has been brought by scientists on it, mainly for its algebraic properties: the Kelvin's notation has not the drawbacks of the Voigt's one, as it will be shown below.

The Kelvin's notation is different from the Voigt's one in that the coefficients 2 affecting $\boldsymbol{\varepsilon}$, Eq. (2.18), are equally distributed over $\boldsymbol{\sigma}$ and $\boldsymbol{\varepsilon}$, in such a way their product still amounts to 2:

$$\{\sigma\}=\left\{\begin{array}{c}\sigma_1=\sigma_{11}\\ \sigma_2=\sigma_{22}\\ \sigma_3=\sigma_{33}\\ \sigma_4=\sqrt{2}\sigma_{23}\\ \sigma_5=\sqrt{2}\sigma_{31}\\ \sigma_6=\sqrt{2}\sigma_{12}\end{array}\right\},\quad \{\varepsilon\}=\left\{\begin{array}{c}\varepsilon_1=\varepsilon_{11}\\ \varepsilon_2=\varepsilon_{22}\\ \varepsilon_3=\varepsilon_{33}\\ \varepsilon_4=\sqrt{2}\varepsilon_{23}\\ \varepsilon_5=\sqrt{2}\varepsilon_{31}\\ \varepsilon_6=\sqrt{2}\varepsilon_{12}\end{array}\right\}. \tag{2.23}$$

In this way, Eqs. (2.20) and (2.21) still hold but there is no difference between $\boldsymbol{\sigma}$ and $\boldsymbol{\varepsilon}$ in the transition from the tensor to the matrix representation; in particular, the way the components of $[C]$ and $[S]$ are deduced from the corresponding ones of $\mathbb{E}$ and $\mathbb{Z}$, are exactly the same (no summation over dummy indexes):

$$[C_{ij}]=\left[\begin{array}{c|c}E_{ppqq} & \sqrt{2}E_{pprs}\\ \hline \sqrt{2}E_{pprs} & 2E_{pqrs}\end{array}\right],\quad [S_{ij}]=\left[\begin{array}{c|c}Z_{ppqq} & \sqrt{2}Z_{pprs}\\ \hline \sqrt{2}Z_{pprs} & 2Z_{pqrs}\end{array}\right]. \tag{2.24}$$

The above symbolic relations can be easily translated in the detailed expressions of C_{ij} and S_{ij}, applying a scheme quite similar to that detailed in Eq. (2.22).

Merhabadi and Cowin (1990), have shown that the Kelvin's notation gives a representation of elasticity by matrices, $[C]$ and $[S]$, representing second-rank symmetric tensors in $\mathbb{R}^6$, which is not the case with the Voigt's notation. Hence, the Kelvin's notation transfers the algebra of elasticity from fourth-rank tensors in $\mathbb{R}^3$, to second-rank tensors in $\mathbb{R}^6$. This fact has some advantages, for instance the rotation of matrices $[C]$ and $[S]$ is made in the same way, unlike with the Voigt's notation. For these reasons, the Kelvin's notation is preferred in this text.

2.1.3 The Mechanical Meaning of the Anisotropic Elastic Constants

In the most general case of a triclinic material, the number of independent elastic moduli is as great as 21; it is important to understand the mechanical meaning of these parameters, because, unlike in the case of an isotropic material, *some unusual, strange mechanical effects can arise in anisotropy*. To discover these effects and connect them to particular elastic parameters, it is worth to use the compliances, i.e. the components of $[S]$ or $\mathbb{Z}$.

Let us consider a cube of a triclinic material, submitted to the only traction σ_1, see Fig. 2.1. In such a case, Eq. (2.21) gives $\varepsilon_k = S_{k1}\sigma_1\ \ \forall k = 1, \ldots, 6$ (or, equivalently, $\varepsilon_{ij} = Z_{ij11}\sigma_{11}\ \ \forall i, j = 1, 2, 3$).

So, while in an isotropic solid only the terms ε_1, corresponding to the direct stretching effect, and $\varepsilon_2 = \varepsilon_3$, corresponding to the *Poisson's effect*, i.e. the deformation in a plane orthogonal to the direction of the normal stress, are not null for a uniaxial traction, in a completely anisotropic body all the components of $\boldsymbol{\varepsilon}$ are not null: a normal stress produces also shearing strains. The coupling effects are hence

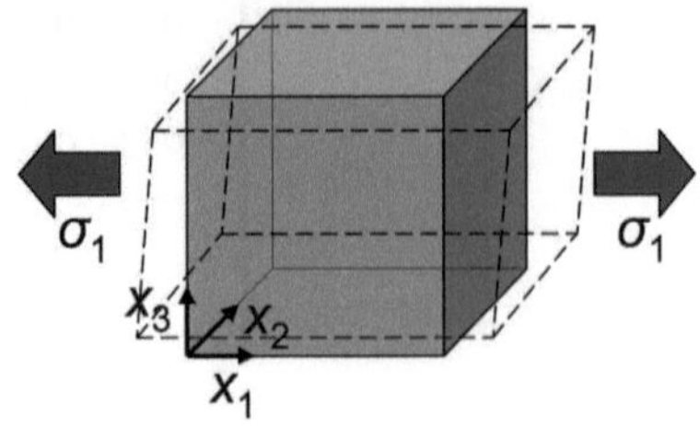

Fig. 2.1 Anisotropic stretched cube

not restricted to the only Poisson's effect, due to the terms $S_{ij},\ i, j = 1, 2, 3,\ i \neq j$: in the anisotropic case, there is also a coupling between normal stress and shear strain, due to terms $S_{kl},\ k = 4, 5, 6,\ l = 1, 2, 3$. In addition, generally speaking $S_{12} \neq S_{23} \neq S_{31}$, so that the Poisson's effect is different in the orthogonal directions, i.e. $\varepsilon_2 \neq \varepsilon_3$. In the same way, usually $S_{4k} \neq S_{5k} \neq S_{6k},\ k = 1, 2, 3$, so that also for the shearing stresses it is $\varepsilon_4 \neq \varepsilon_5 \neq \varepsilon_6$. Finally, the anisotropic cube does not only change its volume under the unique action of a traction, like in isotropic bodies, but it changes also its form: it becomes a prism with no orthogonal faces.

Let us now consider the same cube submitted to a unique shear stress, say σ_5; Eq. (2.21) gives then $\varepsilon_k = S_{k5}\sigma_5 \quad \forall k = 1, \ldots, 6$ (or, equivalently, $\varepsilon_{ij} = Z_{ij31}\sigma_{31} \quad \forall i, j = 1, 2, 3$). This time, we can observe a coupling between shear stresses and extension strains, due to the terms $S_{lk},\ k = 4, 5, 6,\ l = 1, 2, 3$ and also a coupling between a shear stress and the shearing strains in orthogonal planes, due to the terms $S_{ij},\ i, j = 4, 5, 6,\ i \neq j$. This last effect is called the *Chentsov's effect*: it is completely analogous to the Poisson's effect, but it concerns shear stresses and strains in the place of tractions and extensions. Also in this case, the couplings shear stress-extensions and the Chentsov's effect are not necessarily the same in all the planes, because generally speaking $S_{l4} \neq S_{l5} \neq S_{l6},\ l = 1, 2, 3$ and $S_{45} \neq S_{56} \neq S_{64}$. It is then apparent that, submitted to simple shear stress, the cube changes not only its shape, but also its volume, unlike in the case of isotropic bodies.

Finally, the compliance matrix can be subdivided into parts in charge of a particular effect, like in Fig. 2.2. It is immediately recognized that a similar partition is possible also for the stiffness matrix $[C]$.

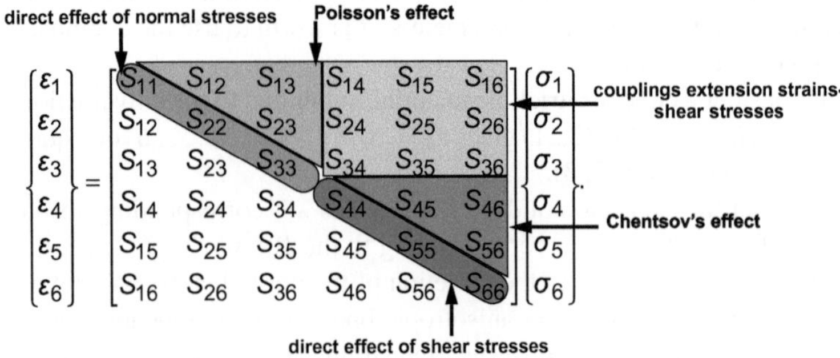

Fig. 2.2 Partition of the compliance matrix by mechanical effects

2.2 Elastic Symmetries

2.2.1 Taking into Account for Elastic Symmetries

Recalling what said in Sect. 1.1, when some kind of symmetry in the behavior is present, then some *equivalent* directions exist, whereon the behavior is the same. The figure formed with these directions is a symmetrical figure allowing all the covering operations of a certain group.

Just because the behavior is the same along equivalent directions, the forms of the elasticity matrix $[C]$[4] and of the strain energy are the same in two frames related by a covering operations. This gives some relations among the components of $[C]$, e.g. some of them are null.[5] Let us sketch the procedure for obtaining such relations:

- the expressions of the strain energy in two frames $\mathscr{R}$ and $\mathscr{R}'$ related by a covering operation are[6]

$$V=\frac{1}{2}\{\varepsilon\}^\top[C]\{\varepsilon\},\quad V'=\frac{1}{2}\{\varepsilon'\}^\top[C]\{\varepsilon'\};\tag{2.25}$$

- the strain tensor $\{\varepsilon'\}$ can be written in the frame $\mathscr{R}$:

$$\{\varepsilon'\}=[R]\{\varepsilon\},\tag{2.26}$$

  with $[R]$ the orthogonal matrix corresponding to the covering operation, i.e. to the symmetry of the material;
- injecting Eq. (2.26) into V', Eq. (2.25), and putting $V=V'$, gives the equation

$$\{\varepsilon\}^\top[C]\{\varepsilon\}=([R]\{\varepsilon\})^\top[C][R]\{\varepsilon\}\quad\forall\{\varepsilon\};\tag{2.27}$$

- this unique scalar equation gives all the relations that must be true for the components of $[C]$ *exactly because it is independent from the strain state*, i.e. because it is true $\forall\{\varepsilon\}$.

Now, the question is: which is the orthogonal matrix $[R]$ corresponding to a given covering operation? This will be the matter of the following Sections.

Before, just a last remark: the procedure sketched above is not the only one; in fact, in place of working with the strain energy, one could directly state that $[C]$,

---

[4]The use of $[C]$ is here preferred to that of $\mathbb{E}$ because it facilitates calculations; of course, the results found for components C_{ij} are immediately valid also for the E_{pqrs}, see Eqs. (2.22) and (2.24).

[5]The same is true for the stress energy; in such a case one can obtain relations among the components of $[S]$ that are exactly the same ones found for $[C]$.

[6]We denote here by a prime a component in $\mathscr{B}'$ or also, for the sake of shortness though with a slight abuse of notation, a vector or tensor whose components are intended to be given in $\mathscr{B}'$. $\mathscr{B}=\{\mathbf{e}_1,\mathbf{e}_2,\mathbf{e}_3\}$ and $\mathscr{B}'=\{\mathbf{e}'_1,\mathbf{e}'_2,\mathbf{e}'_3\}$ are two orthonormal bases of the vector space of translations $\mathscr{V}$ associated with the ordinary euclidean space $\mathscr{E}$, and they are associated with the frames $\mathscr{R}=\{o;\mathscr{B}\}$ and $\mathscr{R}'=\{o';\mathscr{B}'\}$ respectively, $o,o'\in\mathscr{E}$.

or $[S]$, does not change when passing from $\mathscr{R}$ to $\mathscr{R}'$. This approach is practically equivalent to the previous one, but it gives six scalar equations. For instance, for $[C]$ we have:

$$\begin{aligned}&\{\sigma\}=[C]\{\varepsilon\},\ and\ \{\sigma'\}=[C]\{\varepsilon'\}\ \rightarrow\ [R]\{\sigma\}=[C][R]\{\varepsilon\}\ \rightarrow\\&\{\sigma\}=[R]^\top[C][R]\{\varepsilon\}\ \Rightarrow\ [C]=[R]^\top[C][R].\end{aligned} \tag{2.28}$$

2.2.2 Rotation of Axes

Let us consider two orthonormal bases $\mathscr{B}=\{\mathbf{e}_1,\mathbf{e}_2,\mathbf{e}_3\}$ and $\mathscr{B}'=\{\mathbf{e}_1',\mathbf{e}_2',\mathbf{e}_3'\}$ and let us suppose that these two bases are related by the orthogonal tensor $\mathbf{U}$.[7]

We define $\mathbf{U}$ as the tensor such that

$$\mathbf{e}_i=\mathbf{U}\mathbf{e}_i'\ \Rightarrow\ \mathbf{e}_i'=\mathbf{U}^\top\mathbf{e}_i; \tag{2.29}$$

with this definition, it is easy to show that

$$\mathbf{U}=\left[\begin{array}{c}\mathbf{e}_1'\\ \hline \mathbf{e}_2'\\ \hline \mathbf{e}_3'\end{array}\right], \tag{2.30}$$

i.e. the matrix representing $\mathbf{U}$ in the basis $\mathscr{B}'$ has for rows the Cartesian components of the vectors of $\mathscr{B}'$, components expressed in the base $\mathscr{B}$. Algebraically, these components are the director cosines of the angles between two corresponding axes in $\mathscr{B}$ and $\mathscr{B}'$.

Using the above equations, the components in $\mathscr{B}'$ of a tensor of any rank r can be expressed as a linear combination of its components in $\mathscr{B}$, the coefficients of the combination being products of r components of $\mathbf{U}$. In fact, considering that

$$\mathbf{e}_i=\mathbf{U}\mathbf{e}_i'=U_{pq}(\mathbf{e}_p'\otimes\mathbf{e}_q')\mathbf{e}_i'=U_{pq}\delta_{qi}\mathbf{e}_p'=U_{pi}\mathbf{e}_p', \tag{2.31}$$

then, for a vector ($r=1$) it is

$$\mathbf{w}=w_i\mathbf{e}_i=w_iU_{ki}\mathbf{e}_k'\ \Rightarrow\ w_k'=U_{ki}w_i, \tag{2.32}$$

for a second-rank tensor ($r=2$) it is

$$\begin{aligned}&\mathbf{L}=L_{ij}\mathbf{e}_i\otimes\mathbf{e}_j=L_{ij}U_{mi}\mathbf{e}_m'\otimes U_{nj}\mathbf{e}_n'=U_{mi}U_{nj}L_{ij}\mathbf{e}_m'\otimes\mathbf{e}_n'\ \Rightarrow\\&L_{mn}'=U_{mi}U_{nj}L_{ij}\end{aligned} \tag{2.33}$$

[7] $\mathbf{U}$ is not necessarily a proper rotation, because reflections are possible too.

and finally for a fourth-rank tensor ($r = 4$) it is[8]

$$\mathbb{E} = E_{ijkl}\mathbf{e}_i \otimes \mathbf{e}_j \otimes \mathbf{e}_k \otimes \mathbf{e}_l = E_{ijkl}U_{mi}\mathbf{e}'_m \otimes U_{nj}\mathbf{e}'_n \otimes U_{pk}\mathbf{e}'_p \otimes U_{ql}\mathbf{e}'_q = \\ U_{mi}U_{nj}U_{pk}U_{ql}E_{ijkl}\mathbf{e}'_m \otimes \mathbf{e}'_n \otimes \mathbf{e}'_p \otimes \mathbf{e}'_q \;\Rightarrow\; E'_{mnpq} = U_{mi}U_{nj}U_{pk}U_{ql}E_{ijkl}. \tag{2.34}$$

Given $\mathbf{A}, \mathbf{B} \in$ Lin, the *conjugation product* $\mathbf{A} \boxtimes \mathbf{B}$ is the fourth-rank tensor defined by the operation

$$(\mathbf{A} \boxtimes \mathbf{B})\mathbf{C} := \mathbf{A}\mathbf{C}\mathbf{B}^\top \;\; \forall \mathbf{C} \in \text{Lin}. \tag{2.35}$$

It is worth to remark that Eq. (2.35) implies that for the vectors of a basis it is

$$(\mathbf{e}_i \otimes \mathbf{e}_j) \boxtimes (\mathbf{e}_k \otimes \mathbf{e}_l) = \mathbf{e}_i \otimes \mathbf{e}_k \otimes \mathbf{e}_j \otimes \mathbf{e}_l, \tag{2.36}$$

which gives

$$(\mathbf{A} \boxtimes \mathbf{B})_{ijkl} = A_{ik}B_{jl}. \tag{2.37}$$

Once defined the transpose $\mathbb{A}^\top$ of a fourth-rank tensor $\mathbb{A}$ as the unique tensor such that

$$\mathbf{L} \cdot (\mathbb{A}\mathbf{M}) = \mathbf{M} \cdot (\mathbb{A}^\top \mathbf{L}) \;\; \forall \mathbf{L}, \mathbf{M} \in \text{Lin}, \tag{2.38}$$

it is immediate to show that

$$\begin{aligned} (\mathbf{A} \otimes \mathbf{B})^\top &= \mathbf{B} \otimes \mathbf{A}, \\ (\mathbf{A} \boxtimes \mathbf{B})^\top &= \mathbf{A}^\top \boxtimes \mathbf{B}^\top. \end{aligned} \tag{2.39}$$

Like for tensors in Lin, also a tensor $\mathbb{A} \in \mathbb{L}$in is said to be *symmetric* $\iff \mathbb{A} = \mathbb{A}^\top$. It is simple to check that

$$\mathbb{A} = \mathbb{A}^\top \;\Rightarrow\; A^\top_{ijkl} = A_{klij}, \tag{2.40}$$

i.e., the major symmetries of the elastic tensors $\mathbb{E}$ and $\mathbb{Z}$ actually coincide with the definition of symmetric tensor in $\mathbb{L}$in.

For an orthogonal second-rank tensor $\mathbf{U}$, we define its *orthogonal conjugator* $\mathbb{U}$ as

$$\mathbb{U} := \mathbf{U} \boxtimes \mathbf{U}; \tag{2.41}$$

it is not difficult to show that just as $\mathbf{U}$ preserves scalar products of elements in $\mathscr{V}$, its associated orthogonal conjugator $\mathbb{U}$ preserves scalar products in Lin:

$$\mathbb{U}\mathbf{A} \cdot \mathbb{U}\mathbf{B} = \mathbf{A} \cdot \mathbf{B} \;\; \forall \mathbf{A}, \mathbf{B} \in \text{Lin}. \tag{2.42}$$

[8] $\forall\, \mathbf{A}, \mathbf{B}$ and $\mathbf{L} \in$Lin, $\mathbf{A} \otimes \mathbf{B}$ is the fourth-rank tensor defined by the operation $(\mathbf{A} \otimes \mathbf{B})\mathbf{L} := (\mathbf{B} \cdot \mathbf{L})\mathbf{A}$. Applying this rule to the dyads of a basis, we get a fundamental result: $(\mathbf{e}_i \otimes \mathbf{e}_j \otimes \mathbf{e}_k \otimes \mathbf{e}_l)(\mathbf{e}_p \otimes \mathbf{e}_q) = (\mathbf{e}_k \otimes \mathbf{e}_l) \cdot (\mathbf{e}_p \otimes \mathbf{e}_q)(\mathbf{e}_i \otimes \mathbf{e}_j) = \delta_{kp}\delta_{lq}(\mathbf{e}_i \otimes \mathbf{e}_j)$.

In other words, $\mathbb{U}$ is an *orthogonal tensor* in $\mathbb{L}$in (Podio-Guidugli 2000, p. 55). Introducing the *identity* of $\mathbb{L}$in,

$$\mathbb{I} = \mathbf{I} \boxtimes \mathbf{I} \;\Rightarrow\; \mathbb{I} = I_{ijkl}(\mathbf{e}_i \otimes \mathbf{e}_j \otimes \mathbf{e}_k \otimes \mathbf{e}_l) = \delta_{ik}\delta_{jl}(\mathbf{e}_i \otimes \mathbf{e}_j \otimes \mathbf{e}_k \otimes \mathbf{e}_l), \tag{2.43}$$

it is easy to recognize that also for rotations in $\mathbb{L}$in

$$\mathbb{U}\mathbb{U}^\top = \mathbb{U}^\top\mathbb{U} = \mathbb{I}. \tag{2.44}$$

Be $\mathbf{n} \in \mathscr{V}$ a unit vector and let us suppose that $\mathbf{n}$ is orthogonal to a symmetry plane. Then

$$\mathbf{U} := \mathbf{I} - 2\mathbf{n} \otimes \mathbf{n}, \;\; \mathbf{U} = \mathbf{U}^\top, \;\; \mathbb{U} = \mathbf{U} \boxtimes \mathbf{U} = \mathbb{U}^\top, \tag{2.45}$$

is the orthogonal tensor describing the symmetry in the plane whose normal is $\mathbf{n}$. In fact,

$$\begin{aligned} &\mathbf{U}\mathbf{n} = -\mathbf{n} \\ &\mathbf{U}\mathbf{m} = \mathbf{m} \;\; \forall \mathbf{m} \in \mathscr{V} : \; \mathbf{m} \cdot \mathbf{n} = 0, \; |\mathbf{m}| = 1. \end{aligned} \tag{2.46}$$

Thanks to these last definitions, it is possible to give a compact form to results (2.32), (2.33) and (2.34):

$$\begin{aligned} &\mathbf{w}' = \mathbf{U}\mathbf{w}, \\ &\mathbf{L}' = \mathbf{U}\mathbf{L}\mathbf{U}^\top = (\mathbf{U} \boxtimes \mathbf{U})\mathbf{L} = \mathbb{U}\mathbf{L}, \\ &\mathbb{E}' = (\mathbf{U} \boxtimes \mathbf{U})\mathbb{E}(\mathbf{U} \boxtimes \mathbf{U})^\top = \mathbb{U}\mathbb{E}\mathbb{U}^\top. \end{aligned} \tag{2.47}$$

Using Eq. (2.23) and the result of Eq. (2.37), we can now obtain the matrix $[R]$ that corresponds, in the Kelvin's notation, to tensor $\mathbb{U}$; the calculations are rather tedious and a little bit long, but the final result is (Mehrabadi and Cowin 1990),

$$[R] = \begin{bmatrix} U_{11}^2 & U_{12}^2 & U_{13}^2 & \sqrt{2}U_{12}U_{13} & \sqrt{2}U_{13}U_{11} & \sqrt{2}U_{11}U_{12} \\ U_{21}^2 & U_{22}^2 & U_{23}^2 & \sqrt{2}U_{22}U_{23} & \sqrt{2}U_{23}U_{21} & \sqrt{2}U_{21}U_{22} \\ U_{31}^2 & U_{32}^2 & U_{33}^2 & \sqrt{2}U_{32}U_{33} & \sqrt{2}U_{33}U_{31} & \sqrt{2}U_{31}U_{32} \\ \sqrt{2}U_{21}U_{31} & \sqrt{2}U_{22}U_{32} & \sqrt{2}U_{23}U_{33} & U_{23}U_{32} + U_{22}U_{33} & U_{33}U_{21} + U_{31}U_{23} & U_{31}U_{22} + U_{32}U_{21} \\ \sqrt{2}U_{31}U_{11} & \sqrt{2}U_{32}U_{12} & \sqrt{2}U_{33}U_{13} & U_{32}U_{13} + U_{33}U_{12} & U_{31}U_{13} + U_{33}U_{11} & U_{31}U_{12} + U_{32}U_{11} \\ \sqrt{2}U_{11}U_{21} & \sqrt{2}U_{12}U_{22} & \sqrt{2}U_{13}U_{23} & U_{12}U_{23} + U_{13}U_{22} & U_{11}U_{23} + U_{13}U_{21} & U_{11}U_{22} + U_{12}U_{21} \end{bmatrix}. \tag{2.48}$$

The above matrix $[R]$ allows for the change of basis of any second-rank tensor in the Kelvin's notation. In particular for $\boldsymbol{\sigma}$ and $\boldsymbol{\varepsilon}$:

$$\{\sigma'\} = [R]\{\sigma\}, \quad \{\varepsilon'\} = [R]\{\varepsilon\}. \tag{2.49}$$

It can be checked that, when $\mathbf{U}$ is an orthogonal tensor, then

$$[R][R]^\top = [R]^\top[R] = [I], \tag{2.50}$$

i.e., $[R]$ is an orthogonal matrix in $\mathbb{R}^6$; this is not the case with the Voigt's notation. Hence, $[R]$ represents, in the given basis, an orthogonal tensor of Lin over a manifold of dimension six.

It is impossible to put the result of Eq. (2.34) in matrix form, because also in the Kelvin's notation it depends upon four indexes; nevertheless, it is of course possible to express all the components of such an operator, but actually in the most general case these components have so extremely complicate and long expressions that it is practically impossible to write down all of them, so they are omitted here.

2.2.3 *A Tensorial Characterization of Elastic Symmetries*

The results of the previous section give us the possibility of characterizing in an elegant tensorial form the existence of elastic symmetries in a solid (Podio-Guidugli 2000, p. 56).

Let us suppose that a material has a given elastic symmetry and that the two bases $\mathscr{B}$ and $\mathscr{B}'$ correspond to equivalent directions with respect to the symmetry of concern. Physically, this means that it is not possible to detect the change from $\mathscr{B}$ to $\mathscr{B}'$ by experiments measuring stresses, because the behavior is exactly the same in the two cases: $\mathbb{E} = \mathbb{E}'$. Then, applying Eq. $(2.47)_2$ to $\boldsymbol{\sigma}$ and $\boldsymbol{\varepsilon}$,

$$\boldsymbol{\sigma}' = \mathbb{U}\boldsymbol{\sigma}, \quad \boldsymbol{\varepsilon}' = \mathbb{U}\boldsymbol{\varepsilon}, \tag{2.51}$$

and the Hooke's law, Eq. (2.12), we get, because $\mathbb{E} = \mathbb{E}'$,

$$\boldsymbol{\sigma}' = \mathbb{E}\boldsymbol{\varepsilon}' \;\rightarrow\; \mathbb{U}\boldsymbol{\sigma} = \mathbb{E}\mathbb{U}\boldsymbol{\varepsilon} \;\rightarrow \mathbb{U}\mathbb{E}\boldsymbol{\varepsilon} = \mathbb{E}\mathbb{U}\boldsymbol{\varepsilon} \;\Rightarrow\; \mathbb{U}\mathbb{E} = \mathbb{E}\mathbb{U}. \tag{2.52}$$

Hence, an orthogonal transformation $\mathbf{U}$ is in the elastic symmetry group of the material *if and only if* $\mathbb{E}$ *and* $\mathbb{U}$ *commute*, $\mathbb{U}$ being the orthogonal conjugator of $\mathbf{U}$.

The result of Eq. (2.52) constitutes also a way for determining the number and type of independent elastic moduli, i.e. the distinct components of $\mathbb{E}$; this is the way sketched, with reference to matrix $[C]$, in the last paragraph of Sect. 2.2.1, Eq. (2.28); however, as said thereon, the energetic approach is preferred in this text.

2.2.4 *Triclinic Bodies*

A triclinic body has no material symmetries, so Eq. (2.27) cannot be written. As a consequence, it is not possible to reduce the number of independent elastic components, that remains fixed to 21. Matrix $[C]$ appears hence as

$$[C]=\begin{bmatrix} C_{11} & C_{12} & C_{13} & C_{14} & C_{15} & C_{16}\\ & C_{22} & C_{23} & C_{24} & C_{25} & C_{26}\\ & & C_{33} & C_{34} & C_{35} & C_{36}\\ & & & C_{44} & C_{45} & C_{46}\\ & sym & & & C_{55} & C_{56}\\ & & & & & C_{66}\end{bmatrix}. \tag{2.53}$$

2.2.5 *Monoclinic Bodies*

The only symmetry of a monoclinic body is a reflection in a plane. Without loss in generality, we can suppose $x_3 = 0$ to be the symmetry plane. In such a case it is, see Eqs. (2.45) and (2.48),

$$\mathbf{U}=\begin{bmatrix}1&0&0\\0&1&0\\0&0&-1\end{bmatrix}\Rightarrow [R]=\begin{bmatrix}1&0&0&0&0&0\\0&1&0&0&0&0\\0&0&1&0&0&0\\0&0&0&-1&0&0\\0&0&0&0&-1&0\\0&0&0&0&0&1\end{bmatrix}, \tag{2.54}$$

that applied to Eq. (2.27) gives the condition

$$\begin{aligned}&C_{14}\varepsilon_1\varepsilon_4+C_{24}\varepsilon_2\varepsilon_4+C_{34}\varepsilon_3\varepsilon_4+C_{15}\varepsilon_1\varepsilon_5+\\&C_{25}\varepsilon_2\varepsilon_5+C_{35}\varepsilon_3\varepsilon_5+C_{46}\varepsilon_4\varepsilon_6+C_{56}\varepsilon_5\varepsilon_6=0,\end{aligned} \tag{2.55}$$

which is satisfied $\forall\boldsymbol{\varepsilon} \iff$

$$C_{14}=C_{24}=C_{34}=C_{15}=C_{25}=C_{35}=C_{46}=C_{56}=0. \tag{2.56}$$

Hence, a monoclinic body depends upon only 13 distinct elastic moduli:

$$[C]=\begin{bmatrix} C_{11} & C_{12} & C_{13} & 0 & 0 & C_{16}\\ & C_{22} & C_{23} & 0 & 0 & C_{26}\\ & & C_{33} & 0 & 0 & C_{36}\\ & & & C_{44} & C_{45} & 0\\ & sym & & & C_{55} & 0\\ & & & & & C_{66}\end{bmatrix}. \tag{2.57}$$

2.2.6 *Orthotropic Bodies*

Let us now add another plane of symmetry orthogonal to the previous one, say the plane $x_2 = 0$. Following the same procedure, we get successively:

$$\mathbf{U} = \begin{bmatrix} 1 & 0 & 0 \\ 0 & -1 & 0 \\ 0 & 0 & 1 \end{bmatrix} \Rightarrow [R] = \begin{bmatrix} 1 & 0 & 0 & 0 & 0 & 0 \\ 0 & 1 & 0 & 0 & 0 & 0 \\ 0 & 0 & 1 & 0 & 0 & 0 \\ 0 & 0 & 0 & -1 & 0 & 0 \\ 0 & 0 & 0 & 0 & 1 & 0 \\ 0 & 0 & 0 & 0 & 0 & -1 \end{bmatrix}, \tag{2.58}$$

$$\begin{aligned} &(C_{14}\varepsilon_1 + C_{24}\varepsilon_2 + C_{34}\varepsilon_3 + C_{45}\varepsilon_5)\varepsilon_4 + \\ &(C_{16}\varepsilon_1 + C_{26}\varepsilon_2 + C_{36}\varepsilon_3 + C_{56}\varepsilon_5)\varepsilon_6 = 0 \ \forall \boldsymbol{\varepsilon} \iff \\ &C_{14} = C_{24} = C_{34} = C_{45} = C_{16} = C_{26} = C_{36} = C_{56} = 0. \end{aligned} \tag{2.59}$$

So, the existence of the second plane of symmetry has added the four supplementary conditions

$$C_{16} = C_{26} = C_{36} = C_{45} = 0 \tag{2.60}$$

to the previous eight ones, reducing hence to only 9 the number of distinct elastic moduli. Let us now suppose the existence of a third plane of symmetry, orthogonal to the previous ones, the plane $x_1 = 0$. With the same procedure, we get:

$$\mathbf{U} = \begin{bmatrix} -1 & 0 & 0 \\ 0 & 1 & 0 \\ 0 & 0 & 1 \end{bmatrix} \Rightarrow [R] = \begin{bmatrix} 1 & 0 & 0 & 0 & 0 & 0 \\ 0 & 1 & 0 & 0 & 0 & 0 \\ 0 & 0 & 1 & 0 & 0 & 0 \\ 0 & 0 & 0 & 1 & 0 & 0 \\ 0 & 0 & 0 & 0 & -1 & 0 \\ 0 & 0 & 0 & 0 & 0 & -1 \end{bmatrix}, \tag{2.61}$$

$$\begin{aligned} &(C_{15}\varepsilon_1 + C_{25}\varepsilon_2 + C_{35}\varepsilon_3 + C_{45}\varepsilon_4)\varepsilon_5 + \\ &(C_{16}\varepsilon_1 + C_{26}\varepsilon_2 + C_{36}\varepsilon_3 + C_{46}\varepsilon_4)\varepsilon_6 = 0 \ \forall \boldsymbol{\varepsilon} \iff \\ &C_{15} = C_{25} = C_{35} = C_{45} = C_{16} = C_{26} = C_{36} = C_{46} = 0. \end{aligned} \tag{2.62}$$

Rather surprisingly, this new symmetry condition does not give any supplementary condition to those in (2.56) and (2.60). Because the procedure does not depend upon the order of the symmetries, as it is immediately recognized, the only consequence is that the existence of two orthogonal planes of elastic symmetry is physically impossible: only the presence of one or three mutually orthogonal planes of symmetry is admissible. A material having three planes of symmetry is called *orthotropic*. The class of orthotropic materials is very important, because a lot of materials or structures belong to it. An orthotropic material depends hence upon 9 distinct elastic moduli and its matrix [C] looks like

$$[C]=\begin{bmatrix} C_{11} & C_{12} & C_{13} & 0 & 0 & 0\\ & C_{22} & C_{23} & 0 & 0 & 0\\ & & C_{33} & 0 & 0 & 0\\ & & & C_{44} & 0 & 0\\ & sym & & & C_{55} & 0\\ & & & & & C_{66}\end{bmatrix}. \tag{2.63}$$

2.2.7 Axially Symmetric Bodies

We have seen in Sect. 1.4.2 that there are only four possible cases of axial symmetries for crystals: the 2-, 3-, 4- and 6-fold axis of symmetry. In elasticity, there is another possibility, that will be examined in the next Section. Let us then consider the above four cases, in the order, taking as symmetry axis, without loss in generality, the axis x_3.

Let us begin with a 2-fold axis of symmetry; the covering operation corresponds hence to a rotation of π about x_3. In such a case, we have

$$\mathbf{U}=\begin{bmatrix}-1 & 0 & 0\\ 0 & -1 & 0\\ 0 & 0 & 1\end{bmatrix}\Rightarrow [R]=\begin{bmatrix}1&0&0&0&0&0\\0&1&0&0&0&0\\0&0&1&0&0&0\\0&0&0&-1&0&0\\0&0&0&0&-1&0\\0&0&0&0&0&1\end{bmatrix}, \tag{2.64}$$

and we can observe that the matrix $[R]$ is the same of the monoclinic case, Eq. (2.54). Hence, a 2-fold axis of symmetry coincides with a plane of symmetry.

For a 3-fold axis of symmetry, the covering operation corresponds to a rotation of $2\pi/3$ about x_3, which gives

$$\mathbf{U}=\begin{bmatrix}-\frac{1}{2} & \frac{\sqrt{3}}{2} & 0\\ -\frac{\sqrt{3}}{2} & -\frac{1}{2} & 0\\ 0 & 0 & 1\end{bmatrix}\Rightarrow [R]=\begin{bmatrix}\frac{1}{4} & \frac{3}{4} & 0 & 0 & 0 & -\sqrt{\frac{3}{8}}\\ \frac{3}{4} & \frac{1}{4} & 0 & 0 & 0 & \sqrt{\frac{3}{8}}\\ 0&0&1&0&0&0\\ 0&0&0&-\frac{1}{2}&-\frac{\sqrt{3}}{2}&0\\ 0&0&0&\frac{\sqrt{3}}{2}&-\frac{1}{2}&0\\ \sqrt{\frac{3}{8}} & -\sqrt{\frac{3}{8}} & 0&0&0&-\frac{1}{2}\end{bmatrix}; \tag{2.65}$$

in this case, condition (2.27) is very long and omitted here, but finally it gives 14 conditions on the components of $[C]$:

$$\begin{aligned}&C_{16}=C_{26}=C_{34}=C_{35}=C_{36}=C_{45}=0,\\&C_{22}=C_{11},\ \ C_{55}=C_{44},\ \ C_{23}=C_{13},\ \ C_{24}=-C_{14},\\&C_{25}=-C_{15},\ \ C_{56}=\sqrt{2}C_{14},\ \ C_{46}=\sqrt{2}C_{15},\ \ C_{66}=C_{11}-C_{12}.\end{aligned}\tag{2.66}$$

So, there are only 7 distinct elastic moduli:

$$[C]=\begin{bmatrix}C_{11}&C_{12}&C_{13}&C_{14}&C_{15}&0\\&C_{11}&C_{13}&-C_{14}&-C_{15}&0\\&&C_{33}&0&0&0\\&&&C_{44}&0&-\sqrt{2}C_{15}\\&sym&&&C_{44}&\sqrt{2}C_{14}\\&&&&&C_{11}-C_{12}\end{bmatrix}.\tag{2.67}$$

For a 4-fold axis of symmetry, the covering operation corresponds to a rotation of $\pi/2$ about x_3, which gives

$$\mathbf{U}=\begin{bmatrix}0&1&0\\-1&0&0\\0&0&1\end{bmatrix}\Rightarrow[R]=\begin{bmatrix}0&1&0&0&0&0\\1&0&0&0&0&0\\0&0&1&0&0&0\\0&0&0&0&-1&0\\0&0&0&1&0&0\\0&0&0&0&0&-1\end{bmatrix};\tag{2.68}$$

we omit also in this case the long expression of Eq. (2.27), but the final result are 14 conditions different from the (2.66):

$$\begin{aligned}&C_{14}=C_{24}=C_{34}=C_{15}=C_{25}=C_{35}=C_{45}=C_{36}=C_{46}=C_{56}=0,\\&C_{22}=C_{11},\ \ C_{55}=C_{44},\ \ C_{23}=C_{13},\ \ C_{26}=-C_{16},\end{aligned}\tag{2.69}$$

leaving an elastic matrix $[C]$ still depending upon only 7 distinct moduli, but different from the previous case, Eq. (2.67):

$$[C]=\begin{bmatrix}C_{11}&C_{12}&C_{13}&0&0&C_{16}\\&C_{11}&C_{13}&0&0&-C_{16}\\&&C_{33}&0&0&0\\&&&C_{44}&0&0\\&sym&&&C_{44}&0\\&&&&&C_{66}\end{bmatrix}.\tag{2.70}$$

The last case is that of a 6-fold axis of symmetry, with as covering operation a rotation of $\pi/3$ about x_3, which gives

$$\mathbf{U}=\begin{bmatrix}\frac{1}{2} & \frac{\sqrt{3}}{2} & 0\\ -\frac{\sqrt{3}}{2} & \frac{1}{2} & 0\\ 0 & 0 & 1\end{bmatrix}\Rightarrow [R]=\begin{bmatrix}\frac{1}{4} & \frac{3}{4} & 0 & 0 & 0 & \sqrt{\frac{3}{8}}\\ \frac{3}{4} & \frac{1}{4} & 0 & 0 & 0 & -\sqrt{\frac{3}{8}}\\ 0 & 0 & 1 & 0 & 0 & 0\\ 0 & 0 & 0 & \frac{1}{2} & -\frac{\sqrt{3}}{2} & 0\\ 0 & 0 & 0 & \frac{\sqrt{3}}{2} & \frac{1}{2} & 0\\ -\sqrt{\frac{3}{8}} & \sqrt{\frac{3}{8}} & 0 & 0 & 0 & -\frac{1}{2}\end{bmatrix};\tag{2.71}$$

condition (2.27), omitted because too long, gives in this case 16 conditions:

$$\begin{aligned}&C_{14}=C_{24}=C_{34}=C_{15}=C_{25}=C_{35}=\\&C_{45}=C_{16}=C_{26}=C_{36}=C_{46}=C_{56}=0,\\&C_{22}=C_{11},\ \ C_{55}=C_{44},\ \ C_{23}=C_{13},\ \ C_{66}=C_{11}-C_{12},\end{aligned}\tag{2.72}$$

for a final elastic matrix $[C]$ depending upon only 5 moduli:

$$[C]=\begin{bmatrix}C_{11} & C_{12} & C_{13} & 0 & 0 & 0\\ & C_{11} & C_{13} & 0 & 0 & 0\\ & & C_{33} & 0 & 0 & 0\\ & & & C_{44} & 0 & 0\\ & sym & & & C_{44} & 0\\ & & & & & C_{11}-C_{12}\end{bmatrix}.\tag{2.73}$$

2.2.8 *Transversely Isotropic Bodies*

Let us now consider the case of a material with an axis of cylindrical symmetry, i.e. an axis of symmetry where the covering operation is a rotation by any angle θ; such a material is called *transversely isotropic*, and many materials belong to this class, like for instance wood, fiber reinforced composites, laminated steel and so on. Also in this case we can proceed in the usual way. Denoting, for shortening the expressions, $c=\cos\theta,\ s=\sin\theta$, we get:

$$\mathbf{U}=\begin{bmatrix}c & s & 0\\ -s & c & 0\\ 0 & 0 & 1\end{bmatrix}\Rightarrow [R]=\begin{bmatrix}c^2 & s^2 & 0 & 0 & 0 & \sqrt{2}cs\\ s^2 & c^2 & 0 & 0 & 0 & -\sqrt{2}cs\\ 0 & 0 & 1 & 0 & 0 & 0\\ 0 & 0 & 0 & c & -s & 0\\ 0 & 0 & 0 & s & c & 0\\ -\sqrt{2}cs & \sqrt{2}cs & 0 & 0 & 0 & c^2-s^2\end{bmatrix}.\tag{2.74}$$

In this case we obtain exactly the same 16 conditions (2.72); this means that, elastically, the 6-fold axis of symmetry is strictly identical to an axis of cylindrical symmetry. Hence, two such materials cannot be distinguished using only the results of tests on stress or strain. This must not surprise, because this fact is in perfect accordance with the Neumann's principle, as the 6-fold axis of symmetry is contained in the more general case of cylindrical symmetry.

Finally, Eq. (2.73) represents also the elastic matrix of a transversely isotropic material, who has 5 distinct elastic moduli.

2.2.9 *Isotropic Bodies*

Iostropy is the complete symmetry: all the directions are equivalent. The conditions of isotropy could be find following the usual procedure, imposing that Eq. (2.27) is valid for any orthogonal transformation $[R]$. However, this general approach, that can be followed using for instance the Euler angles for expressing a generic $[R]$, results to be very cumbersome and computationally heavy.

A more direct approach is the following one: for a transversely isotropic body, all the directions orthogonal to the axis of symmetry, say x_3, are equivalent. In other words, fixing the axes of x_1 and x_2 is completely arbitrary. Let us then suppose that, besides the equivalence of all the directions in the plane perpendicular to x_3, also x_1 and x_3 are equivalent. We then impose to a material described by a transversely isotropic elastic matrix, Eq. (2.73), this further equivalence, which is described by

$$\mathbf{U}=\begin{bmatrix}0&0&1\\0&1&0\\1&0&0\end{bmatrix}\ \Rightarrow\ [R]=\begin{bmatrix}0&0&1&0&0&0\\0&1&0&0&0&0\\1&0&0&0&0&0\\0&0&0&0&0&1\\0&0&0&0&1&0\\0&0&0&1&0&0\end{bmatrix}. \tag{2.75}$$

This gives three new conditions:

$$C_{13}=C_{12},\quad C_{33}=C_{11},\quad C_{44}=C_{66}, \tag{2.76}$$

which reduce the number of distinct elastic constants from 5 to only 2:

$$[C]=\begin{bmatrix}C_{11}&C_{12}&C_{12}&0&0&0\\&C_{11}&C_{12}&0&0&0\\&&C_{11}&0&0&0\\&&&C_{11}-C_{12}&0&0\\&sym&&&C_{11}-C_{12}&0\\&&&&&C_{11}-C_{12}\end{bmatrix}. \tag{2.77}$$

Because x_1 is any direction, all the directions of the space are equivalent; this can be proved showing that the elastic matrix (2.77) is insensitive to any change of basis leaving x_2 unchanged, i.e.

$$\mathbf{U} = \begin{bmatrix} c & 0 & s \\ 0 & 1 & 0 \\ -s & 0 & c \end{bmatrix} \Rightarrow [R] = \begin{bmatrix} c^2 & 0 & s^2 & 0 & \sqrt{2}cs & 0 \\ 0 & 1 & 0 & 0 & 0 & 0 \\ s^2 & 0 & c^2 & 0 & -\sqrt{2}cs & 0 \\ 0 & 0 & 0 & c & 0 & -s \\ -\sqrt{2}cs & 0 & \sqrt{2}cs & 0 & c^2-s^2 & 0 \\ 0 & 0 & 0 & s & 0 & c \end{bmatrix}, \tag{2.78}$$

which gives as only condition $C_{44} = C_{11} - C_{12}$, a condition already contained in the previous ones, Eq. (2.72) and (2.76). This proves that nothing is added to the previous conditions and hence that all the directions in any meridian plane are equivalent, i.e. that the body is isotropic.

There is another, more elegant and direct way to prove that an isotropic body depends upon only two distinct moduli: because of isotropy, the elastic response is insensitive to a change of frame, so the elastic moduli of an isotropic material cannot be frame-dependent. This means that for an isotropic material, V cannot depend upon the ε_{ij}, that are frame-dependent quantities, but rather on the *invariants* of $\boldsymbol{\varepsilon}$.[9] As a consequence, being V a quadratic function of the ε_{ij}, its general expression is of the type

$$V = \frac{1}{2}c_1 I_1^2 + c_2 I_2, \tag{2.79}$$

with[10]

$$I_1 = \mathrm{tr}\boldsymbol{\varepsilon} = \varepsilon_{ii}, \quad I_2 = \frac{\mathrm{tr}^2\boldsymbol{\varepsilon} - \mathrm{tr}\boldsymbol{\varepsilon}^2}{2} = \frac{\varepsilon_{ii}\,\varepsilon_{ii} - \varepsilon_{ij}\,\varepsilon_{ji}}{2}. \tag{2.80}$$

The third order invariant of $\boldsymbol{\varepsilon}$, i.e. $\det\boldsymbol{\varepsilon}$, cannot enter in the expression of V, because it is a cubic function of the ε_{ij}, while V must be a quadratic function of the ε_{ij}. Then,

$$V = \frac{1}{2}[(c_1 + c_2)\varepsilon_{ii}\,\varepsilon_{ii} - c_2\,\varepsilon_{ij}\,\varepsilon_{ji}], \tag{2.81}$$

so that[11]

$$\begin{aligned} \sigma_{\underline{ii}} &= \frac{\partial V}{\partial \varepsilon_{\underline{ii}}} = (c_1 + c_2)\varepsilon_{ii} - c_2\,\varepsilon_{\underline{ii}}, \\ \sigma_{ij} &= \frac{\partial V}{\partial \varepsilon_{ij}} = -c_2\,\varepsilon_{ji} = -c_2\,\varepsilon_{ij}. \end{aligned} \tag{2.82}$$

---

[9]The elastic potential V is, as any other quantity derived by a scalar product, an invariant, i.e. it is not frame-dependent. Hence, because $[C]$ for an isotropic material is frame independent, the expression of V cannot depend upon frame-dependent quantities, the ε_{ij}, but only upon frame-independent functions of the ε_{ij}: the invariants of $\boldsymbol{\varepsilon}$.

[10]$\boldsymbol{\varepsilon}^2 = \boldsymbol{\varepsilon}\boldsymbol{\varepsilon} = \varepsilon_{ij}\mathbf{e}_i \otimes \mathbf{e}_j\,\varepsilon_{hk}\mathbf{e}_h \otimes \mathbf{e}_k = \varepsilon_{ij}\,\varepsilon_{hk}\,\mathbf{e}_j \cdot \mathbf{e}_h(\mathbf{e}_i \otimes \mathbf{e}_k) = \varepsilon_{ij}\,\varepsilon_{hk}\,\delta_{jh}(\mathbf{e}_i \otimes \mathbf{e}_k) \to \mathrm{tr}\boldsymbol{\varepsilon}^2 = \varepsilon_{ij}\,\varepsilon_{hk}\,\delta_{jh}\mathrm{tr}(\mathbf{e}_i \otimes \mathbf{e}_k) = \varepsilon_{ij}\,\varepsilon_{hk}\,\delta_{jh}\delta_{ik} = \varepsilon_{ij}\,\varepsilon_{ji}$.

[11]Following a common practice, when an index is underlined, it is not a dummy index: no summation over it.

For instance:

$$\begin{aligned}\sigma_{11} &= \frac{\partial V}{\partial \varepsilon_{11}} = (c_1 + c_2)(\varepsilon_{11} + \varepsilon_{22} + \varepsilon_{33}) - c_2\,\varepsilon_{11},\\ \sigma_{12} &= \frac{\partial V}{\partial \varepsilon_{12}} = -c_2\,\varepsilon_{12} \quad \text{etc.}\end{aligned} \tag{2.83}$$

We see hence that in the case of isotropic materials, only two constants are sufficient to characterize the elastic behavior.

2.2.10 Some Remarks About Elastic Symmetries

Some remarks can be done about the results found in the previous Sections. First of all, the results given, in all the cases, for $[C]$ are completely valid also for $[S]$; this is not the case with the Voigt's notation, where for some symmetries, not each S_{ij} has the same expression of the corresponding C_{ij}.

A mechanically important remark is the fact that typically some coupling components disappear in a symmetry basis. The case of orthotropic bodies is emblematic: in the orthotropic frame, the skyline of $[C]$ is exactly the same of an isotropic body and the only coupling is the Poisson's effect. Nevertheless, this is no longer true in any other basis: *in a generic basis, all the anisotropic materials, regardless of their symmetries, behave like a triclinic body, i.e. they have all the coupling terms (generally speaking, their elastic matrix is complete, none of its terms vanishes).*

The only exception to this fact is isotropy; in fact, for an isotropic body the matrices $[C]$ and $[S]$ are completely invariant, i.e. *their only two distinct moduli are tensor invariants* and the only possible coupling is the Poisson's effect. This is the obvious consequence of the fact that all the directions of the space are equivalent. Physically, the fact that the least number of independent elastic constants is two means that in a stressed elastic body there are, in general, at least two distinct and independent deformation effects.

2.2.11 Elasticity of Crystals and Elastic Syngonies

Crystals have an elastic behavior that belongs to one of the cases above or is a combination of these cases. Examining their cases, allows us for entirely defining the ten elastic syngonies introduced in Sect. 1.4.2. In particular, referring to the Voigt's classification, Table 1.1, it is[12]:

[12]We recall that the following classification is based upon the definition of elastic syngony as a class of materials sharing the same number and type of independent elastic moduli, see Sect. 1.4.2.

1. classes 1 and 2 belong to the triclinic case, with 21 constants; their matrix $[C]$ is like in Eq. (2.53) and this crystal syngony corresponds with the *triclinic elastic syngony*;
2. classes 3, 4 and 5 belong to the monoclinic case, with 13 constants; their matrix $[C]$ is like in Eq. (2.57) and this crystal syngony corresponds with the *monoclinic elastic syngony*;
3. classes 6, 7 and 8 of the orthorhombic syngony belong to the orthotropic case, with 9 constants; their matrix $[C]$ is like in Eq. (2.63) and the orthorhombic syngony corresponds hence entirely with the *orthotropic elastic syngony*;
4. classes 12 and 13 of the trigonal syngony belong to the 3-fold rotational symmetry case, with 7 constants; they have a matrix $[C]$ as in Eq. (2.67) and they constitute the *trigonal elastic syngony with 7 constants*;
5. classes 17, 18 and 20 of the tetragonal syngony belong to the 4-fold rotational symmetry case, with 7 constants; their matrix is like in Eq. (2.70) and they constitute the *tetragonal elastic syngony with 7 constants*;
6. classes 9, 10 and 11 of the trigonal syngony are a combination of the 3-fold rotational symmetry and the monoclinic symmetry cases: if the plane of symmetry is the plane $x_1 = 0$, then the usual procedure applied to the matrix (2.67) gives $C_{15} = 0$, and matrix (2.67) becomes

$$[C] = \begin{bmatrix} C_{11} & C_{12} & C_{13} & C_{14} & 0 & 0 \\ & C_{11} & C_{13} & -C_{14} & 0 & 0 \\ & & C_{33} & 0 & 0 & 0 \\ & & & C_{44} & 0 & 0 \\ & sym & & & C_{44} & \sqrt{2}C_{14} \\ & & & & & C_{11}-C_{12} \end{bmatrix}; \tag{2.84}$$

if it is $x_2 = 0$ the plane of symmetry, then it is $C_{14} = 0$ and matrix (2.67) becomes

$$[C] = \begin{bmatrix} C_{11} & C_{12} & C_{13} & 0 & C_{15} & 0 \\ & C_{11} & C_{13} & 0 & -C_{15} & 0 \\ & & C_{33} & 0 & 0 & 0 \\ & & & C_{44} & 0 & -\sqrt{2}C_{15} \\ & sym & & & C_{44} & 0 \\ & & & & & C_{11}-C_{12} \end{bmatrix}; \tag{2.85}$$

these cases constitute the *trigonal elastic syngony with 6 constants*;
7. classes 14, 15, 16 and 19 of the tetragonal syngony are a particular case of the orthotropic symmetry: they have identical elastic properties along the axis x_1 and x_2, which gives the three supplementary conditions $C_{22} = C_{11},\ C_{23} = C_{13},\ C_{55} = C_{44}$, so reducing matrix (2.63) to

$$[C]=\begin{bmatrix} C_{11} & C_{12} & C_{13} & 0 & 0 & 0\\ & C_{11} & C_{13} & 0 & 0 & 0\\ & & C_{33} & 0 & 0 & 0\\ & & & C_{44} & 0 & 0\\ & sym & & & C_{44} & 0\\ & & & & & C_{66}\end{bmatrix}; \quad (2.86)$$

these cases constitute the *tetragonal elastic syngony with 6 constants*;

8. classes of the hexagonal syngony, from the 21 to the 27, belong to the 6-fold rotational symmetry, with 5 constants; together with transversely isotropic materials, that do not exist as crystals, they form the *axe-symmetric elastic syngony*, with $[C]$ as in Eq. (2.73);
9. classes of the cubic syngony, from the 28 to the 32, are a particular case of the orthotropic symmetry: they have identical properties along the three axes, which gives the six supplementary conditions $C_{33}=C_{22}=C_{11},\ C_{23}=C_{13}=C_{12},\ C_{66}=C_{55}=C_{44}$, so reducing matrix (2.63) to

$$[C]=\begin{bmatrix} C_{11} & C_{12} & C_{12} & 0 & 0 & 0\\ & C_{11} & C_{12} & 0 & 0 & 0\\ & & C_{11} & 0 & 0 & 0\\ & & & C_{44} & 0 & 0\\ & sym & & & C_{44} & 0\\ & & & & & C_{44}\end{bmatrix}; \quad (2.87)$$

the cubic crystal syngony corresponds entirely with the *cubic elastic syngony*;

10. the last elastic syngony is the *isotropic elastic syngony*; of course, no crystal syngonies belong to this case; nevertheless, a huge number of materials have an isotropic behavior. Though the texts on crystals and anisotropy usually forget to consider the isotropic case, this one actually exists and for the sake of completeness we prefer here to consider it as an elastic syngony; the isotropic matrix (2.77) can be obtained as a particular case of the cubic one, (2.87), when $C_{44}=C_{11}-C_{12}$.

2.3 The Technical Constants of Elasticity

In practical applications, engineers usually prefer to replace the use of the elastic stiffness matrix components by the so-called *technical elasticity constants* or *engineer moduli*.

Technical constants quantify an effect, a direct or a coupling one, whose mechanical meaning is immediate and that can be easily identified and measured in simple laboratory tests, like for instance unidirectional tensile tests.

Of course, the set of technical constants must be equivalent to the set of independent elastic moduli, which means, on one side, that the number of technical constants and distinct elastic moduli must be the same, i.e. 21, and that the technical constants must represent all the mechanical effects in a stressed body.

Though replacing the components of the stiffness elastic matrix $[C]$, the technical constants are defined as functions of the S_{ij}. Unlike the elastic moduli, only 6 technical constants are *moduli*: they measure a *direct effect* i.e. they correspond to terms on the diagonal of $[C]$, and are homogenous to a stress. The remaining 15 technical constants are *coefficients*: they are dimensionless quantities because constructed as the ratio between two strain components and they measure a *coupling effect*, i.e. they correspond to terms out of the diagonal of $[C]$. Let us introduce all of them.

2.3.1 The Young's Moduli

The three *Young's moduli* generalize to anisotropy the analogous isotropic modulus and are defined in a similar way:

$$E_i := \frac{\sigma_i}{\varepsilon_i}, \quad i = 1, 2, 3, \quad \sigma_i \neq 0, \ \sigma_j = 0 \text{ for } j \neq i, \ j = 1, \ldots, 6. \tag{2.88}$$

As a consequence, from Eq. (2.21) we get the relations (no summation over dummy indexes)

$$S_{ii} = Z_{iiii} = \frac{1}{E_i}, \quad i = 1, 2, 3. \tag{2.89}$$

The mechanical meaning of the Young's moduli is self-evident: each one of them measures the extension stiffness along the direction of one of the frame axes, i.e. the stress to be applied in the direction x_i to stretch the same direction with a strain equal to unity. Generally speaking, the three Young's moduli are different, i.e. in anisotropy the directions of the space have different stiffnesses.

2.3.2 Shear Moduli

Also in this case, the three *shear moduli* generalize to anisotropy the isotropic concept of shear modulus[13]:

$$G_{ij} := \frac{\sigma_k}{2\varepsilon_k}, \quad i, j = 1, 2, 3, \ i \neq j, \ k = 4, 5, 6, \ \sigma_k \neq 0, \ \sigma_h = 0 \text{ for } h \neq k, \ h = 1, \ldots, 6. \tag{2.90}$$

[13]The reader should consider that the definition of the shear moduli normally found in the literature is

$$G_{ij} := \frac{\sigma_{ij}}{\gamma_{ij}},$$

where γ_{ij} is the so-called *technical shear strain*, $\gamma_{ij} := 2\varepsilon_{ij}$. The above equation in the Kelvin's notation just corresponds to Eq. (2.90).

To be remarked that in the literature there is a strange discrepancy in the nomenclature of the G_{ij}s: in fact, the Kelvin notation is used for σ_k and ε_k but in G_{ij} the indexes are those indicating the directions. The correspondences between k and ij are of course those indicated by Eq. (2.19). As a consequence, from Eqs. (2.21) and (2.24) we get the relations (no summation over dummy indexes)

$$2S_{kk} = 4Z_{ijij} = \frac{1}{G_{ij}}, \quad i, j = 1, 2, 3, \ i \neq j, \ k = 4, 5, 6. \tag{2.91}$$

The mechanical meaning of the shear moduli is completely analogous to that of the Young's moduli, but it concerns shear stress and strain, and the same remarks can be done.

2.3.3 Poisson's Coefficients

The definition of the *Poisson's coefficients* or *ratios* in anisotropy is quite similar to that given for isotropic bodies:

$$\nu_{ij} := -\frac{\varepsilon_j}{\varepsilon_i}, \quad i, j = 1, 2, 3, \ \sigma_i \neq 0, \ \sigma_h = 0 \text{ for } h \neq i, \ h = 1, \ldots, 6. \tag{2.92}$$

Like for shear moduli, also in this case the nomenclature makes use, in the same formula, of the Kelvin's notation along with the classical tensorial one.

From the Young's moduli definition, Eq. (2.88), we get

$$\varepsilon_j = -\nu_{ij}\varepsilon_i = -\nu_{ij}\frac{\sigma_i}{E_i}, \quad i, j = 1, 2, 3. \tag{2.93}$$

Through Eq. (2.21) this gives (no summation over dummy indexes)

$$S_{ji} = Z_{jjii} = -\frac{\nu_{ij}}{E_i} \quad \Rightarrow \quad \nu_{ij} = -\frac{S_{ji}}{S_{ii}}, \quad i, j = 1, 2, 3. \tag{2.94}$$

Finally, the symmetry of matrix $[S]$, consequence of the major symmetries of $\mathbb{Z}$, gives the *reciprocity relations*

$$\frac{\nu_{ij}}{E_i} = \frac{\nu_{ji}}{E_j}, \quad i, j = 1, 2, 3, \tag{2.95}$$

which reduce the number of distinct Poisson's coefficients from 6 to only 3.

Some remarks about the Poisson's coefficients: they measure the Poisson's effect, i.e. the deformation in a direction transversal to that of the normal stress. Because, generally speaking, the three Poisson's coefficients are different, the Poisson's effect is different in the different directions. Also, due to the dependence upon the frame orientation, Eq. (2.34), it is possible that in some directions $\nu_{ij} \leq 0$, i.e. directions with null or negative Poisson's coefficients are quite possible in anisotropic elasticity. To end, it is worth to remark that some authors exchange the place of suffixes i and j in the definition of ν_{ij}.

2.3.4 *Chentsov's Coefficients*

The *Chentsov's coefficients* $\mu_{ij,kl}$ play the same role of the Poisson's coefficients with respect to shear stress and strain. They are defined as follows:

$$\begin{aligned} &\mu_{ij,kl} := \frac{\varepsilon_{ij}}{\varepsilon_{kl}}, \ \ i, j, k, l = 1, 2, 3, \ i \neq j, \ k \neq l, \\ &\sigma_{kl} \neq 0, \ \sigma_{pq} = 0 \text{ for } pq \neq kl, \ \ p, q = 1, 2, 3. \end{aligned} \tag{2.96}$$

Hence, coefficient $\mu_{ij,kl}$ measures the Chentsov's effect in the plane ij due to the shear stress σ_{kl}, i.e. the ratio between the shear strain components ε_{ij} and ε_{kl}. By the definition of the G_{ij}s, Eq. (2.90), it follows that (no summation over dummy indexes)

$$\varepsilon_{ij} = \mu_{ij,kl}\varepsilon_{kl} = \mu_{ij,kl}\frac{\sigma_{kl}}{2G_{kl}} \ \ i, j, k, l = 1, 2, 3, \tag{2.97}$$

and through Eqs. (2.21) and (2.24) we get

$$2S_{pq} = 4Z_{ijkl} = \frac{\mu_{ij,kl}}{G_{kl}} \ \Rightarrow \ \mu_{ij,kl} = \frac{S_{pq}}{S_{qq}}, \ i, j, k, l = 1, 2, 3, \ p, q = 4, 5, 6, \tag{2.98}$$

with p that corresponds to the couple ij and q to kl according to the scheme (2.19). The symmetry of $[S]$ gives the reciprocity relations

$$\frac{\mu_{ij,kl}}{G_{kl}} = \frac{\mu_{kl,ij}}{G_{ij}}, \tag{2.99}$$

that, along with the minor symmetries of $\boldsymbol{\sigma}$ and $\boldsymbol{\varepsilon}$ reduce to only three the number of distinct Chentsov's coefficients. Finally, the remarks done for the ν_{ij}s can be rephrased *verbatim* for the $\mu_{ij,kl}$s.

2.3.5 *Coefficients of Mutual Influence of the First Type*

These coefficients characterize the normal strain ε_{ii} due to the shear σ_{jk} (no summation over dummy indexes):

$$\eta_{i,jk} := \frac{\varepsilon_{ii}}{2\varepsilon_{jk}} \quad i, j, k = 1, 2, 3,\ j \neq k,\ \sigma_{jk} \neq 0,\ \sigma_{pq} = 0 \text{ for } pq \neq jk,\ p, q = 1, 2, 3. \tag{2.100}$$

By the definition of the G_{ij}s, Eq. (2.90), it follows that

$$\varepsilon_{ii} = 2\eta_{i,jk}\varepsilon_{jk} = \eta_{i,jk}\frac{\sigma_{jk}}{G_{jk}}, \tag{2.101}$$

and through Eqs. (2.21) and (2.24) we get

$$\sqrt{2}S_{ip} = 2Z_{iijk} = \frac{\eta_{i,jk}}{G_{jk}} \ \Rightarrow\ \eta_{i,jk} = \frac{S_{ip}}{\sqrt{2}S_{pp}},\ i, j, k = 1, 2, 3,\ p = 4, 5, 6, \tag{2.102}$$

with p that corresponds to the couple jk according to the scheme (2.19). For the symmetry of $\boldsymbol{\sigma}$ and $\boldsymbol{\varepsilon}$, the exchange of suffixes j and k has no effects, so the number of distinct coefficients is only 9.

2.3.6 Coefficients of Mutual Influence of the Second Type

These coefficients characterize the shear strain ε_{ij} due to the normal stress σ_{kk} (no summation over dummy indexes):

$$\eta_{ij,k} := \frac{2\varepsilon_{ij}}{\varepsilon_{kk}} \quad i, j, k = 1, 2, 3,\ i \neq j,\ \sigma_{kk} \neq 0,\ \sigma_{pq} = 0 \text{ for } pq \neq kk,\ p, q = 1, 2, 3. \tag{2.103}$$

By the definition of the E_is, Eq. (2.88), it follows that

$$2\varepsilon_{ij} = \eta_{ij,k}\varepsilon_{kk} = \eta_{ij,k}\frac{\sigma_{kk}}{E_k}, \tag{2.104}$$

and through Eqs. (2.21) and (2.24) we get

$$\sqrt{2}S_{pk} = 2Z_{ijkk} = \frac{\eta_{ij,k}}{E_k} \ \Rightarrow\ \eta_{ij,k} = \sqrt{2}\frac{S_{pk}}{S_{kk}},\ i, j, k = 1, 2, 3,\ p = 4, 5, 6, \tag{2.105}$$

with p that corresponds to the couple ij according to the scheme (2.19). Like for the coefficients of the first type, the symmetries of $\boldsymbol{\sigma}$ and $\boldsymbol{\varepsilon}$ reduce the number of distinct coefficients of the second type to only 9.

The coefficients of the second type are not independent from those of the first type; in fact, the symmetry of $[S]$ gives immediately the reciprocity relations

$$\frac{\eta_{ij,k}}{E_k} = \frac{\eta_{k,ij}}{G_{ij}},\ i, j, k = 1, 2, 3. \tag{2.106}$$

So the use of the coefficients of the first or of the second type is arbitrary and equally valid. Also for the coefficients of the first and second type can be repeated almost *verbatim* the remarks done about the other coefficients.

2.3.7 Some Remarks About the Technical Constants

The relations between a technical constant and the corresponding component of $\mathbb{Z}$, given in the previous Sections, are valid regardless of the notation used, i.e. they are the same also with the Voigt's notation. On the contrary, the relations with the components S_{ij} depends upon the notation, and those found above are not completely identical with the Voigt's notation, see for instance (Jones 1999, p. 79).

It is possible, of course, to express also the components of $[C]$ as functions of the technical constants; this necessitates the inversion of $[S]$ and in the most general case it gives so complicate and long expressions that it is impossible to write them.

Nevertheless, in the important case of orthotropic materials the transformation is rather simple. In fact, in the orthotropic frame, the inverse of matrix $[S]$, which is perfectly analogous to matrix (2.63), is given by (no summation on the dummy indexes)

$$
\begin{aligned}
C_{ii} &= \frac{S_{jj}S_{kk}-S_{jk}^2}{S} = \frac{1-\nu_{jk}\nu_{kj}}{\Delta}E_i, \quad i,j,k=1,2,3,\ i\neq j\neq k,\\
C_{ij} &= \frac{S_{ik}S_{kj}-S_{ij}S_{kk}}{S} = \frac{\nu_{ij}+\nu_{ik}\nu_{kj}}{\Delta}E_j, \quad i,j,k=1,2,3,\ i\neq j\neq k,\\
C_{44} &= \frac{1}{S_{44}} = 2G_{23}, \quad C_{55} = \frac{1}{S_{55}} = 2G_{31}, \quad C_{66} = \frac{1}{S_{66}} = 2G_{12}, \quad \text{with}\\
S &= S_{11}S_{22}S_{33} - S_{11}S_{23}^2 - S_{22}S_{13}^2 - S_{33}S_{12}^2 + 2S_{12}S_{23}S_{13},\\
\Delta &= 1-\nu_{12}\nu_{21}-\nu_{23}\nu_{32}-\nu_{31}\nu_{13}-2\nu_{32}\nu_{21}\nu_{13}.
\end{aligned}
\tag{2.107}
$$

In matrix form we have

$$
[C] = \begin{bmatrix}
\frac{S_{22}S_{33}-S_{23}^2}{S} & \frac{S_{13}S_{32}-S_{12}S_{33}}{S} & \frac{S_{12}S_{23}-S_{13}S_{22}}{S} & 0 & 0 & 0\\
 & \frac{S_{11}S_{33}-S_{13}^2}{S} & \frac{S_{21}S_{13}-S_{23}S_{11}}{S} & 0 & 0 & 0\\
 & & \frac{S_{11}S_{22}-S_{12}^2}{S} & 0 & 0 & 0\\
 & & & \frac{1}{S_{44}} & 0 & 0\\
 & sym & & & \frac{1}{S_{55}} & 0\\
 & & & & & \frac{1}{S_{66}}
\end{bmatrix}, \tag{2.108}
$$

and with the technical constants

$$[C] = \begin{bmatrix} \frac{1-\nu_{23}\nu_{32}}{\Delta}E_1 & \frac{\nu_{12}+\nu_{13}\nu_{32}}{\Delta}E_2 & \frac{\nu_{13}+\nu_{12}\nu_{23}}{\Delta}E_3 & 0 & 0 & 0 \\ & \frac{1-\nu_{13}\nu_{31}}{\Delta}E_2 & \frac{\nu_{23}+\nu_{21}\nu_{13}}{\Delta}E_3 & 0 & 0 & 0 \\ & & \frac{1-\nu_{12}\nu_{21}}{\Delta}E_3 & 0 & 0 & 0 \\ & & & 2G_{23} & 0 & 0 \\ & sym & & & 2G_{13} & 0 \\ & & & & & 2G_{12} \end{bmatrix} \tag{2.109}$$

and conversely

$$[S] = \begin{bmatrix} \frac{1}{E_1} & -\frac{\nu_{21}}{E_2} & -\frac{\nu_{31}}{E_3} & 0 & 0 & 0 \\ & \frac{1}{E_2} & -\frac{\nu_{32}}{E_3} & 0 & 0 & 0 \\ & & \frac{1}{E_3} & 0 & 0 & 0 \\ & & & \frac{1}{2G_{23}} & 0 & 0 \\ & sym & & & \frac{1}{2G_{13}} & 0 \\ & & & & & \frac{1}{2G_{12}} \end{bmatrix}. \tag{2.110}$$

It is also worth to specify these results also for the isotropic case

$$[C] = \begin{bmatrix} \frac{(1-\nu)E}{(1-2\nu)(1+\nu)} & \frac{\nu E}{(1-2\nu)(1+\nu)} & \frac{\nu E}{(1-2\nu)(1+\nu)} & 0 & 0 & 0 \\ & \frac{(1-\nu)E}{(1-2\nu)(1+\nu)} & \frac{\nu E}{(1-2\nu)(1+\nu)} & 0 & 0 & 0 \\ & & \frac{(1-\nu)E}{(1-2\nu)(1+\nu)} & 0 & 0 & 0 \\ & & & \frac{E}{1+\nu} & 0 & 0 \\ & sym & & & \frac{E}{1+\nu} & 0 \\ & & & & & \frac{E}{1+\nu} \end{bmatrix}, \tag{2.111}$$

and conversely

$$[S] = \begin{bmatrix} \frac{1}{E} & -\frac{\nu}{E} & -\frac{\nu}{E} & 0 & 0 & 0 \\ & \frac{1}{E} & -\frac{\nu}{E} & 0 & 0 & 0 \\ & & \frac{1}{E} & 0 & 0 & 0 \\ & & & \frac{1+\nu}{E} & 0 & 0 \\ & sym & & & \frac{1+\nu}{E} & 0 \\ & & & & & \frac{1+\nu}{E} \end{bmatrix}. \tag{2.112}$$

To remark that the Voigt's notation can be obtained simply dividing by a factor 2 the components C_{44}, C_{55} and C_{66}, and multiplying by 2 the components S_{44}, S_{55} and S_{66}.

2.4 Bounds on the Elastic Constants

2.4.1 General Conditions and Results

Elastic constants cannot take any value: they are bounded because of the physical fact that the deformation of an elastic body Ω cannot produce energy, i.e. the overall work $\mathcal{L}_{ext}$ done by the applied forces *must be positive*. From the Clapeyron's Theorem

$$\mathcal{L}_{ext} = 2V_{tot} = 2\left(\frac{1}{2}\int_{\Omega} \boldsymbol{\sigma} \cdot \boldsymbol{\varepsilon}\, d\Omega\right), \tag{2.113}$$

we get hence the condition that the total strain energy V_{tot} must be positive. Assuming the strain as independent field over Ω, then the overall condition is

$$V_{tot} = \frac{1}{2}\int_{\Omega} \boldsymbol{\sigma} \cdot \boldsymbol{\varepsilon}\, d\Omega > 0 \ \forall \boldsymbol{\varepsilon} \neq \mathbf{O}. \tag{2.114}$$

The above constraint on the deformation of an elastic body is a strong condition. By a procedure of limit towards small volumes, it is easy to see that it must be true also locally, i.e. $\forall p \in \Omega$; it is just the local form of (2.114) that gives the bounds on the elastic constants of a material. In fact, getting the local form of (2.114) and injecting the Hooke's law (2.12) gives

$$V = \frac{1}{2}\boldsymbol{\sigma} \cdot \boldsymbol{\varepsilon} = \frac{1}{2}\boldsymbol{\varepsilon} \cdot \mathbb{E}\boldsymbol{\varepsilon} > 0 \ \forall \boldsymbol{\varepsilon} \neq \mathbf{O}. \tag{2.115}$$

Equation (2.115) is the mathematical condition corresponding to the thermodynamical fact that no energy can be produced deforming an elastic body: the elasticity stiffness tensor $\mathbb{E}$ must be *positive definite*.

If the $\boldsymbol{\sigma}$ is taken as independent field over Ω in place of $\boldsymbol{\varepsilon}$, we get a similar restriction on the stress energy and finally the condition that the elasticity compliance tensor $\mathbb{Z}$ must be positive definite. Of course, the two approaches give in the end the same results for the elastic constants.

2.4.2 Mathematical Conditions for the Elastic Matrices

It is easier to obtain practical results for the components of matrices $[C]$ and $[S]$ than for tensors $\mathbb{E}$ and $\mathbb{Z}$, so let us rewrite condition (2.115) in its equivalent matrix form:

$$\frac{1}{2}\{\varepsilon\}^{\top}[C]\{\varepsilon\} > 0 \ \forall\{\varepsilon\} \neq \{O\}. \tag{2.116}$$

Of course, the above condition states the positive definiteness of matrix $[C]$, and an analogous condition can be written for matrix $[S]$.

Mathematically, the problem is hence clear: being $[C]$ symmetric, so, by the Spectral Theorem, with real eigenvalues λ_i, condition (2.116) corresponds to impose that all the eigenvalues are positive:

$$\frac{1}{2}\{\varepsilon\}^{\top}[C]\{\varepsilon\} > 0 \ \forall\{\varepsilon\} \neq \{O\} \iff \lambda_i > 0 \ \forall i = 1, \ldots, 6. \tag{2.117}$$

The above result is almost useless, because the Laplace's equation of $[C]$ is an algebraic equation of degree 6. Hence, generally speaking, it is not possible to get an analytic expression of the roots of this equation, the eigenvalues λ_i, for obtaining the searched bounds on the C_{ij}.

Nevertheless, a first qualitative result is that the number of conditions on the C_{ij}s is 6. As distinct components are, in the most general case, 21, the conditions on the C_{ij}s are not necessarily simple bounds but at least some of them are necessarily relations among some of the components. Also, for the hexagonal, cubic and isotropic syngonies the number of conditions is redundant with respect to the distinct elastic constants, so some of them have lower and upper bounds and/or some of the bounds are redundant (this, anyway, can be true also for other syngonies).

Though the approach by eigenvalues is practically impossible, there is another mathematical approach which is completely general and feasible. To this purpose, let us introduce the following definitions and theorems of matrix algebra.

A *principal minor* of a matrix $[A]$ is the determinant of the sub-matrix extracted from $[A]$ removing an equal number of rows and columns having the same indexes, i.e. preserving the leading diagonal.

A *leading principal minor of order r* is the determinant of a principal $r \times r$ sub-matrix whose rows and columns are the first r rows and columns of $[A]$. Hence, a $n \times n$ matrix has n leading principal minors.

Theorem 1 (Necessary condition for a symmetric matrix to be positive definite) *All the principal minors of a positive definite $n \times n$ symmetric matrix $[A]$ are positive.*

Proof By the hypothesis,

$$\{x\}^{\top}[A]\{x\} > 0 \ \forall\{x\} \neq \{0\}. \tag{2.118}$$

Then, for a principal $r \times r$ sub-matrix $[A^r]$ extracted from $[A]$ deleting $n - r$ rows and columns with the same indexes, we have

$$\{x^r\}^{\top}[A^r]\{x^r\} = \{x\}^{\top}[A]\{x\} > 0 \ \forall\{x\}, \{x^r\} \neq \{0\} \tag{2.119}$$

where $\{x\}$ is any vector whose components corresponding to the removed rows of $[A]$ are null and with at least one of the other components different from zero, while $\{x^r\}$ is the $r-$dimensional vector obtained removing from $\{x\}$ the components corresponding to the removed rows of $[A]$.

Hence, $[A^r]$ is positive definite and because it is symmetric, its eigenvalues are positive, see Eq. (2.117). Then, its determinant, which is the product of its eigenvalues, is positive too. This proves that $\det[A^r] > 0$ and hence the theorem.

Theorem 2 (Necessary and sufficient condition for a symmetric matrix to be positive definite) *For a $n \times n$ symmetric matrix $[A]$ to be positive definite it is necessary and sufficient that its n leading principal minors are all positive.*

The proof of this theorem is non trivial and the reader is referred to (Hohn 1958, p. 340).

The six principal minors of $[C]$ are

$$M_1 = C_{11}, \quad M_2 = \begin{vmatrix} C_{11} & C_{12} \\ C_{12} & C_{22} \end{vmatrix}, \quad M_3 = \begin{vmatrix} C_{11} & C_{12} & C_{13} \\ C_{12} & C_{22} & C_{23} \\ C_{13} & C_{23} & C_{33} \end{vmatrix},$$
$$M_4 = \begin{vmatrix} C_{11} & C_{12} & C_{13} & C_{14} \\ C_{12} & C_{22} & C_{23} & C_{24} \\ C_{13} & C_{23} & C_{33} & C_{34} \\ C_{14} & C_{24} & C_{34} & C_{44} \end{vmatrix}, \quad M_5 = \begin{vmatrix} C_{11} & C_{12} & C_{13} & C_{14} & C_{15} \\ C_{12} & C_{22} & C_{23} & C_{24} & C_{25} \\ C_{13} & C_{23} & C_{33} & C_{34} & C_{35} \\ C_{14} & C_{24} & C_{34} & C_{44} & C_{45} \\ C_{15} & C_{25} & C_{35} & C_{45} & C_{55} \end{vmatrix}, \tag{2.120}$$
$$M_6 = \det[C].$$

Contrarily to the eigenvalues, it is always possible to explicit the above expressions and hence the 6 conditions

$$M_i > 0, \; i = 1, \ldots, 6. \tag{2.121}$$

That is why the use of Theorem 2 is more interesting than condition (2.117), though to write down the 6 conditions in the most general case of a triclinic material gives so long expressions that they are omitted here.

We can, however, consider the different elastic syngonies and because they have a simpler form of $[C]$ than in the triclinic case, also conditions (2.121) will be simpler. In particular, let us consider here some cases whose results are particularly simple (the bounds are written for matrix $[C]$, but similar results can be written for $[S]$ too; redundant bounds have been omitted):

- orthotropic elastic syngony, Eq. (2.63):

$$\begin{aligned} &C_{ii} > 0, \quad i = 1, 4, 5, 6, \\ &C_{11}C_{22} - C_{12}^2 > 0, \\ &C_{11}C_{22}C_{33} - C_{33}C_{12}^2 - C_{11}C_{23}^2 - C_{22}C_{13}^2 + 2C_{12}C_{13}C_{23} > 0; \end{aligned} \tag{2.122}$$

- tetragonal elastic syngony with 6 constants, Eq. (2.86):

$$\begin{aligned} &C_{44} > 0,\\ &C_{66} > 0,\\ &C_{11}^2 - C_{12}^2 > 0,\\ &(C_{11} - C_{12})\left[C_{33}(C_{11} + C_{12}) - 2C_{13}^2\right] > 0; \end{aligned} \tag{2.123}$$

- axially symmetric elastic syngony, Eq. (2.73):

$$\begin{aligned} &C_{44} > 0,\\ &C_{11}^2 - C_{12}^2 > 0,\\ &(C_{11} - C_{12})\left[C_{33}(C_{11} + C_{12}) - 2C_{13}^2\right] > 0; \end{aligned} \tag{2.124}$$

- cubic elastic syngony, Eq. (2.87)

$$\begin{aligned} &C_{44} > 0,\\ &C_{11} - C_{12} > 0,\\ &C_{11} + 2C_{12} > 0; \end{aligned} \tag{2.125}$$

- isotropic elastic syngony, Eq. (2.77):

$$\begin{aligned} &C_{11} - C_{12} > 0,\\ &C_{11} + 2C_{12} > 0. \end{aligned} \tag{2.126}$$

2.4.3 A Mechanical Approach

The bounds on the elastic constants can be found also by a direct mechanical approach, based upon the fact that the strain energy must be positive *for each possible choice of the strain field* $\boldsymbol{\varepsilon}$. This allows for choosing particularly simple strain fields, giving some direct, simple results. Let us see how (no summation over dummy indexes): choose a field $\{\varepsilon\}$ with only one component $\varepsilon_i \neq 0$. Then,

$$V > 0 \iff C_{ii} > 0,\ i = 1, \ldots, 6; \tag{2.127}$$

we get hence six conditions. Unfortunately, each one of them is only a *necessary* condition for the strain energy be positive, so the (2.127) *do not constitute a set of necessary and sufficient* conditions for the positiveness of V. Nevertheless, they give us a precious information: *all the moduli responsible for the direct effects are strictly positive*. Using the stress energy instead of the strain energy, it is immediately recognized that it is also:

$$S_{ii} > 0\ \forall i = 1, \ldots, 6. \tag{2.128}$$

2.4.4 Bounds on the Technical Constants

The results of Eqs. (2.89), (2.91) and (2.128) give immediately

$$E_i > 0, \quad G_{ij} > 0 \ \forall i, j = 1, 2, 3: \tag{2.129}$$

all the Young's and shear moduli are strictly positive quantities, result that is valid for any kind of elastic syngony.

To these necessary conditions some other relations for the technical constants can be added. First of all, let us consider a spherical state of stress; it is then easy to see that

$$\begin{aligned} &\{\sigma\} = \sigma\{I\} \ \Rightarrow \ \{\sigma\}^\top[S]\{\sigma\} > 0 \iff \\ &S_{11} + S_{22} + S_{33} + 2(S_{13} + S_{32} + S_{21}) > 0. \end{aligned} \tag{2.130}$$

Replacing in the above result the expressions of the S_{ij}s from Eqs. (2.89) and (2.94) gives the condition

$$\frac{1 - 2\nu_{12}}{E_1} + \frac{1 - 2\nu_{23}}{E_2} + \frac{1 - 2\nu_{31}}{E_3} > 0. \tag{2.131}$$

This result is valid regardless of the elastic syngony; for the cubic and isotropic syngonies it becomes the well known bound $\nu < 1/2$ on the Poisson's coefficient. A simpler but rougher estimation can be obtained from bound (2.131), (Lekhnitskii 1950, p. 85):

$$\begin{aligned} &\frac{3 - 2(\nu_{12} + \nu_{23} + \nu_{31})}{\min\{E_1, E_2, E_3\}} > \frac{1 - 2\nu_{12}}{E_1} + \frac{1 - 2\nu_{23}}{E_2} + \frac{1 - 2\nu_{31}}{E_3} > 0 \Rightarrow \\ &\nu_{12} + \nu_{23} + \nu_{31} < \frac{3}{2}. \end{aligned} \tag{2.132}$$

Some other necessary conditions can be given expressing the C_{ii} in terms of the technical parameters. This is impossible in the most general case of the triclinic syngony, for the calculations are too complicate. However, this can be done for the orthotropic syngony; the supplementary bounds can be found expressing the (2.127) as functions of the technical constants through Eq. (2.107) and taking into account the positivity of the Young's moduli, Eq. (2.129):

$$\begin{aligned} &1 - \nu_{ij}\nu_{ji} > 0 \ \forall i, j = 1, 2, 3; \\ &\Delta = 1 - \nu_{12}\nu_{21} - \nu_{23}\nu_{32} - \nu_{31}\nu_{13} - 2\nu_{32}\nu_{21}\nu_{13} > 0. \end{aligned} \tag{2.133}$$

Condition $(2.133)_2$ can be transformed to

$$\nu_{32}\nu_{21}\nu_{13} < \frac{1}{2}\left(1 - \nu_{32}^2\frac{E_2}{E_3} - \nu_{21}^2\frac{E_1}{E_2} - \nu_{13}^2\frac{E_3}{E_1}\right) < \frac{1}{2}. \tag{2.134}$$

Through the reciprocity conditions on the Poisson's coefficients, Eq. (2.95), conditions $(2.133)_1$ can be written also as

$$|\nu_{ij}| < \sqrt{\frac{E_i}{E_j}} \quad \forall i, j = 1, 2, 3, \tag{2.135}$$

or equivalently

$$|S_{ij}| < \sqrt{S_{ii} S_{jj}} \quad \forall i, j = 1, 2, 3. \tag{2.136}$$

Some remarks to end this part; first of all, the bounds concern frame dependent quantities, and of course they are more easily written in a frame composed by symmetry directions. Then, the only, general, necessary and sufficient conditions are the (2.121), that can always be written and used in numerical applications, e.g. for checking the validity of the results of experimental tests.

In the case of orthotropic materials, a set of conditions on the technical constants can be easily written, but it is still questionable whether or not it constitutes a set of necessary *and* sufficient conditions for the positivity of the strain energy, a point never treated in the literature. Finally, bounds on the Chentsov's and mutual influence coefficients are apparently unknown in the literature.

In the case of isotropic materials, the conditions of positivity of the strain energy reduce to the well known three bounds on E and ν

$$E > 0, \quad -1 < \nu < \frac{1}{2}, \tag{2.137}$$

but when the bounds are written for the two distinct components of $[C]$, C_{11} and C_{12}, then rather surprisingly the bounds are only two, see Eq. (2.126):

$$C_{11} - C_{12} > 0, \quad C_{11} + 2C_{12} > 0. \tag{2.138}$$

Also when the isotropic constitutive law is written under the form of the Lamé's equations

$$\boldsymbol{\sigma} = 2\mu\boldsymbol{\varepsilon} + \lambda \operatorname{tr}\boldsymbol{\varepsilon}\, \mathbf{I}, \tag{2.139}$$

it is easy to show that the only two bounds on the Lamé's constants λ and μ are

$$\mu > 0, \quad 2\mu + 3\lambda > 0, \tag{2.140}$$

that corresponds exactly to bounds (2.138). This fact shows that the number of necessary and sufficient conditions for the strain energy to be positive depends upon the choice of the elastic constants and that, anyway, it is quite hard to establish a priori its value, whose maximum remains however 6.

A last remark: all the bounds and conditions written in this Section are written on *frame dependent quantities*, apart those written for the isotropic case, Eq. (2.126), of course. In particular, conditions (2.122) to (2.125) are valid exclusively in the symmetry frame where the respective matrices $[C]$ have been written. In the plane case, we will see that it is possible, with the polar formalism, to give *completely invariant bounds*, i.e. bounds established on tensor invariants, which are not yet known for the general 3D case.

2.5 An Observation About the Decomposition of the Strain Energy

Let us consider a point which is true at least for isotropic materials but often thought as generally true also for other elastic syngonies: is it possible to decompose the strain, or stress, energy into *spherical* and *deviatoric* parts? In other words, we ponder whether or not it is *always* possible to write

$$V = V_{sph} + V_{dev}, \tag{2.141}$$

where V_{sph}, the *spherical* part of V is produced *exclusively* by the spherical part of $\boldsymbol{\varepsilon}$ and by its corresponding part of $\boldsymbol{\sigma}$, i.e.

$$V_{sph} = \frac{1}{2}\boldsymbol{\varepsilon}_{sph} \cdot \mathbb{E}\boldsymbol{\varepsilon}_{sph}, \tag{2.142}$$

and V_{dev}, the *deviatoric* part of V is produced *exclusively* by the deviatoric part of $\boldsymbol{\varepsilon}$ and by its corresponding part of $\boldsymbol{\sigma}$, i.e.

$$V_{dev} = \frac{1}{2}\boldsymbol{\varepsilon}_{dev} \cdot \mathbb{E}\boldsymbol{\varepsilon}_{dev}. \tag{2.143}$$

Mechanically, such a decomposition means that V can be considered as the sum of two parts: one, V_{sph}, due to volume changes not accompanied by shape changes, the other one, V_{dev}, produced by isochoric shape changes. This decomposition is, for instance, at the basis of the Hüber–Hencky–von Mises criterion, where the only V_{dev} is considered to be responsible of yielding.

We recall that it is always possible to decompose $\boldsymbol{\sigma}$ and $\boldsymbol{\varepsilon}$ into a spherical and a deviatoric part

$$\begin{aligned} &\boldsymbol{\sigma} = \boldsymbol{\sigma}_{sph} + \boldsymbol{\sigma}_{dev}, \ \ \boldsymbol{\sigma}_{sph} = \frac{1}{3}\text{tr}\boldsymbol{\sigma}\ \mathbf{I}, \ \ \boldsymbol{\sigma}_{dev} = \boldsymbol{\sigma} - \boldsymbol{\sigma}_{sph}, \\ &\boldsymbol{\varepsilon} = \boldsymbol{\varepsilon}_{sph} + \boldsymbol{\varepsilon}_{dev}, \ \ \boldsymbol{\varepsilon}_{sph} = \frac{1}{3}\text{tr}\boldsymbol{\varepsilon}\ \mathbf{I}, \ \ \boldsymbol{\varepsilon}_{dev} = \boldsymbol{\varepsilon} - \boldsymbol{\varepsilon}_{sph}, \end{aligned} \tag{2.144}$$

and that any spherical part is orthogonal to any deviatoric part:

$$\begin{aligned}\boldsymbol{\sigma}_{sph}\cdot\boldsymbol{\varepsilon}_{dev}&=\frac{1}{3}\mathrm{tr}\boldsymbol{\sigma}\,\mathbf{I}\cdot\left(\boldsymbol{\varepsilon}-\frac{1}{3}\mathrm{tr}\boldsymbol{\varepsilon}\,\mathbf{I}\right)=\frac{1}{3}\mathrm{tr}\boldsymbol{\varepsilon}\;\mathrm{tr}\boldsymbol{\sigma}-\frac{1}{3}\mathrm{tr}\boldsymbol{\varepsilon}\;\mathrm{tr}\boldsymbol{\sigma}=0,\\ \boldsymbol{\sigma}_{dev}\cdot\boldsymbol{\varepsilon}_{sph}&=\left(\boldsymbol{\sigma}-\frac{1}{3}\mathrm{tr}\boldsymbol{\sigma}\,\mathbf{I}\right)\cdot\frac{1}{3}\mathrm{tr}\boldsymbol{\varepsilon}\,\mathbf{I}=\frac{1}{3}\mathrm{tr}\boldsymbol{\varepsilon}\;\mathrm{tr}\boldsymbol{\sigma}-\frac{1}{3}\mathrm{tr}\boldsymbol{\varepsilon}\;\mathrm{tr}\boldsymbol{\sigma}=0.\end{aligned}\tag{2.145}$$

Using decomposition (2.144) we have

$$\begin{aligned}V=&\frac{1}{2}\boldsymbol{\varepsilon}\cdot\mathbb{E}\boldsymbol{\varepsilon}=\frac{1}{2}(\boldsymbol{\varepsilon}_{sph}+\boldsymbol{\varepsilon}_{dev})\cdot\mathbb{E}(\boldsymbol{\varepsilon}_{sph}+\boldsymbol{\varepsilon}_{dev})=\\ &\frac{1}{2}\boldsymbol{\varepsilon}_{sph}\cdot\mathbb{E}\boldsymbol{\varepsilon}_{sph}+\frac{1}{2}\boldsymbol{\varepsilon}_{dev}\cdot\mathbb{E}\boldsymbol{\varepsilon}_{dev}+\frac{1}{2}\boldsymbol{\varepsilon}_{sph}\cdot\mathbb{E}\boldsymbol{\varepsilon}_{dev}+\frac{1}{2}\boldsymbol{\varepsilon}_{dev}\cdot\mathbb{E}\boldsymbol{\varepsilon}_{sph}.\end{aligned}\tag{2.146}$$

For the decomposition (2.141) to be true, it is necessary and sufficient that

$$\boldsymbol{\varepsilon}_{sph}\cdot\mathbb{E}\boldsymbol{\varepsilon}_{dev}=0\ \Rightarrow\ \mathrm{tr}\left[\boldsymbol{\varepsilon}_{sph}^{\top}(\mathbb{E}\boldsymbol{\varepsilon}_{dev})\right]=0\ \ \forall\boldsymbol{\varepsilon}.\tag{2.147}$$

In fact, whenever Eq. (2.147) is satisfied, for definition (2.38) it is

$$\boldsymbol{\varepsilon}_{dev}\cdot\mathbb{E}\boldsymbol{\varepsilon}_{sph}=\mathbb{E}^{\top}\boldsymbol{\varepsilon}_{dev}\cdot\boldsymbol{\varepsilon}_{sph}=\boldsymbol{\varepsilon}_{sph}\cdot\mathbb{E}\boldsymbol{\varepsilon}_{dev},\tag{2.148}$$

because of the symmetry of $\mathbb{E}$, i.e. for its major symmetries. This result shows that the two mixed terms in (2.146) are identical.

Through (2.144), condition (2.147) can be written as

$$\mathrm{tr}\left[\frac{1}{3}\mathrm{tr}\boldsymbol{\varepsilon}\;\mathbf{I}(\mathbb{E}\boldsymbol{\varepsilon}_{dev})\right]=0\ \ \forall\boldsymbol{\varepsilon}\quad\Longleftrightarrow\quad\mathrm{tr}(\mathbb{E}\boldsymbol{\varepsilon}_{dev})=0.\tag{2.149}$$

The components of $\mathbb{E}$ must satisfy Eq. (2.149) for the decomposition (2.141) to be possible. It can be rewritten as

$$\mathrm{tr}\left[\mathbb{E}\left(\boldsymbol{\varepsilon}-\frac{1}{3}\mathrm{tr}\boldsymbol{\varepsilon}\;\mathbf{I}\right)\right]=0\ \Rightarrow\ 3\mathrm{tr}(\mathbb{E}\boldsymbol{\varepsilon})-\mathrm{tr}\boldsymbol{\varepsilon}\;\mathrm{tr}(\mathbb{E}\mathbf{I})=0\ \ \forall\boldsymbol{\varepsilon}.\tag{2.150}$$

Actually, it is easy to check that condition (2.147) corresponds to impose that

$$\boldsymbol{\sigma}_{dev}=\mathbb{E}\boldsymbol{\varepsilon}_{dev},\quad\boldsymbol{\sigma}_{sph}=\mathbb{E}\boldsymbol{\varepsilon}_{sph}.\tag{2.151}$$

Condition (2.150) can be written by components:

$$E_{hhkk}\varepsilon_{ii}-3E_{jjpq}\varepsilon_{pq}=0\ \ \forall\varepsilon_{mn},\ \ i,j,h,k,p,q,m,n=1,2,3.\tag{2.152}$$

Generally speaking, this quantity does not vanish for any possible choice of $\boldsymbol{\varepsilon}$. As a consequence, *for a generic anisotropic material decomposition of the strain energy into a spherical and deviatoric part is not possible*. Nevertheless, it can be checked that for the cubic syngony Eq. (2.152) is always satisfied. In fact, for an orthotropic material condition (2.152) becomes

$$\begin{aligned}&\frac{1}{3}[E_{1111}(2\varepsilon_{11}-\varepsilon_{22}-\varepsilon_{33})+E_{2222}(2\varepsilon_{22}-\varepsilon_{11}-\varepsilon_{33})+\\&E_{3333}(2\varepsilon_{33}-\varepsilon_{22}-\varepsilon_{11})]+\\&\frac{2}{3}[E_{1122}(\varepsilon_{11}+\varepsilon_{22}-2\varepsilon_{33})+E_{1133}(\varepsilon_{11}+\varepsilon_{33}-2\varepsilon_{22})+\\&E_{2233}(\varepsilon_{22}+\varepsilon_{33}-2\varepsilon_{11})]=0,\end{aligned} \tag{2.153}$$

condition which is not yet satisfied, generally speaking, but which is always satisfied when

$$E_{1111}=E_{2222}=E_{3333},\quad E_{1122}=E_{2233}=E_{1133}, \tag{2.154}$$

i.e. by cubic materials. Hence, for materials of the cubic syngony, decomposition (2.141) is always possible and, a fortiori, by isotropic materials too, as well known, because they can be considered as a special case of the cubic syngony, see Sect. 2.2.11, point 10.

2.6 Determination of Symmetry Planes

The classification in elastic syngonies presupposes that, for a given material, the existing equivalent directions are known, so as to write $\mathbb{E}$, or equivalently $[C]$, in a symmetry frame, which makes some of the E_{ijkl}s, and the corresponding C_{ij}s, vanish.

But when a material is completely unknown, the independent measures to be done in experimental tests to characterize the material are as much as 21; practically, it is very complicate to do all of these tests. Nevertheless, the existence of possible symmetry planes remains unknown also once all the C_{ij}s are known, if $[C]$ is a full matrix.

The problem is hence the following one: given a general matrix $[C]$, is it possible to determine if some planes of symmetry exist and which they are? We will see in Sect. 4.1 that in the planar case it is very simple to determine whether or not the material has some kind of elastic symmetry and the symmetry directions using the polar formalism. In the three dimensional case, the problem is much more complicate; it has been solved by Cowin and Mehrabadi in two works, (Cowin and Mehrabadi 1987; Cowin 1989), successively completed by Ting (1996). We give here a brief account of these results.

Be $\mathbf{n}$ and $\mathbf{m}$ two vectors such that $||\mathbf{n}||=||\mathbf{m}||=1,\ \mathbf{m}\cdot\mathbf{n}=0$, with $\mathbf{n}$ orthogonal to a symmetry plane for a material whose elastic tensor is $\mathbb{E}$. Consider the following second-rank symmetric tensors: $\mathbf{V}=\mathbb{E}\mathbf{I}$, $\mathbf{W}$ the acoustic[14] tensor relative to the basis

[14]The *acoustic* or *Green-Christoffel tensor* $\mathbf{A_u}$ relative to the direction $\mathbf{u}$ is the unique tensor such that $\mathbf{A_u}\mathbf{w}=\mathbb{E}(\mathbf{w}\otimes\mathbf{u})\mathbf{u}\ \forall\mathbf{w}\in\mathscr{V}$.

direction $\mathbf{e}_p$, $\mathbf{X}$ and $\mathbf{Y}$ the acoustic tensors relative to $\mathbf{n}$ and $\mathbf{m}$, respectively.[15] We can now state the following:

Theorem 3 *The following statements are equivalent ($\lambda_i \in \mathbb{R},\ i = 1, \ldots, 6$):*

1. *the material has a plane of symmetry whose normal is* $\mathbf{n}$*;*
2. $\mathbf{Vn} = \lambda_1 \mathbf{Yn} = \lambda_2 \mathbf{n}$*;*
3. $\mathbf{Wn} = \lambda_3 \mathbf{Yn} = \lambda_4 \mathbf{n}$*;*
4. $\mathbf{Xn} = \lambda_5 \mathbf{Yn} = \lambda_6 \mathbf{n}$.

Proof Without loss of generality, let us suppose that $\mathbf{n} = \mathbf{e}_1$ and $\mathbf{m} = \cos\theta \mathbf{e}_2 + \sin\theta \mathbf{e}_3$. When $\mathbf{n}$ is an eigenvector of $\mathbf{V}$, $\mathbf{W}$, $\mathbf{X}$ or $\mathbf{Y}$ then

$$\begin{aligned}
&\mathbf{Vn} = \lambda_v \mathbf{n} \ \rightarrow\ E_{i1qq}\mathbf{e}_i = \lambda_v \mathbf{e}_1,\\
&\mathbf{Wn} = \lambda_w \mathbf{n} \ \rightarrow\ E_{ip1p}\mathbf{e}_i = \lambda_w \mathbf{e}_1,\\
&\mathbf{Xn} = \lambda_x \mathbf{n} \ \rightarrow\ E_{i111}\mathbf{e}_i = \lambda_x \mathbf{e}_1,\\
&\mathbf{Yn} = \lambda_y \mathbf{n} \ \rightarrow\\
&\left[E_{i212}\cos^2\theta + E_{i313}\sin^2\theta + (E_{i213} + E_{i312})\sin\theta\cos\theta\right]\mathbf{e}_i = \lambda_y \mathbf{e}_1\ \forall\theta.
\end{aligned} \tag{2.155}$$

For $i = 1$, the above results give the values of the respective eigenvalues, but for $i = 2, 3$ we get, respectively,

$$\begin{aligned}
&E_{21qq} = E_{31qq} = 0,\\
&E_{2p1p} = E_{3p1p} = 0,\\
&E_{2111} = E_{3111} = 0,\\
&E_{2212}\cos^2\theta + E_{2313}\sin^2\theta + (E_{2213} + E_{2312})\sin\theta\cos\theta =\\
&E_{3212}\cos^2\theta + E_{3313}\sin^2\theta + (E_{3213} + E_{3312})\sin\theta\cos\theta = 0\ \forall\theta.
\end{aligned} \tag{2.156}$$

Passing to the C_{ij}s for the sake of convenience, and writing down in the order all the above relations, we get

$$\begin{aligned}
&C_{15} + C_{25} + C_{35} = C_{16} + C_{26} + C_{36} = 0,\\
&C_{15} + C_{35} + \frac{C_{46}}{\sqrt{2}} = C_{16} + C_{26} + \frac{C_{45}}{\sqrt{2}} = 0,\\
&C_{15} = C_{16} = 0,\\
&C_{25} = C_{26} = C_{35} = C_{36} = C_{45} = C_{46} = 0.
\end{aligned} \tag{2.157}$$

[15] It is simple to verify that

$$\begin{aligned}
&\mathbf{V} = \mathbb{E}\mathbf{I} = E_{ikqq}\mathbf{e}_i \otimes \mathbf{e}_k,\\
&\mathbf{W} = E_{ipkp}\mathbf{e}_i \otimes \mathbf{e}_k,\\
&\mathbf{X} = E_{ilkm}n_l n_m \mathbf{e}_i \otimes \mathbf{e}_k,\\
&\mathbf{Y} = E_{ijkh}m_j m_h \mathbf{e}_i \otimes \mathbf{e}_k.
\end{aligned}$$

If the material has $x_1 = 0$ as unique plane of symmetry, it belongs to the monoclinic syngony and its matrix $[C]$ is given by, see Sect. 2.2.5,

$$[C] = \begin{bmatrix} C_{11} & C_{12} & C_{13} & C_{14} & 0 & 0 \\ & C_{22} & C_{23} & C_{24} & 0 & 0 \\ & & C_{33} & C_{34} & 0 & 0 \\ & & & C_{44} & 0 & 0 \\ & sym & & & C_{55} & C_{56} \\ & & & & & C_{66} \end{bmatrix}, \tag{2.158}$$

that is:

$$C_{15} = C_{16} = C_{25} = C_{26} = C_{35} = C_{36} = C_{45} = C_{46} = 0. \tag{2.159}$$

It is then clear that conditions $(2.157)_{1,4}$, $(2.157)_{2,4}$ or $(2.157)_{3,4}$ imply (2.159) and vice-versa, which proves the theorem.

This theorem states that the material has a plane of symmetry whose normal is **n** if and only if **n** is the eigenvector of **Y** and of at least another tensor among **V**, **W** or **X**.

2.6.1 *Physical Interpretations*

A physical interpretation of Theorem 3 is possible in the frame of the acoustics theory, see (Ting 1996, p. 61): tensor **X** is the acoustic tensor for the elastic waves that propagate in the direction of **n**. An elastic wave is a *longitudinal wave* whenever **n** is an eigenvector of **X**; in such a case, **n** is called a *specific direction* of **X**. It has been proved by Kolodner that in an anisotropic material there exist always at least three different specific directions (Kolodner 1966). When **n** is an eigenvector of **Y**, then the wave is *transversal*, **m** is the direction of the wave propagation and **n** is called the *specific axis*.

Then conditions $(2.157)_{3,4}$, i.e. when **n** is an eigenvector of **X** and **Y**, are equivalent to say that **n** is at the same time a specific direction and a specific axis, i.e. it is simultaneously the direction of propagation of longitudinal waves and the transversal direction of transversal waves propagating along the direction of **m** orthogonal to **n**.

A statical interpretation has also been given by Hayes and Norris. It traduces the above acoustics conditions into equivalent statical conditions. They have been resumed in the following

Theorem 4 *A material has a plane of symmetry if and only if at least two orthogonal planes of pure shear exist, sharing a common shear direction which is the normal to the plane of symmetry.*

For the proof of this Theorem, rather articulated, we address the reader to the original paper (Hayes and Norris 1991).

2.7 Curvilinear Anisotropy

When in a body there are directions that are not parallel but mechanically equivalent, then the body possesses a *curvilinear anisotropy*. It is still possible to write the Hooke's law in a rectangular coordinate system, as done until now. However, in doing so, the components of $[C]$ or $[S]$ are no more constants, but vary with the position according to the variation of the coordinate directions with respect to the equivalent directions.

Be $\{\xi, \eta, \zeta\}$ the coordinate directions of the curvilinear coordinates that coincide with the mechanically equivalent directions. With self-evident meaning of the symbols, the Hooke's law can be written in the curvilinear coordinate system as (Lekhnitskii 1950, p. 64),

$$\left\{\begin{array}{c}\sigma_{\xi\xi}\\ \sigma_{\eta\eta}\\ \sigma_{\zeta\zeta}\\ \sqrt{2}\sigma_{\eta\zeta}\\ \sqrt{2}\sigma_{\zeta\xi}\\ \sqrt{2}\sigma_{\xi\eta}\end{array}\right\}=\left[\begin{array}{cccccc}C_{11}&C_{12}&C_{13}&C_{14}&C_{15}&C_{16}\\ &C_{22}&C_{23}&C_{24}&C_{25}&C_{26}\\ &&C_{33}&C_{34}&C_{35}&C_{36}\\ &&&C_{44}&C_{45}&C_{46}\\ &sym&&&C_{55}&C_{56}\\ &&&&&C_{66}\end{array}\right]\left\{\begin{array}{c}\varepsilon_{\xi\xi}\\ \varepsilon_{\eta\eta}\\ \varepsilon_{\zeta\zeta}\\ \sqrt{2}\varepsilon_{\eta\zeta}\\ \sqrt{2}\varepsilon_{\zeta\xi}\\ \sqrt{2}\varepsilon_{\xi\eta}\end{array}\right\}, \qquad (2.160)$$

where the C_{ij}s are constants. In some cases of non homogenous bodies, the C_{ij}s can depend upon the coordinates $\{\xi, \eta, \zeta\}$. Of course, if some type of elastic symmetry is present in the body, then some of the C_{ij}s can be null, as in the ordinary cases of the elastic syngonies.

A special case of curvilinear anisotropy is that of *cylindrical anisotropy*: the body has an axis of symmetry, not necessarily inside the body itself, all the directions crossing this axis at right angles are equivalent, as well as all the directions parallel to the axis and the directions orthogonal to the first two directions.

Using a customary set of cylindrical coordinates $\{r, \theta, z\}$, with z the axis of symmetry, then the Hooke's law can be written as

$$\left\{\begin{array}{c}\sigma_{rr}\\ \sigma_{\theta\theta}\\ \sigma_{zz}\\ \sqrt{2}\sigma_{\theta z}\\ \sqrt{2}\sigma_{zr}\\ \sqrt{2}\sigma_{r\theta}\end{array}\right\}=\left[\begin{array}{cccccc}C_{11}&C_{12}&C_{13}&C_{14}&C_{15}&C_{16}\\ &C_{22}&C_{23}&C_{24}&C_{25}&C_{26}\\ &&C_{33}&C_{34}&C_{35}&C_{36}\\ &&&C_{44}&C_{45}&C_{46}\\ &sym&&&C_{55}&C_{56}\\ &&&&&C_{66}\end{array}\right]\left\{\begin{array}{c}\varepsilon_{rr}\\ \varepsilon_{\theta\theta}\\ \varepsilon_{zz}\\ \sqrt{2}\varepsilon_{\theta z}\\ \sqrt{2}\varepsilon_{zr}\\ \sqrt{2}\varepsilon_{r\theta}\end{array}\right\}. \qquad (2.161)$$

A special case of cylindrical anisotropy is that of *cylindrical orthotropy*: each plane which is radial, tangential or orthogonal to the symmetry axis is a plane of symmetry. In such a case matrix $[C]$ in Eq. (2.161) is simplified:

$$\begin{Bmatrix} \sigma_{rr} \\ \sigma_{\theta\theta} \\ \sigma_{zz} \\ \sqrt{2}\sigma_{\theta z} \\ \sqrt{2}\sigma_{zr} \\ \sqrt{2}\sigma_{r\theta} \end{Bmatrix} = \begin{bmatrix} C_{11} & C_{12} & C_{13} & 0 & 0 & 0 \\ & C_{22} & C_{23} & 0 & 0 & 0 \\ & & C_{33} & 0 & 0 & 0 \\ & & & C_{44} & 0 & 0 \\ & sym & & & C_{55} & 0 \\ & & & & & C_{66} \end{bmatrix} \begin{Bmatrix} \varepsilon_{rr} \\ \varepsilon_{\theta\theta} \\ \varepsilon_{zz} \\ \sqrt{2}\varepsilon_{\theta z} \\ \sqrt{2}\varepsilon_{zr} \\ \sqrt{2}\varepsilon_{r\theta} \end{Bmatrix}. \quad (2.162)$$

It is worth noting that cylindrical orthotropy is *not* equivalent to transverse isotropy (that in fact depends upon only 5 constants, not upon 9). Actually, transverse isotropy is a special case of cylindrical orthotropy, because not only the radial and tangential directions are equivalent, but all the directions lying in a plane orthogonal to the symmetry axis are equivalent directions.

Some examples of cylindrical anisotropy are a block of wood with regular yearly cylindrical layers, or metallic pipes, for their manufacturing process, or a circular reinforced concrete slab with steel bars disposed radially and circumferentially, a bicycle wheel, when homogenized, a circular stone arch and so on.

Another, less common, type of curvilinear anisotropy is that of *spherical anisotropy*: there is a center of symmetry, not necessarily belonging to the body, and all the rays emanating from it are equivalent directions. Also, the tangents to the meridians and to the parallels are equivalent directions too. Using a standard spherical coordinate systems $\{\rho, \theta, \varphi\}$, where the directions of the coordinate axes coincide with the equivalent directions, Eq. (2.160) becomes

$$\begin{Bmatrix} \sigma_{\rho\rho} \\ \sigma_{\theta\theta} \\ \sigma_{\varphi\varphi} \\ \sqrt{2}\sigma_{\theta\varphi} \\ \sqrt{2}\sigma_{\varphi\rho} \\ \sqrt{2}\sigma_{\rho\theta} \end{Bmatrix} = \begin{bmatrix} C_{11} & C_{12} & C_{13} & C_{14} & C_{15} & C_{16} \\ & C_{22} & C_{23} & C_{24} & C_{25} & C_{26} \\ & & C_{33} & C_{34} & C_{35} & C_{36} \\ & & & C_{44} & C_{45} & C_{46} \\ & sym & & & C_{55} & C_{56} \\ & & & & & C_{66} \end{bmatrix} \begin{Bmatrix} \varepsilon_{\rho\rho} \\ \varepsilon_{\theta\theta} \\ \varepsilon_{\varphi\varphi} \\ \sqrt{2}\varepsilon_{\theta\varphi} \\ \sqrt{2}\varepsilon_{\varphi\rho} \\ \sqrt{2}\varepsilon_{\rho\theta} \end{Bmatrix}. \quad (2.163)$$

The case of *spherical orthotropy* is get when each meridian and tangential plane is a plane of symmetry as well as each plane orthogonal to these two planes. Then, Eq. (2.163) becomes

$$\begin{Bmatrix} \sigma_{\rho\rho} \\ \sigma_{\theta\theta} \\ \sigma_{\varphi\varphi} \\ \sqrt{2}\sigma_{\theta\varphi} \\ \sqrt{2}\sigma_{\varphi\rho} \\ \sqrt{2}\sigma_{\rho\theta} \end{Bmatrix} = \begin{bmatrix} C_{11} & C_{12} & C_{13} & 0 & 0 & 0 \\ & C_{22} & C_{23} & 0 & 0 & 0 \\ & & C_{33} & 0 & 0 & 0 \\ & & & C_{44} & 0 & 0 \\ & sym & & & C_{55} & 0 \\ & & & & & C_{66} \end{bmatrix} \begin{Bmatrix} \varepsilon_{\rho\rho} \\ \varepsilon_{\theta\theta} \\ \varepsilon_{\varphi\varphi} \\ \sqrt{2}\varepsilon_{\theta\varphi} \\ \sqrt{2}\varepsilon_{\varphi\rho} \\ \sqrt{2}\varepsilon_{\rho\theta} \end{Bmatrix}. \quad (2.164)$$

To remark the difference between isotropy and spherical orthotropy: isotropy is a special case of spherical orthotropy, because all the directions are equivalent, not only those emanating from the centre of symmetry. This reduces the number of independent elastic constants from 9 to only 2.

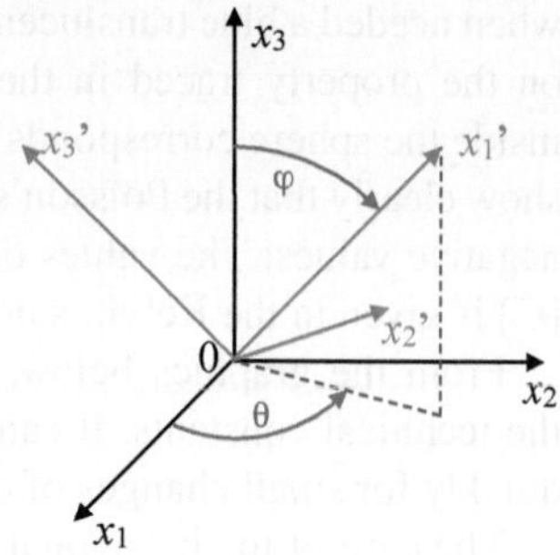

Fig. 2.3 Scheme of the frame rotation for tracing the elastic constants 3D-graphics

2.8 Some Examples of Anisotropic Materials

To end this Chapter, we give in this Section some examples of anisotropic materials, showing the matrix $[C]$ (in GPa) and the 3D-directional diagrams of some of the technical constants. These last have been obtained as the value get by the constant on the axis of x_1' of a frame $\{x_1', x_2', x_3'\}$ rotated with respect to the frame $\{x_1, x_2, x_3\}$ where the matrix $[C]$ is known, see Fig. 2.3. The rotation matrix $[R]$ is obtained according to Eq. (2.48), with a rotation tensor $\mathbf{U}$ that is

$$\mathbf{U} = \begin{bmatrix} \sin\varphi\cos\theta & \sin\varphi\sin\theta & \cos\varphi \\ -\sin\theta & \cos\theta & 0 \\ -\cos\varphi\cos\theta & -\cos\varphi\sin\theta & \sin\varphi \end{bmatrix}. \tag{2.165}$$

So, with this choice axis x_2 lies always in the horizontal plane.

The compliance matrix $[S']$ in the rotated frame can be obtained using the inverse of relations (2.49):

$$\begin{gathered} \{\varepsilon\} = [S]\{\sigma\} \rightarrow [R]^\top\{\varepsilon'\} = [S][R]^\top\{\sigma'\} \rightarrow \\ \{\varepsilon'\} = [R][S][R]^\top\{\sigma'\} \Rightarrow [S'] = [R][S][R]^\top. \end{gathered} \tag{2.166}$$

This result can be applied to $[C]$ too, and it is the matrix corresponding of Eq. (2.34). Once the S_{ij}s known, the technical constants can be easily calculated using the results of Sect. 2.3.

Through Eqs. (2.165) and (2.166) it can be shown that for the materials of the hexagonal elastic syngony it is always

$$S_{14} = S_{16} = S_{24} = S_{26} = S_{34} = S_{36} = S_{45} = S_{56} = 0. \tag{2.167}$$

For these materials, the only Chentsov's and mutual influence coefficients that are not identically null are $\mu_{23,12}$, $\eta_{1,31}$, $\eta_{2,31}$, $\eta_{3,31}$, $\eta_{31,1}$, $\eta_{31,2}$ and $\eta_{31,3}$.

Different cases are considered below; for each one of them, the directional 3D-graphics of E_1, G_{12}, ν_{12}, $\mu_{23,12}$, $\eta_{1,31}$ and $\eta_{31,1}$ are traced. For the four last constants, when needed a blue translucent sphere is also traced: it represents the surface whereon the property traced in the graphics vanishes. So, the part of the 3D-graphics inside the sphere corresponds to negative values of the property itself. The graphics show clearly that the Poisson's, Chentsov's and mutual influence coefficients can get negative values. The values of the C_{ij}s, E_1 and G_{12} are in GPa. To remember that $[C]$ is given in the Kelvin's notation, Eq. (2.24).

From the graphics below, one can appreciate the extreme variety of forms of the technical constants. It can be remarked how anisotropy properties change very quickly for small changes of direction (Fig. 2.9, 2.10, 2.11, 2.12, 2.13 and 2.14).

The case of the hexagonal syngony is very articulated, and it can be shown that there are as much as 8 possible different profiles of $E_1(\varphi)$, (Vannucci 2015) (of course E_1, like all the other properties, does not depend upon θ, because the hexagonal elastic syngony is equivalent to transverse isotropy).

- Anorthite ($CaAl_2Si_2O_8$)
  Crystal syngony: Monoclinic, $N = 13$, plane of symmetry: $x_2 = 0$.
  Source: Evans and Grove (2004)

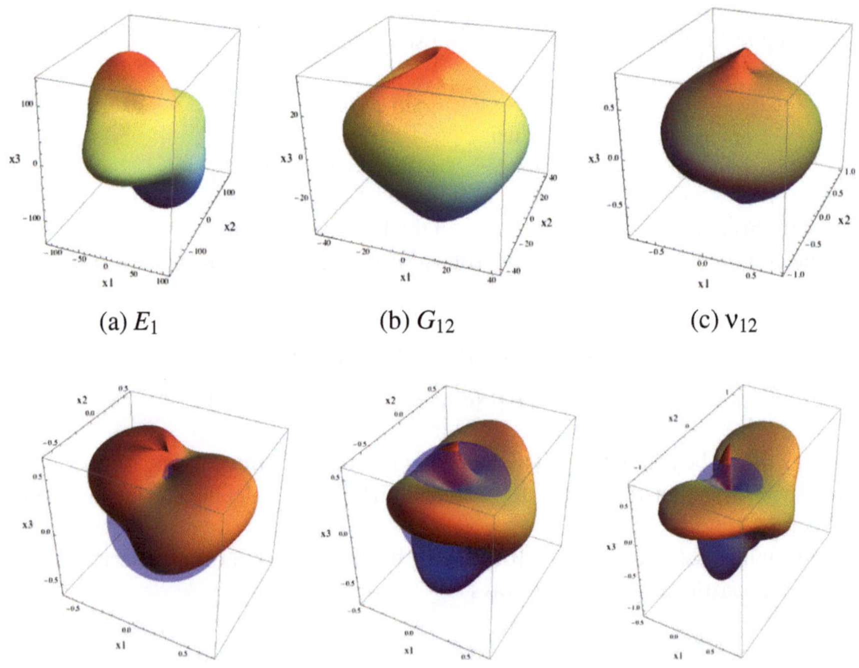

(a) E_1 (b) G_{12} (c) ν_{12}

(d) $\mu_{23,12}$ (e) $\eta_{1,31}$ (f) $\eta_{31,1}$

Fig. 2.4 Directional 3D-graphics of Anorthite

$$[C] = \begin{bmatrix} 124 & 66 & 50 & 0 & -26.9 & 0 \\ & 205 & 42 & 0 & -9.9 & 0 \\ & & 156 & 0 & -25.4 & 0 \\ & & & 48 & 0 & -2 \\ & sym & & & 80 & 0 \\ & & & & & 84 \end{bmatrix}$$

- Perovskite ($CaTiO_3$)
  Crystal syngony: Orthorhombic, $N = 9$.
  Source: Evans and Grove (2004)

$$[C] = \begin{bmatrix} 515 & 117 & 117 & 0 & 0 & 0 \\ & 525 & 139 & 0 & 0 & 0 \\ & & 435 & 0 & 0 & 0 \\ & & & 48 & 0 & 0 \\ & sym & & & 404 & 0 \\ & & & & & 350 \end{bmatrix}$$

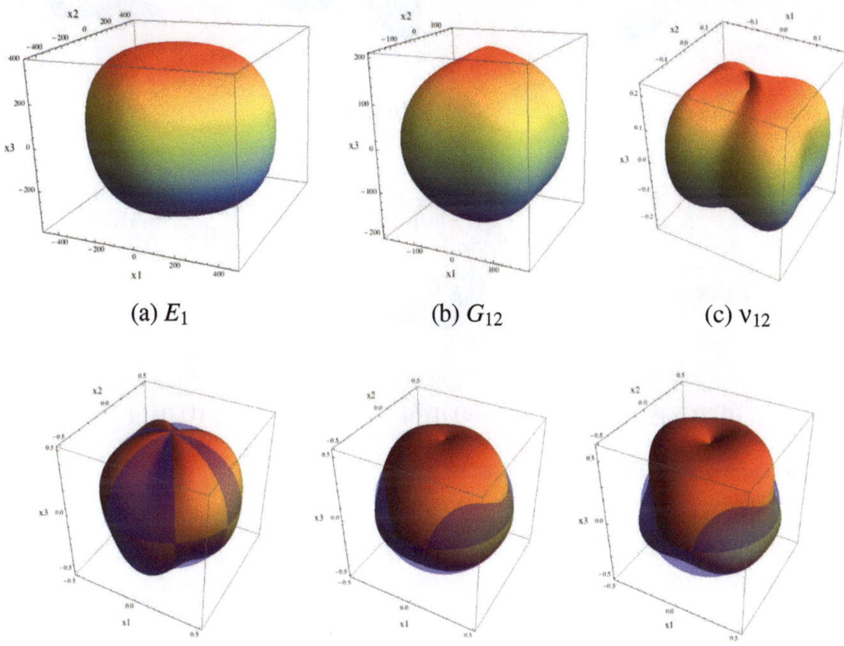

(a) E_1 (b) G_{12} (c) ν_{12}

(d) $\mu_{23,12}$ (e) $\eta_{1,31}$ (f) $\eta_{31,1}$

Fig. 2.5 Directional 3D-graphics of Perovskite

- Dolomite ($CaMg(CO_3)_2$) (*estimated)
  Crystal syngony: Trigonal, $N = 7$.
  Source: Bakri and Zaoui (2011)

$$[C] = \begin{bmatrix} 196.6 & 64.4 & 54.7 & 31.7 & 25.3^* & 0 \\ & 196.6 & 54.7 & -31.7 & -25.3^* & 0 \\ & & 110 & 0 & 0 & 0 \\ & & & 83.2 & 0 & -35.84 \\ & sym & & & 83.2 & 44.8 \\ & & & & & 132.2 \end{bmatrix}$$

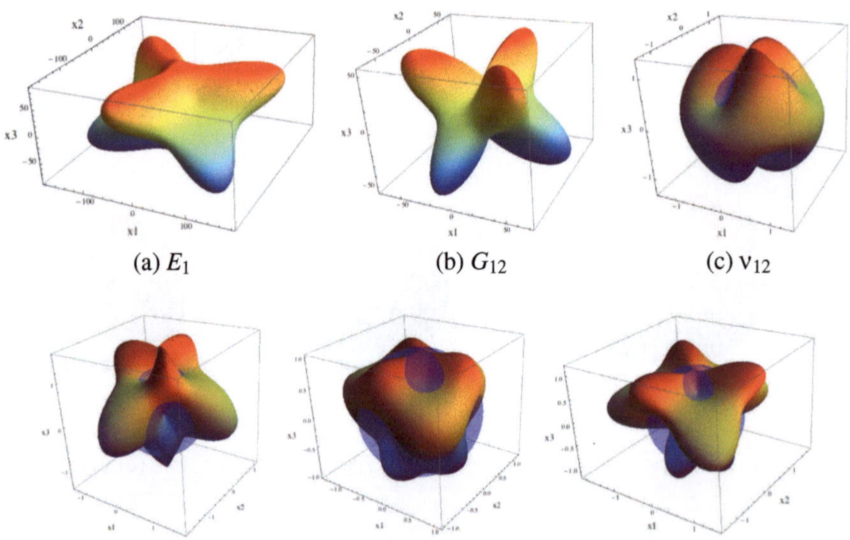

(a) E_1 (b) G_{12} (c) ν_{12}

(d) $\mu_{23,12}$ (e) $\eta_{1,31}$ (f) $\eta_{31,1}$

Fig. 2.6 Directional 3D-graphics of Dolomite

- Calcium Tungstate ($CaWO_4$)
  Crystal syngony: Tetragonal, $N = 7$.
  Source: Landolt and Börnstein (1992).

$$[C] = \begin{bmatrix} 141 & 61 & 41 & 0 & 0 & 1.9 \\ & 141 & 41 & 0 & 0 & -1.9 \\ & & 125 & 0 & 0 & 0 \\ & & & 67.4 & 0 & 0 \\ & sym & & & 67.4 & 0 \\ & & & & & 81.4 \end{bmatrix}$$

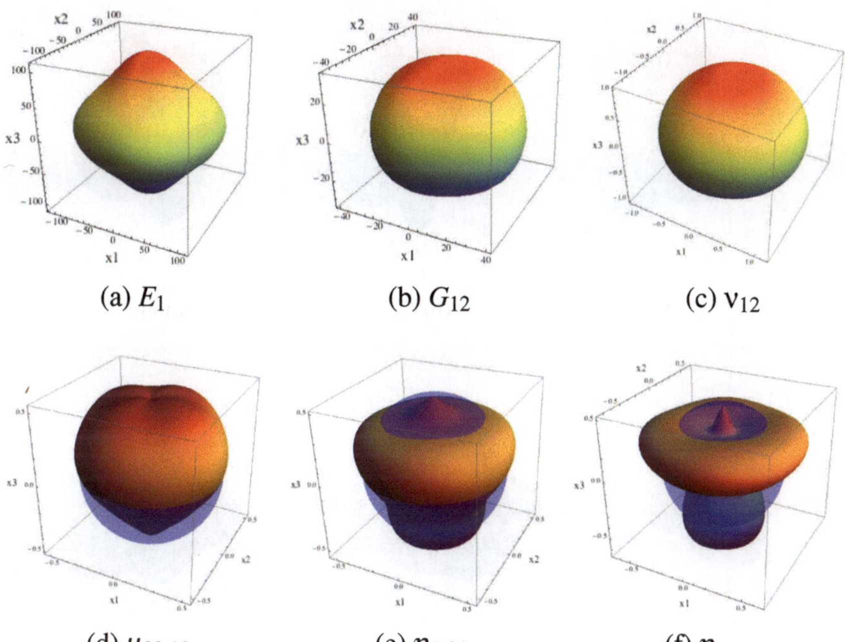

(a) E_1 (b) G_{12} (c) ν_{12}

(d) $\mu_{23,12}$ (e) $\eta_{1,31}$ (f) $\eta_{31,1}$

Fig. 2.7 Directional 3D-graphics of Calcium Tungstate

- Quartz (SiO_2)
  Crystal syngony: Trigonal, $N = 6$.
  Source: Landolt and Börnstein (1992)

$$[C] = \begin{bmatrix} 86.8 & 7.1 & 14.4 & 24.3 & 0 & 0 \\ & 86.8 & 14.4 & -24.3 & 0 & 0 \\ & & 107.5 & 0 & 0 & 0 \\ & & & 116.4 & 0 & 0 \\ & sym & & & 116.4 & 34.4 \\ & & & & & 79.7 \end{bmatrix}$$

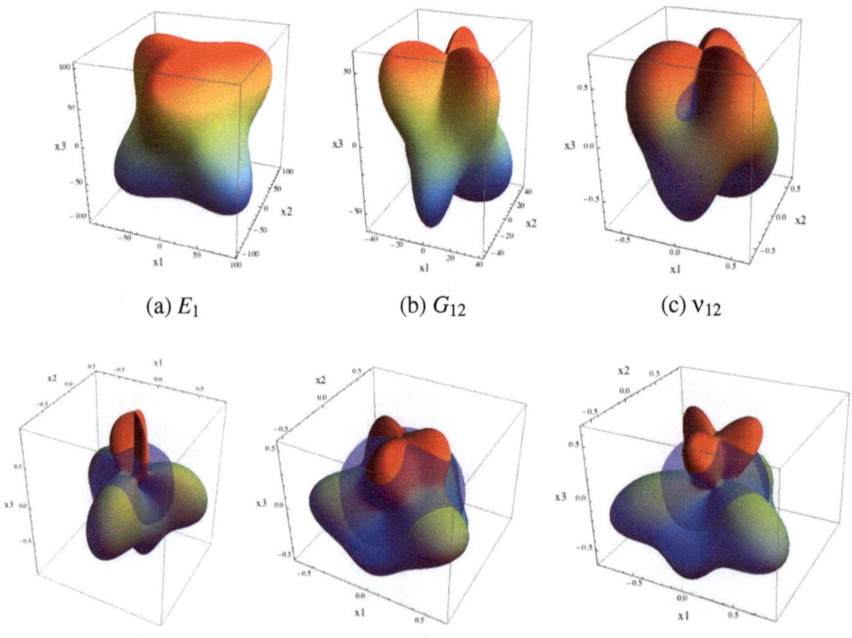

(a) E_1 (b) G_{12} (c) ν_{12}

(d) $\mu_{23,12}$ (e) $\eta_{1,31}$ (f) $\eta_{31,1}$

Fig. 2.8 Directional 3D-graphics of Quartz

- Zircon ($ZrSiO_4$)
  Crystal syngony: Tetragonal, $N = 6$.
  Source: Evans and Grove (2004)

$$[C] = \begin{bmatrix} 424 & 70 & 149 & 0 & 0 & 0 \\ & 424 & 149 & 0 & 0 & 0 \\ & & 489 & 0 & 0 & 0 \\ & & & 262 & 0 & 0 \\ & sym & & & 262 & 0 \\ & & & & & 96 \end{bmatrix}$$

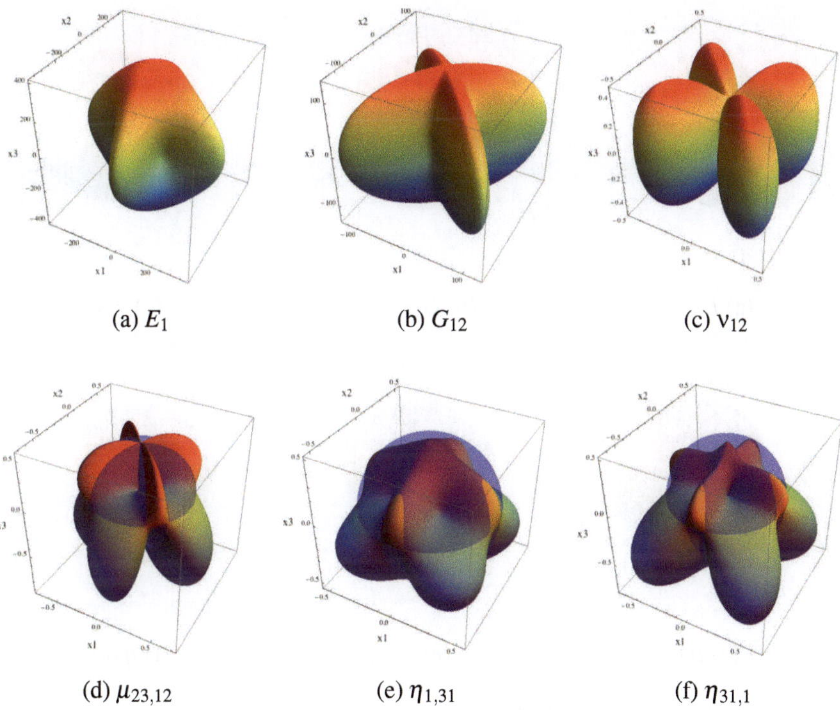

(a) E_1 (b) G_{12} (c) ν_{12}

(d) $\mu_{23,12}$ (e) $\eta_{1,31}$ (f) $\eta_{31,1}$

Fig. 2.9 Directional 3D-graphics of Zircon

- Ice (H_2O)
  Crystal syngony: Hexagonal, $N = 5$.
  Source: Evans and Grove (2004)

$$[C] = \begin{bmatrix} 13.5 & 6.5 & 6 & 0 & 0 & 0 \\ & 13.5 & 6 & 0 & 0 & 0 \\ & & 15 & 0 & 0 & 0 \\ & & & 6 & 0 & 0 \\ & sym & & & 6 & 0 \\ & & & & & 7 \end{bmatrix}$$

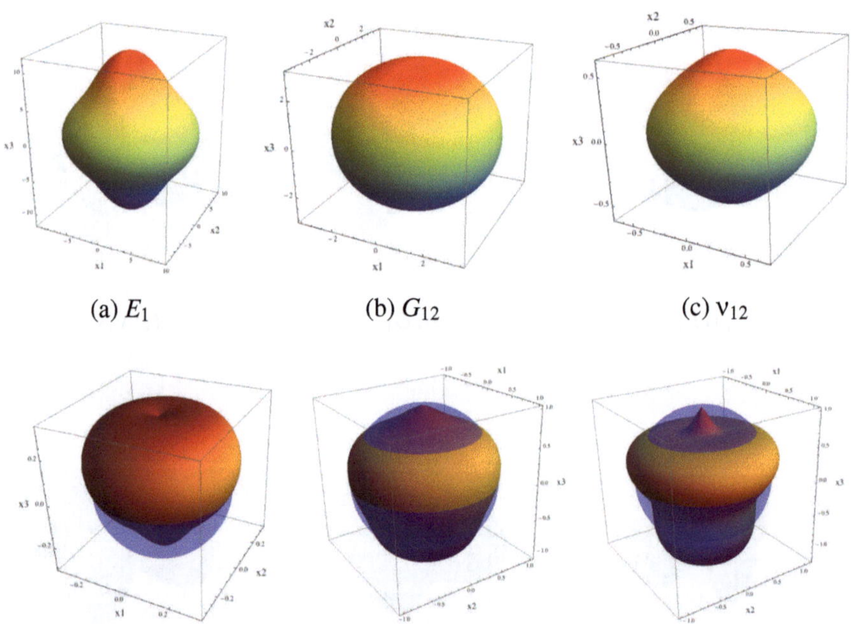

(a) E_1 (b) G_{12} (c) ν_{12}

(d) $\mu_{23,12}$ (e) $\eta_{1,31}$ (f) $\eta_{31,1}$

Fig. 2.10 Directional 3D-graphics of Ice

- Titanium Boride (TiB_2)
  Crystal syngony: Hexagonal, $N = 5$.
  Source: Landolt and Börnstein (1992)

$$[C] = \begin{bmatrix} 648.3 & 404.2 & 317.7 & 0 & 0 & 0 \\ & 648.3 & 317.7 & 0 & 0 & 0 \\ & & 439.3 & 0 & 0 & 0 \\ & & & 500 & 0 & 0 \\ & sym & & & 500 & 0 \\ & & & & & 244.1 \end{bmatrix}$$

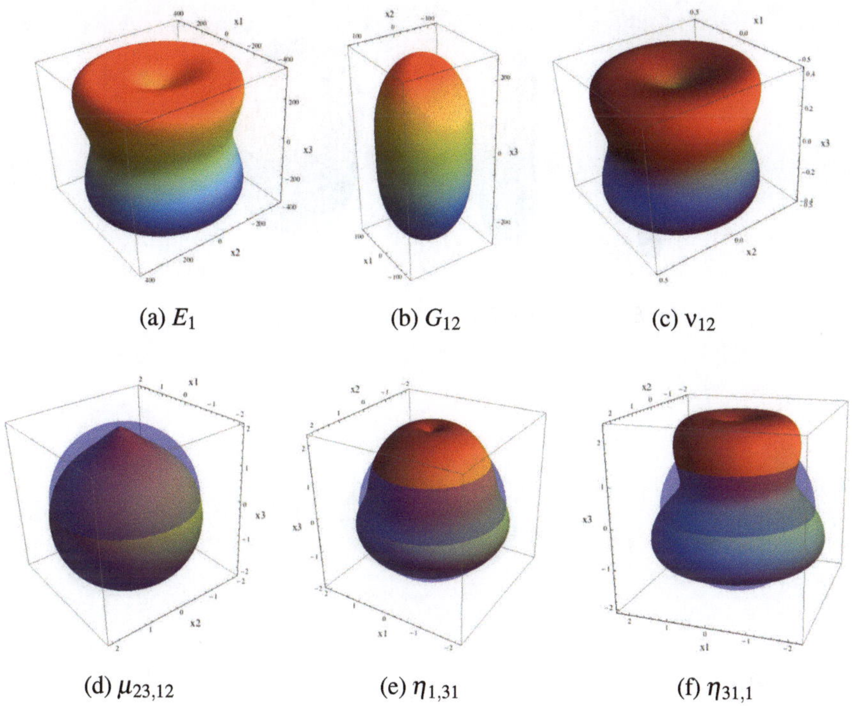

(a) E_1 (b) G_{12} (c) ν_{12}

(d) $\mu_{23,12}$ (e) $\eta_{1,31}$ (f) $\eta_{31,1}$

Fig. 2.11 Directional 3D-graphics of Titanium Boride

- Pine Wood
  Transversely isotropic, $N = 5$.
  Source: Lekhnitskii (1950)

$$[C] = \begin{bmatrix} 0.45 & 0.11 & 0.13 & 0 & 0 & 0 \\ & 0.45 & 0.13 & 0 & 0 & 0 \\ & & 10.1 & 0 & 0 & 0 \\ & & & 1.5 & 0 & 0 \\ & sym & & & 1.5 & 0 \\ & & & & & 0.34 \end{bmatrix}$$

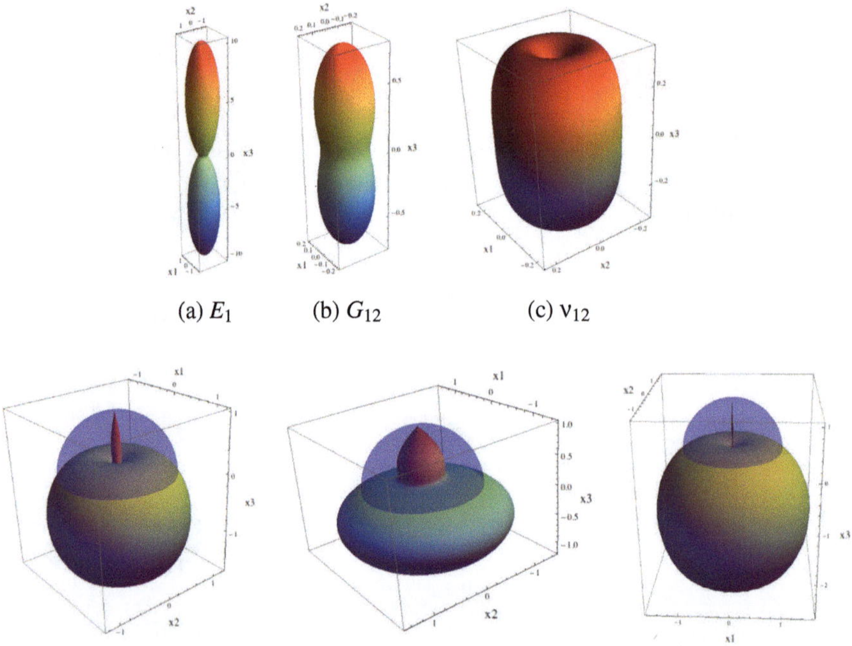

(a) E_1 (b) G_{12} (c) ν_{12}

(d) $\mu_{23,12}$ (e) $\eta_{1,31}$ (f) $\eta_{31,1}$

Fig. 2.12 Directional 3D-graphics of Pine Wood

- Gold (Au)
  Crystal syngony: Cubic, $N = 3$.
  Source: Evans and Grove (2004)

$$[C] = \begin{bmatrix} 191 & 162 & 162 & 0 & 0 & 0 \\ & 191 & 162 & 0 & 0 & 0 \\ & & 191 & 0 & 0 & 0 \\ & & & 84 & 0 & 0 \\ & sym & & & 84 & 0 \\ & & & & & 84 \end{bmatrix}$$

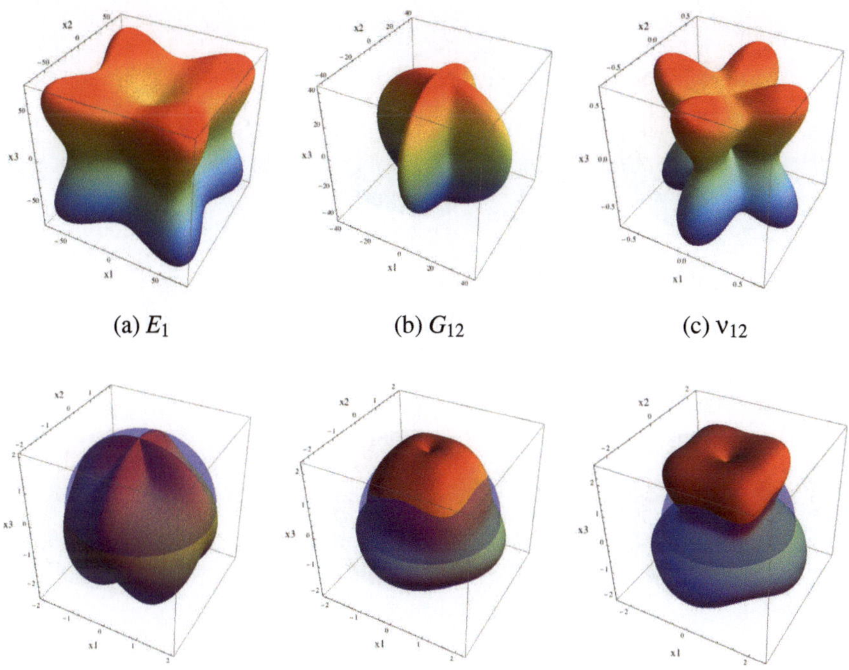

(a) E_1 (b) G_{12} (c) ν_{12}

(d) $\mu_{23,12}$ (e) $\eta_{1,31}$ (f) $\eta_{31,1}$

Fig. 2.13 Directional 3D-graphics of Gold

- Diamond (C)
  Crystal syngony: Cubic, $N = 3$.
  Source: Evans and Grove (2004)

$$[C] = \begin{bmatrix} 1079 & 124 & 124 & 0 & 0 & 0 \\ & 1079 & 124 & 0 & 0 & 0 \\ & & 1079 & 0 & 0 & 0 \\ & & & 1156 & 0 & 0 \\ & sym & & & 1156 & 0 \\ & & & & & 1156 \end{bmatrix}$$

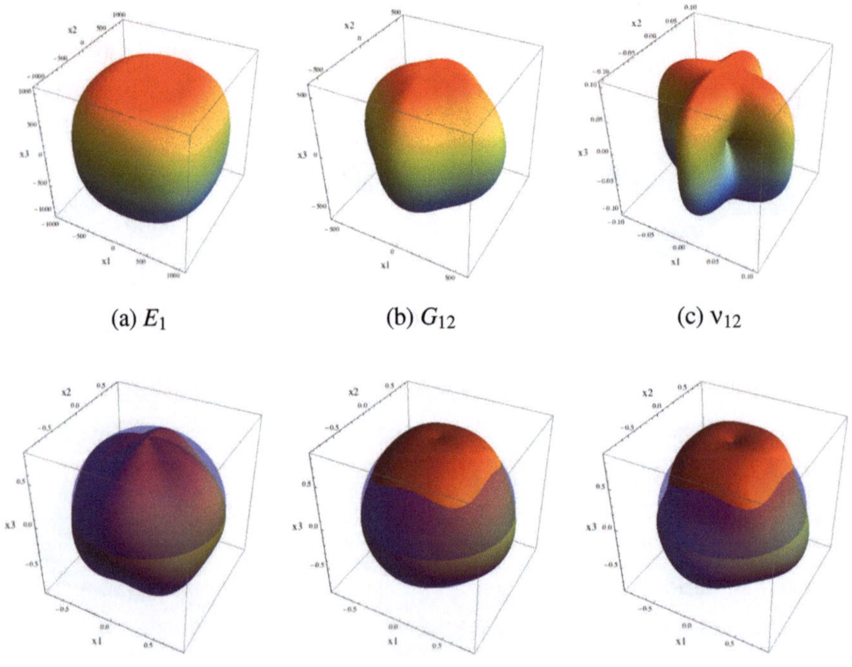

(a) E_1 (b) G_{12} (c) ν_{12}

(d) $\mu_{23,12}$ (e) $\eta_{1,31}$ (f) $\eta_{31,1}$

Fig. 2.14 Directional 3D-graphics of Diamond

References

Z. Bakri, A. Zaoui, Structural and mechanical properties of dolomite rock under high pressure conditions: a first-principles study. Phys. Status Solidi **248**, 1894–1900 (2011)

S.C. Cowin, Properties of the anisotropic elasticity tensor. Q. J. Mech. Appl. Math. **42**, 249–266 (1989)

S.C. Cowin, M.M. Mehrabadi, On the identification of material symmetry for anisotropic elastic materials. Q. J. Mech. Appl. Math. **40**, 451–476 (1987)

J. Evans, T. Grove, *Structure of Earth Materials - MIT Open CourseWare* (MIT, Boston, 2004), http://ocw.mit.edu/courses/earth-atmospheric-and-planetary-sciences/12-108-structure-of-earth-materials-fall-2004/lecture-notes/lec20.pdf

G. Green, On the laws of reflexion and refraction of light at the common surface of two non-crystallized media. Camb. Philos. Soc. Trans. **7** (1839)

M.E. Gurtin, *An Introduction to Continuum Mechanics* (Academic Press Inc., New York, 1981)

M.A. Hayes, A.N. Norris, Static implications of the existence of a plane of symmetry in an anisotropic elastic solid. Q. J. Mech. Appl. Math. **45**, 141–147 (1991)

F.E. Hohn, *Elementary Matrix Algebra* (MacMillan, New York, 1958)

R.M. Jones, *Mechanics of Composite Materials*, 2nd edn. (Taylor & Francis, Philadelphia, 1999)

I.I. Kolodner, Existence of longitudinal waves in anisotropic media. J. Acoust. Soc. Am. **40**, 730–731 (1966)

H.H. Landolt, R. Börnstein, *Second and Higher Order Elastic Constants*, Numerical Data and Functional Relationships in Science and Technology III, vol. 29/a (Springer, Berlin, 1992)

S.G. Lekhnitskii, in *Theory of Elasticity of an Anisotropic Elastic Body*. English translation by P. Fern (Holden-Day, San Francisco, 1963), p. 1950

A.E.H. Love, *A Treatise on the Mathematical Theory of Elasticity* (Dover, New York, 1944)

M.M. Mehrabadi, S.C. Cowin, Eigentensors of linear anisotropic elastic materials. Q. J. Mech. Appl. Math. **43**, 15–41 (1990)

P. Podio-Guidugli, A primer in elasticity. J. Elast. **58**, 1–104 (2000)

W. Thomson Lord Kelvin, Elements of a mathematical theory of elasticity. Philos. Trans. R. Soc. **146**, 481–498 (1856)

W. Thomson Lord Kelvin, Mathematical theory of elasticity. Encycl. Br. **7**, 819–825 (1878)

T.C.T. Ting, *Anisotropic Elasticity* (Oxford University Press, Oxford, 1996)

P. Vannucci, A note on the computation of the extrema of Young's modulus for hexagonal materials: an approach by planar tensor invariants. Appl. Math. Comput. (2015)

W. Voigt, *Lehrbuch der Kristallphysik* (B.G. Teubner, Leipzig, 1910)

Chapter 3
Plane Anisotropic Elasticity

Abstract This Chapter is an introduction to plane anisotropic elasticity. Some classical topics are detailed, starting from the general concepts of stress functions. Then a deep analysis of the concepts of plane states is proposed: the cases of plane strain, plane stress and generalized plane stress are detailed, with an analysis of the differences that can be found in the literature on this subject. The mechanical consistency of plane stress and a comparison between the above three cases is then given. Successively, the reader is introduced to the Lekhnitskii theory, which is presented in detail. The last part of the Chapter concerns a brief presentation of the Stroh formalism.

3.1 Introduction

The results found in the previous Chapter are quite general and constitute the basis for the mechanical analysis of anisotropic elastic bodies. Nevertheless, in a great number of situations the problem can be reduced from a 3D to a planar one, because of its geometry and loading conditions. This reduction can considerably simplify the problem and also open the way to the use of special mathematical techniques, like for instance complex variables.

For this reason, it is interesting to specialize to the planar case the results found in Chap. 2 and develop special theories for plane anisotropic elasticity.

Actually, different cases can be considered; to this purpose, it is worth to make a distinction between *plane tensors* and *plane fields*. A plane tensor, for instance $\boldsymbol{\sigma}$ or $\boldsymbol{\varepsilon}$ or also the displacement vector $\mathbf{u}$, considered as a tensor of the first rank, is a tensor whose components orthogonal to a given plane, say the plane $x_3 = 0$, are all null (i.e. $\sigma_{13} = \sigma_{23} = \sigma_{33} = 0$, $\varepsilon_{13} = \varepsilon_{23} = \varepsilon_{33} = 0$, or $u_3 = 0$), while a plane field is a tensor function whose components are scalar functions independent of x_3: $\sigma_{ij} = \sigma_{ij}(x_1, x_2)$, $\varepsilon_{ij} = \varepsilon_{ij}(x_1, x_2)$ or $u_i = u_i(x_1, x_2)$, $\forall i, j = 1, 2, 3$. A plane field is, hence, not necessarily a plane tensor and viceversa, and cases are possible, depending on the assumptions, where in the same problem one of the tensors is not plane nor a plane field, while the others are plane tensors and/or plane fields. The possible combinations are different, and the literature is not always completely clear

P. Vannucci, *Anisotropic Elasticity*, Lecture Notes in Applied and Computational Mechanics 85, DOI 10.1007/978-981-10-5439-6_3

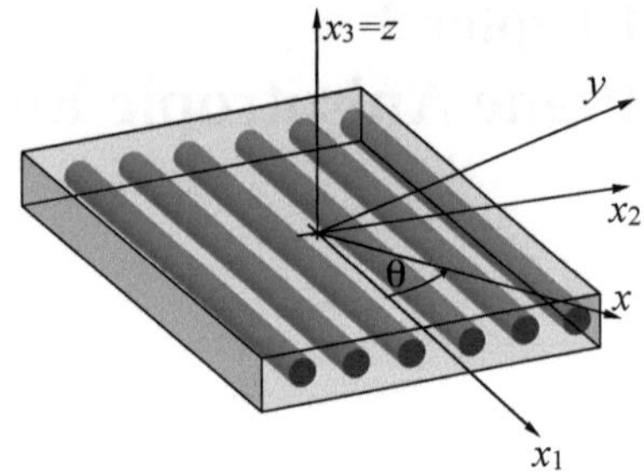

Fig. 3.1 General sketch for the planar case

about this topic. We have tried to give an exposition as complete as possible about this subject, considering the different approaches and the possible definitions existing in the literature. So, the cases of plane strain, plane stress, generalized plane stress are presented, and different points of view in the literature are exposed. Then, the bases of the Lekhnitskii theory are presented and a short account of the Stroh formalism given. The question of the representation of elasticity in planar problems is only rapidly considered in Sect. 3.3.2, while the polar formalism, a subject too vast to be bounded to a Section, is introduced in Chap. 4.

Figure 3.1 shows the general sketch that will be used in the following for the planar case: without loss in generality, we will consider that the structure belongs to the plane $x_3 = 0$ and we will use two bases: the first one, $\mathscr{B} = \{\mathbf{e}_1,\ \mathbf{e}_2,\ \mathbf{e}_3\}$, with axes $\{x_1, x_2, x_3\}$, that we will call the *material basis*, because in this basis the properties of the material are known. It is typically the case of orthotropic materials, whose elastic moduli are customarily given with respect to the orthotropic basis, like the case in Fig. 3.1, where an orthotropic layer composed by an isotropic matrix reinforced by aligned fibers, a typical unidirectional composite ply, is ideally represented. The second basis, $\mathscr{B}' = \{\mathbf{e}_1',\ \mathbf{e}_2',\ \mathbf{e}_3'\}$, with axes $\{x_1', x_2', x_3'\}$, is a generic basis, rotated counterclockwise through an angle θ about the axis $x_3 = x_3'$.

3.2 Stress Functions

Before introducing plane states, it is worth to recall a classical mathematical technique for plane problems, based upon the use of complex variables: the *stress function*, first introduced in 1862 by the astronomer Airy (1862), then completed in a subsequent paper (Airy 1863). Airy noticed that in 2D problems the equilibrium equations of a body subjected to only surface tractions (i.e. with a null body vector) indicate that the σ_{ij} can be regarded as the second-order partial derivatives of a single scalar function, the *Airy's stress function.*[1] We give here the most general approach

[1]Later, Maxwell generalized this result to the 3D case, introducing three of such scalar stress functions (Maxwell 1870).

to the Airy's stress function, valid regardless of the type of material and including also the presence of a body vector **b**, Eq. (1.28) (Milne-Thomson 1960, p. 5).

Consider a plane system defined by the conditions

$$\sigma_{ij} = \sigma_{ij}(x_1, x_2), \quad \sigma_{23} = \sigma_{31} = 0, \tag{3.1}$$

which implies that the equilibrium equations (1.27) reduce to

$$\sigma_{\alpha\beta,\beta} = b_\alpha, \quad \alpha, \beta = 1, 2, \tag{3.2}$$

and that $b_3 = 0$. For such a plane problem, we introduce the complex variable (i denotes the imaginary unit and a bar the complex conjugate)

$$z = x_1 + ix_2 \rightarrow \overline{z} = x_1 - ix_2, \tag{3.3}$$

and conversely

$$x_1 = \frac{1}{2}(z + \overline{z}), \quad x_2 = -\frac{1}{2}i(z - \overline{z}). \tag{3.4}$$

For the differential operators we have then the following equivalences

$$\begin{cases} \dfrac{\partial}{\partial x_1} = \dfrac{\partial}{\partial z} + \dfrac{\partial}{\partial \overline{z}}, \\ \dfrac{\partial}{\partial x_2} = i\dfrac{\partial}{\partial z} - i\dfrac{\partial}{\partial \overline{z}}, \end{cases} \quad \begin{cases} 2\dfrac{\partial}{\partial z} = \dfrac{\partial}{\partial x_1} - i\dfrac{\partial}{\partial x_2}, \\ 2\dfrac{\partial}{\partial \overline{z}} = \dfrac{\partial}{\partial x_1} + i\dfrac{\partial}{\partial x_2}. \end{cases} \tag{3.5}$$

If $(3.5)_1$ is injected into (3.2) we get

$$\begin{aligned} \frac{\partial \sigma_{11}}{\partial z} + \frac{\partial \sigma_{11}}{\partial \overline{z}} + i\left(\frac{\partial \sigma_{12}}{\partial z} - \frac{\partial \sigma_{12}}{\partial \overline{z}}\right) &= b_1, \\ \frac{\partial \sigma_{21}}{\partial z} + \frac{\partial \sigma_{21}}{\partial \overline{z}} + i\left(\frac{\partial \sigma_{22}}{\partial z} - \frac{\partial \sigma_{22}}{\partial \overline{z}}\right) &= b_2; \end{aligned} \tag{3.6}$$

multiplying the second equation by $-i$ and adding the result to the first equation gives

$$\frac{\partial \Theta}{\partial z} - \frac{\partial \Phi}{\partial \overline{z}} = b_1 - ib_2, \tag{3.7}$$

where

$$\Theta = \sigma_{11} + \sigma_{22}, \quad \Phi = \sigma_{22} - \sigma_{11} + 2i\sigma_{12}, \tag{3.8}$$

are the *Kolosov's fundamental stress combinations* (Kolosov 1909).

Be Θ_0, Φ_0 a particular solution of (3.7) corresponding to the action of the body vector, i.e. such that

$$\frac{\partial\Theta_0}{\partial z} - \frac{\partial\Phi_0}{\partial\overline{z}} = b_1 - ib_2; \tag{3.9}$$

then, the general solution of (3.7) is

$$\begin{aligned}\Theta &= \Theta_0 + 4\frac{\partial^2\chi}{\partial z\partial\overline{z}},\\ \Phi &= \Phi_0 + 4\frac{\partial^2\chi}{\partial z^2}.\end{aligned} \tag{3.10}$$

The arbitrary real valued function

$$\chi = \chi(x_1, x_2) = \chi(z, \overline{z}) \tag{3.11}$$

is the *Airy's stress function.*

The solution of the stress problem is hence reduced to the knowledge of the Airy's function. In fact, from Eqs. (3.8) and (3.10) we get

$$\begin{aligned}\sigma_{11} &= \frac{1}{2}\Theta - \frac{1}{4}(\Phi + \overline{\Phi}) = \sigma_{11}^0 + \frac{\partial^2\chi}{\partial x_2^2},\\ \sigma_{22} &= \frac{1}{2}\Theta + \frac{1}{4}(\Phi + \overline{\Phi}) = \sigma_{22}^0 + \frac{\partial^2\chi}{\partial x_1^2},\\ \sigma_{12} &= -\frac{1}{4}i(\Phi - \overline{\Phi}) = \sigma_{12}^0 - \frac{\partial^2\chi}{\partial x_1\partial x_2},\end{aligned} \tag{3.12}$$

where

$$\begin{aligned}\sigma_{11}^0 &= \frac{1}{2}\Theta_0 - \frac{1}{4}(\Phi_0 + \overline{\Phi}_0),\\ \sigma_{22}^0 &= \frac{1}{2}\Theta_0 + \frac{1}{4}(\Phi_0 + \overline{\Phi}_0),\\ \sigma_{12}^0 &= -\frac{1}{4}i(\Phi_0 - \overline{\Phi}_0),\end{aligned} \tag{3.13}$$

are a particular solution of the equilibrium equations (3.2) accounting for the body vector. When this last is null, i.e. when the body is acted upon uniquely by surface tractions, Eq. (3.12) becomes simply

$$\begin{aligned}\sigma_{11} &= \frac{\partial^2\chi}{\partial x_2^2},\\ \sigma_{22} &= \frac{\partial^2\chi}{\partial x_1^2},\\ \sigma_{12} &= -\frac{\partial^2\chi}{\partial x_1\partial x_2}.\end{aligned} \tag{3.14}$$

It is an easy task to check that for the equilibrium in case of volume forces depending upon a potential $U, \mathbf{f} = \nabla U$, then Eq. (3.14) becomes

$$\sigma_{11} = \frac{\partial^2 \chi}{\partial x_2^2} - U,$$
$$\sigma_{22} = \frac{\partial^2 \chi}{\partial x_1^2} - U, \tag{3.15}$$
$$\sigma_{12} = -\frac{\partial^2 \chi}{\partial x_1 \partial x_2}.$$

The knowledge of the Airy's stress function gives hence the stress components that *automatically satisfy the equilibrium equations.*

The Airy's function can be introduced also without making use of complex variables, by a technique that lets appear also another stress function used in the following. To this purpose, let us introduce the

Theorem 5 *Be $f_1(x_1, x_2)$ and $f_2(x_1, x_2)$ two scalar plane functions such that*

$$\frac{\partial f_1}{\partial x_1} + \frac{\partial f_2}{\partial x_2} = 0; \tag{3.16}$$

then a potential function $\Psi(x_1, x_2)$ exists such that

$$f_1 = -\frac{\partial \Psi}{\partial x_2}, \quad f_2 = \frac{\partial \Psi}{\partial x_1}. \tag{3.17}$$

Proof It is immediate to check that Eq. (3.17) satisfy Eq. (3.16). Then, let us suppose that f_1 and f_2 satisfy Eq. (3.16) and consider the function

$$\Psi(x_1, x_2) = \int_{a_1}^{x_1} f_2(\xi, x_2)\, d\xi - \int_{a_2}^{x_2} f_1(x_1, \eta)\, d\eta. \tag{3.18}$$

Upon differentiation and after using Eq. (3.16) the (3.17) is readily found.

The equilibrium equations of a system subjected to only surface tractions and where $\sigma_{ij} = \sigma_{ij}(x_1, x_2)$ are in the form of (3.16):

$$\sigma_{11,1} + \sigma_{12,2} = 0,$$
$$\sigma_{21,1} + \sigma_{22,2} = 0, \tag{3.19}$$
$$\sigma_{31,1} + \sigma_{32,2} = 0,$$

so for Theorem 5 there exist scalar functions $\varphi_i(x_1, x_2)$ such that

$$\sigma_{i1} = -\varphi_{i,2}, \quad \sigma_{i2} = \varphi_{i,1}. \tag{3.20}$$

Because $\sigma_{12}=\sigma_{21}$,

$$\varphi_{1,1}+\varphi_{2,2}=0, \tag{3.21}$$

which once more is in the form of (3.16), so it exists a scalar function $\chi(x_1,x_2)$ such that

$$\varphi_1=-\chi_{,2}\,,\quad \varphi_2=\chi_{,1}\,. \tag{3.22}$$

Then, by (3.20), we get the (3.14). Putting $\varphi_3=-\Psi$, called *stress function*[2] (Ting 1996, p. 119), we get also

$$\sigma_{23}=-\Psi_{,1}\,,\quad \sigma_{31}=\Psi_{,2}\,. \tag{3.23}$$

We remark that in this approach we did not make the assumption $\sigma_{23}=\sigma_{13}=0$; as a result, we get a second stress function, Ψ.

When the stresses are represented through χ and Ψ by Eqs. (3.14) and (3.23), then the equilibrium equations are automatically satisfied. To remark that σ_{33} cannot be determined by this way. We will see the use of functions χ and Ψ in Sect. 3.9.

3.3 Simplifying the General Relations

3.3.1 Rotation of Axes

It is possible, in the plane case, to give a synthetic, matrix expression of the transformation of the Cartesian components from the basis $\mathscr{B}=\{x_1,x_2\}$ to the basis $\mathscr{B}'=\{x_1',x_2'\}$, see Fig. 3.1. In particular, we will give the explicit form of the transformations (2.28), (2.49) and (2.166).

The change of axes sketched in Fig. 3.1, is represented by the orthogonal tensor, Eq. (2.30),

$$\mathbf{U}=\begin{bmatrix} c & s & 0\\ -s & c & 0\\ 0 & 0 & 1\end{bmatrix},\quad c=\cos\theta,\ s=\sin\theta, \tag{3.24}$$

which gives the rotation matrix, Eq. (2.48),

$$[R]=\begin{bmatrix} c^2 & s^2 & 0 & 0 & 0 & \sqrt{2}cs\\ s^2 & c^2 & 0 & 0 & 0 & -\sqrt{2}cs\\ 0 & 0 & 1 & 0 & 0 & 0\\ 0 & 0 & 0 & c & -s & 0\\ 0 & 0 & 0 & s & c & 0\\ -\sqrt{2}cs & \sqrt{2}cs & 0 & 0 & 0 & c^2-s^2\end{bmatrix}. \tag{3.25}$$

[2] In the literature, also χ is called stress function, which is a little bit ambiguous.

If the only plane components of $\{\varepsilon\}$ are considered, extracting from matrix (3.25) the relevant components, Eq. (2.49) becomes

$$\begin{Bmatrix} \varepsilon'_1 \\ \varepsilon'_2 \\ \varepsilon'_6 \end{Bmatrix} = \begin{bmatrix} c^2 & s^2 & \sqrt{2}cs \\ s^2 & c^2 & -\sqrt{2}cs \\ -\sqrt{2}cs & \sqrt{2}cs & c^2 - s^2 \end{bmatrix} \begin{Bmatrix} \varepsilon_1 \\ \varepsilon_2 \\ \varepsilon_6 \end{Bmatrix}. \tag{3.26}$$

When matrix (3.25) is injected into Eq. (2.66), then, rearranging in a vector the components of $[S]$, we get the matrix relation

$$\begin{Bmatrix} S'_{11} \\ S'_{16} \\ S'_{12} \\ S'_{66} \\ S'_{26} \\ S'_{22} \end{Bmatrix} = \begin{bmatrix} c^4 & 2\sqrt{2}c^3 s & 2c^2s^2 & 2c^2s^2 & 2\sqrt{2}cs^3 & s^4 \\ -\sqrt{2}c^3 s & c^4 - 3c^2s^2 & \sqrt{2}cs(c^2 - s^2) & \sqrt{2}cs(c^2 - s^2) & 3c^2s^2 - s^4 & \sqrt{2}cs^3 \\ c^2s^2 & \sqrt{2}cs(s^2 - c^2) & c^4 + s^4 & -2c^2s^2 & \sqrt{2}cs(c^2 - s^2) & c^2s^2 \\ 2c^2s^2 & 2\sqrt{2}cs(s^2 - c^2) & -4c^2s^2 & (c^2 - s^2)^2 & 2\sqrt{2}cs(c^2 - s^2) & 2c^2s^2 \\ -\sqrt{2}cs^3 & 3c^2s^2 - s^4 & \sqrt{2}cs(s^2 - c^2) & \sqrt{2}cs(s^2 - c^2) & c^4 - 3c^2s^2 & \sqrt{2}c^3 s \\ s^4 & -2\sqrt{2}cs^3 & 2c^2s^2 & 2c^2s^2 & -2\sqrt{2}c^3 s & c^4 \end{bmatrix} \begin{Bmatrix} S_{11} \\ S_{16} \\ S_{12} \\ S_{66} \\ S_{26} \\ S_{22} \end{Bmatrix}. \tag{3.27}$$

Equations (3.25), (3.26) and (3.27) are valid also for $\{\sigma\}$ and $[C]$ respectively, because they are written in the Kelvin's notation; e.g., if Eq. (3.25) is used in Eq. (2.28) then the same transformation matrix of Eq. (3.27) is obtained for $[C]$. When the Voigt's notation is used, this is not true and the transformations for $\{\sigma\}$ and $[C]$ are different from those for $\{\varepsilon\}$ and $[S]$.

3.3.2 *The Tsai and Pagano Parameters*

Equation (3.27) is to be used each time that a change of frame has to be done, which is the case, e.g., in several computations concerning the mechanics of laminated plates. The drawback of such equation is that it is rather cumbersome to be used. So, with the aim to obtain a simpler equation for the axes rotation of the stiffness matrix of an anisotropic layer, Tsai and Pagano (1968) proposed a transformation of Eq. (3.27), obtained exclusively using standard trigonometric identities. The transformation introduced by Tsai and Pagano is

$$\begin{Bmatrix} Q'_{11} \\ Q'_{12} \\ Q'_{16} \\ Q'_{22} \\ Q'_{26} \\ Q'_{66} \end{Bmatrix} = \begin{bmatrix} 1 & \cos 2\theta & \cos 4\theta & 0 & 0 & 2\sin 2\theta & \sin 4\theta \\ 0 & 0 & -\cos 4\theta & 1 & 0 & 0 & -\sin 4\theta \\ 0 & \frac{\sqrt{2}}{2}\sin 2\theta & \sqrt{2}\sin 4\theta & 0 & 0 & \sqrt{2}\cos 2\theta & \sqrt{2}\cos 4\theta \\ 1 & -\cos 2\theta & \cos 4\theta & 0 & 0 & -2\sin 2\theta & \sin 4\theta \\ 0 & \frac{\sqrt{2}}{2}\sin 2\theta & -\sqrt{2}\sin 4\theta & 0 & 0 & \sqrt{2}\cos 2\theta & -\sqrt{2}\cos 4\theta \\ 0 & 0 & -2\cos 4\theta & 0 & 2 & 0 & -2\sin 4\theta \end{bmatrix} \begin{Bmatrix} U_1 \\ U_2 \\ U_3 \\ U_4 \\ U_5 \\ U_6 \\ U_7 \end{Bmatrix}. \tag{3.28}$$

The above relation has been written, as usually done, for the reduced stiffness matrix of a plane stress state $[Q]$, Eq. (3.62), the situation normally considered for the mechanics of laminates. The Kelvin's notation has been used, that is why it is slightly different from the original one, written for the Voigt's notation, and as such valid exclusively for stiffness, while Eq. (3.28) can be applied to the compliance matrix $[S]$ as well.

The quantities U_is are the so-called *Tsai and Pagano parameters*. They are linear combinations of the components of the matrix in the original frame:

$$\begin{aligned}
U_1 &= \frac{1}{8}(3Q_{11} + 2Q_{12} + 3Q_{22} + 2Q_{66}),\\
U_2 &= \frac{1}{2}(Q_{11} - Q_{22}),\\
U_3 &= \frac{1}{8}(Q_{11} - 2Q_{12} + Q_{22} - 2Q_{66}),\\
U_4 &= \frac{1}{8}(Q_{11} + 6Q_{12} + Q_{22} - 2Q_{66}),\\
U_5 &= \frac{1}{8}(Q_{11} - 2Q_{12} + Q_{22} + 2Q_{66}),\\
U_6 &= \frac{1}{2\sqrt{2}}(Q_{16} + Q_{26}),\\
U_7 &= \frac{1}{2\sqrt{2}}(Q_{16} - Q_{26}).
\end{aligned} \tag{3.29}$$

In the literature, the U_is are often called *invariants*, like in the same title of the original publication, but this is not completely correct. In fact, not all of them are tensor invariants, in particular it can be checked that U_2, U_3, U_6 and U_7 are frame dependent quantities. To remark that Tsai and Pagano make use of 7 quantities to express 6 other functions. As a consequence, the U_is are not all independent, and in fact it is simple to see that, for instance,

$$U_5 = \frac{U_1 - U_4}{2}. \tag{3.30}$$

Th U_is have not a direct and clear physical meaning, nor they are immediately linked to the anisotropic properties or to the elastic symmetries. In the literature, their use is exclusively bounded to the design of laminates.

3.3.3 Plane and Antiplane States and Tensors

The reduction from a 3D to the planar case sketched in Fig. 3.1 can be done in the two different cases of *plane strain* or *plane stress*, see *infra*. There are substantial differences between the two cases, but a common algebraic basis can be given for both of them.

To this purpose, let us rewrite the Hooke's law (2.20) like (p stands for *plane* and a for *antiplane*)

$$\begin{cases} \{\sigma^p\} = [C1]\{\varepsilon^p\} + [C2]\{\varepsilon^a\}, \\ \{\sigma^a\} = [C2]^\top\{\varepsilon^p\} + [C3]\{\varepsilon^a\}, \end{cases} \tag{3.31}$$

and its inverse (2.21) like

$$\begin{cases} \{\varepsilon^p\} = [S1]\{\sigma^p\} + [S2]\{\sigma^a\}, \\ \{\varepsilon^a\} = [S2]^\top\{\sigma^p\} + [S3]\{\sigma^a\}. \end{cases} \tag{3.32}$$

In Eqs. (3.31) and (3.32) it is:

$$\{\sigma^p\} = \begin{Bmatrix} \sigma_1 \\ \sigma_2 \\ \sigma_6 \end{Bmatrix}, \quad \{\sigma^a\} = \begin{Bmatrix} \sigma_3 \\ \sigma_4 \\ \sigma_5 \end{Bmatrix}, \tag{3.33}$$

$$\{\varepsilon^p\} = \begin{Bmatrix} \varepsilon_1 \\ \varepsilon_2 \\ \varepsilon_6 \end{Bmatrix}, \quad \{\varepsilon^a\} = \begin{Bmatrix} \varepsilon_3 \\ \varepsilon_4 \\ \varepsilon_5 \end{Bmatrix}, \tag{3.34}$$

$$\begin{aligned} &[C1] = \begin{bmatrix} C_{11} & C_{12} & C_{16} \\ & C_{22} & C_{26} \\ sym & & C_{66} \end{bmatrix}, \quad [C2] = \begin{bmatrix} C_{13} & C_{14} & C_{15} \\ C_{23} & C_{24} & C_{25} \\ C_{36} & C_{46} & C_{56} \end{bmatrix}, \\ &[C3] = \begin{bmatrix} C_{33} & C_{34} & C_{35} \\ & C_{44} & C_{45} \\ sym & & C_{55} \end{bmatrix}, \end{aligned} \tag{3.35}$$

$$\begin{aligned} &[S1] = \begin{bmatrix} S_{11} & S_{12} & S_{16} \\ & S_{22} & S_{26} \\ sym & & S_{66} \end{bmatrix}, \quad [S2] = \begin{bmatrix} S_{13} & S_{14} & S_{15} \\ S_{23} & S_{24} & S_{25} \\ S_{36} & S_{46} & S_{56} \end{bmatrix}, \\ &[S3] = \begin{bmatrix} S_{33} & S_{34} & S_{35} \\ & S_{44} & S_{45} \\ sym & & S_{55} \end{bmatrix}. \end{aligned} \tag{3.36}$$

The above results are the common algebraic basis for developing, separately but dually, the two cases of plane strain and plane stress. Generalizing what established in Sect. 3.1, we will call, in short, *plane tensors* all those with the superscript p and *antiplane* all those with the superscript a, i.e. it is antiplane any component out of the plane $x_3 = 0$.

3.4 Plane Strain

We define *plane strain* a state for which the displacement vector $\mathbf{u} = (u_1, u_2, u_3)$ is such that

$$u_3 = 0,\ u_\alpha = u_\alpha(x_1, x_2),\ \alpha = 1, 2. \tag{3.37}$$

Through the strain-displacement relations (1.25), Eq. (3.37) gives

$$\begin{aligned}&\varepsilon_3 = u_{3,3} = 0,\ \varepsilon_4 = \frac{u_{2,3} + u_{3,2}}{2} = 0,\ \varepsilon_5 = \frac{u_{1,3} + u_{3,1}}{2} = 0 \rightarrow\\ &\{\varepsilon^a\} = \{0\}, \quad \{\varepsilon^p\} = \{\varepsilon^p(x_1, x_2)\},\end{aligned} \tag{3.38}$$

which justifies the name *plane strain*: the *antiplane strain* $\{\varepsilon^a\}$ *is null* and the plane strain $\{\varepsilon^p\}$ is a *plane field*.

From Eqs. (3.31) and (3.32) we get hence, for the in plane tensors,

$$\begin{aligned}\{\sigma^p\} &= [C1]\{\varepsilon^p\},\\ \{\varepsilon^p\} &= [\Sigma]\{\sigma^p\},\end{aligned} \tag{3.39}$$

while for the antiplane tensors it is

$$\begin{aligned}&\{\sigma^a\} = [C2]^\top\{\varepsilon^p\} = -[S3]^{-1}[S2]^\top\{\sigma^p\} = [C2]^\top[\Sigma]\{\sigma^p\},\\ &\{\varepsilon^a\} = \{0\},\end{aligned} \tag{3.40}$$

with

$$[\Sigma] = [C1]^{-1} = [S1] - [S2][S3]^{-1}[S2]^\top, \tag{3.41}$$

the *reduced compliance matrix*. We can observe that in plane strain, the stiffness of the in-plane part, $[C1]$, does not change with respect to the 3D case, while the in-plane compliance changes: $[\Sigma] \neq [S1]$. Also, unlike $\{\varepsilon^a\}$, $\{\sigma^a\} \neq \{0\}$: the antiplane stress is not null in plane strain, generally speaking.

It is in principle possible to detail the components of $[\Sigma]$ in the general case of triclinic material, but the result is rather cumbersome. Bounding ourselves to the interesting case of monoclinic material, with $x_3 = 0$ plane of symmetry, gives, see Eqs. (2.56) and (2.57),

$$\Sigma_{ij} = S_{ij} - \frac{S_{i3}S_{j3}}{S_{33}},\quad i, j = 1, 2, 6, \tag{3.42}$$

and

$$\{\sigma^a\} = \begin{Bmatrix}\sigma_3\\ \sigma_4\\ \sigma_5\end{Bmatrix} = \begin{Bmatrix}C_{13}\varepsilon_1 + C_{23}\varepsilon_2 + C_{36}\varepsilon_6\\ 0\\ 0\end{Bmatrix}. \tag{3.43}$$

Through Eq. (3.40) we get also

$$\sigma_3 = -\frac{S_{13}\sigma_1 + S_{23}\sigma_2 + S_{36}\sigma_6}{S_{33}}. \tag{3.44}$$

It is hence apparent that in the case of monoclinic material with $x_3 = 0$ plane of symmetry, the transverse shear components σ_4 and σ_5 vanish in plane strain. This is not the case for σ_3.

For an orthotropic material with $\{x_1, x_2, x_3\}$ the orthotropic frame, because $C_{i6} = S_{i6} = 0\ \forall i = 1, 2, 3$, Eq. (2.63), the previous results can be further simplified:

$$[\Sigma] = \begin{bmatrix} S_{11} - \dfrac{S_{13}^2}{S_{33}} & S_{12} - \dfrac{S_{13}S_{23}}{S_{33}} & 0 \\ & S_{22} - \dfrac{S_{23}^2}{S_{33}} & 0 \\ sym & & S_{66} \end{bmatrix}, \tag{3.45}$$

$$\sigma_3 = C_{13}\varepsilon_1 + C_{23}\varepsilon_2 = -\frac{S_{13}\sigma_1 + S_{23}\sigma_2}{S_{33}}. \tag{3.46}$$

We can give the above result also as function of the components of $\mathbb{Z}$: because in the Kelvin notation Eq. (2.24) is to be considered, we obtain

$$[\Sigma] = \begin{bmatrix} Z_{1111} - \dfrac{Z_{1133}^2}{Z_{3333}} & Z_{1122} - \dfrac{Z_{1133}Z_{2233}}{Z_{3333}} & 0 \\ & Z_{2222} - \dfrac{Z_{2233}^2}{Z_{3333}} & 0 \\ sym & & 2Z_{1212} \end{bmatrix}. \tag{3.47}$$

Finally, if the S_{ij}s in Eq. (3.45) are expressed as functions of the technical constants, cf. Sect. 2.3, then we obtain

$$[\Sigma] = \begin{bmatrix} \dfrac{1 - \nu_{13}\nu_{31}}{E_1} & -\dfrac{\nu_{12}(\nu_{21} + \nu_{23}\nu_{31})}{\nu_{21}E_1} & 0 \\ & \dfrac{1 - \nu_{23}\nu_{32}}{E_2} & 0 \\ sym & & \dfrac{1}{2G_{12}} \end{bmatrix}, \tag{3.48}$$

$$\sigma_3 = \frac{(\nu_{13} + \nu_{12}\nu_{23})\varepsilon_1 + (\nu_{23} + \nu_{21}\nu_{13})\varepsilon_2}{1 - \nu_{12}\nu_{21} - \nu_{13}\nu_{31} - \nu_{23}\nu_{32} - 2\nu_{13}\nu_{32}\nu_{21}}E_3 = \nu_{31}\sigma_1 + \nu_{32}\sigma_2. \tag{3.49}$$

Adapting the above results to an isotropic body is immediate:

$$[\Sigma]=\begin{bmatrix}\frac{1-\nu^2}{E} & -\frac{\nu(1+\nu)}{E} & 0\\ & \frac{1-\nu^2}{E} & 0\\ sym & & \frac{1+\nu}{E}\end{bmatrix}, \tag{3.50}$$

$$\sigma_3=\frac{\nu(\varepsilon_1+\varepsilon_2)}{(1-2\nu)(1+\nu)}E=\nu(\sigma_1+\sigma_2). \tag{3.51}$$

Three remarks: first, condition (3.37) implies not only that $\{\varepsilon^p\}$ is a plane function, Eq. (3.38), but even, through the Hooke's law, that $\{\sigma\}$ is a plane function too:

$$\sigma_i=\sigma_i(x_1,x_2),\ \forall i=1,\ldots,6. \tag{3.52}$$

Then, plane strain is typical of infinitely long cylindrical bodies subjected to loadings that do not depend upon x_3, the longitudinal axis (e.g. a pipe with internal or/and external pressure, a rail under its own weight etc.). In such cases, the assumption (3.37) is plausible.

Finally, generally speaking $\sigma_3(x_1,x_2)\neq 0$. Hence, a plane strain is possible, for finite cylinders, only when appropriate actions are applied at the bases of the cylinder, in order to ensure the existence of $\sigma_3(x_1,x_2)\neq 0$ and that $u_3=0$.

3.4.1 The Concept of Plane Strain in the Literature

The concept of plane strain can get different definitions in the literature; the definition given here, Eq. (3.37), is the same one given by Love (1944, p. 137), Muskhelishvili (1953, p. 89), and by Rand and Rovenski (2005, p. 87).

A general and rigorous definition, valid not only for infinitely long cylinders, is given by Milne-Thomson (1960, p. 11):

> [...] a state of *plane deformation* is said to exist if the following conditions are satisfied: (i) one of the principal directions of deformation is the same at every point of the material; (ii) apart from a rigid body movement of the material as a whole, particles which occupy planes perpendicular to the fixed principal direction prior to the deformation continue to occupy the same planes after the deformation.

To remark that Milne-Thomson makes use of the term *plane deformation* and not of *plane strain*. Of course, the above definition implies that there is no warping of the material planes orthogonal to the invariable principal direction and that

$$u_3=0,\ \ \varepsilon_3=0, \tag{3.53}$$

but not the second assumption of (3.37), i.e. $u_\alpha=u_\alpha(x_1,x_2),\ \alpha=1,2$. Actually, Milne-Thomson shows that this is a consequence of the definition of plane strain and of the strain-displacement relation in non-linear elasticity, i.e. taking the Green-Lagrange tensor as measure of deformation, Eq. (1.22):

$$\varepsilon_{ij} = \frac{1}{2}(u_{i,j} + u_{j,i} + u_{k,i}u_{k,j}) \rightarrow \varepsilon_3 = u_{3,3} + \frac{1}{2}(u_{1,3}^2 + u_{2,3}^2 + u_{3,3}^2), \quad (3.54)$$

which, through Eq. (3.53), brings to

$$u_{1,3}^2 + u_{2,3}^2 = 0 \Rightarrow u_{1,3} = u_{2,3} = 0 \Rightarrow u_\alpha = u_\alpha(x_1, x_2),\ \alpha = 1, 2. \quad (3.55)$$

This result, along with $(3.53)_1$, gives

$$\begin{aligned}
\varepsilon_1 &= u_{1,1} + \frac{1}{2}(u_{1,1}^2 + u_{2,1}^2),\\
\varepsilon_2 &= u_{2,2} + \frac{1}{2}(u_{1,2}^2 + u_{2,2}^2),\\
\varepsilon_6 &= \frac{1}{\sqrt{2}}(u_{1,2} + u_{2,1} + u_{1,1}u_{1,2} + u_{2,1}u_{2,2}),\\
\varepsilon_3 &= \varepsilon_4 = \varepsilon_5 = 0.
\end{aligned} \quad (3.56)$$

Of course, the results of Eq. (3.56) reduce to the classical infinitesimal strain tensor (1.25) in the limit (1.24). As a consequence, also $\{\sigma\} = \{\sigma(x_1, x_2)\}$, i.e. the stress tensor is a plane field too, so giving what Milne-Thomson calls *a plane system*, see Sect. 3.5.

Ting (1996, p. 65) introduces the argument as *antiplane deformations* and then he develops, substantially, the theory described in this Section and later on in Sect. 3.7. Then, (Ting 1996, p. 134) he introduces another category of plane deformation problems, those where the only basic assumption is

$$u_i = u_i(x_1, x_2)\ \forall i = 1, 2, 3. \quad (3.57)$$

There is a substantial difference between this plane case and the one developed in this Section or that defined by Milne-Thomson, because now $u_3 \neq 0$. Ting calls this type of plane deformation *the Stroh formalism*, and the fundamentals of such theory will be introduced in Sect. 3.10.

Green and Zerna (1954, p. 184) introduce the concept of plane strain as a system where the displacement and strain components are independent from x_3, so substantially the same definition given by Ting for the Stroh formalism, but then they develop all the theory in the framework of nonlinear elasticity, which is far beyond the scope of this text.

3.5 Plane Stress

By definition, we assume that an elastic body is in a *plane stress* state when the antiplane stress $\{\sigma^a\}$ is null and the plane stress $\{\sigma^p\}$ is a plane function.

$$\begin{aligned}&\{\sigma^a\}=\{0\}\ \rightarrow\ \sigma_3=\sigma_4=\sigma_5=0,\\&\{\sigma^p\}=\{\sigma^p(x_1,x_2)\},\ \rightarrow\ \sigma_i=\sigma_i(x_1,x_2)\ \forall i=1,2,6.\end{aligned}\tag{3.58}$$

As a consequence of (3.58) and of the equation of motion, (1.27), also the body vector is a plane function: $\mathbf{b}=\mathbf{b}(x_1,x_2)$.

The case of plane stress is completely analogous to the previous one of plane strain: because of the symmetry of relations (3.31) and (3.32), the developments for plane stress can be obtained repeating *verbatim* those for plane strain, simply replacing the strains with stresses, the compliances with stiffnesses:

$$\begin{aligned}&\{\sigma^p\}=[Q]\{\varepsilon^p\},\\&\{\varepsilon^p\}=[S1]\{\sigma^p\},\end{aligned}\tag{3.59}$$

and for the antiplane tensors

$$\begin{aligned}&\{\sigma^a\}=0,\\&\{\varepsilon^a\}=[S2]^\top\{\sigma^p\}=-[C3]^{-1}[C2]^\top\{\varepsilon^p\}=[S2]^\top[Q]\{\varepsilon^p\},\end{aligned}\tag{3.60}$$

with

$$[Q]=[S1]^{-1}=[C1]-[C2][C3]^{-1}[C2]^\top,\tag{3.61}$$

the *reduced stiffness matrix*. In a dual manner with respect to the results of plane strain, in case of plane stress the compliance of the in-plane part, $[S1]$, does not change with respect to the 3D case, while the in-plane stiffness changes: $[Q]\neq[C1]$. Also, unlike $\{\sigma^a\}$, $\{\varepsilon^a\}\neq\{0\}$: the antiplane strain is not null in plane stress, generally speaking.

For a monoclinic material we obtain

$$Q_{ij}=C_{ij}-\frac{C_{i3}C_{j3}}{C_{33}},\quad i,j=1,2,6,\tag{3.62}$$

and

$$\{\varepsilon^a\}=\left\{\begin{matrix}\varepsilon_3\\\varepsilon_4\\\varepsilon_5\end{matrix}\right\}=\left\{\begin{matrix}S_{13}\sigma_1+S_{23}\sigma_2+S_{36}\sigma_6\\0\\0\end{matrix}\right\}.\tag{3.63}$$

Through Eq. (3.60) we get also

$$\varepsilon_3=-\frac{C_{13}\varepsilon_1+C_{23}\varepsilon_2+C_{36}\varepsilon_6}{C_{33}}.\tag{3.64}$$

So, in the case of monoclinic material with $x_3 = 0$ plane of symmetry, the transverse shear deformations ε_4 and ε_5 vanish in plane stress, but not ε_3.

For an orthotropic material with $\{x_1, x_2, x_3\}$ the orthotropic frame, like in Fig. 3.1, we get:

$$[Q] = \begin{bmatrix} C_{11} - \dfrac{C_{13}^2}{C_{33}} & C_{12} - \dfrac{C_{13}C_{23}}{C_{33}} & 0 \\ & C_{22} - \dfrac{C_{23}^2}{C_{33}} & 0 \\ sym & & C_{66} \end{bmatrix}, \tag{3.65}$$

$$\varepsilon_3 = S_{13}\sigma_1 + S_{23}\sigma_2 = -\frac{C_{13}\varepsilon_1 + C_{23}\varepsilon_2}{C_{33}}. \tag{3.66}$$

Using the fact that $[Q] = [S1]^{-1}$, Eq. (3.61), we get also

$$[Q] = \begin{bmatrix} \dfrac{S_{22}}{S_{11}S_{22} - S_{12}^2} & -\dfrac{S_{12}}{S_{11}S_{22} - S_{12}^2} & 0 \\ & \dfrac{S_{11}}{S_{11}S_{22} - S_{12}^2} & 0 \\ sym & & \dfrac{1}{S_{66}} \end{bmatrix}; \tag{3.67}$$

this result gives a bound on the Young's moduli. In fact, recalling Eqs. (2.89), (2.94) and (2.95), along with the positiveness of the Young's moduli, we obtain (no summation over dummy indexes)

$$Q_{ii} > E_i, \ i = 1, 2. \tag{3.68}$$

We can give the above result also as function of the components of $\mathbb{E}$: remembering that in the Kelvin notation we need to consider Eq. (2.24), we get easily

$$[Q] = \begin{bmatrix} E_{1111} - \dfrac{E_{1133}^2}{E_{3333}} & E_{1122} - \dfrac{E_{1133}E_{2233}}{E_{3333}} & 0 \\ & E_{2222} - \dfrac{E_{2233}^2}{E_{3333}} & 0 \\ sym & & 2E_{1212} \end{bmatrix}. \tag{3.69}$$

Finally, if the S_{ij}s in Eq. (3.67) are expressed as functions of the technical constants, cf. Sect. 2.3, then we obtain

$$[Q]=\begin{bmatrix}\frac{E_1}{1-\nu_{12}\nu_{21}} & \frac{\nu_{12}E_2}{1-\nu_{12}\nu_{21}} & 0\\ & \frac{E_2}{1-\nu_{12}\nu_{21}} & 0\\ sym & & 2G_{12}\end{bmatrix}, \tag{3.70}$$

$$\varepsilon_3=-\nu_{13}\frac{\sigma_1}{E_1}-\nu_{23}\frac{\sigma_2}{E_2}=-\frac{\varepsilon_1(\nu_{13}+\nu_{12}\nu_{23})+\varepsilon_2(\nu_{23}+\nu_{13}\nu_{21})}{1-\nu_{12}\nu_{21}}. \tag{3.71}$$

It is immediate to adapt the above results to an isotropic body:

$$[Q]=\begin{bmatrix}\frac{E}{1-\nu^2} & \frac{\nu E}{1-\nu^2} & 0\\ & \frac{E}{1-\nu^2} & 0\\ sym & & \frac{E}{1+\nu}\end{bmatrix}, \tag{3.72}$$

$$\varepsilon_3=-\frac{\nu}{E}(\sigma_1+\sigma_2)=-\frac{\nu}{1-\nu}(\varepsilon_1+\varepsilon_2). \tag{3.73}$$

A remark about the displacement vector $\mathbf{u}=(u_1,u_2,u_3)$: generally speaking, it is not a plane function:

$$\mathbf{u}=\mathbf{u}(x_1,x_2,x_3), \tag{3.74}$$

i.e., the problem is not plane for the displacements. This is a fundamental difference with plane strain; in fact, for plane strain, $\mathbf{u}$, $\boldsymbol{\varepsilon}$ and $\boldsymbol{\sigma}$ are all plane fields.

The state of plane stress is typical of thin, flat bodies, like plates or slabs. A plate is *thin* when its thickness is much smaller than its typical in-plane dimension. Be $x_3=0$ the mid-plane of the plate, like in Fig. 3.1. If the plate is submitted to only in-plane loadings, then, because of the small thickness of the plate and assuming a continuous distribution of the σ_{ij}s through the plate's thickness, assumptions (3.58) are a good approximation of reality.

3.5.1 The Concept of Plane Stress in the Literature

Let us briefly consider the notion of plane stress in the literature. The definition given here, Eq. (3.58), is rather classical, and it is, for instance, that given by Love (1944, p. 137), or by Rand and Rovenski (2005, p. 90). Milne-Thomson (1960, p. 5), gives perhaps the most general definition:

> [...] a *plane system* is one for which there exists a plane such that the stress tensor is the same at all material points of any normal to this plane as at the material point in which that normal meets the plane.

To remark the use of the term *plane system* and not of *plane stress* by Milne-Thomson. Also, his definition is not completely identical to that given in (3.58), because it is not required that condition $(3.58)_1$ be satisfied. Nevertheless, the same author immediately after considers only plane systems with $\sigma_4 = \sigma_5 = 0$. This implies that, for the third equation of motion, the body vector $\mathbf{b}$ is planar: $b_3 = 0$. So, all the actions are parallel to the plane of the system.

Lekhnitskii (1963, p. 103) analyzes exactly the general case of plane system as defined by Milne-Thomson and Ting (1996, p. 118) calls explicitly such a system *the Lekhnitskii Formalism*. Its theoretical implications are analyzed in Sect. 3.9.

3.6 Generalized Plane Stress

The concept of *generalized plane stress* was first introduced by Filon (1903) and successively developed by Love (1944, p. 138), Muskhelishvili (1953, p. 92) and by Lekhnitskii (1963, p. 134), as a special case of his plane theory, see Sect. 3.9.11.

Let us consider a thin plate whose thickness is $2t$, acted upon by only loadings parallel to the mid-plane $x_3 = 0$ and with the two surfaces unloaded:

$$\sigma_3 = \sigma_4 = \sigma_5 = 0 \text{ at } x_3 = \pm t. \tag{3.75}$$

For a triclinic material, the plane stress $\{\sigma^p\}$ will generate also antiplane strains, $\{\varepsilon^a\} \neq \{0\}$, which implies that $u_3(x_1, x_2, 0) \neq 0$: the mid-plane of the plate will warp under in-plane loadings. To exclude this possibility, we will consider only anisotropic materials with at least

$$C_{14} = C_{15} = C_{24} = C_{25} = C_{34} = C_{35} = C_{46} = C_{56} = 0. \tag{3.76}$$

The most general materials satisfying such requirements, are those of the monoclinic syngony with $x_3 = 0$ as plane of symmetry, Eq. (2.56).

Now, we introduce the *average displacements*

$$\widehat{u}_i = \frac{1}{2t} \int_{-t}^{+t} u_i \, dx_3 \ \forall i = 1, 2, 3, \tag{3.77}$$

and

$$[u_i] = \frac{u_i(x_1, x_2, t) - u_i(x_1, x_2, -t)}{2t} \ \forall i = 1, 2, 3. \tag{3.78}$$

We make the further assumption that all the applied forces are symmetrically distributed with respect to the mid-plane of the plate, so that the stresses are symmetric with respect to this plane. As a consequence, also the displacements will be symmetric and, by (3.77) and (3.78), it will be

$$[u_1] = [u_2] = 0, \ \ \widehat{u}_3 = 0, \tag{3.79}$$

which gives

$$\frac{1}{2t}\int_{-t}^{+t} u_{i,j}\, dx_3 = \begin{cases} \widehat{u}_{i,j} \ \ \forall i, j \neq 3, \\ \widehat{\varepsilon}_3 \ \text{if } i = j = 3, \\ 0 \ \text{otherwise.} \end{cases} \tag{3.80}$$

This result means that the average displacement is a plane vector and also a plane field:

$$\widehat{u}_3 = 0, \ \ \widehat{u}_\alpha = \widehat{u}_\alpha(x_1, x_2), \ \ \alpha = 1, 2, \tag{3.81}$$

and that for the average strain it is

$$\widehat{\varepsilon}_4 = \widehat{\varepsilon}_5 = 0, \ \ \widehat{\varepsilon}_3 \neq 0, \ \ \{\widehat{\varepsilon}\} = \{\widehat{\varepsilon}(x_1, x_2)\}, \tag{3.82}$$

i.e. the strain tensor in not plane but it is a plane field.

As a consequence, considering the requirements (3.76), integrating the Hooke's law over the thickness gives

$$\widehat{\sigma}_i = C_{i1}\widehat{\varepsilon}_1 + C_{i2}\widehat{\varepsilon}_2 + C_{i3}\widehat{\varepsilon}_3 + C_{i6}\widehat{\varepsilon}_6. \tag{3.83}$$

Applying the third equilibrium equation, see (1.27),

$$\sigma_{5,1} + \sigma_{4,2} + \sqrt{2}\sigma_{3,3} = 0, \tag{3.84}$$

at the plate's surfaces, $x_3 = \pm t$, for the (3.75) we get

$$\sigma_{3,3} = 0. \tag{3.85}$$

The consequence of (3.75) and of the last result is that we can assume

$$\sigma_3 \simeq 0 \ \ \forall x_3 \in [-t, t] \ \Rightarrow \ \widehat{\sigma}_3 = 0. \tag{3.86}$$

Then, writing the (3.83) for $\widehat{\sigma}_{33}$ gives the condition

$$\widehat{\varepsilon}_3 = -\frac{1}{C_{33}}\left(C_{31}\widehat{\varepsilon}_1 + C_{32}\widehat{\varepsilon}_2 + C_{36}\widehat{\varepsilon}_6\right), \tag{3.87}$$

that injected back into (3.83) gives

$$\{\widehat{\sigma}\} = [\widehat{C}]\{\widehat{\varepsilon}\}, \tag{3.88}$$

where the components of the *reduced elastic stiffness matrix* $[\widehat{C}]$ are given by

$$\widehat{C}_{ij} = C_{ij} - \frac{C_{i3}C_{j33}}{C_{33}}. \tag{3.89}$$

The above components define the reduced elastic stiffness matrix exactly as $[Q]$, see Eq. (3.62). Nevertheless, the difference with plane stress is that in generalized plane stress all the equations are satisfied on the average, not locally. If t is very small compared to the other relevant dimensions of the plate then generalized plane stress is a good approximation.

To notice that, through Eqs. (3.76) and (3.82) it is

$$\widehat{\sigma}_4 = \widehat{\sigma}_5 = 0. \tag{3.90}$$

Let us now integrate the equilibrium equations (1.27) on the thickness of the plate; then, Eq. (3.75) gives

$$\begin{aligned}\sqrt{2}\widehat{\sigma}_{1,1} + \widehat{\sigma}_{6,2} &= 0,\\ \widehat{\sigma}_{6,1} + \sqrt{2}\widehat{\sigma}_{2,2} &= 0,\\ \widehat{\sigma}_{5,1} + \widehat{\sigma}_{4,2} &= 0.\end{aligned} \tag{3.91}$$

3.7 Mechanical Consistency of Plane States

We have introduced in Sect. 3.4 plane strain and in Sect. 3.5 plane stress; we ponder now on their mechanical consistency, i.e. if such states are physically possible.

Let us begin with plane strain, Eq. (3.37); injecting the Hooke's law in the equilibrium equations (1.27) of a body submitted only to loadings on its boundary, these reduce to

$$\begin{aligned}&E_{i111}\varepsilon_{11,1} + 2E_{i112}\varepsilon_{12,1} + E_{i122}\varepsilon_{22,1} +\\ &E_{i211}\varepsilon_{11,2} + 2E_{i212}\varepsilon_{12,2} + E_{i222}\varepsilon_{22,2} = 0\end{aligned} \quad \forall i = 1, 2, 3. \tag{3.92}$$

The coefficients of the third equation are

$$\begin{aligned}&E_{3111} = C_{15},\ E_{3121} = E_{3112} = C_{56},\ E_{3212} = E_{3221} = C_{46},\\ &E_{3222} = C_{24},\ E_{3211} = C_{14},\ E_{3122} = C_{25}.\end{aligned} \tag{3.93}$$

All of these coefficients are null for a monoclinic material with $x_3 = 0$ as plane of symmetry, Eq. (2.56).

Let us now consider the *antiplane deformations*

$$u_1 = u_2 = 0,\ \ u_3 = u_3(x_1, x_2)\ \rightarrow\ u_{3,3} = \varepsilon_{33} = 0,\ \{\varepsilon^p\} = \{0\}. \tag{3.94}$$

Now the three equations of equilibrium reduce to

$$E_{i113}\varepsilon_{13,1} + E_{i123}\varepsilon_{23,1} + E_{i213}\varepsilon_{13,2} + E_{i223}\varepsilon_{23,2} = 0\ \forall i = 1, 2, 3. \tag{3.95}$$

The coefficients of the two first Eq. (3.95) are exactly the (3.93). Hence, a monoclinic body satisfies automatically, for each applied loading on the boundary, the third plane equilibrium equation and the two first antiplane equations: *plane and antiplane deformations are uncoupled*. The monoclinic condition is not the minimal requirement: the true necessary conditions are the (3.93) to be null, while for a monoclinic material, required for generalized plane stress, it is *also* $C_{34} = C_{35} = 0$.

This result is obviously valid also for the other elastic syngonies that satisfy the same conditions, namely for the orthotropic, tetragonal, axially-symmetric, cubic and isotropic ones. For all such materials, the plane strain state is a possible situation and it is an *exact theory*. To remark that in this circumstance, it is also $\sigma_4 = \sigma_5 = 0$.

For a triclinic or trigonal body, or for any other syngony not correctly oriented (i.e. for which $x_3 = 0$ is not one of the symmetry planes), a plane strain deformation or an antiplane one cannot exist, generally speaking: also in the case where the three components of displacement u_i depend upon only x_1 and x_2, all of them are coupled, so that u_3 does not vanish, in general. Such a state is called a *generalized plane strain*: $u_3 \neq 0$, but $\varepsilon_3 = 0$ because nothing is function of x_3, so that $u_{3,3} = 0$. This case is analyzed in Sect. 3.9.9.

The compatibility equations (1.29) give an equation for the Airy's stress function (3.14). In fact, with the assumptions (3.38) all the compatibility equations are automatically satisfied but the first, that in the Kelvin's notation reads like

$$\sqrt{2}\varepsilon_{6,12} = \varepsilon_{1,22} + \varepsilon_{2,11}. \tag{3.96}$$

Using Eq. $(3.39)_2$ and expressing the stress components by the (3.14), remembering that $\sigma_6 = \sqrt{2}\sigma_{12}$, we get the following homogenized biharmonic equation for the Airy's stress function χ:

$$\nabla_1^4 \chi = 0, \tag{3.97}$$

where

$$\begin{aligned}\nabla_1^4 = \Sigma_{22}\frac{\partial^4}{\partial x_1^4} - 2\sqrt{2}\Sigma_{26}\frac{\partial^4}{\partial x_1^3 \partial x_2} + 2(\Sigma_{12} + \Sigma_{66})\frac{\partial^4}{\partial x_1^2 \partial x_2^2} - \\ 2\sqrt{2}\Sigma_{16}\frac{\partial^4}{\partial x_1 \partial x_2^3} + \Sigma_{11}\frac{\partial^4}{\partial x_2^4}\end{aligned} \tag{3.98}$$

is the *generalized biharmonic differential operator* occurring in the present case and in the Kelvin's notation[3].

Let us now turn the attention on plane stress. If the Airy's function is used in the first compatibility equation (3.96), and proceeding like in the previous case, but now with the strain-stress relation $(3.59)_2$, we get the biharmonic equation for χ

$$\nabla_2^4 \chi = 0, \tag{3.99}$$

[3]Changing of material syngony or of plane state, other operators can be introduced; for instance, it can be easily checked that for an orthotropic material, $\Sigma_{16} = \Sigma_{26} = 0$, so that ∇_1^4 has a simpler form, while for an isotropic material we get $\nabla_1^4 = \frac{1-\nu^2}{E}\nabla^4$, where ∇^4 is the customary double laplacian.

with now

$$\nabla_2^4 = S_{22}\frac{\partial^4}{\partial x_1^4} - 2\sqrt{2}S_{26}\frac{\partial^4}{\partial x_1^3\partial x_2} + 2(S_{12}+S_{66})\frac{\partial^4}{\partial x_1^2\partial x_2^2} - \\ 2\sqrt{2}S_{16}\frac{\partial^4}{\partial x_1\partial x_2^3} + S_{11}\frac{\partial^4}{\partial x_2^4} \tag{3.100}$$

the generalized biharmonic operator for the present case. Formally, ∇_2^4 is identical to ∇_1^4, but the components of the compliance tensor $[S]$ are to be used in place of those of the reduced compliance $[\Sigma]$.

The other compatibility equations, for a strain tensor that is a plane field but *not* a plane tensor, because generally speaking in plane stress $\{\varepsilon^a\} \neq \{0\}$, are, in the Kelvin's notation,

$$\varepsilon_{3,11} = 0, \;\; \varepsilon_{3,12} = 0, \;\; \varepsilon_{3,22} = 0, \;\; \varepsilon_{4,11} = \varepsilon_{5,12}, \;\; \varepsilon_{4,12} = \varepsilon_{5,22}. \tag{3.101}$$

Also considering materials that are at least monoclinic, for which $\varepsilon_4 = \varepsilon_5 = 0$, so that the two last equations are automatically satisfied, the first three equations are left unsatisfied, unless ε_3 is a linear function of x_1, x_2:

$$\varepsilon_3(x_1, x_2) = c_0 + c_1 x_1 + c_2 x_2. \tag{3.102}$$

In all the other cases, the plane stress analysis is not exact, and can be considered as accurate only in the limit of thin plates acted upon by surface tractions parallel to the mid-plane of the plate.

3.8 Comparison of Plane States

We can now compare the three cases of plane strain, plane stress and generalized plane stress.

Plane Strain:

- The displacement is a plane vector and also a plane field: $u_3 = 0, u_\alpha = u_\alpha(x_1, x_2)$, $\alpha = 1, 2$;
- the strain tensor is plane and also a plane field: $\{\varepsilon\} = \{\varepsilon^p(x_1, x_2)\}, \;\; \{\varepsilon^a\} = \{0\}$;
- the stress tensor is *not* plane but it is a plane field: $\{\sigma\} = \{\sigma(x_1, x_2)\}$; for a material with the moduli (3.93) null, it is also $\sigma_4 = \sigma_5 = 0$, but $\sigma_3 \neq 0$;
- the equilibrium equations in case of null body vector, for a material with the moduli (3.93) null, reduce to $\sigma_{ij,j} = 0 \;\; j = 1, 2, \;\forall i = 1, 2, 3$; the third of these equations corresponds to the antiplane state, uncoupled from the plane one;
- the Hooke's law does not change with respect to the 3D case: $\{\sigma\} = [C]\{\varepsilon^p\}$;
- the inverse Hooke's law becomes: $\{\varepsilon^p\} = [\Sigma]\{\sigma^p\}$, with $[\Sigma]$ the reduced compliance matrix whose components are given by Eq. (3.42) for a material at least monoclinic;

- the theory of plane strain is exact.

Plane Stress

- The displacement is *not* a plane vector *nor* a plane field: $u_i = u_i(x_1, x_2, x_3),\ \forall i = 1, 2, 3$;
- the strain tensor is *not* plane but it is a plane field: $\{\varepsilon\} = \{\varepsilon(x_1, x_2)\}$; for a material at least monoclinic, it is also $\varepsilon_4 = \varepsilon_5 = 0$, but $\varepsilon_3 \neq 0$;
- the stress tensor is plane and also a plane field: $\{\sigma\} = \{\sigma^p(x_1, x_2)\}$;
- the equilibrium equations for a null body vector reduce to $\sigma_{ij,j} = 0\ \ i, j = 1, 2$, *regardless of the material*;
- the Hooke's law becomes: $\{\sigma^p\} = [Q]\{\varepsilon^p\}$, with $[Q]$ the reduced stiffness matrix whose components for a material at least monoclinic are given by Eq. (3.62);
- the inverse Hooke's law does not change with respect to the 3D case: $\{\varepsilon\} = [S]\{\sigma^p\}$;
- the theory of plane stress is not exact.

Generalized Plane Stress

- All the relations are given on the average, i.e. as average values on the thickness of the plate, not locally;
- the theory is valid for thin plates of a material at least monoclinic, with $\sigma_3 = \sigma_4 = \sigma_5 = 0$ on the plate's surfaces and submitted uniquely to loadings parallel to the plate's mid-plane;
- the average displacement is a plane vector and also a plane field: $\widehat{u}_3 = 0,\ \ \widehat{u}_\alpha = \widehat{u}_\alpha(x_1, x_2),\ \alpha = 1, 2$;
- the average strain tensor is *not* plane, because $\widehat{\varepsilon}_4 = \widehat{\varepsilon}_5 = 0$, but $\widehat{\varepsilon}_3 \neq 0$; nevertheless, it is a plane field: $\{\widehat{\varepsilon}\} = \{\widehat{\varepsilon}(x_1, x_2)\}$;
- the average stress tensor is *not* plane but it is a plane field: $\{\widehat{\sigma}\} = \{\widehat{\sigma}^p(x_1, x_2)\}$;
- the equilibrium equations reduce to $\widehat{\sigma}_{ij,j} = 0\ \ j = 1, 2\ \forall i = 1, 2, 3$;
- the average Hooke's law becomes: $\{\widehat{\sigma}\} = [\widehat{C}]\{\widehat{\varepsilon}\}$, with $[\widehat{C}] = [Q]$ the reduced stiffness matrix whose components are given by Eq. (3.62);
- the theory of generalized plane stress is exact, on the average, only if $\widehat{\sigma}_3$ is exactly zero everywhere in the plate.

It appears hence that plane strain and generalized plane stress are formally identical, provided that the stiffness matrix of plane strain is replaced by the reduced stiffness matrix for generalized plane stress, and of course considering that in generalized plane stress all the relations are valid on the average. Nevertheless, some differences remain, for instance $\sigma_3 \neq 0$ and $\varepsilon_3 = 0$ in plane strain, while it is assumed that $\widehat{\sigma}_3 = 0$ and $\varepsilon_3 \neq 0$ in generalized plane stress. The case of plane stress is not formally identical to plane strain or to generalized plane stress because the displacement vector is not plane nor a plane field, besides the fact that $\varepsilon_3 \neq 0$ and $\sigma_3 = 0$.

3.9 The Lekhnitskii Theory

3.9.1 The General Lekhnitskii Problem

We will name a *Lekhnitskii Problem* every problem of the elastic equilibrium of an anisotropic body whose stress field is constrained to satisfy *uniquely* the condition of plane field:

$$\boldsymbol{\sigma} = \boldsymbol{\sigma}(x_1, x_2), \tag{3.103}$$

hence, generally speaking, with $\{\sigma^a\} \neq \{0\}$, i.e. the stress is *not necessarily a plane tensor*. The same properties, by the reverse Hooke's law, are true for the strain too, but not for the displacement, see Sect. 3.5:

$$\boldsymbol{\varepsilon} = \boldsymbol{\varepsilon}(x_1, x_2), \quad \mathbf{u} = \mathbf{u}(x_1, x_2, x_3). \tag{3.104}$$

The *Lekhnitskii theory* or *formalism* (Lekhnitskii 1963, p. 103; Ting 1996, p. 118), is the mathematical theory for obtaining a general formulation of the solution to the Lekhnitskii Problem. It is based upon the use of the stress functions χ and Ψ, Eqs. (3.15), (3.23). We will see that the Lekhnitskii theory comprehends, as special cases, plane deformation, generalized plane strain and generalized plane stress.

We follow here the original approach of Lekhnitskii, considering the general case of an anisotropic body belonging to any possible elastic syngony, submitted to surface tractions on the boundaries and to volume forces depending upon a potential U,

$$\mathbf{f} = \nabla U. \tag{3.105}$$

The loading and the geometrical conditions generate a stress state defined by a plane field like in Eq. (3.103). This is namely the case of a body bounded by a cylindrical surface, submitted to body forces and tractions on the boundary, provided that these loads act orthogonally to and do not vary along the cylinder axis.[4] For a cylinder of finite length, it is assumed that the stresses reduce to an equivalent finite force and moment acting upon the bases of the cylinder.

Ting (1996, p. 118), does not consider the presence of body forces and for the representation of the general solution, he considers only the case where $\varepsilon_3 = 0$, a simpler case treated apart in Sect. 3.9.9. Milne-Thomson (1960, p. 174), considers a plane deformation, Sect. 3.4, of a material with $x_3 = 0$ as plane of symmetry. Hence, his basic assumptions are

$$\begin{gathered} u_3 = 0, \quad \varepsilon_3 = \varepsilon_4 = \varepsilon_5 = 0, \quad \sigma_4 = \sigma_5 = 0, \\ \mathbf{u} = \mathbf{u}(x_1, x_2), \quad \boldsymbol{\varepsilon} = \boldsymbol{\varepsilon}(x_1, x_2), \quad \boldsymbol{\sigma} = \boldsymbol{\sigma}(x_1, x_2). \end{gathered} \tag{3.106}$$

[4]Lekhnitskii names explicitly this theory: *The state of stress of a homogeneous anisotropic body bounded by a cylindrical surface in which the stress do not vary along the generator.*

So, the problem considered by Milne-Thomson is still a Lekhnitskii problem, because the basic assumption (3.103) is still satisfied, but it is a simpler case: $\boldsymbol{\varepsilon}$ and $\mathbf{u}$ are not only plane fields, but also plane tensors. Moreover, $\boldsymbol{\sigma}$ is simpler, because $\sigma_4 = \sigma_5 = 0$ (actually, Milne-Thomson considers only the case of plane deformation, a special case of the Lekhnitskii problem considered in Sect. 3.9.10). The approach is nevertheless similar, but Milne-Thomson develops a rigorous method based upon the use of complex variables.

All the following developments are done with the Kelvin's notation, that is why some equations are slightly different from those that can be found in the cited texts.

3.9.2 *The Decomposition of the Displacement Field*

The displacement vector $\mathbf{u}(x_1, x_2, x_3)$ is decomposed into a plane vector field[5]

$$\mathbf{u}^\pi = \mathbf{u}^\pi(x_1, x_2) \tag{3.107}$$

and a field complementary to the plane one, depending also upon x_3. This can be done in the following way: in the Kelvin's notation, it is

$$\begin{aligned} &\varepsilon_1 = u_{1,1}, \quad \varepsilon_2 = u_{2,2}, \quad \varepsilon_3 = u_{3,3}, \\ &\varepsilon_4 = \frac{u_{2,3} + u_{3,2}}{\sqrt{2}}, \quad \varepsilon_5 = \frac{u_{1,3} + u_{3,1}}{\sqrt{2}}, \quad \varepsilon_6 = \frac{u_{1,2} + u_{2,1}}{\sqrt{2}}. \end{aligned} \tag{3.108}$$

Because the stress is a plane field, for the Hooke's law it is also

$$\varepsilon_i = \varepsilon_i(x_1, x_2) \ \forall i = 1, \ldots, 6; \tag{3.109}$$

as a consequence, in Eq. (3.108) the right-hand sides are independent of x_3. The most general expression for the components of $\mathbf{u}(x_1, x_2, x_3)$ is hence

$$\begin{aligned} u_1(x_1, x_2, x_3) &= u_1^\pi(x_1, x_2) + u(x_2, x_3), \\ u_2(x_1, x_2, x_3) &= u_2^\pi(x_1, x_2) + v(x_1, x_3), \\ u_3(x_1, x_2, x_3) &= u_3^\pi(x_1, x_2) + x_3 w(x_1, x_2). \end{aligned} \tag{3.110}$$

Injecting Eq. (3.110) into Eq. $(3.108)_{4,5,6}$ gives

$$\varepsilon_4 = \frac{u^\pi_{3,2} + v_{,3} + x_3 w_{,2}}{\sqrt{2}}, \quad \varepsilon_5 = \frac{u^\pi_{3,1} + u_{,3} + x_3 w_{,1}}{\sqrt{2}}, \quad \varepsilon_6 = \frac{u^\pi_{1,2} + u^\pi_{2,1} + u_{,2} + v_{,1}}{\sqrt{2}}, \tag{3.111}$$

[5]It is worth noting that the symbol π denotes only a plane field, not a plane vector.

and because of Eqs. (3.107) and (3.109), the quantities

$$v_{,3} + x_3 w_{,2}, \quad u_{,3} + x_3 w_{,1}, \quad u_{,2} + v_{,1} \tag{3.112}$$

cannot depend upon x_3. Then, $u_{,3}$ and $v_{,3}$ must be linear in x_3, while $w_{,1}$ is a function of x_1 and $w_{,2}$ of x_2. Then, we can put

$$\begin{aligned} u(x_2, x_3) &= -\frac{1}{2}x_3^2(A + D\,x_2) + x_3 f(x_2),\\ v(x_1, x_3) &= -\frac{1}{2}x_3^2(B + D\,x_1) + x_3 g(x_1),\\ w(x_1, x_2) &= A\,x_1 + B\,x_2 + C + D\,x_1 x_2, \quad A, B, C, D \in \mathbb{R}. \end{aligned} \tag{3.113}$$

Injecting (3.113) into (3.112)$_3$ leads to

$$-D\,x_3^2 + x_3\frac{df(x_2)}{dx_2} + x_3\frac{dg(x_1)}{dx_1}, \tag{3.114}$$

a quantity that must be independent of x_3, which gives

$$D = 0, \quad f(x_2) = -(\omega x_2 + \gamma_2), \quad g(x_1) = \omega x_1 + \gamma_1, \quad \omega, \gamma_1, \gamma_2 \in \mathbb{R}. \tag{3.115}$$

Hence, the displacement field has the expression

$$\begin{aligned} u_1(x_1, x_2, x_3) &= u_1^\pi(x_1, x_2) - \frac{1}{2}A\,x_3^2 - \omega x_2 x_3 - \gamma_2 x_3,\\ u_2(x_1, x_2, x_3) &= u_2^\pi(x_1, x_2) - \frac{1}{2}B\,x_3^2 + \omega x_1 x_3 + \gamma_1 x_3,\\ u_3(x_1, x_2, x_3) &= u_3^\pi(x_1, x_2) + x_3(A\,x_1 + B\,x_2 + C). \end{aligned} \tag{3.116}$$

Any rigid displacement can be added to $\mathbf{u}(x_1, x_2, x_3)$ without altering the strain and stress fields; we can hence add the displacement $\boldsymbol{\delta}(x_1, x_2, x_3)$ corresponding to an infinitesimal rigid rotation[6] θ around the axis $\boldsymbol{\gamma} = (\gamma_1, \gamma_2, \gamma_3)$,

$$\mathbf{R}_\gamma = \mathbf{I} + \boldsymbol{\Gamma}, \tag{3.117}$$

[6]It can be shown (Podio-Guidugli 2000, p. 7), that any rotation $\mathbf{R}$ can be represented as

$$\mathbf{R} = \mathbf{I} + \sin\theta\mathbf{W} + (1 - \cos\theta)\mathbf{W}^2,$$

where θ is the rotation angle and $\mathbf{W}$ is the axial tensor relative to the unit vector $\mathbf{w}$ that determines the rotation axis. For infinitesimal rotations, i.e. for $\theta \to 0$,

$$\lim_{\theta\to 0} \mathbf{R} = \mathbf{I} + \theta\mathbf{W}.$$

Equation (3.117) is then obtained incorporating θ into the components of $\boldsymbol{\Gamma}$.

with $\boldsymbol{\Gamma}$ the axial tensor corresponding to $\boldsymbol{\gamma}$:

$$\boldsymbol{\Gamma} = \begin{bmatrix} 0 & -\gamma_3 & \gamma_2 \\ \gamma_3 & 0 & -\gamma_1 \\ -\gamma_2 & \gamma_1 & 0 \end{bmatrix}; \tag{3.118}$$

hence

$$\boldsymbol{\delta}(x_1, x_2, x_3) = \mathbf{R}_\gamma \mathbf{x} - \mathbf{x} = \left\{ \begin{array}{c} \gamma_2 x_3 - \gamma_3 x_2 \\ \gamma_3 x_1 - \gamma_1 x_3 \\ \gamma_1 x_2 - \gamma_2 x_1 \end{array} \right\}. \tag{3.119}$$

Once the displacement (3.119) added to the displacement (3.116) and the terms depending upon x_1 and x_2 incorporated in the $u_i^\pi(x_1, x_2)$, we get the final expression for $\mathbf{u}(x_1, x_2, x_3)$:

$$\begin{aligned} u_1(x_1, x_2, x_3) &= u_1^\pi(x_1, x_2) - \frac{1}{2} A\, x_3^2 - \omega x_2 x_3, \\ u_2(x_1, x_2, x_3) &= u_2^\pi(x_1, x_2) - \frac{1}{2} B\, x_3^2 + \omega x_1 x_3, \\ u_3(x_1, x_2, x_3) &= u_3^\pi(x_1, x_2) + x_3(A\, x_1 + B\, x_2 + C). \end{aligned} \tag{3.120}$$

The terms in (3.120) depending upon x_3 accounts for the difference between the two assumptions of plane stress field or plane displacement field (but *not* of plane strain, when the assumption $u_3 = 0$ is also done, see Sect. 3.4), and finally between the Lekhnitskii and the Stroh theories, see Sect. 3.10.

3.9.3 Strain Field and Compatibility Equations

With the components (3.120), Eq. (3.108) becomes

$$\begin{aligned} &\varepsilon_1 = u_{1,1}^\pi, \quad \varepsilon_2 = u_{2,2}^\pi, \quad \varepsilon_3 = A\, x_1 + B\, x_2 + C, \\ &\varepsilon_4 = \frac{u_{3,2}^\pi + \omega x_1}{\sqrt{2}}, \quad \varepsilon_5 = \frac{u_{3,1}^\pi - \omega x_2}{\sqrt{2}}, \quad \varepsilon_6 = \frac{u_{1,2}^\pi + u_{2,1}^\pi}{\sqrt{2}}. \end{aligned} \tag{3.121}$$

Because ε_3 is linear in x_1 and x_2, then the deformation corresponds to a bending about the line $A\, x_1 + B\, x_2 + C = 0$; the deformation determined by ω is clearly, on its side, a torsion about the axis x_3.

With the above expressions for the ε_is, it is readily checked that the only two compatibility equations that are not identically satisfied are Eq. $(1.29)_{1,4}$, that with the Kelvin's notation become

$$\begin{aligned} \varepsilon_{1,22} + \varepsilon_{2,11} &= \sqrt{2}\, \varepsilon_{6,12}, \\ \varepsilon_{4,1} - \varepsilon_{5,2} &= \sqrt{2}\, \omega. \end{aligned} \tag{3.122}$$

These relations will give the two differential equations to be satisfied by the stress functions χ and Ψ.

3.9.4 Differential Equations for χ and Ψ

The components ε_i are linked to the stresses σ_i by the Hooke's reverse law:

$$\varepsilon_i = S_{ij}\sigma_j, \quad i, j = 1, \ldots, 6, \tag{3.123}$$

the S_{ij}s being the components of the elastic compliance matrix $[S]$ in the Kelvin's notation, Eq. (2.24)$_2$.

The two stress functions χ and Ψ cannot determine σ_3. Hence, if a solution is looked for in terms of χ and Ψ, σ_3 must be eliminated. This can be done deducing σ_3 from Eq. (3.123):

$$\sigma_3 = \frac{\varepsilon_3}{S_{33}} - \frac{1}{S_{33}} \sum_{\substack{j=1 \\ j \neq 3}}^{6} S_{3j}\sigma_j, \tag{3.124}$$

and injecting it back into the (3.123), so obtaining

$$\varepsilon_i = S'_{i1}\sigma_1 + S'_{i2}\sigma_2 + S'_{i4}\sigma_4 + S'_{i5}\sigma_5 + S'_{i6}\sigma_6 + S^*_{i3}\varepsilon_3, \quad i = 1, 2, 4, 5, 6. \tag{3.125}$$

with

$$S'_{ij} = S_{ij} - \frac{S_{i3}S_{j3}}{S_{33}}, \quad S^*_{i3} = \frac{S_{i3}}{S_{33}}, \quad i, j = 1, \ldots, 6. \tag{3.126}$$

The components S'_{ij} are called *reduced elastic compliances*, and they are exactly equal to the components Σ_{ij}, Eq. (3.42), also called reduced compliances. This is rather surprisingly, because the Σ_{ij}s arise in a plane strain problem, quite different from the Lekhnitskii theory, where the only assumption is a plane field for stress. Actually, there are important differences between the S'_{ij}s and the Σ_{ij}s: while the S'_{ij}s are valid for each elastic syngony, the Σ_{ij}s in the form of Eq. (3.42) are correct only for a material at least monoclinic with $x_3 = 0$ as plane of symmetry. Moreover, the S'_{ij}s are defined for the 3D case, while the Σ_{ij}s define only plane components. Actually, though the S'_{ij}s are equal to the Σ_{ij}s, they are deduced in a completely different way, which explains why in a problem with a plane stress field, which however is *not* a plane stress state, there are reduced compliances and not reduced stiffnesses, Sect. 3.5.

To remark that, with definition (3.126),

$$S'_{ij} = S'_{ji}, \tag{3.127}$$

and

$$S'_{i3} = S'_{3i} = 0 \;\; \forall i = 1, \ldots, 6. \tag{3.128}$$

We now express the σ_{ij}s using the stress functions χ and Ψ,

$$\begin{aligned}\sigma_1 &= \chi_{,22} - U,\\ \sigma_2 &= \chi_{,11} - U,\\ \sigma_6 &= -\sqrt{2}\,\chi_{,12},\\ \sigma_4 &= -\sqrt{2}\,\Psi_{,1},\\ \sigma_5 &= \sqrt{2}\,\Psi_{,2}.\end{aligned} \tag{3.129}$$

Substituting these relations into Eq. (3.125) gives

$$\begin{aligned}\varepsilon_i = S'_{i1}(\chi_{22} - U) + S'_{i2}(\chi_{11} - U) - \sqrt{2}S'_{i4}\Psi_{,1} + \sqrt{2}S'_{i5}\Psi_{,2} -\\ \sqrt{2}S'_{i6}\chi_{,12} + S^*_{i3}\varepsilon_3, \quad i = 1, 2, 4, 5, 6.\end{aligned} \tag{3.130}$$

The derivatives of the ε_is can now be calculated and injected into the compatibility equations (3.122); remembering the expression of ε_3, Eq. (3.121)$_3$, some standard passages leads to the following result:

$$\begin{aligned}\nabla_1^4\chi + \nabla_1^3\Psi = C_1,\\ \nabla_1^3\chi + \nabla_1^2\Psi = C_2,\end{aligned} \tag{3.131}$$

where the known terms at the right-hand side C_1 and C_2 are

$$\begin{aligned}C_1 &= (S'_{12} + S'_{22})U_{,11} - \sqrt{2}(S'_{16} + S'_{26})U_{,12} + (S'_{11} + S'_{12})U_{,22},\\ C_2 &= -2\omega + \sqrt{2}\left[S^*_{34}A - S^*_{35}B - (S'_{14} + S'_{24})U_{,1} + (S'_{15} + S'_{25})U_{,2}\right].\end{aligned} \tag{3.132}$$

The differential operators are

$$\begin{aligned}\nabla_1^2 =& 2\left(S'_{44}\frac{\partial^2}{\partial x_1^2} - 2S'_{45}\frac{\partial^2}{\partial x_1\partial x_2} + S'_{55}\frac{\partial^2}{\partial x_2^2}\right),\\ \nabla_1^3 =& \sqrt{2}\left[-S'_{24}\frac{\partial^3}{\partial x_1^3} + (S'_{25} + \sqrt{2}S'_{46})\frac{\partial^3}{\partial x_1^2\partial x_2} - \right.\\ &\left.(S'_{14} + \sqrt{2}S'_{56})\frac{\partial^3}{\partial x_1\partial x_2^2} + S'_{15}\frac{\partial^3}{\partial x_2^3}\right],\\ \nabla_1^4 =& S'_{22}\frac{\partial^4}{\partial x_1^4} - 2\sqrt{2}S'_{26}\frac{\partial^4}{\partial x_1^3\partial x_2} + 2(S'_{12} + S'_{66})\frac{\partial^4}{\partial x_1^2\partial x_2^2} - \\ &2\sqrt{2}S'_{16}\frac{\partial^4}{\partial x_1\partial x_2^3} + S'_{11}\frac{\partial^4}{\partial x_2^4}.\end{aligned} \tag{3.133}$$

∇_1^4 is not only formally identical to the operator given by Eq. (3.98), but, because of the above mentioned identity of the S_{ij}s and Σ_{ij}s, they are exactly the same operator; that is why we have indicated with the same symbol both of them.

Equation (3.131) are a system of non-homogeneous differential equations for χ and Ψ; together with the appropriate boundary conditions, they define a boundary value problem reduced to the knowledge of the scalar two-dimensional functions χ and Ψ. The Lekhnitskii theory has hence transformed a 3D problem into a two-dimensional one, the dependence upon x_3 being however recovered in the above relations for the ε_i and $\mathbf{u}$.

The equations in (3.131) can be rearranged for uncoupling χ and Ψ and for obtaining a homogeneous problem. To this end, let us pose

$$\begin{aligned} \chi &= \chi^h + \chi^p, \\ \Psi &= \Psi^h + \Psi^p, \end{aligned} \tag{3.134}$$

where the superscript h denotes the solutions of the associated homogeneous equations

$$\begin{aligned} \nabla_1^4 \chi^h + \nabla_1^3 \Psi^h &= 0, \\ \nabla_1^3 \chi^h + \nabla_1^2 \Psi^h &= 0, \end{aligned} \tag{3.135}$$

and p a particular solution of Eq. (3.131) depending upon the known terms (3.132) and usually rather simple to be found. So, we concentrate on the homogeneous equations (3.135) and first of all we uncouple χ and Ψ as follows:

$$\begin{aligned} \nabla_1^2(\nabla_1^4 \chi^h + \nabla_1^3 \Psi^h) &= 0, \\ \nabla_1^3(\nabla_1^3 \chi^h + \nabla_1^2 \Psi^h) &= 0, \end{aligned} \tag{3.136}$$

and subtracting the second equation from the first one gives

$$(\nabla_1^2 \nabla_1^4 - \nabla_1^3 \nabla_1^3)\chi^h = 0. \tag{3.137}$$

The same can be done for Ψ^h: applying the operator ∇_1^3 to Eq. (3.135)$_1$ and ∇_1^4 to Eq. (3.135)$_2$, then subtracting the first equation from the second one, the result is exactly the same:

$$(\nabla_1^2 \nabla_1^4 - \nabla_1^3 \nabla_1^3)\Psi^h = 0. \tag{3.138}$$

Equations (3.137) and (3.138) are two uncoupled sixth-order differential equations for χ and Ψ.

3.9.5 General Solution of the Homogeneous Equations

Let us consider Eq. (3.137); the sixth-order differential operator $\nabla_1^2\nabla_1^4 - \nabla_1^3\nabla_1^3$ can be decomposed into six linear operators:

$$(\nabla_1^2\nabla_1^4 - \nabla_1^3\nabla_1^3)\chi^h = D_6D_5D_4D_3D_2D_1\chi^h, \tag{3.139}$$

where the linear operators D_k are given by

$$D_k = \frac{\partial}{\partial x_2} - \mu_k\frac{\partial}{\partial x_1} \quad \forall k = 1, \ldots, 6. \tag{3.140}$$

In the above equation, μ_k is a solution of the sextic characteristic algebraic equation associated to (3.137):

$$\delta_4(\mu)\delta_2(\mu) - \delta_3^2(\mu) = 0, \tag{3.141}$$

where

$$\begin{aligned}
&\delta_2(\mu) = 2\left(S'_{44} - 2S'_{45}\mu + S'_{55}\mu^2\right),\\
&\delta_3(\mu) = \sqrt{2}\left[-S'_{24} + (S'_{25} + \sqrt{2}S'_{46})\mu - (S'_{14} + \sqrt{2}S'_{56})\mu^2 + S'_{15}\mu^3\right],\\
&\delta_4(\mu) = S'_{22} - 2\sqrt{2}S'_{26}\mu + 2(S'_{12} + S'_{66})\mu^2 - 2\sqrt{2}S'_{16}\mu^3 + S'_{11}\mu^4.
\end{aligned} \tag{3.142}$$

The integration of a sixth-order differential equation is hence reduced to the integration of six equations of the first order: posing

$$D_1\chi^h = \varphi_2,\quad D_2\varphi_2 = \varphi_3,\quad D_3\varphi_3 = \varphi_4,\quad D_4\varphi_4 = \varphi_5,\quad D_5\varphi_5 = \varphi_6, \tag{3.143}$$

φ_6 satisfies to equation

$$D_6\varphi_6 = 0. \tag{3.144}$$

The general integral is a function of the argument $x_1 + \mu_6x_2$:

$$\varphi_6 = f_6^V(x_1 + \mu_6x_2). \tag{3.145}$$

Then, φ_5 satisfies a first-order non-homogeneous equation:

$$D_5\varphi_5 = f_6^V(x_1 + \mu_6x_2), \tag{3.146}$$

whose integral is

$$\varphi_5 = f_5^{IV}(x_1 + \mu_5x_2) + \frac{f_6^{IV}(x_1 + \mu_6x_2)}{\mu_6 - \mu_5}. \tag{3.147}$$

Repeating the same procedure until χ^h gives

$$\chi^h = \sum_{k=1}^{6} \frac{f_k(x_1 + \mu_k x_2)}{\prod_{j=1}^{k-1}(\mu_k - \mu_j)}. \tag{3.148}$$

For Eq. (3.138), repeating *verbatim* the same procedure for Ψ^h gives a similar result:

$$\Psi^h = \sum_{k=1}^{6} \frac{g_k(x_1 + \mu_k x_2)}{\prod_{j=1}^{k-1}(\mu_k - \mu_j)}. \tag{3.149}$$

It can be proved that the roots of (3.141) are all complex, see Sect. 3.9.6. Hence, posing

$$\mu_k = \alpha_k + i\beta_k, \quad \mu_{k+3} = \overline{\mu}_k, \quad \alpha_k, \beta_k \in \mathbb{R}, \ \beta_k > 0 \ \forall k = 1, 2, 3, \tag{3.150}$$

$$\left.\begin{aligned} \chi_k^h &= \frac{f_k(z_k)}{\prod_{j=1}^{k-1}(\mu_k - \mu_j)}, \\ \Psi_k^h &= \frac{g_k(z_k)}{\prod_{j=1}^{k-1}(\mu_k - \mu_j)}, \end{aligned}\right. \quad k = 1, \ldots, 6, \tag{3.151}$$

with the complex variables z_k defined by

$$z_k = x_1 + \mu_k x_2, \quad k = 1, \ldots, 6, \tag{3.152}$$

we have

$$\begin{aligned} \chi^h &= 2\sum_{k=1}^{3} Re\left[\chi_k^h(z_k)\right], \\ \Psi^h &= 2\sum_{k=1}^{3} Re\left[\Psi_k^h(z_k)\right]. \end{aligned} \tag{3.153}$$

Because χ^h and Ψ^h are not only given by Eqs. (3.137) and (3.138) but they also satisfy Eq. (3.135), it can be checked that it is also (prime denotes differentiation)

$$\Psi_k^h(z_k) = \lambda_k \, \chi_k^{h'}(z_k), \quad k = 1, 2, 3, \tag{3.154}$$

with

$$\lambda_k = -\frac{\delta_3(\mu_k)}{\delta_2(\mu_k)} = -\frac{\delta_4(\mu_k)}{\delta_3(\mu_k)}, \quad k = 1, 2, 3. \tag{3.155}$$

Hence, Eq. $(3.153)_2$ can also be put in the form

$$\Psi^h = 2\sum_{k=1}^{3} Re\left[\lambda_k \chi_k^{h'}(z_k)\right]. \tag{3.156}$$

Equations (3.153)$_1$ and (3.156) are the general form of the solution to the homogeneous problem (3.135).

To obtain the final expression for χ and Ψ, besides the particular solution χ^p and Ψ^p, the boundary conditions must be taken into account, see Sect. (3.9.8).

3.9.6 Roots of the Characteristic Equation

We show in this Section that the roots of the characteristic equation (3.141) cannot be real. To this purpose, let us range the S'_{ij}s in a 5×5 matrix $[S']$ as follows:

$$[S'] = \left[\begin{array}{c|c} [S4'] & [S3'] \\ \hline [S3']^\top & [S2'] \end{array}\right] = \left[\begin{array}{ccc|cc} S'_{11} & S'_{12} & S'_{16} & S'_{14} & S'_{15} \\ & S'_{22} & S'_{26} & S'_{24} & S'_{25} \\ & & S'_{66} & S'_{64} & S'_{65} \\ \hline & sym & & S'_{44} & S'_{45} \\ & & & & S'_{55} \end{array}\right]. \tag{3.157}$$

The first step is to use the positiveness of the strain energy V to prove the following

Theorem 6 *The matrices* $[S']$, $[S2']$ *and* $[S4']$ *are positive definite.*

Proof Let us introduce the 5×5 matrix

$$\left[C^5\right] = \begin{bmatrix} C_{11} & C_{12} & C_{14} & C_{15} & C_{16} \\ & C_{22} & C_{24} & C_{25} & C_{26} \\ & & C_{44} & C_{45} & C_{46} \\ & sym & & C_{55} & C_{56} \\ & & & & C_{66} \end{bmatrix}, \tag{3.158}$$

obtained removing the third column and row from $[C]$. Being a principal minor of $[C]$, $[C^5]$ is positive definite, as a consequence of the positiveness of the strain energy, see Theorem 1.

We now prove that the matrix

$$[S^5] = \begin{bmatrix} S'_{11} & S'_{12} & S'_{14} & S'_{15} & S'_{16} \\ & S'_{22} & S'_{24} & S'_{25} & S'_{26} \\ & & S'_{44} & S'_{45} & S'_{46} \\ & sym & & S'_{55} & S'_{56} \\ & & & & S'_{66} \end{bmatrix}, \tag{3.159}$$

whose components S'_{ij} are defined by Eq. (3.126), is the inverse of $[C^5]$:

$$[C^5][S^5] = [I]. \tag{3.160}$$

This can be proved noticing that each component of $[C^5]$ can be written as

$$C^5_{ij} = C_{ij} - \delta_{i3}C_{3j} - C_{i3}\delta_{3j} + \delta_{i3}C_{33}\delta_{3j}, \ \ i, j = 1, 2, 3. \tag{3.161}$$

Using Eq. (3.126), it is simple to check that

$$C^5_{ik}S^5_{kj} = \delta_{ij} - \delta_{i3}\delta_{3j}, \ \ i, j = 1, 2, 3; \tag{3.162}$$

Regarding at matrices $[C^5]$, $[S^5]$ and $[\delta_{ij} - \delta_{i3}\delta_{3j}]$ as 6×6 matrices, they all have the third row and column whose elements are zero. Hence, if these matrices are reduced to 5×5 matrices eliminating the third row and column, then Eq. (3.162) is exactly equal to Eq. (3.160), i.e. $[S^5]$ is equal to $[C^5]^{-1}$. As a consequence, also $[S^5]$ is positive definite.[7]

Matrix $[S']$ is just the matrix $[S^5]$, with rows and columns 3 and 5 swapped. Of course, this does not change the positive definiteness, which is preserved also for $[S']$. Finally, matrices $[S2']$ and $[S4']$ are just principal sub-matrices of $[S']$, and as such they are positive definite too, see Theorem 1.

We can now prove the

Theorem 8 *The roots of the characteristic equation cannot be real.*

Proof Consider the vector

$$\{u\} = \left\{\frac{\{v\}}{\lambda\{w\}}\right\} = \left\{\begin{array}{c}\mu^2\\ 1\\ -\sqrt{2}\,\mu\\ \hline -\sqrt{2}\,\lambda\\ \sqrt{2}\,\lambda\mu\end{array}\right\}, \ \ \lambda = -\frac{\delta_3(\mu)}{\delta_2(\mu)}; \tag{3.163}$$

vectors $\{u\}$, $\{v\}$ and $\{w\}$ are not null; it is then

$$\{w\}^\top[S2']\{w\} = \delta_2(\mu) > 0 \tag{3.164}$$

because $[S2']$ is positive definite. Moreover,

---

[7] To prove this fact, we have the

Theorem 7 *The inverse* $[A]^{-1}$ *of a symmetric positive definite matrix* $[A]$ *is positive definite.*

Proof Because it is symmetric, $[A]$ can be diagonalized and has on the diagonal real eigenvalues. Then, $[A]^{-1}$ has on the diagonal the reciprocal of the eigenvalues of $[A]$: $A_{ij} = \delta_{ij}\lambda_i \Rightarrow A^{-1}_{ij} = \delta_{ij}/\lambda_i$ (no summation over i). Because $[A]$ is positive definite, $\lambda_i > 0 \ \forall i$, Eq. (2.117). So, the eigenvalues $1/\lambda_i$ of $[A]^{-1}$ are all positive too, which proves that also $[A]^{-1}$ is positive definite.

$$\{u\}^\top[S']\{u\} = \delta_4(\mu) + 2\lambda\delta_3(\mu) + \lambda^2\delta_2(\mu) = \frac{\delta_2(\mu)\delta_4(\mu) - \delta_3^2(\mu)}{\delta_2(\mu)} > 0 \tag{3.165}$$

because $[S']$ is positive definite. Hence,

$$\delta_2(\mu)\delta_4(\mu) - \delta_3^2(\mu) > 0 \ \forall\mu, \tag{3.166}$$

which proves that the characteristic equation (3.141) cannot have real roots: $\mu_k \in \mathbb{C}\ \forall k = 1, \ldots, 6$.

Because the characteristic equation has real coefficients, its six roots are complex conjugate. That is why the six roots can be put in the form of Eq. (3.150).

Something more can be said about the six roots if the elastic syngony is specified. Namely:

- for a monoclinic material with $x_1 = 0$ plane of symmetry, $S_{i5} = S_{i6} = 0\ \forall i = 1, \ldots, 4 \Rightarrow S'_{15} = S'_{25} = S'_{16} = S'_{26} = S'_{45} = S'_{46} = 0 \rightarrow$ the characteristic equation is a cubic polynomial of μ^2. Then, if the three roots of μ^2 are real, they are negative, because $\mu_k \in \mathbb{C}$. So, in such a case $\mu_k = i\beta_k,\ \beta_k > 0\ \forall k = 1, 2, 3$. The other possibility, is that one of the roots is real, say $\mu_3^2 \in \mathbb{R}$, and the two other ones complex conjugate: $\mu_1 = \alpha + i\beta, \mu_2 = -\alpha + i\beta, \mu_3 = i\beta_3,\ \alpha, \beta, \beta_3 \in \mathbb{R},\ \beta > 0, \beta_3 > 0$. A similar result is obtained also in the case of $x_2 = 0$ plane of symmetry;
- for a monoclinic material with $x_3 = 0$ plane of symmetry, it turns out to be $S'_{14} = S'_{15} = S'_{24} = S'_{25} = S'_{46} = S'_{56} = 0 \Rightarrow \delta_3(\mu) = 0$ identically. The roots of the characteristic equation are hence those of the equations

$$\begin{aligned} \delta_2(\mu) &= 0, \\ \delta_4(\mu) &= 0. \end{aligned} \tag{3.167}$$

Let μ_1 and μ_2 be the roots of Eq. $(3.167)_2$:

$$\begin{aligned} \mu_1 &= \alpha_1 + i\beta_1, \\ \mu_2 &= \alpha_2 + i\beta_2, \end{aligned} \quad \alpha_i, \beta_i \in \mathbb{R},\ \ \beta_i > 0\ \forall i = 1, 2, \tag{3.168}$$

while if μ_3 is the root of Eq. $(3.167)_1$ then[8]

$$\mu_3 = \frac{S'_{45} + i\sqrt{S'_{44}S'_{55} - S'^2_{45}}}{S'_{55}}. \tag{3.169}$$

[8]For this case, $\lambda_3 \rightarrow \infty$, see Eq. (3.155). To avoid this situation, in the literature the term $\chi_3^{h'}$ is everywhere replaced by the term $\chi_3^{h'}/\lambda_3$ for μ_3. We prefer to preserve here the general form of the equations given in the previous Sections, because formally simpler and because we will see in Sect. 3.9.10 that, in the present case, the third term in all the summations giving stresses and displacements vanishes.

- for transversely isotropic materials, with x_3 as axis of symmetry, the characteristic equation reduces to

$$(\mu^2+1)^3=0 \Rightarrow \mu_1=\mu_2=\mu_3=i. \tag{3.170}$$

The same result is true for isotropic materials.
- Ting (1996, p. 125) has shown that any material with a matrix $[S']$ like

$$[S']=\begin{bmatrix} S'_{11} & \eta^2 S'_{12} & \eta S'_{14} & S'_{15} & \eta S'_{16} \\ & \eta^4 S'_{22} & \eta^3 S'_{24} & \eta^2 S'_{25} & \eta^3 S'_{26} \\ & & \eta^2 S'_{44} & \eta S'_{45} & \eta^2 S'_{46} \\ & sym & & S'_{55} & \eta S'_{56} \\ & & & & \eta^2 S'_{66} \end{bmatrix}, \quad \eta \in \mathbb{R},\ \eta>0, \tag{3.171}$$

has also three identical imaginary roots:

$$\mu_1=\mu_2=\mu_3=i\eta. \tag{3.172}$$

Moreover, rotating through an angle θ such a material about the axis x_3 transforms the above purely imaginary identical roots into three identical complex roots:

$$\mu_1=\mu_2=\mu_3=\frac{i\eta\cos\theta-\sin\theta}{i\eta\sin\theta+\cos\theta}. \tag{3.173}$$

Because any complex number is uniquely determined by two real numbers and the above roots depend upon the two parameters η and θ, it is always possible to obtain an anisotropic material with three given identical roots μ_k by rotating about the axis of x_3 a material whose matrix $[S']$ is like in Eq. (3.171).

3.9.7 General Expressions for Stresses and Displacements

The components of stress can be recovered injecting Eqs. (3.153)$_1$ and (3.156) into Eq. (3.129), and remembering Eq. (3.134). For the component σ_3, Eq. (3.121)$_3$ must be injected into Eq. (3.124). Some standard passages lead to the following result

$$\begin{aligned} \sigma_1&=2\sum_{k=1}^{3} Re\left[\mu_k^2\,\chi_k^{h''}(z_k)\right]+\frac{\partial^2\chi^p}{\partial x_2^2}-U, \\ \sigma_2&=2\sum_{k=1}^{3} Re\left[\chi_k^{h''}(z_k)\right]+\frac{\partial^2\chi^p}{\partial x_1^2}-U, \\ \sigma_3&=\frac{1}{S_{33}}\left(Ax_1+Bx_2+C-S_{13}\sigma_1-S_{23}\sigma_2-S_{43}\sigma_4-S_{53}\sigma_5-S_{63}\sigma_6\right), \end{aligned} \tag{3.174}$$

$$\sigma_4 = -2\sqrt{2}\sum_{k=1}^{3} Re\left[\lambda_k\, \chi_k^{h''}(z_k)\right] - \sqrt{2}\frac{\partial \Psi^p}{\partial x_1},$$

$$\sigma_5 = 2\sqrt{2}\sum_{k=1}^{3} Re\left[\lambda_k\, \mu_k\, \chi_k^{h''}(z_k)\right] + \sqrt{2}\frac{\partial \Psi^p}{\partial x_2},$$

$$\sigma_6 = -2\sqrt{2}\sum_{k=1}^{3} Re\left[\mu_k\, \chi_k^{h''}(z_k)\right] - \sqrt{2}\frac{\partial^2 \chi^p}{\partial x_1 \partial x_2}.$$

The displacement components (3.120) can be obtained integrating the expressions of the strains. First, the stresses (3.174) are injected into the strains ε_1 and ε_2 given by Eq. (3.125). Then, see Eq. $(3.121)_{1,2}$, integrating ε_1 with respect to x_1 and ε_2 with respect to x_2 gives $u_1^{\pi}(x_1, x_2)$ and $u_2^{\pi}(x_1, x_2)$ respectively, that injected into Eq. $(3.120)_{1,2}$ give finally

$$\begin{aligned} u_1(x_1, x_2, x_3) &= 2\sum_{k=1}^{3} Re\left[\xi_k\, \chi_k^{h'}(z_k)\right] + \tilde{u}_1(x_1, x_2) - \frac{1}{2}Ax_3^2 - \omega x_2 x_3, \\ u_2(x_1, x_2, x_3) &= 2\sum_{k=1}^{3} Re\left[\eta_k\, \chi_k^{h'}(z_k)\right] + \tilde{u}_2(x_1, x_2) - \frac{1}{2}Bx_3^2 + \omega x_1 x_3, \end{aligned} \tag{3.175}$$

where $\tilde{u}_1(x_1, x_2)$ and $\tilde{u}_2(x_1, x_2)$ are the particular solutions corresponding to χ^p, Ψ^p and U, while

$$\begin{aligned} \xi_k &= \mu_k^2 S'_{11} + S'_{12} - \sqrt{2}\mu_k S'_{16} - \sqrt{2}\lambda_k S'_{14} + \sqrt{2}\lambda_k \mu_k S'_{15}, \\ \eta_k &= \mu_k S'_{12} + \frac{S'_{22}}{\mu_k} - \sqrt{2}S'_{26} - \sqrt{2}\frac{\lambda_k}{\mu_k}S'_{24} + \sqrt{2}\lambda_k S'_{25}. \end{aligned} \tag{3.176}$$

For u_3, from Eq. $(3.121)_5$ we get

$$u_{3,1}^{\pi} = \sqrt{2}\varepsilon_5 + \omega x_2, \tag{3.177}$$

with ε_5 given by Eq. (3.125) for $i = 5$; integrating with respect to x_1 and injecting the result into Eq. $(3.120)_3$ gives finally

$$u_3(x_1, x_2, x_3) = 2\sum_{k=1}^{3} Re\left[\zeta_k\, \chi_k^{h'}(z_k)\right] + \tilde{u}_3(x_1, x_2) + x_3(Ax_1 + B_2 + C), \tag{3.178}$$

with

$$\zeta_k = \sqrt{2}\mu_k^2 S'_{15} + \sqrt{2}S'_{25} - 2\mu_k S'_{65} - 2\lambda_k S'_{45} + 2\lambda_k \mu_k S'_{55} \tag{3.179}$$

and $\tilde{u}_3(x_1, x_2)$ the particular solution corresponding to χ^p, Ψ^p and U.

Strains are recovered from Eqs. $(3.121)_3$ and (3.125).

3.9.8 Boundary Conditions

The solution of the boundary value problem (3.131) needs to specify the boundary conditions. Let us consider a cylinder submitted to body forces and on its boundary only to surface tractions, both of them orthogonal to and independent of x_3, the direction of the cylinder's axis. In addition, we still suppose that the body forces depend upon a potential, Eq. (3.105).

Let us first consider the boundary conditions on the lateral surface $\partial\Omega$. Then, if $\mathbf{n} = (\cos\theta, \sin\theta, 0)$, with θ the angle that $\mathbf{n}$ forms with the axis x_1, is the unit outward normal to the cross-section contour, and $\mathbf{t} = (t_1, t_2, 0)$ is the surface traction vector, on the boundary it must be

$$\mathbf{t} = \boldsymbol{\sigma}\,\mathbf{n} \rightarrow \begin{cases} t_1 = \sigma_1 \cos\theta + \dfrac{\sigma_6}{\sqrt{2}} \sin\theta, \\ t_2 = \dfrac{\sigma_6}{\sqrt{2}} \cos\theta + \sigma_2 \sin\theta, \\ 0 = \sigma_5 \cos\theta + \sigma_4 \sin\theta, \end{cases} \quad \text{at } \partial\Omega. \tag{3.180}$$

The above conditions can be simplified if $\cos\theta$ and $\sin\theta$ are expressed in terms of the arc-length s of the contour,[9]

$$\cos\theta = \frac{dx_2}{ds}, \quad \sin\theta = -\frac{dx_1}{ds}, \tag{3.181}$$

and if Eq. (3.129) is used to express the stresses, then we easily get

$$\begin{aligned} \chi_{,1} &= \int_0^s \left(\mp t_2 + U\frac{dx_1}{ds} \right) + c_1, \\ \chi_{,2} &= \int_0^s \left(\pm t_1 + U\frac{dx_2}{ds} \right) + c_2, \\ \Psi &= c_3, \end{aligned} \quad \text{at } \partial\Omega, \quad c_1,\ c_2,\ c_3 \in \mathbb{R}, \tag{3.182}$$

the upper signs being for the outer contour, the lower for the inner one.

Using in the above relations Eqs. (3.153)$_1$, (3.156), (3.175) and (3.178) gives

[9] As customarily done, we consider as positive the direction s that leaves the region of the cross-section on the left of the contour; as a consequence, the sign of $\cos\theta$ and $\sin\theta$ in Eq. (3.181) change on the interior contours, when they exist.

$$
\begin{aligned}
&2\sum_{k=1}^{3} Re\left[\chi_k^{h'}(z_k)\right] = \int_0^s \left(\mp t_2 + U\frac{dx_1}{ds}\right) - \chi_{,1}^p + c_1,\\
&2\sum_{k=1}^{3} Re\left[\mu_k\, \chi_k^{h'}(z_k)\right] = \int_0^s \left(\pm t_1 + U\frac{dx_2}{ds}\right) - \chi_{,2}^p + c_2, \quad \text{at } \partial\Omega.\\
&2\sum_{k=1}^{3} Re\left[\lambda_k \chi_k^{h'}(z_k)\right] = -\Psi^p + c_3,
\end{aligned}
\tag{3.183}
$$

We turn now the attention on the conditions at the ends of the cylinder; the stresses on each basis are statically equivalent to a force and to a moment resultant (Ω is the cross-section domain):

$$
\begin{aligned}
\mathbf{F} &= \left(\int_\Omega \sigma_5 d\Omega, \int_\Omega \sigma_4 d\Omega, \int_\Omega \sigma_3 d\Omega\right),\\
\mathbf{M} &= \left(\int_\Omega \sigma_3 x_2 d\Omega, \int_\Omega \sigma_3 x_1 d\Omega, \int_\Omega (\sigma_4 x_1 - \sigma_5 x_2) d\Omega\right).
\end{aligned}
\tag{3.184}
$$

It can be proved (Lekhnitskii 1963, p. 113), that the first two components of **F** are identically null. The third component is equivalent to an axial force, say N, while the three components of **M** are, generally speaking, not null, say M_1, M_2 and M_3. This quantities, that are the same in any cross-section of the cylinder because stresses do not vary with x_3, are actually the conditions to be given at the ends of the cylinder. Once these four quantities known, the four constants A, B, C and ω can be calculated.

3.9.9 Generalized Plane Strain

The concept of generalized plane strain has been introduced in Sect. 3.7: in such a state, all the tensors are independent of x_3, so they are plane fields but generally speaking, they are not plane, in particular, $u_3 \neq 0$.

Generalized plane strain is a special case of the Lekhnitskii problem, which corresponds to put

$$
A = B = C = \omega = 0. \tag{3.185}
$$

In fact, as a consequence of this assumption and of Eqs. (3.121)$_3$ and (3.120),

$$
\varepsilon_3 = 0, \quad u_k = u_k^\pi(x_1, x_2) \;\; \forall k = 1, 2, 3 : \tag{3.186}
$$

also **u**, like already $\boldsymbol{\sigma}$ and $\boldsymbol{\varepsilon}$, is a plane field; nevertheless, generally speaking, ε_4 and ε_5 are not null.

Then, in this case we get simpler expressions for σ_3:

$$\sigma_3 = -\frac{1}{S_{33}}\left(S_{13}\sigma_1 + S_{23}\sigma_2 + S_{43}\sigma_4 + S_{53}\sigma_5 + S_{63}\sigma_6\right), \tag{3.187}$$

and for the u_ks:

$$\begin{aligned} u_1(x_1, x_2) &= 2\sum_{k=1}^{3} Re\left[\xi_k\, \chi_k^{h'}(z_k)\right] + \tilde{u}_1(x_1, x_2),\\ u_2(x_1, x_2) &= 2\sum_{k=1}^{3} Re\left[\eta_k\, \chi_k^{h'}(z_k)\right] + \tilde{u}_2(x_1, x_2),\\ u_3(x_1, x_2) &= 2\sum_{k=1}^{3} Re\left[\zeta_k\, \chi_k^{h'}(z_k)\right] + \tilde{u}_3(x_1, x_2), \end{aligned} \tag{3.188}$$

but also for the particular integrals χ^p and Ψ^p, because the non-homogeneous term C_2 in Eq. (3.132) is simpler:

$$C_2 = \sqrt{2}\left[(S'_{14} + S'_{24})U_{,1} + (S'_{15} + S'_{25})U_{,2}\right]. \tag{3.189}$$

The case of generalized plane strain is typical of an infinitely long cylinder, whose material is generally anisotropic, acted upon as specified in Sect. 3.9.1. In such a case the boundary conditions are uniquely those in Eq. (3.183).

3.9.10 Plane Deformation

Let us consider a particular case of generalized plain strain, as introduced in the previous Section: an infinitely long cylinder whose material has at each point the plane $x_3 = 0$ as plane of symmetry, x_3 being the cylinder's axis. We have already shown, Sect. 3.7, that in such a case plane and antiplane deformations are uncoupled and that, finally, the problem is reduced to a case of plane deformation, i.e.:

$$\begin{aligned} &u_3 = 0,\ u_\alpha = u_\alpha(x_1, x_2),\ \alpha = 1, 2,\\ &\varepsilon_3 = \varepsilon_4 = \varepsilon_5 = 0,\ \{\varepsilon^p\} = \{\varepsilon^p(x_1, x_2)\}. \end{aligned} \tag{3.190}$$

For a material having at least $x_3 = 0$ as plane of symmetry, it is also, Sect. 3.9.6,

$$\delta_3(\mu) = 0, \tag{3.191}$$

which implies,[10] through Eq. (3.155),

$$\lambda_k = 0\ \forall k = 1, 2, 3, \tag{3.192}$$

[10] Actually, for this particular case $\lambda_3 \to \infty$, see Note 8, and $\chi_3^{h'}$ should be replaced by $\chi_3^{h'}/\lambda_3$. In the end, the result is the same: the third term in all the summations vanishes everywhere,

and, Eqs. (2.57) and (3.126),

$$\begin{aligned}&S_{14}=S_{15}=S_{24}=S_{25}=S_{34}=S_{35}=S_{46}=S_{56}=0 \Rightarrow\\&S'_{14}=S'_{15}=S'_{24}=S'_{25}=S'_{46}=S'_{56}=0.\end{aligned} \tag{3.193}$$

If the material is orthotropic, in addition it is also

$$S_{16}=S_{26}=S_{36}=S_{45}=0 \Rightarrow S'_{16}=S'_{26}=S'_{45}=0. \tag{3.194}$$

Considering all the above conditions, we have then

$$\Psi=0,\ \ C_2=0, \tag{3.195}$$

so χ remains the only stress function and finally we get

$$\begin{aligned}u_1(x_1,x_2)&=2\sum_{k=1}^{2} Re\left[\xi_k\, \chi_k^{h'}(z_k)\right]+\tilde{u}_1(x_1,x_2),\\u_2(x_1,x_2)&=2\sum_{k=1}^{2} Re\left[\eta_k\, \chi_k^{h'}(z_k)\right]+\tilde{u}_2(x_1,x_2),\\u_3(x_1,x_2)&=0;\end{aligned} \tag{3.196}$$

$$\begin{aligned}\xi_k&=\mu_k^2 S'_{11}+S'_{12}-\sqrt{2}\mu_k S'_{16},\\\eta_k&=\mu_k S'_{12}+\frac{S'_{22}}{\mu_k}-\sqrt{2}S'_{26},\qquad k=1,2;\\\zeta_k&=0,\end{aligned} \tag{3.197}$$

$$\begin{aligned}\sigma_1&=2\sum_{k=1}^{2} Re\left[\mu_k^2\, \chi_k^{h''}(z_k)\right]+\frac{\partial^2\chi^p}{\partial x_2^2}-U,\\\sigma_2&=2\sum_{k=1}^{2} Re\left[\chi_k^{h''}(z_k)\right]+\frac{\partial^2\chi^p}{\partial x_1^2}-U,\\\sigma_3&=-\frac{1}{S_{33}}\left(S_{13}\sigma_1+S_{23}\sigma_2+S_{63}\sigma_6\right),\\\sigma_4&=0,\\\sigma_5&=0,\\\sigma_6&=-2\sqrt{2}\sum_{k=1}^{2} Re\left[\mu_k\, \chi_k^{h''}(z_k)\right]-\sqrt{2}\frac{\partial^2\chi^p}{\partial x_1\partial x_2};\end{aligned} \tag{3.198}$$

(Footnote 10 continued)
see also Eq. (3.204), that is why we prefer to assume immediately that all the λ_k are null: the final result is the same.

$$\begin{aligned}
\varepsilon_1 &= S'_{11}\sigma_1 + S'_{11}\sigma_2 + S'_{16}\sigma_6,\\
\varepsilon_2 &= S'_{12}\sigma_1 + S'_{22}\sigma_2 + S'_{26}\sigma_6,\\
\varepsilon_3 &= 0,\\
\varepsilon_4 &= 0,\\
\varepsilon_5 &= 0,\\
\varepsilon_6 &= S'_{16}\sigma_1 + S'_{26}\sigma_2 + S'_{66}\sigma_6;
\end{aligned} \tag{3.199}$$

$$\left.\begin{aligned}
2\sum_{k=1}^{2} Re\left[\chi_k^{h'}(z_k)\right] &= \int_0^s \left(\mp t_2 + U\frac{dx_1}{ds}\right) - \chi^p_{,1} + c_1,\\
2\sum_{k=1}^{2} Re\left[\mu_k\, \chi_k^{h'}(z_k)\right] &= \int_0^s \left(\pm t_1 + U\frac{dx_2}{ds}\right) - \chi^p_{,2} + c_2,
\end{aligned}\right\} \text{ at } \partial\Omega,\quad c_1, c_2 \in \mathbb{R}. \tag{3.200}$$

To remark that in the absence of body forces, $U = 0$, all the particular integrals are identically null: $\chi^p = 0$, $\tilde{u}_\alpha = 0\ \forall\alpha = 1, 2$.

In this case, where the stress function Ψ is absent, the differential equation for χ^h reduces to

$$\nabla_1^4 \chi^h = 0; \tag{3.201}$$

it is a fourth-order differential equation, whose characteristic associated algebraic equation is

$$\delta_4(\mu) = S'_{22} - 2\sqrt{2}S'_{26}\mu + 2(S'_{12} + S'_{66})\mu^2 - 2\sqrt{2}S'_{16}\mu^3 + S'_{11}\mu^4 = 0. \tag{3.202}$$

Also in this case, the four roots μ_k are complex conjugate. In fact, $[S4']$ is positive definite and it is easy to check that, Eq. (3.163),

$$\{v\}^\top[S4']\{v\} = \delta_4(\mu) > 0. \tag{3.203}$$

So, the four roots of Eq. (3.202), assuming that they are distinct,[11] are

$$\begin{aligned}
\mu_1 &= \alpha_1 + i\beta_1,\\
\mu_2 &= \alpha_2 + i\beta_2,\\
\mu_3 &= \overline{\mu}_1,\\
\mu_4 &= \overline{\mu}_2.
\end{aligned} \qquad \alpha_i, \beta_i \in \mathbb{R},\ \ \beta_i > 0\ \forall i = 1, 2, \tag{3.204}$$

The case named by Lekhnitskii plane deformation is the one considered by Milne-Thomson and it is a special case of a plane strain state, because $\sigma_4 = \sigma_5 = 0$. Nevertheless, recalling what said in Sect. 3.7, this is also the only case when the

[11] In case of identical roots, $\mu_1 = \mu_2$, it can be shown that the problem can be reduced to that of an isotropic body (Lekhnitskii 1963, p. 137).

plane strain state is an exact theory. So, we can establish that plane strain coincides with the case of plane deformation of the Lekhnitskii theory.

3.9.11 Generalized Plane Stress

The case of generalized plane stress has already been introduced in Sect. 3.6. With the assumptions that define this case, we remark that the situation for the average stresses and displacements is exactly that described in the previous section, i.e. of plane deformation. There is hence a precise link between these two cases, as already anticipated in Sect. 3.8.

In particular, the average stresses $\widehat{\sigma_1}$, $\widehat{\sigma_2}$ and $\widehat{\sigma_6}$ still depend upon the stress function χ as in Eq. (3.129):

$$\begin{aligned}\widehat{\sigma_1} &= \chi_{,22} - U,\\ \widehat{\sigma_2} &= \chi_{,11} - U,\\ \widehat{\sigma_6} &= -\sqrt{2}\,\chi_{,12}.\end{aligned} \tag{3.205}$$

Remembering that one of the basic assumptions of generalized plane stress is a material that is at least monoclinic with $x_3 = 0$ as plane of symmetry, Eq. (2.56), and the fact that $\widehat{\sigma_3} = \sigma_4 = \sigma_5 = 0$, see Eqs. (3.86) and (3.199), then

$$\begin{aligned}\widehat{\varepsilon_i} &= S_{i1}\widehat{\sigma_1} + S_{i2}\widehat{\sigma_2} + S_{i6}\widehat{\sigma_6}, \quad i = 1, 2, 3, 6,\\ \widehat{\varepsilon_4} &= \widehat{\varepsilon_5} = 0.\end{aligned} \tag{3.206}$$

In this case, no need to introduce the reduced compliances S'_{ij} for eliminating $\widehat{\sigma_3}$, because this last is already null. Hence, the solution given in the previous section for the case of plane deformation is still valid in the case of generalized plane stress, provided that the reduced compliances S'_{ij} are everywhere replaced by the compliances S_{ij}.

In this case, the characteristic equation (3.202) becomes

$$\delta_4(\mu) = S_{22} - 2\sqrt{2}S_{26}\mu + 2(S_{12} + S_{66})\mu^2 - 2\sqrt{2}S_{16}\mu^3 + S_{11}\mu^4 = 0, \tag{3.207}$$

and it is elementary to remark that its roots are still complex, because the matrix $[S4]$, corresponding of matrix $[S4']$ but with the S_{ij}s replacing the S'_{ij}s, is obviously positive definite.

3.9.12 A Final Consideration

In the literature, the problems of plane deformation and of generalized plane stress are often combined and called *the plane problem of the theory of elasticity*. We can hence remark that the Lekhnitskii theory is a general frame where generalized plane strain, plane strain and generalized plane stress are special cases. Nevertheless, the case of plane stress, as defined in Sect. 3.5 is not comprehended in the Lekhnitskii theory.

3.10 The Stroh Theory

3.10.1 The General Stroh Problem

We will name a *Stroh Problem* every problem of the elastic equilibrium of an anisotropic body whose displacement field is constrained to satisfy *uniquely* the condition of plane field[12]:

$$\mathbf{u} = \mathbf{u}(x_1, x_2). \tag{3.208}$$

The same properties are obviously true for the strain too, and, through the Hooke's law, for the stress:

$$\boldsymbol{\varepsilon} = \boldsymbol{\varepsilon}(x_1, x_2), \quad \boldsymbol{\sigma} = \boldsymbol{\sigma}(x_1, x_2). \tag{3.209}$$

To remark that a consequence of assumption (3.208) is that $\varepsilon_{33} = 0$, but *not* that $\sigma_{33} = 0$.

The *Stroh theory* or *formalism* (Ting 1996, p. 134), is the mathematical theory for obtaining a general formulation of the solution to the Stroh Problem. We will see that there are several similarities with the Lekhnitskii theory.

Since the basic condition of the Stroh problem is put on $\mathbf{u}(x_1, x_2)$, it is worth to write the Hooke's law directly as a function of the u_is; remembering the symmetry of $\mathbb{E}$, it is

$$\boldsymbol{\sigma} = \mathbb{E}\nabla\mathbf{u} \quad \rightarrow \quad \sigma_{ij} = E_{ijkl}u_{k,l}, \quad i, j, k = 1, 2, 3, \ l = 1, 2. \tag{3.210}$$

As a consequence, the equation of motion can be rewritten as

$$\text{div}(\mathbb{E}\nabla\mathbf{u}) = \mathbf{b} \rightarrow E_{ijkl}u_{k,lj} = b_i \quad i, k = 1, 2, 3, \ j, l = 1, 2, \tag{3.211}$$

which shows that also the body vector is necessarily a plane field.

We pose

$$\mathbf{u}(x_1, x_2) = \mathbf{u}^h(x_1, x_2) + \mathbf{u}^p(x_1, x_2), \tag{3.212}$$

where $\mathbf{u}^p(x_1, x_2)$ is a particular solution that satisfies

[12] We prefer to use, throughout all Sect. 3.10, the standard tensor notation for vectors and matrices, indicating them by boldfaced uppercase and lowercase letters, respectively. This is to remind the fact that all the theory is developed for tensor components, not for the matrix Kelvin's notation, with the exception of Sect. 3.10.3, where the Kelvin's notation is used because more suitable for analyzing the effect of the elastic syngonies.

$$E_{ijkl}u^p_{k,lj} = b_i \quad i,k = 1,2,3, \ \ j,l = 1,2, \tag{3.213}$$

and $\mathbf{u}^h(x_1, x_2)$ the solution of the associated homogeneous equation

$$E_{ijkl}u^h_{k,lj} = 0 \quad i,k = 1,2,3, \ \ j,l = 1,2. \tag{3.214}$$

The solution $\mathbf{u}^p(x_1, x_2)$ depends upon the particular distribution of the body vector $\mathbf{b}(x_1, x_2)$ and is usually simple to be found. So, we will concentrate on the solution $\mathbf{u}^h(x_1, x_2)$, i.e. on the case of absence of body forces.

3.10.2 The Theory of Eshelby, Read and Shockley

The theoretical developments of the Stroh formalism can be traced back to a work of Eshelby, Read and Shockley (Eshelby et al. 1953), successively developed by Stroh in two papers (Stroh 1958, 1962).

We can put

$$\mathbf{u}^h = \mathbf{a} f(z) \ \ \rightarrow \ \ u^h_k = a_k f(z), \ \ k = 1,2,3, \tag{3.215}$$

where the variable z is a linear combination of x_1 and x_2; without lack of generality, we set

$$z = x_1 + p x_2; \tag{3.216}$$

the a_ks and p are four constants to be determined, while $f(z)$ is an arbitrary function.

Let us differentiate Eq. (3.215):

$$\begin{aligned} u^h_{k,l} &= a_k \frac{df}{dz}\frac{\partial z}{\partial x_l} = (\delta_{l1} + p\delta_{l2}) a_k f'(z), \\ u^h_{k,lj} &= (\delta_{l1} + p\delta_{l2}) a_k \frac{df}{dz}\frac{\partial z}{\partial x_j} = (\delta_{l1} + p\delta_{l2})(\delta_{j1} + p\delta_{j2}) a_k f''(z). \end{aligned} \tag{3.217}$$

Injecting Eq. (3.217) into Eq. (3.214) gives an equation which is satisfied $\forall z$ if and only if

$$E_{ijkl}(\delta_{l1} + p\delta_{l2})(\delta_{j1} + p\delta_{j2}) a_k = 0 \quad i,k = 1,2,3, \ \ j,l = 1,2. \tag{3.218}$$

Equation (3.218) can be transformed to

$$\left[E_{i1k1} + p\left(E_{i1k2} + E_{i2k1}\right) + p^2 E_{i2k2}\right] a_k = 0 \quad i,k = 1,2,3, \ \ j,l = 1,2, \tag{3.219}$$

or in matrix form

$$\mathbf{W}\mathbf{a} = \mathbf{O}, \ \ \mathbf{W} = \mathbf{Q} + p\left(\mathbf{R} + \mathbf{R}^\top\right) + p^2\mathbf{T}, \tag{3.220}$$

where the 3 × 3 matrices **Q**, **R** and **T** are

$$Q_{ik} = E_{i1k1} \rightarrow \mathbf{Q} = \begin{bmatrix} E_{1111} & E_{1121} & E_{1131} \\ & E_{2121} & E_{2131} \\ sym & & E_{3131} \end{bmatrix}; \tag{3.221}$$

$$R_{ik} = E_{i1k2} \rightarrow \mathbf{R} = \begin{bmatrix} E_{1112} & E_{1122} & E_{1132} \\ E_{2112} & E_{2122} & E_{2132} \\ E_{3112} & E_{3122} & E_{3132} \end{bmatrix}; \tag{3.222}$$

$$T_{ik} = E_{i2k2} \rightarrow \mathbf{T} = \begin{bmatrix} E_{1212} & E_{1222} & E_{1232} \\ & E_{2222} & E_{2232} \\ sym & & E_{3232} \end{bmatrix}. \tag{3.223}$$

Finally, matrix **W** is,

$$\mathbf{W} = \begin{bmatrix} E_{1111} + 2E_{1112}p + E_{1212}p^2 & E_{1112} + (E_{1122} + E_{1212})p + E_{2212}p^2 & E_{1113} + (E_{1123} + E_{1312})p + E_{2312}p^2 \\ & E_{1212} + 2E_{2212}p + E_{2222}p^2 & E_{1312} + (E_{2213} + E_{2312})p + E_{2223}p^2 \\ sym & & E_{1313} + E_{2313}p + E_{2323}p^2 \end{bmatrix}. \tag{3.224}$$

To remark that matrices **Q** and **T** are principal minors of the stiffness matrix; hence, because this last is positive definite by virtue of the positive definiteness of the strain energy, they are positive definite too, Theorem 1.

Equation (3.220) has a non trivial solution, $\mathbf{a} = \mathbf{O}$, if and only if

$$\det \mathbf{W} = 0. \tag{3.225}$$

We need, hence, to look for the values of p that satisfy Eq. (3.225), that are eigenvalues depending upon the elastic moduli E_{ijkl}; to this purpose, we have the

Theorem 9 *The six eigenvalues p, roots of the characteristic equation (3.225) cannot be real.*

Proof Making the scalar product of Eq. (3.218) and vector **a** gives

$$E_{ijkl}(\delta_{l1} + p\delta_{l2})a_k(\delta_{j1} + p\delta_{j2})a_i = 0 \quad i, k = 1, 2, 3, \ j, l = 1, 2, \tag{3.226}$$

and putting

$$\begin{aligned} (\delta_{j1} + p\delta_{j2})a_i &= u^h_{i,j}, \\ (\delta_{l1} + p\delta_{l2})a_k &= u^h_{k,l}, \end{aligned} \tag{3.227}$$

we get

$$E_{ijkl}u^h_{i,j}u^h_{k,l} = E_{ijkl}\varepsilon_{ij}\varepsilon_{kl} = 2V = 0, \tag{3.228}$$

which cannot be true, unless $\mathbf{u}^h = \mathbf{O}$, because of the positive definiteness of the strain energy V. Hence, the eigenvalues p cannot be real.

The eigenvalues p are hence complex and in particular complex conjugate, because the coefficients of the sextic equation (3.225) are all real valued. As a consequence, also the associated eigenvectors $\mathbf{a}_j$, given by Eq. (3.220) once known the eigenvalues p_j, are complex conjugate. Following the same procedure already used for the Lekhnitskii problem, we pose

$$p_j = \alpha_j + i\beta_j,\ \ p_{j+3} = \overline{p}_j,\ \ \alpha_j, \beta_j \in \mathbb{R},\ \ \beta_j > 0,\ \ j = 1, 2, 3. \tag{3.229}$$

We assume that the eigenvalues are distinct; then, the general solution is obtained superposing the six solutions of the type (3.215):

$$\mathbf{u} = \sum_{j=1}^{3} \left[\mathbf{a}_j f_j(z_j) + \overline{\mathbf{a}}_j f_{j+3}(\overline{z}_j)\right], \tag{3.230}$$

where the $f_j(z_j)$ are six arbitrary functions whose respective arguments are

$$z_j = x_1 + p_j x_2,\ \ j = 1, \ldots, 6. \tag{3.231}$$

The stress components can be recovered injecting Eq. (3.208) into Eq. (3.210):

$$\sigma_{ij} = E_{ijkl}(\delta_{1l} + p\delta_{2l})a_k f'(z)\ \ \rightarrow\ \ \begin{cases} \sigma_{i1} = (Q_{ij} + pR_{ij})a_j f'(z), \\ \sigma_{i2} = (R_{ji} + pT_{ij})a_j f'(z), \end{cases} \tag{3.232}$$

and hence, through Eqs. (3.219)–(3.223) and (3.229),

$$\begin{aligned} \begin{Bmatrix} \sigma_{11} \\ \sigma_{12} \\ \sigma_{13} \end{Bmatrix} &= \sum_{j=1}^{3} \left[\left(\mathbf{Q} + p_j \mathbf{R}\right)\mathbf{a}_j f'_j(z_j) + \left(\mathbf{Q} + \overline{p}_j \mathbf{R}\right)\overline{\mathbf{a}}_j f'_{j+3}(\overline{z}_j)\right], \\ \begin{Bmatrix} \sigma_{21} \\ \sigma_{22} \\ \sigma_{23} \end{Bmatrix} &= \sum_{j=1}^{3} \left[\left(\mathbf{R}^\top + p_j \mathbf{T}\right)\mathbf{a}_j f'_j(z_j) + \left(\mathbf{R}^\top + \overline{p}_j \mathbf{T}\right)\overline{\mathbf{a}}_j f'_{j+3}(\overline{z}_j)\right]. \end{aligned} \tag{3.233}$$

The component σ_{33} is then deduced from the condition

$$\varepsilon_{33} = Z_{33hk}\sigma_{hk} = 0\ \Rightarrow\ \sigma_{33} = -\frac{Z_{33hk}\sigma_{hk}}{Z_{3333}},\ \ hk \neq 33. \tag{3.234}$$

3.10.3 The Eigenvalues p_j and the Elastic Syngony

The above results are quite general; nevertheless, they can be simplified for the different elastic syngonies, because some of the elastic moduli are null or depending upon other moduli. To this purpose, it is worth to rewrite matrices **Q**, **R**, **T** and **W** using the Kelvin's notation:

$$[Q]=\begin{bmatrix} C_{11} & \frac{C_{16}}{\sqrt{2}} & \frac{C_{15}}{\sqrt{2}} \\ & \frac{C_{66}}{2} & \frac{C_{56}}{2} \\ sym & & \frac{C_{55}}{2}\end{bmatrix}; \tag{3.235}$$

$$[R]=\begin{bmatrix} \frac{C_{16}}{\sqrt{2}} & C_{12} & \frac{C_{14}}{\sqrt{2}} \\ \frac{C_{66}}{2} & \frac{C_{26}}{\sqrt{2}} & \frac{C_{46}}{2} \\ \frac{C_{56}}{2} & \frac{C_{25}}{\sqrt{2}} & \frac{C_{45}}{2}\end{bmatrix}; \tag{3.236}$$

$$[T]=\begin{bmatrix} \frac{C_{66}}{2} & \frac{C_{26}}{\sqrt{2}} & \frac{C_{46}}{2} \\ & C_{22} & \frac{C_{24}}{\sqrt{2}} \\ sym & & \frac{C_{44}}{2}\end{bmatrix}; \tag{3.237}$$

$$[W]=\begin{bmatrix} C_{11}+\sqrt{2}C_{16}p+\frac{C_{66}}{2}p^2 & \frac{C_{16}}{\sqrt{2}}+(C_{12}+\frac{C_{66}}{2})p+\frac{C_{26}}{\sqrt{2}}p^2 & \frac{C_{15}}{\sqrt{2}}+(\frac{C_{14}}{\sqrt{2}}+\frac{C_{56}}{2})p+\frac{C_{46}}{2}p^2 \\ & \frac{C_{66}}{2}+\sqrt{2}C_{26}p+C_{22}p^2 & \frac{C_{56}}{2}+(\frac{C_{25}}{\sqrt{2}}+\frac{C_{46}}{2})p+\frac{C_{24}}{\sqrt{2}}p^2 \\ sym & & \frac{C_{55}}{2}+C_{45}p+\frac{C_{44}}{2}p^2\end{bmatrix}. \tag{3.238}$$

Let us now consider the effect of some elastic syngonies on the eigenvalues p_j:

- for a monoclinic material with $x_1=0$ plane of symmetry, $C_{i5}=C_{i6}=0\ \forall i=1,\ldots,4$; matrix $[W]$ is hence reduced to

$$[W]=\frac{1}{2}\begin{bmatrix} 2C_{11}+C_{66}p^2 & (2C_{12}+C_{66})p & (\sqrt{2}C_{14}+C_{56})p \\ & C_{66}+2C_{22}p^2 & C_{56}+\sqrt{2}C_{24}p^2 \\ sym & & C_{55}+C_{44}p^2\end{bmatrix}. \tag{3.239}$$

  In this case, the sextic equation is of degree 3 in p^2; two are the possible cases: (i) the three roots of $p^2\in\mathbb{R}$; because $p_j\in\mathbb{C}$, then $p_j^2<0\ \forall j=1,2,3$; so $p_j=i\beta_j,\ \beta_j>0\ \forall j=1,2,3$; (ii) one of the roots of p^2 is real and the two others are complex conjugate: $p_1=\alpha+i\beta,\ p_2=-\alpha+i\beta,\ p_3=i\beta_3,\ \alpha,\beta,\beta_3\in\mathbb{R},\ \beta,\beta_3>0$. An analogous result is obtained for a monoclinic material with $x_2=0$ plane of symmetry.
- For a monoclinic material with $x_3=0$ plane of symmetry, $C_{i4}=C_{i5}=0\ \forall i=1,\ldots,3,\ C_{46}=C_{56}=0$; matrix $[W]$ becomes

$$[W]=\frac{1}{2}\begin{bmatrix}2C_{11}+2\sqrt{2}C_{16}p+C_{66}p^2 & \sqrt{2}C_{16}+(2C_{12}+C_{66})p+\sqrt{2}C_{26}p^2 & 0\\ & C_{66}+2\sqrt{2}C_{26}p+2C_{22}p^2 & 0\\ sym & & C_{55}+2C_{45}p+C_{44}p^2\end{bmatrix}. \quad (3.240)$$

The sextic equation is decomposed into a quadratic and a quartic one; the solutions are $p_1=\alpha_1+i\beta_1,\ p_2=\alpha_2+i\beta_2,\ \alpha_h,\beta_h\in\mathbb{R},\ \beta_h>0\ \forall h=1,2$ and $p_3=\frac{1}{C_{44}}\left(-C_{45}+i\sqrt{C_{44}C_{55}-C_{45}^2}\right)$.

- For a transversely isotropic material with x_3 as axis of symmetry, $C_{i4}=C_{i5}=C_{i6}=0\ \forall i=1,2,3,\ C_{45}=C_{46}=C_{56}=0,\ C_{22}=C_{11},\ C_{23}=C_{13},\ C_{55}=C_{44}$ and $C_{66}=C_{11}-C_{12}$; matrix $[W]$ becomes

$$[W]=\frac{1}{2}\begin{bmatrix}2C_{11}+(C_{11}-C_{12})p^2 & (C_{11}+C_{12})p & 0\\ & C_{11}-C_{12}+2C_{11}p^2 & 0\\ sym & & C_{44}(1+p^2)\end{bmatrix}. \quad (3.241)$$

for such a matrix $[W]$ the eigenvalues are $p_1=p_2=p_3=i$. Then, rank$[W]=1$, so there are only two independent eigenvectors $\mathbf{a}$ for the triple eigenvalue.

- For an isotropic material, it is also $C_{33}=C_{11},\ C_{13}=C_{12}$ and $C_{44}=C_{11}-C_{12}$, so matrix $[W]$ becomes

$$[W]=\frac{1}{2}\begin{bmatrix}2C_{11}+(C_{11}-C_{12})p^2 & (C_{11}+C_{12})p & 0\\ & C_{11}-C_{12}+2C_{11}p^2 & 0\\ sym & & (C_{11}-C_{12})(1+p^2)\end{bmatrix}. \quad (3.242)$$

It is then evident that the isotropic case coincides with the previous one, with the only difference that now C_{44} is no more an independent modulus. This situation is not exclusively limited to isotropic or transversely isotropic materials, but it is shared also by infinitely many other anisotropic materials, see Sect. 3.9.6 (Ting 1996, p. 125).

3.10.4 The Sextic Formalism of Stroh

We introduce now a further transformation of the results from Eshelby, Read and Shockley, Sect. 3.10.2, obtained through the introduction of a vector stress function.

Let us rewrite the stress components, Eq. (3.232), as

$$\sigma_{i1}=-pb_i f'(z),\quad \sigma_{i2}=b_i f'(z), \quad (3.243)$$

where, using Eq. (3.220),

$$\mathbf{b} = \left(\mathbf{R}^\top + p\mathbf{T}\right)\mathbf{a} = -\frac{1}{p}\left(\mathbf{Q} + p\mathbf{R}\right)\mathbf{a}. \tag{3.244}$$

Introducing the *vector stress function*

$$\boldsymbol{\varphi} = \mathbf{b}f(z), \tag{3.245}$$

Equation (3.243) can be rewritten as

$$\sigma_{i1} = -\varphi_{i,2}, \quad \sigma_{i2} = \varphi_{i,1}; \tag{3.246}$$

the general solution for $\boldsymbol{\varphi}$ is get superposing six solutions of the form (3.245) associated with the six eigenvalues p_j:

$$\boldsymbol{\varphi} = \sum_{j=1}^{3}\left[\mathbf{b}_j f_j(z_j) + \overline{\mathbf{b}}_j f_{j+3}(\overline{z}_j)\right]. \tag{3.247}$$

Equations (3.230) and (3.247) constitute the *sextic formalism of Stroh*, and the vectors $\mathbf{a}_j$ and $\mathbf{b}_j$, $j = 1, \ldots, 6$, are the *Stroh eigenvalues*, related by Eq. (3.244). Once the stress function $\boldsymbol{\varphi}$ known, stresses are calculated through Eq. (3.246), while strains are get differentiating $\mathbf{u}$. The stress component σ_{33} is still given by Eq. (3.234) or, in the Kelvin's notation,

$$\varepsilon_3 = S_{3k}\sigma_k = 0 \;\Rightarrow\; \sigma_3 = -\frac{S_{3k}\sigma_k}{S_{33}}, \; k \neq 3. \tag{3.248}$$

Some other forms can be given to the Stroh formalism. First of all, functions $f_j(z_j)$ have usually the same form; they can hence be put in the form

$$f_j(z_j) = f(z_j)q_j, \quad f_{j+3}(\overline{z}_j) = \overline{f}(\overline{z}_j)\overline{q}_j, \quad q_j \in \mathbb{C}, \; j = 1, 2, 3 \tag{3.249}$$

Introducing the matrices

$$\mathbf{A} = [\mathbf{a}_1, \mathbf{a}_2, \mathbf{a}_3], \;\; \mathbf{B} = [\mathbf{b}_1, \mathbf{b}_2, \mathbf{b}_3], \tag{3.250}$$

and the diagonal function matrix

$$\mathbf{F} = \begin{bmatrix} f(z_1) & 0 & 0 \\ 0 & f(z_2) & 0 \\ 0 & 0 & f(z_3) \end{bmatrix}, \tag{3.251}$$

then

$$\{u\} = 2\mathrm{Re}(\mathbf{AFq}), \quad \{\varphi\} = 2\mathrm{Re}(\mathbf{BFq}). \tag{3.252}$$

So, everything is reduced to determine $\mathbf{F}$ and $\mathbf{q}$.

Another representation can be obtained as follows: once introduced the vectors

$$\mathbf{f1} = \begin{Bmatrix} f_1(z_1) \\ f_2(z_2) \\ f_3(z_3) \end{Bmatrix}, \quad \mathbf{f2} = \begin{Bmatrix} f_4(\overline{z}_1) \\ f_5(\overline{z}_2) \\ f_6(\overline{z}_3) \end{Bmatrix}, \tag{3.253}$$

Equations (3.230) and (3.247) can be put in the symbolic form

$$\left\{\begin{array}{c} \mathbf{u} \\ \hline \boldsymbol{\varphi} \end{array}\right\} = \left[\begin{array}{c|c} \mathbf{A} & \overline{\mathbf{A}} \\ \hline \mathbf{B} & \overline{\mathbf{B}} \end{array}\right] \left\{\begin{array}{c} \mathbf{f1} \\ \hline \mathbf{f2} \end{array}\right\}. \tag{3.254}$$

We will show in Sect. (3.10.5) that

$$\left[\begin{array}{c|c} \mathbf{B}^\top & \mathbf{A}^\top \\ \hline \overline{\mathbf{B}}^\top & \overline{\mathbf{A}}^\top \end{array}\right] \left[\begin{array}{c|c} \mathbf{A} & \overline{\mathbf{A}} \\ \hline \mathbf{B} & \overline{\mathbf{B}} \end{array}\right] = \left[\begin{array}{c|c} \mathbf{I} & \mathbf{O} \\ \hline \mathbf{O} & \mathbf{I} \end{array}\right], \tag{3.255}$$

so that Eq. (3.254) can be rewritten as

$$\left[\begin{array}{c|c} \mathbf{B}^\top & \mathbf{A}^\top \\ \hline \overline{\mathbf{B}}^\top & \overline{\mathbf{A}}^\top \end{array}\right] \left\{\begin{array}{c} \mathbf{u} \\ \hline \boldsymbol{\varphi} \end{array}\right\} = \left\{\begin{array}{c} \mathbf{f1} \\ \hline \mathbf{f2} \end{array}\right\}, \tag{3.256}$$

which gives the representation

$$\mathbf{A}^\top \boldsymbol{\varphi} + \mathbf{B}^\top \mathbf{u} = \mathbf{f1}. \tag{3.257}$$

A question is the relation existing between the vector stress function $\boldsymbol{\varphi}$ and the Airy's function χ. To this purpose, we observe that, by Eq. (3.246),

$$\sigma_{12} = \sigma_{21} \quad \Rightarrow \quad \varphi_{1,1} + \varphi_{2,2} = 0. \tag{3.258}$$

This relation is of the type (3.16); so, it exists a scalar function χ, which is precisely the Airy's function, such that

$$\varphi_1 = -\chi_{,2}, \quad \varphi_2 = \chi_{,1}, \tag{3.259}$$

and hence, see Eq. (3.245),

$$\chi = \int_c^z \varphi_2(t)dt = b_2 \int_c^z f(t)dt, \tag{3.260}$$

with c an arbitrary constant.

It is possible to give a physical interpretation to the vector stress function $\boldsymbol{\varphi}$. Consider the surface traction $\mathbf{t}$ on the boundary $\partial\Omega$ of an anisotropic elastic body Ω; the Cauchy's Theorem states that

$$\mathbf{t} = \boldsymbol{\sigma}\mathbf{n}, \tag{3.261}$$

n being the outward unit normal to $\partial\Omega$. Let us consider now the curve γ, intersection between the plane $x_3 = 0$ and $\partial\Omega$. Over γ,

$$\mathbf{n} = \left(-\frac{dx_2}{ds}; \frac{dx_1}{ds}; 0\right), \tag{3.262}$$

with s a curvilinear abscissa on γ chosen in such a way that for increasing s the domain Ω is left on the right side. Then, by Eqs. (3.246) and (3.262) we get

$$\frac{d\varphi_i}{ds} = \varphi_{i,1}\frac{dx_1}{ds} + \varphi_{i,2}\frac{dx_2}{ds} = \sigma_{i1}n_1 + \sigma_{i2}n_2 = \sigma_{ij}n_j = t_i \quad \rightarrow \quad \mathbf{t} = \frac{d\boldsymbol{\varphi}}{ds}. \tag{3.263}$$

If s_1 and s_2 are two points on γ, then the resultant of the surface tractions on the arc $s_1 - s_2$ of γ is

$$\mathbf{r} = \int_{s_1}^{s_2} \mathbf{t}(s)ds = \boldsymbol{\varphi}(s_2) - \boldsymbol{\varphi}(s_1) = \Delta\boldsymbol{\varphi}. \tag{3.264}$$

If $\mathbf{e}_3$ is the unit vector of the axis of x_3, the moment about this axis due to the surface tractions **t** on the arc $s_1$divs_2 of γ is, see Eq. (3.260),

$$\begin{aligned}
\mathbf{m} &= \int_{s_1}^{s_2} (x_1t_2 - x_2t_1)\, ds\, \mathbf{e}_3 = \int_{s_1}^{s_2} (x_1d\varphi_2 - x_2d\varphi_1)\, \mathbf{e}_3 = \\
&= \left[\int_{s_1}^{s_2} d\,(x_1\varphi_2 - x_2\varphi_1) - \int_{s_1}^{s_2} (\varphi_2dx_1 - \varphi_1dx_2)\right] \mathbf{e}_3 = \\
&= \int_{s_1}^{s_2} d\,(x_1\varphi_2 - x_2\varphi_1 - \chi)\, \mathbf{e}_3 = \Delta\,(x_1\varphi_2 - x_2\varphi_1 - \chi)\, \mathbf{e}_3 = \\
&= \left[\Delta\,(x_1\varphi_2 - x_2\varphi_1) - \int_{s_1}^{s_2} \varphi_2(t)dt\right] \mathbf{e}_3.
\end{aligned} \tag{3.265}$$

3.10.5 Algebraic Questions

The two relations in Eq. (3.244) can be rewritten in the matrix form

$$\left[\begin{array}{c|c} -\mathbf{Q} & \mathbf{O} \\ \hline -\mathbf{R}^\top & \mathbf{I} \end{array}\right] \left\{\begin{array}{c} \mathbf{a} \\ \hline \mathbf{b} \end{array}\right\} = p \left[\begin{array}{c|c} \mathbf{R} & \mathbf{I} \\ \hline \mathbf{T} & \mathbf{O} \end{array}\right] \left\{\begin{array}{c} \mathbf{a} \\ \hline \mathbf{b} \end{array}\right\}, \tag{3.266}$$

and because **T** is positive definite, its inverse $\mathbf{T}^{-1}$ exists and

$$\left[\begin{array}{c|c} \mathbf{O} & \mathbf{T}^{-1} \\ \hline \mathbf{I} & -\mathbf{R}\mathbf{T}^{-1} \end{array}\right] \left[\begin{array}{c|c} \mathbf{R} & \mathbf{I} \\ \hline \mathbf{T} & \mathbf{O} \end{array}\right] = \left[\begin{array}{c|c} \mathbf{I} & \mathbf{O} \\ \hline \mathbf{O} & \mathbf{I} \end{array}\right]. \tag{3.267}$$

Then, if we multiply on the left both sides of Eq. (3.266) by the first matrix in Eq. (3.267), we get the equation

$$\mathbf{N}\mathbf{v} = p\mathbf{v}, \tag{3.268}$$

where

$$\mathbf{N} = \left[\begin{array}{c|c} \mathbf{N1} & \mathbf{N2} \\ \hline \mathbf{N3} & \mathbf{N1}^\top \end{array}\right], \quad \mathbf{v} = \left\{\begin{array}{c} \mathbf{a} \\ \hline \mathbf{b} \end{array}\right\}, \tag{3.269}$$

with

$$\mathbf{N1} = -\mathbf{T}^{-1}\mathbf{R}^\top, \quad \mathbf{N2} = \mathbf{T}^{-1}, \quad \mathbf{N3} = \mathbf{R}\mathbf{T}^{-1}\mathbf{R}^\top - \mathbf{Q}. \tag{3.270}$$

Matrix **N** is the *fundamental elasticity matrix* (Ingebrigtsen and Tonning 1969); it is easy to check that **N2** and **N3** are symmetric and, as already noticed, **N2** is also positive definite; moreover, **N3** is a stiffness matrix, **N2** is a compliance matrix and **N1** is a dimensionless matrix.

Equation (3.268) clearly shows that the composed vector **v** is an eigenvector of matrix **N** relative to the eigenvalue p. In particular, because **N** is not symmetric, **v** is a right eigenvector[13] while a left eigenvector **w** satisfies

$$\mathbf{N}^\top\mathbf{w} = p\mathbf{w}. \tag{3.271}$$

We introduce the matrix

$$\mathbf{J} = \left[\begin{array}{c|c} \mathbf{O} & \mathbf{I} \\ \hline \mathbf{I} & \mathbf{O} \end{array}\right] \Rightarrow \mathbf{J} = \mathbf{J}^\top = \mathbf{J}^{-1} \Rightarrow \mathbf{J}\mathbf{N} = (\mathbf{J}\mathbf{N})^\top = \mathbf{N}^\top\mathbf{J}; \tag{3.272}$$

then, applying **J** to Eq. (3.268), we get

$$\mathbf{J}\mathbf{N}\mathbf{v} = p\mathbf{J}\mathbf{v} \Rightarrow \mathbf{N}^\top(\mathbf{J}\mathbf{v}) = p(\mathbf{J}\mathbf{v}), \tag{3.273}$$

i.e. the left eigenvector **w** relative to the same eigenvalue p is

$$\mathbf{w} = \mathbf{J}\mathbf{v} = \left\{\begin{array}{c} \mathbf{b} \\ \hline \mathbf{a} \end{array}\right\}. \tag{3.274}$$

Let us assume that all the eigenvalues p are distinct; we normalise the eigenvectors **v** in such a way that

---

[13]Given a matrix **A**, **x** is a *right eigenvector* relative to the eigenvalue λ if $\mathbf{A}\mathbf{x} = \lambda\mathbf{x}$, while **y** is a *left eigenvector* relative to the eigenvalue μ if $\mathbf{A}^\top\mathbf{y} = \mu\mathbf{y}$. The eigenvalues are the roots of the *characteristic equations* $\det(\mathbf{A} - \lambda\mathbf{I}) = 0$ and $\det(\mathbf{A}^\top - \mu\mathbf{I}) = 0$. Because $\mathbf{I} = \mathbf{I}^\top$ and the determinant is the same for a tensor and its transpose, $\lambda = \mu$, i.e. the eigenvalues are the same for the right and left eigenvectors. Let us now consider two eigenvalues, $\lambda_j \neq \lambda_k$ and the respective right and left eigenvectors: $\mathbf{A}\mathbf{x}_j = \lambda_j\mathbf{x}_j$ and $\mathbf{A}^\top\mathbf{y}_k = \lambda_k\mathbf{y}_k$. Then $\lambda_j\mathbf{x}_j \cdot \mathbf{y}_k = \mathbf{A}\mathbf{x}_j \cdot \mathbf{y}_k = \mathbf{A}^\top\mathbf{y}_k \cdot \mathbf{x}_j = \lambda_k\mathbf{x}_j \cdot \mathbf{y}_k \rightarrow (\lambda_j - \lambda_k)\mathbf{y}_k \cdot \mathbf{x}_j = 0 \Rightarrow \mathbf{x}_j \cdot \mathbf{y}_k = 0$ because $\lambda_j \neq \lambda_k$: right and left eigenvectors of distinct eigenvalues are mutually orthogonal. By the same definition, if **A** is symmetric, right and left eigenvectors coincide and the distinction is meaningless.

$$\mathbf{v}_i \cdot \mathbf{w}_j = \delta_{ij} \ \Rightarrow \mathbf{a}_i \cdot \mathbf{b}_j + \mathbf{b}_i \cdot \mathbf{a}_j = \delta_{ij}. \tag{3.275}$$

By the definition of matrices **A** and **B**, Eq. (3.250), this last equation corresponds to Eq. (3.255), known also as *orthogonality relation*, so proving it. Of course, the matrix product in Eq. (3.255) commutes, so we can write also

$$\left[\begin{array}{c|c} \mathbf{A} & \overline{\mathbf{A}} \\ \hline \mathbf{B} & \overline{\mathbf{B}} \end{array}\right] \left[\begin{array}{c|c} \mathbf{B}^\top & \mathbf{A}^\top \\ \hline \overline{\mathbf{B}}^\top & \overline{\mathbf{A}}^\top \end{array}\right] = \left[\begin{array}{c|c} \mathbf{I} & \mathbf{O} \\ \hline \mathbf{O} & \mathbf{I} \end{array}\right], \tag{3.276}$$

known as *closure relation*. Developing this equation, we easily find that

$$\mathrm{Re}\left(\mathbf{A}\mathbf{B}^\top\right) = \frac{1}{2}\mathbf{I}, \ \ \mathrm{Re}\left(\mathbf{A}\mathbf{A}^\top\right) = \mathrm{Re}\left(\mathbf{B}\mathbf{B}^\top\right) = \mathbf{O}. \tag{3.277}$$

As a consequence, the matrices **S**, **H** and **L** defined by

$$\mathbf{S} = i(2\mathbf{A}\mathbf{B}^\top - \mathbf{I}), \ \ \mathbf{H} = 2i\mathbf{A}\mathbf{A}^\top, \ \ \mathbf{L} = -2i\mathbf{B}\mathbf{B}^\top, \tag{3.278}$$

are real-valued; it can be shown, through a rotation about the axis x_3 (Ting 1996, p. 209), that **S, H** and **L** are rank-two tensors, called the *Barnett-Lothe tensors* (Barnett and Lothe 1973). It is clear that

$$\mathbf{H} = \mathbf{H}^\top, \ \ \mathbf{L} = \mathbf{L}^\top, \tag{3.279}$$

and it can be shown that they are also positive definite, which means that **A** and **B** are non-singular. The importance of the Barnett-Lothe tensors is that, being real, they give the possibility of obtaining the solution to many problems in a real form.

Introducing the matrix

$$\mathbf{P} = \begin{bmatrix} p_1 & 0 & 0 \\ 0 & p_2 & 0 \\ 0 & 0 & p_3 \end{bmatrix}, \tag{3.280}$$

Equation (3.268) for each one of the eigenvalues p can be rewritten in a unique equation as

$$\mathbf{N}\left[\begin{array}{c|c} \mathbf{A} & \overline{\mathbf{A}} \\ \hline \mathbf{B} & \overline{\mathbf{B}} \end{array}\right] = \left[\begin{array}{c|c} \mathbf{A} & \overline{\mathbf{A}} \\ \hline \mathbf{B} & \overline{\mathbf{B}} \end{array}\right] \left[\begin{array}{c|c} \mathbf{P} & \mathbf{O} \\ \hline \mathbf{O} & \overline{\mathbf{P}} \end{array}\right], \tag{3.281}$$

and hence by Eq. (3.276),

$$\mathbf{N} = \left[\begin{array}{c|c} \mathbf{A} & \overline{\mathbf{A}} \\ \hline \mathbf{B} & \overline{\mathbf{B}} \end{array}\right] \left[\begin{array}{c|c} \mathbf{P} & \mathbf{O} \\ \hline \mathbf{O} & \overline{\mathbf{P}} \end{array}\right] \left[\begin{array}{c|c} \mathbf{B}^\top & \mathbf{A}^\top \\ \hline \overline{\mathbf{B}}^\top & \overline{\mathbf{A}}^\top \end{array}\right], \tag{3.282}$$

which diagonalizes **N**.

The above relations have been obtained in the case of distinct eigenvectors. When some of the eigenvectors of **N** coincide, the above developments are no more valid. The relative developments and the structure of the matrices of the Stroh formalism go beyond the scope of this text; the reader is addressed to the detailed treatise of Ting (1996), for a deeper insight in the matter.

3.11 Plane States: Nomenclature

Before going on, it is worth to precise something about the nomenclature of the tensors. We have seen that there are different interesting cases, in applications, of plane problems in elasticity, such as plane strain, plane stress, generalized plane stress etc.

Nevertheless, one can imagine to be in a *plane world* with only two dimensions, and state all the equations in this hypothetic world. Of course, such a situation can represent different practical situations, like plane strain or plane stress and so on.

In other words, we can continue to work with the classical equations of elasticity in a plane situation, without necessarily specifying in which state actually we are. In such a case, we will continue to use the customary nomenclature for the Hooke's law:

$$\begin{aligned}\{\sigma\} &= [C]\{\varepsilon\},\\ \{\varepsilon\} &= [S]\{\sigma\}.\end{aligned} \tag{3.283}$$

Every time that we will state an equation in a general sense, without the need for specifying to which state it is referred to, we will use the above symbols, namely for the stiffness and compliance tensors.

Whenever the situation is that of plane strain, then we will write

$$\begin{aligned}\{\sigma\} &= [C]\{\varepsilon\},\\ \{\varepsilon\} &= [\Sigma]\{\sigma\},\end{aligned} \tag{3.284}$$

and in case of plane stress or generalized plane stress

$$\begin{aligned}\{\sigma\} &= [Q]\{\varepsilon\},\\ \{\varepsilon\} &= [S]\{\sigma\}.\end{aligned} \tag{3.285}$$

In other words, in case of plane strain and stress we will use the reduced compliance and stiffness tensors respectively, $[\Sigma]$ and $[Q]$. In all the cases, we will omit, for the sake of simplicity, the superscript p for indicating the plane case: it will be tacitly understood.

References

G.B. Airy, On the strains in the interior of beams. Britannic Association Reports, 32th meeting, pp. 82–86 (1862)

G.B. Airy, On the strains in the interior of beams. Philos. Trans. R. Soc. **153**, 49–79 (1863)

D.M. Barnett, J. Lothe, Synthesis of the sextic and the integral formalism for dislocations, Green's functions and surface waves in anisotropic elastic solids. Phys. Nor. **7**, 13–19 (1973)

J.D. Eshelby, W.T. Read, W. Shockley, Anisotropic elasticity with applications to dislocation theory. Acta Met. **1**, 251–259 (1953)

L.N.G. Filon, On the approximate solution for the bending of a beam of rectangular cross-section. Philos. Trans. R. Soc. **201**, 63–155 (1903)

A. Green, W. Zerna, *Theoretical Elasticity* (Clarendon, Oxford, 1954)

K.A. Ingebrigtsen, A. Tonning, Elastic surface waves in crystals. Phys. Rev. **184**, 942–951 (1969)

G.V. Kolosov, On the application of complex function theory to a plane problem of the mathematical theory of elasticity. Technical report, Dorpat (Yuriev) University (now Tartu University), Tartu, Estonia (1909)

S.G. Lekhnitskii, in *Theory of Elasticity of an Anisotropic Elastic Body*, English translation (1963) by P. Fern (Holden-Day, San Francisco, 1950)

A.E.H. Love, *A Treatise on the Mathematical Theory of Elasticity* (Dover, New York, 1944)

J.C. Maxwell, On reciprocal figure, frames and diagrams of forces. Trans. R. Soc. Edinburgh **26** (1870)

L.M. Milne-Thomson, *Plane Elastic Systems* (Springer, Berlin, 1960)

N.I. Muskhelishvili, *Some Basic Problems of the Mathematical Theory of Elasticity* (P. Noordhoff Ltd., The Netherlands, 1953)

P. Podio-Guidugli, A primer in elasticity. J. Elast. **58**, 1–104 (2000)

O. Rand, V. Rovenski, *Analytical Methods in Anisotropic Elasticity* (Birkhäuser, Boston, 2005)

A.N. Stroh, Dislocations and cracks in anisotropic elasticity. Philosophical Magazine **3**, 625–646 (1958)

A.N. Stroh, Steady state problems in anisotropic elasticity. J. Math. Phys. **41**, 77–103 (1962)

T.C.T. Ting, *Anisotropic Elasticity* (Oxford University Press, Oxford, 1996)

S.W. Tsai, N.J. Pagano, Invariant properties of composite materials, in *Composite Materials Workshop*, ed. by S.W. Tsai, J.C. Halpin, N.J. Pagano, Technomic, Stamford, CT (1968)

Chapter 4
The Polar Formalism

Abstract The polar formalism, a mathematical technique used to represent plane tensors by invariants and angles, is introduced in this chapter. The theory is fully developed in detail, starting from the pioneer, founding cworks of Verchery to the more recent developments. The algebra of the method is completely given and different topics are developed: the decomposition of the strain energy and the bounds on the polar invariants, a full analysis of all the possible elastic symmetries in plane elasticity, the cases of special plane anisotropic materials, the theory of polar projectors, some cases of interaction between geometry and anisotropy, plane piezoelectricity, anisotropy induced by damage, the polar invariant formulation of strength criteria for anisotropic layers. The chapter ends with different examples of plane anisotropic materials.

4.1 Introduction: Why the Polar Formalism?

In 1979 G. Verchery presented a memory about the invariants of an elasticity-type tensor (Verchery 1982). This short paper marks the birth of the *polar formalism* or *method*.

We have seen that for anisotropic materials the Cartesian components of a tensor describing a given property all depend upon the direction; moreover, this dependence is rather cumbersome, Sect. 2.2.2, also in the plane case, Eq. (3.27). Hence, when the Cartesian components are used for representing an anisotropic tensor, *none of these components are an intrinsic*[1] *quantity: all of them are frame-dependent parameters.* In addition, if a privileged direction linked to the anisotropic property exist, *it does not appear explicitly.*

Basically, the polar formalism is an *algebraic technique to represent a plane tensor using only tensor invariants and angles* (that is why the method is called *polar*). Hence, the intrinsic quantities describing a given anisotropic property and the direc-

[1]We will use the word *intrinsic* as synonymous of *invariant*. While invariant has a clear and precise mathematical meaning, a tensorial quantity whose value is preserved under frame changes, the word intrinsic has a more physical signification: it indicates a quantity that characterizes intrinsically a physical property, that belongs, in some sense, to it.

P. Vannucci, *Anisotropic Elasticity*, Lecture Notes in Applied and Computational Mechanics 85, DOI 10.1007/978-981-10-5439-6_4

tion directly and explicitly appear in the equations. It is exactly the use of *invariants and angles that makes the polar method interesting for analyzing anisotropic phenomena*: on one side, the invariants are not linked to the particular choice of the axes, so they give an intrinsic representation of the property. On the other side, the explicit use of angles makes appear directly one of the fundamental aspects of anisotropy: the direction. This is possible because, unlike other tensor representations, the polar method does not use exclusively polynomial invariants.

Moreover, we will see that the invariants used in the polar formalism are linked to the symmetries of the property, i.e. for elasticity to the elastic symmetries: actually, *they represent in an invariant way the symmetries*. In addition, the polar method allows for obtaining *much simpler formulae for the rotation of the axes* than Eq. (3.27). The entire method is based upon the use of a special complex variable transformation, that is why it can be used only for representing plane tensors.

For its characteristics, the polar formalism is well suited for design problems and for theoretical analyses. The possibility of working directly with tensor invariants gives in fact some mathematical advantages in certain transformations.

Because the polar invariants represent intrinsically the symmetries, the polar formalism opens the way to a new approach to the analysis of the material symmetries. While in a traditional approach the analysis of the symmetries is essentially *geometric*, in the polar formalism it is strictly *algebraic*.

In fact, with the traditional approach, one analyzes the effects that a geometric symmetry of the material behavior has on the Cartesian tensor components. Typically, some of them vanishes in a particular frame, the symmetry frame, i.e. the frame whose axes coincide with the equivalent directions of the given material symmetry. So, this approach gives a typical structure of the tensor *but exclusively in the symmetry frame*: the algebraic effects of this analysis are evident only in this special frame, and vanishes in a general frame, at least apparently.

In the polar formalism, the approach is quite the opposite one: a material symmetry is intrinsically detected by a special value get by one or more polar invariants, and this, of course, *regardless of the frame in which the Cartesian components are written*. The point of view is hence clearly algebraic: the symmetry is seen as an algebraic property, and more important than its geometric description, is the effect that the invariants have on the Cartesian components and the property they represent when these invariants get the values corresponding to a symmetry. This approach focuses hence on the *algebraic effects* of the symmetry; it has allowed to discover some planar elastic symmetries unknown in the past and, studying the anisotropy of complex materials, the links that exist between the tensorial symmetries and the elastic symmetries, etc.

With the polar formalism, the classification of the elastic symmetries is strictly based upon the algebraic properties of the tensor polar invariants, not upon the geometric symmetries: then, mechanical aspects assume a greater importance than the geometric ones. This point of view lets appear a fundamental fact: *to the same material symmetry, classified according to a merely geometric criterion, can belong different algebraic symmetries which have different mechanical properties*.

This chapter contains the fundamentals of the polar formalism, while some of its applications can be found in the following chapters. The subject treated in this chapter is an organic presentation of different results, that can be found principally in two publications (Vannucci 2002a, p. 65; Vannucci 2005), besides the original memory from Verchery and some other papers, recalled in the text when needed.

The polar formalism apply directly to tensor components; that is why we prefer to develop the entire theory continuing to use them in place of switching immediately to the Kelvin's notation. The final forms of the main results in the Kelvin's notation are given in Sect. 4.13.

4.2 The Transformation of Verchery

The polar formalism, as already said, is an algebraic technique based upon the use of a complex variable change. However, unlike what done in other approaches, namely in the works of Muskhelishvili (1953), Green and Zerna (1954) or Milne-Thomson (1960), Verchery introduces a different transformation. The reason is that, as we will see, this transformation allows for obtaining particularly simple matrices, namely diagonal matrices for the rotations and anti-diagonal matrices for mirror symmetries.[2] In short, the transformation of Verchery has better algebraic properties than the one usually introduced in the literature.

Just like Green and Zerna, Verchery introduces a complex variable change, interpreted as a change of frame. Let us consider a vector $\mathbf{x} = (x_1, x_2)$, and the transformation

$$\mathsf{X}^1 = \frac{1}{\sqrt{2}}\overline{k}z, \quad \mathsf{X}^2 = \overline{\mathsf{X}}^1, \quad k = e^{i\frac{\pi}{4}}, \tag{4.1}$$

giving the contravariant components of $\mathsf{X}^{cont} = (X^1, X^2)$, the transformed of $\mathbf{x}$ (the transformation is not orthogonal). Equation (4.1) is the *transformation of Verchery*; z is the complex variable

$$z = x_1 + ix_2. \tag{4.2}$$

The transformation (4.1) can be applied not only to rank-1 tensors, the vectors, but also to tensors of any rank. To this purpose, it is worth to write Eq. (4.1) in a matrix form[3]:

$$\mathsf{X}^{cont} = \mathbf{m}_1\mathbf{x}, \quad \rightarrow \quad \mathbf{m}_1 = \frac{1}{\sqrt{2}}\begin{bmatrix} \overline{k} & k \\ k & \overline{k} \end{bmatrix} = \frac{1}{2}\begin{bmatrix} 1-i & 1+i \\ 1+i & 1-i \end{bmatrix}. \tag{4.3}$$

[2]Any plane symmetry in 3D corresponds to a symmetry with respect to a straight line in 2D; for the sake of simplicity, and for recalling that we are not in 3D, we will call *mirror symmetry* any symmetry with respect to a straight line.

[3]For the sake of simplicity, we continue to use the standard notation of tensors also when these are considered as matrices, like in Sect. 3.10. As we use the tensorial components also for matrices, i.e. we do not use the Kelvin's notation, there is no risk of mistakes.

The covariant components can be easily obtained using the metric tensor $\mathbf{g}$:

$$\mathsf{X}_{cov} = \mathbf{g}_{cov}\, \mathsf{X}^{cont}, \tag{4.4}$$

whose components can be found expressing the length ds of an infinitesimal arc:

$$\begin{aligned} ds^2 &= dx_1^2 + dx_2^2 = dz d\overline{z} = 2\, \mathsf{dX}^1 \mathsf{dX}^2, \\ ds^2 &= \mathsf{dX}^{cont} \cdot \mathbf{g}_{cov} \mathsf{dX}^{cont} = g_{ij} \mathsf{dX}^i \mathsf{dX}^j, \end{aligned} \quad \rightarrow \quad \mathbf{g}_{cov} = \begin{bmatrix} 0 & 1 \\ 1 & 0 \end{bmatrix}. \tag{4.5}$$

Hence

$$\mathbf{m}_1^{-1} = \mathbf{g}_{cov}\, \mathbf{m}_1 \tag{4.6}$$

and, considering Eq. (4.1),

$$\mathsf{X}_{cov} = (\mathsf{X}_1, \mathsf{X}_2) = (\mathsf{X}^2, \mathsf{X}^1) = (\overline{\mathsf{X}}^1, \overline{\mathsf{X}}^2) \;\rightarrow\; \mathsf{X}_{cov} = \overline{\mathsf{X}}^{cont} = \mathbf{m}_1^{-1} \mathbf{x}. \tag{4.7}$$

This fact is typical of the transformation of Verchery: *all the covariant components are equal to the contravariant components that are obtained swapping indexes 1 and 2, or, equivalently, they are the complex conjugate of the corresponding contravariant components, and vice-versa.* A further result for this transformation is that

$$ds^2 = \mathsf{dX}_{cov} \cdot \mathbf{g}^{cont} \mathsf{dX}_{cov} = g^{ij} \mathsf{dX}_i \mathsf{dX}_j \;\rightarrow\; \mathbf{g}^{cont} = \mathbf{g}_{cov}^{-1} = \mathbf{g}_{cov} := \mathbf{g}. \tag{4.8}$$

Matrix $\mathbf{m}_1$ operates the transformation of rank-1 tensors, and it has some remarkable algebraic properties, that can be readily found. It is important to notice that these properties are shared by all the matrices $\mathbf{m}_j$ that operate the transformation for rank-j tensors. Such properties, easy to be checked for $\mathbf{m}_1$, are:

$$\begin{aligned} \mathbf{m}_j^\top &= \mathbf{m}_j, \\ \overline{\mathbf{m}}_j^\top &\neq \mathbf{m}_j, \qquad\qquad \forall j \geq 1. \\ \mathbf{m}_j^{-1} &= \overline{\mathbf{m}}_j^\top = \overline{\mathbf{m}}_j, \end{aligned} \tag{4.9}$$

Hence, matrices $\mathbf{m}_j$ are unitary, but not Hermitian because of Eq. $(4.9)_2$, symmetric with respect to both the diagonals and the inverse coincides with the complex conjugate.[4]

[4]To make a comparison, the transformation normally used, cf. Green and Zerna, is defined by the equations $X^1 = z$, $X^2 = \overline{z}$. Following the same procedure used here for the Verchery's transformation, it is easy to check that in this case all the listed properties are no more valid.

The matrix $\mathbf{m}_2$ for the transformation of rank-2 tensors can be computed in the following way:

$$\mathbf{m}_2=\left[\begin{array}{c|c} m_1^{11}\mathbf{m}_1 & m_1^{12}\mathbf{m}_1 \\ \hline m_1^{21}\mathbf{m}_1 & m_1^{22}\mathbf{m}_1 \end{array}\right]=\frac{1}{2}\begin{bmatrix} -i & 1 & 1 & i \\ 1 & -i & i & 1 \\ 1 & i & -i & 1 \\ i & 1 & 1 & 0 \end{bmatrix}. \tag{4.10}$$

It is not too hard to check that $\mathbf{m}_2$ has the properties (4.9). If we represent a second-rank tensor $\mathbf{L}$ as a column vector,[5] then

$$\mathsf{L}^{cont}=\mathbf{m}_2\mathbf{L}\ \rightarrow\ \begin{Bmatrix} \mathsf{L}^{11} \\ \mathsf{L}^{12} \\ \mathsf{L}^{21} \\ \mathsf{L}^{22} \end{Bmatrix}=\frac{1}{2}\begin{bmatrix} -i & 1 & 1 & i \\ 1 & -i & i & 1 \\ 1 & i & -i & 1 \\ i & 1 & 1 & -i \end{bmatrix}\begin{Bmatrix} L_{11} \\ L_{12} \\ L_{21} \\ L_{22} \end{Bmatrix}. \tag{4.11}$$

As already happened for X^{cont}, we can notice that only two complex components of L^{cont} are sufficient to define $\mathbf{L}$, because

$$\mathsf{L}^{21}=\overline{\mathsf{L}}^{12},\quad \mathsf{L}^{22}=\overline{\mathsf{L}}^{11}. \tag{4.12}$$

This is a consequence of the Verchery's transformation, valid for tensors of any rank. In addition, it is also

$$\operatorname{tr}\mathbf{L}=\mathsf{L}^{12}+\mathsf{L}^{21} \tag{4.13}$$

and, once put, see Eq. (2.47)$_2$,

$$\mathbb{G}=\mathbf{g}\boxtimes\mathbf{g}, \tag{4.14}$$

we get also

$$\mathsf{L}_{cov}=\mathbf{g}\mathsf{L}^{cont}\mathbf{g}^{\top}=\mathbb{G}\mathsf{L}^{cont},\ \rightarrow\ \mathsf{L}_{ij}=g_{im}g_{jn}\mathsf{L}^{mn}\ \rightarrow\ \mathsf{L}_{cov}=\begin{bmatrix} \mathsf{L}^{22} & \mathsf{L}^{21} \\ \mathsf{L}^{12} & \mathsf{L}^{11} \end{bmatrix}, \tag{4.15}$$

confirming what said above about the relation between covariant and contravariant components. Remembering Eqs. (4.9)$_3$ and (4.12), we then have also

$$\mathsf{L}_{cov}=\overline{\mathsf{L}}^{cont}=\begin{bmatrix} \overline{\mathsf{L}}^{11} & \overline{\mathsf{L}}^{12} \\ \overline{\mathsf{L}}^{21} & \overline{\mathsf{L}}^{22} \end{bmatrix}\ \rightarrow\ \mathsf{L}_{cov}=\mathbf{m}_2^{-1}\mathbf{L}. \tag{4.16}$$

---

[5]The order in which the components of a tensor appear in the column is not arbitrary, but obeys to the following rule: the first component is that whose indexes are all 1 and the successive components increase the indexes starting from the right: 1111, 1112, 1121, 1122, 1211, 1212, 1221, 1222 and so on.

In the case, interesting for us, of a symmetric second-rank tensor, eliminating the component 21, Eq. (4.11) becomes

$$\begin{Bmatrix} \mathsf{L}^{11} \\ \mathsf{L}^{12} \\ \mathsf{L}^{22} \end{Bmatrix} = \frac{1}{2} \begin{bmatrix} -i & 2 & i \\ 1 & 0 & 1 \\ i & 2 & -i \end{bmatrix} \begin{Bmatrix} L_{11} \\ L_{12} \\ L_{22} \end{Bmatrix}. \tag{4.17}$$

Let us consider now the case of a fourth-rank tensor $\mathbb{T}$; the transformation matrix $\mathbf{m}_4$ is computed as

$$\mathbf{m}_4 = \left[\begin{array}{c|c|c|c} m_2^{11}\mathbf{m}_2 & m_2^{12}\mathbf{m}_2 & m_2^{13}\mathbf{m}_2 & m_2^{14}\mathbf{m}_2 \\ \hline m_2^{21}\mathbf{m}_2 & m_2^{22}\mathbf{m}_2 & m_2^{23}\mathbf{m}_2 & m_2^{24}\mathbf{m}_2 \\ \hline m_2^{31}\mathbf{m}_2 & m_2^{32}\mathbf{m}_2 & m_2^{33}\mathbf{m}_2 & m_2^{34}\mathbf{m}_2 \\ \hline m_2^{41}\mathbf{m}_2 & m_2^{42}\mathbf{m}_2 & m_2^{43}\mathbf{m}_2 & m_2^{44}\mathbf{m}_2 \end{array}\right]. \tag{4.18}$$

The contravariant components of $\mathbb{T}$ can be computed as usual:

$$\mathsf{T}^{cont} = \mathbf{m}_4 \mathbb{T}, \tag{4.19}$$

and writing $\mathbb{T}$ in the form of a column vector we get, after some rather lengthy computations,

$$\begin{Bmatrix} \mathsf{T}^{1111} \\ \mathsf{T}^{1112} \\ \mathsf{T}^{1121} \\ \mathsf{T}^{1122} \\ \mathsf{T}^{1211} \\ \mathsf{T}^{1212} \\ \mathsf{T}^{1221} \\ \mathsf{T}^{1222} \\ \mathsf{T}^{2111} \\ \mathsf{T}^{2112} \\ \mathsf{T}^{2121} \\ \mathsf{T}^{2122} \\ \mathsf{T}^{2211} \\ \mathsf{T}^{2212} \\ \mathsf{T}^{2221} \\ \mathsf{T}^{2222} \end{Bmatrix} = \frac{1}{4} \begin{bmatrix}
-1 & -i & -i & 1 & -i & 1 & 1 & i & -i & 1 & 1 & i & 1 & i & i & -1 \\
-i & -1 & 1 & -i & 1 & -i & i & 1 & 1 & -i & i & 1 & i & 1 & -1 & i \\
-i & 1 & -1 & -i & 1 & i & -i & 1 & 1 & i & -i & 1 & i & -1 & 1 & i \\
1 & -i & -i & -1 & i & 1 & 1 & -i & i & 1 & 1 & -i & -1 & i & i & 1 \\
-i & 1 & 1 & i & -1 & -i & -i & 1 & 1 & i & i & -1 & -i & 1 & 1 & i \\
1 & -i & i & 1 & -i & -1 & 1 & -i & i & 1 & -1 & i & 1 & -i & i & 1 \\
1 & i & -i & 1 & -i & 1 & -1 & -i & i & -1 & 1 & i & 1 & i & -i & 1 \\
i & 1 & 1 & -i & 1 & -i & -i & -1 & -1 & i & i & 1 & i & 1 & 1 & -i \\
-i & 1 & 1 & i & 1 & i & i & -1 & -1 & -i & i & 1 & -i & 1 & 1 & i \\
1 & -i & i & 1 & i & 1 & -1 & i & -i & -1 & 1 & -i & 1 & -i & i & 1 \\
1 & i & -i & 1 & i & -1 & 1 & i & -i & 1 & -1 & -i & 1 & i & -i & 1 \\
i & 1 & 1 & -i & -1 & i & i & 1 & 1 & -i & -i & -1 & i & 1 & 1 & -i \\
1 & i & i & -1 & -i & 1 & 1 & i & -i & 1 & 1 & i & -1 & -i & -i & 1 \\
i & 1 & -1 & i & 1 & -i & i & 1 & 1 & -i & i & 1 & -i & -1 & 1 & -i \\
i & -1 & 1 & i & 1 & i & -i & 1 & 1 & i & -i & 1 & -i & 1 & -1 & -i \\
-1 & i & i & 1 & i & 1 & 1 & -i & i & 1 & 1 & -i & 1 & -i & -i & -1
\end{bmatrix} \begin{Bmatrix} T_{1111} \\ T_{1112} \\ T_{1121} \\ T_{1122} \\ T_{1211} \\ T_{1212} \\ T_{1221} \\ T_{1222} \\ T_{2111} \\ T_{2112} \\ T_{2121} \\ T_{2122} \\ T_{2211} \\ T_{2212} \\ T_{2221} \\ T_{2222} \end{Bmatrix}. \tag{4.20}$$

To check that $\mathbf{m}_4$ has the properties (4.9) is still rather straightforward, despite the size, 16×16, of the matrix.

Once more, only eight complex components T^{ijkl} are needed, because

$$
\begin{aligned}
&\mathsf{T}^{2111}=\overline{\mathsf{T}}^{1222},\mathsf{T}^{2112}=\overline{\mathsf{T}}^{1221},\mathsf{T}^{2121}=\overline{\mathsf{T}}^{1212},\mathsf{T}^{2122}=\overline{\mathsf{T}}^{1211},\\
&\mathsf{T}^{2211}=\overline{\mathsf{T}}^{1122},\mathsf{T}^{2212}=\overline{\mathsf{T}}^{1121},\mathsf{T}^{2221}=\overline{\mathsf{T}}^{1112},\mathsf{T}^{2222}=\overline{\mathsf{T}}^{1111}.
\end{aligned}
\tag{4.21}
$$

Also for the covariant components of $\mathbb{T}$ we get, see Eq. (2.47)$_3$,

$$
\begin{aligned}
&\mathsf{T}_{cov}=\mathbb{G}\mathsf{T}^{cont}\mathbb{G}^\top \;\rightarrow\; \mathsf{T}_{ijkl}=g_{im}g_{jn}g_{kp}g_{lq}\mathsf{T}^{mnpq},\\
&\mathsf{T}_{cov}=\overline{\mathsf{T}}^{cont},\;\; \mathsf{T}_{cov}=\mathbf{m}_4^{-1}\mathbb{T}\;\rightarrow\\
&\mathsf{T}_{1111}=\mathsf{T}^{2222}=\overline{\mathsf{T}}^{1111},\;\mathsf{T}_{1112}=\mathsf{T}^{2221}=\overline{\mathsf{T}}^{1112},\;\mathsf{T}_{1121}=\mathsf{T}^{2212}=\overline{\mathsf{T}}^{1121},\;\text{etc.}
\end{aligned}
\tag{4.22}
$$

We are specially interested in elasticity-type tensors, i.e. tensors whose components satisfy the minor and major symmetries, Eqs. (2.13) and (2.11). For a plane tensors, these symmetries give the following ten conditions

$$
\begin{aligned}
&T_{1112}=T_{1121}=T_{1211}=T_{2111},\quad T_{1122}=T_{2211},\\
&T_{1212}=T_{2112}=T_{2121}=T_{1221},\quad T_{1222}=T_{2122}=T_{2212}=T_{2221}.
\end{aligned}
\tag{4.23}
$$

As a consequence, there are only six independent components for a plane elasticity-type tensor. Considering the above relations, Eq. (4.20) for the case of a tensor of the elasticity type is considerably simplified:

$$
\begin{Bmatrix}\mathsf{T}^{1111}\\ \mathsf{T}^{1112}\\ \mathsf{T}^{1122}\\ \mathsf{T}^{1212}\\ \mathsf{T}^{1222}\\ \mathsf{T}^{2222}\end{Bmatrix}=\frac{1}{4}\begin{bmatrix}-1 & -4i & 2 & 4 & 4i & -1\\ -i & 2 & 0 & 0 & 2 & i\\ 1 & 0 & -2 & 4 & 0 & 1\\ 1 & 0 & 2 & 0 & 0 & 1\\ i & 2 & 0 & 0 & 2 & -i\\ -1 & 4i & 2 & 4 & -4i & -1\end{bmatrix}\begin{Bmatrix}T_{1111}\\ T_{1112}\\ T_{1122}\\ T_{1212}\\ T_{1222}\\ T_{2222}\end{Bmatrix}.
\tag{4.24}
$$

It can be noticed that only four components of T^{cont} are sufficient to know $\mathbb{T}$: T^{1111}, T^{1112}, which are complex, T^{1122} and T^{1212}, which are real.

4.3 Tensor Rotation

We consider a new frame $\{x_1', x_2'\}$, rotated through an angle θ with respect to the initial frame $\{x_1, x_2\}$, see Fig. 3.1. We pose

$$
r=e^{-i\theta},
\tag{4.25}
$$

so that in the new frame the complex variable is

$$
z'=r\,z.
\tag{4.26}
$$

If we apply the Verchery's transformation (4.1) we get the new contravariant components of $\mathbf{x}$:

$$\begin{aligned}\mathsf{X}^{1'} &= \frac{1}{\sqrt{2}}\overline{k}\,z' = \frac{1}{\sqrt{2}}\overline{k}\,r\,z = r\,\mathsf{X}^1,\\ \mathsf{X}^{2'} &= \frac{1}{\sqrt{2}}k\,\overline{z}' = \frac{1}{\sqrt{2}}k\,\overline{r}\,\overline{z} = \overline{r}\,\mathsf{X}^2,\end{aligned} \tag{4.27}$$

so that we can write

$$\mathsf{X}^{cont'} = \mathbf{R}_1\mathsf{X}^{cont} \quad\rightarrow\quad \begin{Bmatrix}\mathsf{X}^{1'}\\ \mathsf{X}^{2'}\end{Bmatrix} = \begin{bmatrix} r & 0\\ 0 & \overline{r}\end{bmatrix}\begin{Bmatrix}\mathsf{X}^1\\ \mathsf{X}^2\end{Bmatrix}. \tag{4.28}$$

The rotation matrix has a characteristic that is common to all the rotation matrices, for any tensor rank: *it is diagonal*. This is a fundamental result of the Verchery's transformation because, as we will see below, it is just this property that allows for easily find tensor invariants.

The direct transformation of the real Cartesian components can be obtained using Eqs. (4.3) and (4.28):

$$\mathbf{x}' = \mathbf{m}_1^{-1}\mathsf{X}^{cont'} = \mathbf{m}_1^{-1}\mathbf{R}_1\mathsf{X}^{cont} = \mathbf{m}_1^{-1}\mathbf{R}_1\mathbf{m}_1\mathbf{x}. \tag{4.29}$$

Developing the calculations, one obtains

$$\mathbf{x}' = \mathbf{r}_1\mathbf{x},\ \ \mathbf{r}_1 = \mathbf{m}_1^{-1}\mathbf{R}_1\mathbf{m}_1 = \begin{bmatrix} c & s\\ -s & c\end{bmatrix};\ \ c = \cos\theta,\ s = \sin\theta. \tag{4.30}$$

It can be noticed that $\mathbf{r}_1$ is the classical matrix for the rotation of tensors in $\mathbb{R}^2$.

The rotation matrix $\mathbf{R}_2$ for rank-two tensors can be constructed with the same rule used for $\mathbf{m}_2$, Eq. (4.10), for finally obtaining

$$\mathsf{L}^{cont'} = \mathbf{R}_2\mathsf{L}^{cont} \quad\rightarrow\quad \begin{Bmatrix}\mathsf{L}^{11'}\\ \mathsf{L}^{12'}\\ \mathsf{L}^{21'}\\ \mathsf{L}^{22'}\end{Bmatrix} = \begin{bmatrix} r^2 & 0 & 0 & 0\\ 0 & 1 & 0 & 0\\ 0 & 0 & 1 & 0\\ 0 & 0 & 0 & \overline{r}^2\end{bmatrix}\begin{Bmatrix}\mathsf{L}^{11}\\ \mathsf{L}^{12}\\ \mathsf{L}^{21}\\ \mathsf{L}^{22}\end{Bmatrix}. \tag{4.31}$$

For symmetric tensors, the above equation reduces to

$$\begin{Bmatrix}\mathsf{L}^{11'}\\ \mathsf{L}^{12'}\\ \mathsf{L}^{22'}\end{Bmatrix} = \begin{bmatrix} r^2 & 0 & 0\\ 0 & 1 & 0\\ 0 & 0 & \overline{r}^2\end{bmatrix}\begin{Bmatrix}\mathsf{L}^{11}\\ \mathsf{L}^{12}\\ \mathsf{L}^{22}\end{Bmatrix}. \tag{4.32}$$

Also in this case, we can find the matrix $\mathbf{r}_2$ for the rotation of the real Cartesian components:

$$\mathbf{L}' = \mathbf{m}_2^{-1}\mathsf{L}^{cont'} = \mathbf{m}_2^{-1}\mathbf{R}_2\mathsf{L}^{cont} = \mathbf{m}_2^{-1}\mathbf{R}_2\mathbf{m}_2\mathbf{L} \rightarrow$$

$$\mathbf{L}' = \mathbf{r}_2\mathbf{L}, \quad \mathbf{r}_2 = \mathbf{m}_2^{-1}\mathbf{R}_2\mathbf{m}_2 = \begin{bmatrix} c^2 & sc & sc & s^2 \\ -sc & c^2 & -s^2 & sc \\ -sc & -s^2 & c^2 & sc \\ s^2 & -sc & -sc & c^2 \end{bmatrix}, \tag{4.33}$$

which is the classical rotation matrix for rank-two tensors in the plane. If matrix $\mathbf{r}_2$ is written in the Kelvin's notation, once eliminated the term L_{21}, for taking into account for the symmetry, and rearranged, it coincides with the rotation matrix of Eq. (3.26).

For tensors of the fourth rank, the procedure is exactly the same:

$$\mathsf{T}^{cont'} = \mathbf{R}_4\mathsf{T}^{cont}, \tag{4.34}$$

which, after some lengthy calculations, reads like (we have omitted all the zero components, to better remark the diagonal character of the matrix)

$$\begin{Bmatrix} \mathsf{T}^{1111'} \\ \mathsf{T}^{1112'} \\ \mathsf{T}^{1121'} \\ \mathsf{T}^{1122'} \\ \mathsf{T}^{1211'} \\ \mathsf{T}^{1212'} \\ \mathsf{T}^{1221'} \\ \mathsf{T}^{1222'} \\ \mathsf{T}^{2111'} \\ \mathsf{T}^{2112'} \\ \mathsf{T}^{2121'} \\ \mathsf{T}^{2122'} \\ \mathsf{T}^{2211'} \\ \mathsf{T}^{2212'} \\ \mathsf{T}^{2221'} \\ \mathsf{T}^{2222'} \end{Bmatrix} = \begin{bmatrix} r^4 &&&&&&&&&&&&&&& \\ & r^2 &&&&&&&&&&&&&& \\ && r^2 &&&&&&&&&&&&& \\ &&& 1 &&&&&&&&&&&& \\ &&&& r^2 &&&&&&&&&&& \\ &&&&& 1 &&&&&&&&&& \\ &&&&&& 1 &&&&&&&&& \\ &&&&&&& \bar{r}^2 &&&&&&&& \\ &&&&&&&& r^2 &&&&&&& \\ &&&&&&&&& 1 &&&&&& \\ &&&&&&&&&& 1 &&&&& \\ &&&&&&&&&&& \bar{r}^2 &&&& \\ &&&&&&&&&&&& 1 &&& \\ &&&&&&&&&&&&& \bar{r}^2 && \\ &&&&&&&&&&&&&& \bar{r}^2 & \\ &&&&&&&&&&&&&&& \bar{r}^4 \end{bmatrix} \begin{Bmatrix} \mathsf{T}^{1111} \\ \mathsf{T}^{1112} \\ \mathsf{T}^{1121} \\ \mathsf{T}^{1122} \\ \mathsf{T}^{1211} \\ \mathsf{T}^{1212} \\ \mathsf{T}^{1221} \\ \mathsf{T}^{1222} \\ \mathsf{T}^{2111} \\ \mathsf{T}^{2112} \\ \mathsf{T}^{2121} \\ \mathsf{T}^{2122} \\ \mathsf{T}^{2211} \\ \mathsf{T}^{2212} \\ \mathsf{T}^{2221} \\ \mathsf{T}^{2222} \end{Bmatrix}, \tag{4.35}$$

which for an elasticity-type tensor[6] becomes

[6]We continue to indicate a fourth-rank tensor, even in the case of an elastic tensor, by the letter $\mathbb{T}$ to maintain a wide generality, because the polar representation is valid for *any* tensor, not only for the stiffness elasticity tensor.

$$\begin{Bmatrix} \mathsf{T}^{1111'} \\ \mathsf{T}^{1112'} \\ \mathsf{T}^{1122'} \\ \mathsf{T}^{1212'} \\ \mathsf{T}^{1222'} \\ \mathsf{T}^{2222'} \end{Bmatrix} = \begin{bmatrix} r^4 & & & & & \\ & r^2 & & & & \\ & & 1 & & & \\ & & & 1 & & \\ & & & & \overline{r}^2 & \\ & & & & & \overline{r}^4 \end{bmatrix} \begin{Bmatrix} \mathsf{T}^{1111} \\ \mathsf{T}^{1112} \\ \mathsf{T}^{1122} \\ \mathsf{T}^{1212} \\ \mathsf{T}^{1222} \\ \mathsf{T}^{2222} \end{Bmatrix}. \tag{4.36}$$

Also in this case, for the rotation of the real Cartesian components we get

$$\begin{aligned} &\mathbb{T}' = \mathbf{m}_4^{-1}\mathsf{T}^{cont'} = \mathbf{m}_4^{-1}\mathbf{R}_4\mathsf{T}^{cont} = \mathbf{m}_4^{-1}\mathbf{R}_4\mathbf{m}_4\mathbb{T} \ \rightarrow \\ &\mathbb{T}' = \mathbf{r}_4\mathbb{T}, \ \ \mathbf{r}_4 = \mathbf{m}_4^{-1}\mathbf{R}_4\mathbf{m}_4. \end{aligned} \tag{4.37}$$

We explicit the matrix $\mathbf{r}_4$ only for the case of elasticity-like tensors:

$$\mathbf{r}_4 = \begin{bmatrix} c^4 & 4sc^3 & 2s^2c^2 & 4s^2c^2 & 4s^3c & s^4 \\ -sc^3 & c^4-3s^2c^2 & sc^3-s^3c & 2(sc^3-s^3c) & 3s^2c^2-s^4 & s^3c \\ s^2c^2 & 2(s^3c-sc^3) & c^4+s^4 & -4s^2c^2 & 2(sc^3-s^3c) & s^2c^2 \\ s^2c^2 & 2(s^3c-sc^3) & -2s^2c^2 & (c^2-s^2)^2 & 2(sc^3-s^3c) & s^2c^2 \\ -s^3c & 3s^2c^2-s^4 & s^3c-sc^3 & 2(s^3c-sc^3) & c^4-3s^2c^2 & sc^3 \\ s^4 & -4s^3c & 2s^2c^2 & 4s^2c^2 & -4sc^3 & c^4 \end{bmatrix}, \tag{4.38}$$

which is the classical rotation matrix for elasticity tensors in the plane. If matrix $\mathbf{r}_4$ is written in the Kelvin's notation, it coincides with the rotation matrix of Eq. (3.27).

4.4 Tensor Invariants Under Frame Rotations

The results of the previous section allows us for obtaining rather easily the tensor invariants under frame rotations; this is particularly simple thanks to the fact that all the rotation tensors $\mathbf{R}_j$ for the contravariant complex components are diagonal, which is far to be the case for the rotation tensors $\mathbf{r}_j$ of the real Cartesian components.This fact is the major algebraic effect of the Verchery's transformation, and motivates the method and the passage to contravariant complex components.

For better understanding the procedure, let us start with the simpler case, that of vectors; looking at Eq. (4.28), one can see immediately that the only invariant quantity, i.e. *the only quantity that can be formed using the contravariant components and whose transformation to another frame does not depend upon* r, is $\mathsf{X}^1\mathsf{X}^2$. In fact,

$$\mathsf{X}^{1'}\mathsf{X}^{2'} = r\mathsf{X}^1\,\overline{r}\mathsf{X}^2 = \mathsf{X}^1\mathsf{X}^2. \tag{4.39}$$

A vector has hence only a quadratic tensor invariant; using Eq. (4.1), we get

$$\mathsf{X}^1\mathsf{X}^2 = \frac{1}{\sqrt{2}}kz\frac{1}{\sqrt{2}}k\bar{z} = \frac{x_1^2 + x_2^2}{2}, \tag{4.40}$$

which is half the square of the norm of **x**, the only invariant quantity in a vector.

The same procedure can be applied to the other tensors. For **L**, Eq. (4.31) gives two complex conjugate linear invariants, L^{12}, L^{21}, and a quadratic one, $\mathsf{L}^{11}\mathsf{L}^{22}$:

$$\begin{aligned}
\mathsf{L}^{12} &= \frac{1}{2}\left[L_{11} + L_{22} - i\left(L_{12} - L_{21}\right)\right], \\
\mathsf{L}^{21} = \overline{\mathsf{L}}^{12} &= \frac{1}{2}\left[L_{11} + L_{22} + i\left(L_{12} - L_{21}\right)\right], \\
\mathsf{L}^{11}\mathsf{L}^{22} &= \frac{1}{4}\left[\left(L_{11} - L_{22}\right)^2 + \left(L_{12} + L_{21}\right)^2\right],
\end{aligned} \tag{4.41}$$

and hence the three independent real invariants of **L** are

$$\begin{aligned}
l_1 &= \mathrm{Re}\left(\mathsf{L}^{12}\right) = \mathrm{Re}\left(\mathsf{L}^{21}\right) = \frac{1}{2}\left(L_{11} + L_{22}\right) = \frac{1}{2}\mathrm{tr}\,\mathbf{L}, \\
l_2 &= \mathrm{Im}\left(\mathsf{L}^{12}\right) = \mathrm{Im}\left(\mathsf{L}^{21}\right) = \frac{1}{2}\left(L_{12} - L_{21}\right), \\
q_1 &= \frac{1}{4}\left[\left(L_{11} - L_{22}\right)^2 + \left(L_{12} + L_{21}\right)^2\right],
\end{aligned} \tag{4.42}$$

which for a symmetric tensor become only two, a linear, l_1, and a quadratic one, q_1:

$$\begin{aligned}
l_1 &= \mathsf{L}^{12} = \mathsf{L}^{21} = \frac{1}{2}\left(L_{11} + L_{22}\right) = \frac{1}{2}\mathrm{tr}\,\mathbf{L}, \\
q_1 &= \mathsf{L}^{11}\mathsf{L}^{22} = \frac{1}{4}\left[\left(L_{11} - L_{22}\right)^2 + 4L_{12}^2\right],
\end{aligned} \tag{4.43}$$

while $l_2 = 0$ identically.

For a fourth-rank tensor $\mathbb{T}$ Eq. (4.35) gives 43 invariants on the whole, of which 6 are linear, 17 quadratics and 20 cubics. Nevertheless, they cannot be all independent. In fact, there can be at most 15 independent invariants for $\mathbb{T}$, because it has 16 components.[7] So, 28 syzygies necessarily exist among the 43 invariants.[8] To determine all the independent invariants of a fourth-rank general tensor in $\mathbb{R}^2$ is very long and at present, it is still to be done.

We are, however, interested in elasticity-type tensors. In such a case, we have only 6 independent components, which means that there must be 5 tensor independent

---

[7] In $\mathbb{R}^2$, a frame is fixed by the choice of a unique parameter: an angle measured from a direction chosen conventionally. Hence, if a tensor has n distinct components, it can have at most $n - 1$ independent invariants.

[8] A *syzygy* is a relation between two or more tensor invariants. The search for syzygies is a crucial point in determining which are the dependent invariants; unfortunately, no general method exists for finding the syzygies.

invariants for an elasticity tensor in $\mathbb{R}^2$. Scrutiny of Eq. (4.121) is much simpler and it gives the following six real invariants:

$$\begin{aligned} L_1 &= \mathsf{T}^{1122}, \\ L_2 &= \mathsf{T}^{1212}, \\ Q_1 &= \mathsf{T}^{1111}\mathsf{T}^{2222}, \\ Q_2 &= \mathsf{T}^{1112}\mathsf{T}^{1222}, \\ C_1 + iC_2 &= \mathsf{T}^{1111}\left(\mathsf{T}^{1222}\right)^2. \end{aligned} \tag{4.44}$$

L_1 and L_2 are linear, Q_1 and Q_2 quadratic and C_1 and C_2 cubic. Because there can be only five independent invariants, one syzygy must exist. This can be readily found observing that

$$\begin{aligned} C_1^2 + C_2^2 &= (C_1 + iC_2)(C_1 - iC_2) = \mathsf{T}^{1111}\left(\mathsf{T}^{1222}\right)^2 \overline{\mathsf{T}}^{1111}\left(\overline{\mathsf{T}}^{1222}\right)^2 = \\ &= \mathsf{T}^{1111}\left(\mathsf{T}^{1222}\right)^2 \mathsf{T}^{2222}\left(\mathsf{T}^{1112}\right)^2 = Q_1 Q_2^2. \end{aligned} \tag{4.45}$$

In obtaining this result, we have used $\mathsf{T}^{2111} = \mathsf{T}^{1112}$ and Eq. (4.21). The Cartesian expression of these invariants can be found using Eq. (4.24):

$$\begin{aligned} L_1 &= \frac{1}{4}\left(T_{1111} - 2T_{1122} + 4T_{1212} + T_{2222}\right), \\ L_2 &= \frac{1}{4}\left(T_{1111} + 2T_{1122} + T_{2222}\right), \\ Q_1 &= \frac{1}{16}\left(T_{1111} - 2T_{1122} - 4T_{1212} + T_{2222}\right)^2 + \left(T_{1112} - T_{1222}\right)^2, \\ Q_2 &= \frac{1}{16}\left(T_{1111} - T_{2222}\right)^2 + \frac{1}{4}\left(T_{1112} + T_{1222}\right)^2, \\ C_1 &= \frac{1}{64}\left(T_{1111} - 2T_{1122} - 4T_{1212} + T_{2222}\right)\left[\left(T_{1111} - T_{2222}\right)^2 - \right. \\ &\left. -4\left(T_{1112} + T_{1222}\right)^2\right] + \frac{1}{4}\left(T_{1112}^2 - T_{1222}^2\right)\left(T_{1111} - T_{2222}\right), \\ C_2 &= \frac{1}{16}\left(T_{1112} - T_{1222}\right)\left[\left(T_{1111} - T_{2222}\right)^2 - 4\left(T_{1112} + T_{1222}\right)^2\right] - \\ &- \frac{1}{16}\left(T_{1112} + T_{1222}\right)\left(T_{1111} - T_{2222}\right)\left(T_{1111} - 2T_{1122} - 4T_{1212} + T_{2222}\right). \end{aligned} \tag{4.46}$$

This result shows how it should be difficult to find the tensor invariants directly using the real Cartesian components, while it has been rather straightforward using the transformation of Verchery.

4.5 The Polar Components

Using the tensor invariants, we may now introduce the *polar components*. The polar components are in the same number of the independent Cartesian components, i.e. they are equal to the number of the invariants plus one: this last parameter introduces the frame orientation. Unlike the invariants, not all the polar components are polynomial functions of the Cartesian components. This happens for two reasons: on one side, the need for having the same dimensions for all the moduli. On the other side, for introducing the effect of the direction: polynomials are not fitted for well representing angles. There is not a unique way to introduce the polar components, operation that remains somewhat arbitrary: other choices could be made, always preserving the same number of invariant components. We follow here the original approach of Verchery.

4.5.1 *Second-Rank Symmetric Tensors*

The polar components of a symmetric second-rank tensor are introduced posing

$$\begin{aligned} \mathsf{L}^{11} &= Re^{2i\left(\Phi-\frac{\pi}{4}\right)}, \\ \mathsf{L}^{12} &= T. \end{aligned} \tag{4.47}$$

T and R are real quantities. They are *moduli*, in the sense that they are quantities having the same dimensions of the tensor they represent (e.g. the dimensions of a stress for tensor $\boldsymbol{\sigma}$). For what concerns T, from Eq. (4.43)$_1$ we have that

$$T = \frac{L_{11} + L_{22}}{2} = \frac{1}{2}\text{tr}\,\mathbf{L}. \tag{4.48}$$

Being the modulus of a complex quantity, R is non negative. In particular, for Eqs. (4.12)$_2$ and (4.17), it is

$$Re^{2i\Phi} = \frac{L_{11} - L_{22}}{2} + iL_{12} \tag{4.49}$$

and

$$R = \sqrt{\mathsf{L}^{11}\overline{\mathsf{L}}^{11}} = \sqrt{\mathsf{L}^{11}\mathsf{L}^{22}} = \sqrt{\left(\frac{L_{11} - L_{22}}{2}\right)^2 + L_{12}^2} \geq 0. \tag{4.50}$$

$\mathsf{L}^{11}\mathsf{L}^{22}$ is an invariant, (4.43)$_2$; as a consequence, both T and R are invariant quantities.

Φ is to be interpreted as an angle; from Eq. (4.49),

$$\tan 2\Phi = \frac{2L_{12}}{L_{11} - L_{22}}. \tag{4.51}$$

Because L_{12} and $L_{11} - L_{22}$ are not invariant quantities, Φ depends upon the value these Cartesian components take in the chosen frame: Φ is not an invariant, and it entirely determines the frame orientation. Actually, Φ measures the angle between the direction whereupon $L_{12} = 0$ and the axis x_1 of the frame in which the components L_{ij} are given.

Equations (4.48), (4.49) and (4.51) define the three polar components, T, R and Φ, of $\mathbf{L}$ as functions of its Cartesian components L_{ij}. It is easy to obtain the reverse equations, that give the L_{ij}s as functions of the polar components:

$$\begin{aligned} L_{11} &= T + R\cos 2\Phi, \\ L_{12} &= R\sin 2\Phi, \\ L_{22} &= T - R\cos 2\Phi. \end{aligned} \tag{4.52}$$

T represents the spherical part of $\mathbf{L}$ and R the deviatoric one, in the sense that

$$\begin{aligned} \mathbf{L}_{sph} &= T\mathbf{I} \quad \rightarrow \quad \|\mathbf{L}_{sph}\| = \sqrt{2}T, \\ \mathbf{L}_{dev} &= \mathbf{L} - \mathbf{L}_{sph} \quad \rightarrow \quad \|\mathbf{L}_{dev}\| = \sqrt{2}R. \end{aligned} \tag{4.53}$$

4.5.2 Elasticity Tensors

We have seen that for the description of an elasticity tensor, two complex and two real contravariant components are sufficient. Then, the polar components of a fourth-rank elasticity-type tensor are introduced putting:

$$\begin{aligned} \mathsf{T}^{1111} &= 2R_0 e^{4i(\Phi_0 - \frac{\pi}{4})}, \\ \mathsf{T}^{1112} &= 2R_1 e^{2i(\Phi_1 - \frac{\pi}{4})}, \\ \mathsf{T}^{1122} &= 2T_0, \\ \mathsf{T}^{1212} &= 2T_1. \end{aligned} \tag{4.54}$$

The six parameters $T_0, T_1, R_0, R_1, \Phi_0$ and Φ_1 are the *polar components of* $\mathbb{T}$. In particular, T_0, T_1, R_0 and R_1 are *polar moduli*, i.e. they have the same dimensions of the tensor they represent: the dimensions of a stress, if $\mathbb{T}$ is a stiffness tensor, the dimensions of the reciprocal of a stress, if $\mathbb{T}$ is a compliance tensor and so on. Moreover,

$$R_0 \geq 0, \quad R_1 \geq 0, \tag{4.55}$$

because they are proportional to the modulus of a complex quantity. Φ_0 and Φ_1 are to be interpreted as *polar angles*; we see hence that *the polar formalism gives a representation of elasticity using exclusively moduli and angles*. In this sense, it is quite different from the classical Cartesian representation, where only moduli are used, and from the representation by technical constants, which makes use of moduli and coefficients.

Using Eqs. (4.21) and (4.44), it is simple to show that

$$\begin{aligned}
&L_1 = 2T_0,\\
&L_2 = 2T_1,\\
&Q_1 = 4R_0^2,\\
&Q_2 = 4R_1^2,\\
&C_1 + iC_2 = 8R_0R_1^2 e^{4i(\Phi_0-\Phi_1)} \Rightarrow \begin{aligned} C_1 &= 8R_0R_1^2\cos 4(\Phi_0-\Phi_1),\\ C_2 &= 8R_0R_1^2\sin 4(\Phi_0-\Phi_1).\end{aligned}
\end{aligned} \tag{4.56}$$

This result shows that T_0, T_1, R_0, R_1 and $\Phi_0 - \Phi_1$ *are tensor invariants*. They constitute a complete set of independent invariants for $\mathbb{T}$. In particular, T_0 and T_1 are linear invariants, R_0 and R_1 are functions of quadratic invariants and $\Phi_0 - \Phi_1$ is a function of a cubic invariant, that is hence represented by a difference of angles.

Injecting Eq. (4.54) into (4.24) we obtain the Cartesian expression of the polar components:

$$\begin{aligned}
8T_0 &= T_{1111} - 2T_{1122} + 4T_{1212} + T_{2222},\\
8T_1 &= T_{1111} + 2T_{1122} + T_{2222},\\
8R_0e^{4i\Phi_0} &= T_{1111} - 2T_{1122} - 4T_{1212} + T_{2222} + 4i(T_{1112} - T_{1222}),\\
8R_1e^{2i\Phi_1} &= T_{1111} - T_{2222} + 2i\left(T_{1112} + T_{1222}\right),
\end{aligned} \tag{4.57}$$

or, more explicitly,

$$\begin{aligned}
T_0 &= \frac{1}{8}(T_{1111} - 2T_{1122} + 4T_{1212} + T_{2222}),\\
T_1 &= \frac{1}{8}(T_{1111} + 2T_{1122} + T_{2222}),\\
R_0 &= \frac{1}{8}\sqrt{(T_{1111} - 2T_{1122} - 4T_{1212} + T_{2222})^2 + 16(T_{1112} - T_{1222})^2},\\
R_1 &= \frac{1}{8}\sqrt{(T_{1111} - T_{2222})^2 + 4(T_{1112} + T_{1222})^2},\\
\tan 4\Phi_0 &= \frac{4(T_{1112} - T_{1222})}{T_{1111} - 2T_{1122} - 4T_{1212} + T_{2222}},\\
\tan 2\Phi_1 &= \frac{2\left(T_{1112} + T_{1222}\right)}{T_{1111} - T_{2222}}.
\end{aligned} \tag{4.58}$$

It is apparent that the polar angles Φ_0 and Φ_1 are functions of the Cartesian components of $\mathbb{T}$. They are, by consequence, frame dependent, though their difference is an invariant. Hence, the value of one of them depends upon the other one: only one of the two polar angles if free, and its choice corresponds to fix a frame. The choice usually done is to put

$$\Phi_1 = 0, \tag{4.59}$$

which corresponds to have the highest value of the component T_{1111} in correspondence of the axis of x_1.

The reverse relations of Eq. (4.57) can be obtained injecting Eq. (4.54) into the reverse of Eq. (4.24), easy to be obtained because of Eq. (4.9)$_3$: the inverse of matrix $\mathbf{m}_4$ is obtained simply changing the sign of the imaginary unit i. We get

$$\begin{aligned}
T_{1111}&=T_0+2T_1+R_0\cos 4\Phi_0+4R_1\cos 2\Phi_1,\\
T_{1112}&=R_0\sin 4\Phi_0+2R_1\sin 2\Phi_1,\\
T_{1122}&=-T_0+2T_1-R_0\cos 4\Phi_0,\\
T_{1212}&=T_0-R_0\cos 4\Phi_0,\\
T_{1222}&=-R_0\sin 4\Phi_0+2R_1\sin 2\Phi_1,\\
T_{2222}&=T_0+2T_1+R_0\cos 4\Phi_0-4R_1\cos 2\Phi_1.
\end{aligned} \tag{4.60}$$

4.6 Change of Frame

Let us consider now a change of frame from the original one $\{x_1, x_2\}$ to a frame $\{x'_1, x'_2\}$ rotated counterclockwise through an angle θ, like in Fig. 3.1. Then, Eq. (4.32) gives

$$\mathsf{L}^{11'} = r^2\mathsf{L}^{11} = Re^{2i(\Phi-\theta-\frac{\pi}{4})}, \tag{4.61}$$

while L^{12} does not change because it is an invariant. So, inserting this result in Eq. (4.17) and using again Eq. (4.12)$_2$ we obtain, following the same procedure used in Sect. 4.5.1,

$$Re^{2i(\Phi-\theta)} = \frac{L_{11}(\theta)-L_{22}(\theta)}{2} + iL_{12}(\theta), \tag{4.62}$$

and for the reverse equations

$$\begin{aligned}
L_{11}(\theta) &= T + R\cos 2(\Phi-\theta),\\
L_{12}(\theta) &= R\sin 2(\Phi-\theta),\\
L_{22}(\theta) &= T - R\cos 2(\Phi-\theta).
\end{aligned} \tag{4.63}$$

The above equations are an analytical representation of the *Mohr's circle*: T is the abscissa of the centre of the circle, while R is the radius and Φ the orientation of the principal components, i.e. the orientation of the frame. In fact, the equation of the Mohr's circle for $\mathbf{L}$ is

$$\left(\xi-\frac{L_{11}+L_{22}}{2}\right)^2+\eta^2=\left(\frac{L_{11}-L_{22}}{2}\right)^2+L_{12}^2, \tag{4.64}$$

then any point for which $(\xi,\eta)=(L_{11}(\theta),L_{12}(\theta))$ or $(\xi,\eta)=(L_{22}(\theta),L_{12}(\theta))$ belong to the circle.

For what concerns an elasticity tensor, we follow the same procedure: using Eqs. (4.121) and (4.54) we get

$$\begin{aligned}\mathsf{T}^{1111'}&=r^4\mathsf{T}^{1111}=2r^4R_0e^{4i(\Phi_0-\theta-\frac{\pi}{4})},\\ \mathsf{T}^{1112'}&=r^2\mathsf{T}^{1112}=2r^2R_1e^{2i(\Phi_1-\theta-\frac{\pi}{4})},\end{aligned} \tag{4.65}$$

that used in Eq. (4.24) gives

$$\begin{aligned}8T_0&=T_{1111}(\theta)-2T_{1122}(\theta)+4T_{1212}(\theta)+T_{2222}(\theta),\\ 8T_1&=T_{1111}(\theta)+2T_{1122}(\theta)+T_{2222}(\theta),\\ 8R_0e^{4i(\Phi_0-\theta)}&=T_{1111}(\theta)-2T_{1122}(\theta)-4T_{1212}(\theta)+T_{2222}(\theta)+\\ &\quad+4i\left[T_{1112}(\theta)-T_{1222}(\theta)\right],\\ 8R_1e^{2i(\Phi_1-\theta)}&=T_{1111}(\theta)-T_{2222}(\theta)+2i\left[T_{1112}(\theta)+T_{1222}(\theta)\right],\end{aligned} \tag{4.66}$$

and for the reverse equations

$$\begin{aligned}T_{1111}(\theta)&=T_0+2T_1+R_0\cos4\left(\Phi_0-\theta\right)+4R_1\cos2\left(\Phi_1-\theta\right),\\ T_{1112}(\theta)&=R_0\sin4\left(\Phi_0-\theta\right)+2R_1\sin2\left(\Phi_1-\theta\right),\\ T_{1122}(\theta)&=-T_0+2T_1-R_0\cos4\left(\Phi_0-\theta\right),\\ T_{1212}(\theta)&=T_0-R_0\cos4\left(\Phi_0-\theta\right),\\ T_{1222}(\theta)&=-R_0\sin4\left(\Phi_0-\theta\right)+2R_1\sin2\left(\Phi_1-\theta\right),\\ T_{2222}(\theta)&=T_0+2T_1+R_0\cos4\left(\Phi_0-\theta\right)-4R_1\cos2\left(\Phi_1-\theta\right).\end{aligned} \tag{4.67}$$

It is possible to give a graphical construction corresponding to Eq. (4.67), presented in Fig. 4.1. This construction is called *generalized Mohr's circles*.

Equations (4.63) and (4.67), when compared with matrices (4.33) and (4.38), show one of the greatest advantages of the polar formalism: *the Cartesian components in the new frame are obtained simply subtracting the angle θ from the polar angles.* The operation of the change of frame is hence particularly simple when the Cartesian components are given as functions of the polar parameters.

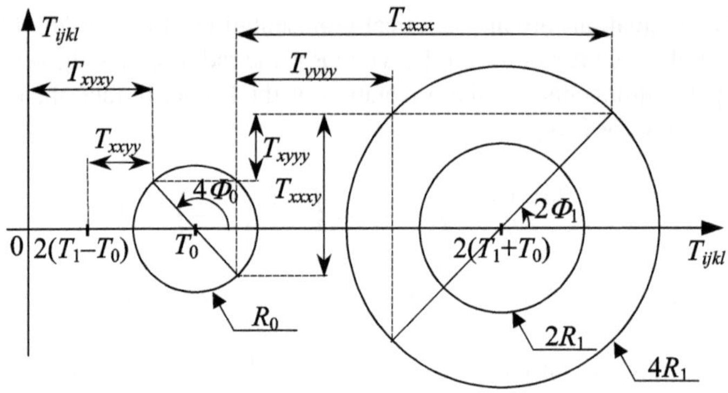

Fig. 4.1 Generalized Mohr's circles

4.7 Harmonic Interpretation of the Polar Formalism

The results obtained in the previous section deserve some commentary. Let us first consider the case of an elastic tensor, Eq. (4.67). Each Cartesian component is the sum of different terms, and in the most general case, that of $T_{1111}(\theta)$ and $T_{2222}(\theta)$, we have:

- an invariant term, $T_0 + 2T_1$, which represents the mean value of the components and its isotropic part; T_0 and T_1 are hence the *isotropic polar invariants*;
- a term which is a circular function of 2θ whose amplitude is proportional to R_1;
- a term which is a circular function of 4θ whose amplitude is proportional to R_0;
- these two terms are shifted of an angle $\Phi_0 - \Phi_1$; this term is an invariant, so we have its physical meaning: it represents the angular shifting of the two anisotropic phases;
- R_0, R_1 and $\Phi_0 - \Phi_1$ are hence the *anisotropic polar invariants*;
- R_0 and R_1 represent, to within a factor, the amplitude of the anisotropic phases, that are directional fluctuations around the isotropic average.

We have hence a new interpretation of anisotropic elasticity in $\mathbb{R}^2$: the anisotropic elastic behavior can be regarded as a finite sum of harmonics: a constant term, the isotropic phase, and two anisotropic phases, one varying with 2θ, the other one with 4θ. The amplitude of all of these phases and the phase offset of the anisotropic phases are tensor invariants, i.e. they are *intrinsic properties of the material.*

The above considerations give the physical meaning of the polar invariants. We will see in the next sections that the elastic symmetries are determined by some special values taken by them and the role played by these invariants in the strain energy decomposition.

4.8 Polar Parameters of the Inverse Tensor

Equation (4.67) is valid for any elastic tensor, hence also for $\mathbb{T}^{-1}$, that we will indicate, for the sake of generality, by $\mathbb{S}$. We denote the polar components of $\mathbb{S}$ by lower-case letters: t_0, t_1, r_0, r_1 and $\varphi_0 - \varphi_1$. These can be found expressing the Cartesian components of $\mathbb{S}$ as functions of those of $\mathbb{T}$, and these last by their polar components, Eq. (4.67). Comparing the result so found with Eq. (4.67) written for $\mathbb{S}$, gives $t_0, t_1, r_0, r_1, \varphi_0$ and φ_1. The calculations are rather heavy and only the final result is given here:

$$\begin{aligned} t_0 &= \frac{2}{\Delta}\left(T_0 T_1 - R_1^2\right), \\ t_1 &= \frac{1}{2\Delta}\left(T_0^2 - R_0^2\right), \\ r_0 e^{4i\varphi_0} &= \frac{2}{\Delta}\left(R_1^2 e^{4i\Phi_1} - T_1 R_0 e^{4i\Phi_0}\right), \\ r_1 e^{2i\varphi_1} &= \frac{R_1 e^{2i\Phi_1}}{\Delta}\left[R_0 e^{4i(\Phi_0-\Phi_1)} - T_0\right]. \end{aligned} \tag{4.68}$$

From the above equations, we obtain also

$$\begin{aligned} r_0 =& \frac{2}{\Delta}\sqrt{\left(R_1^2\cos 4\Phi_1 - T_1 R_0 \cos 4\Phi_0\right)^2 + \left(R_1^2 \sin 4\Phi_1 - T_1 R_0 \sin 4\Phi_0\right)^2}, \\ r_1 =& \frac{R_1}{\Delta}\{[T_0\cos 2\Phi_1 - R_0\cos(4(\Phi_0-\Phi_1)+2\Phi_1)]^2 + \\ & [T_0\sin 2\Phi_1 - R_0\sin(4(\Phi_0-\Phi_1)+2\Phi_1)]^2\}^{\frac{1}{2}}, \end{aligned} \tag{4.69}$$

and

$$\begin{aligned} \tan 4\varphi_0 &= \frac{R_1^2\sin 4\Phi_1 - T_1 R_0 \sin 4\Phi_0}{R_1^2\cos 4\Phi_1 - T_1 R_0\cos 4\Phi_0}, \\ \tan 2\varphi_1 &= \frac{R_0\sin[4(\Phi_0-\Phi_1)+2\Phi_1] - T_0\sin 2\Phi_1}{R_0\cos[4(\Phi_0-\Phi_1)+2\Phi_1] - T_0\cos 2\Phi_1}. \end{aligned} \tag{4.70}$$

Δ is an invariant quantity, defined by

$$\begin{aligned} \Delta &= 8T_1\left(T_0^2 - R_0^2\right) - 16R_1^2\left[T_0 - R_0\cos 4\left(\Phi_0 - \Phi_1\right)\right] = \\ &= \det\begin{bmatrix} T_{1111} & T_{1122} & T_{1112} \\ & T_{2222} & T_{1222} \\ sym & & T_{1212}\end{bmatrix}. \end{aligned} \tag{4.71}$$

It will be proved in Sect. 4.11 that Δ is a positive quantity.

Equations (4.68) being symmetric, i.e. we can switch $\mathbb{T}$ and $\mathbb{S}$, we notice that

$$R_1 = 0 \Leftrightarrow r_1 = 0, \qquad R_0 = 0 \not\Leftrightarrow r_0 = 0. \tag{4.72}$$

Equation (4.72) has a considerable importance in the determination of all the elastic symmetries, analyzed in Sect. 4.12.

4.9 Technical Constants and Polar Invariants

We can now express the technical constants as functions of the polar invariants. To this purpose, we express first the components of the compliance tensor $\mathbb{S}$ as functions of the compliance polar invariants:

$$\begin{aligned}
S_{1111}(\theta)&=t_0+2t_1+r_0\cos 4(\varphi_0-\theta)+4r_1\cos 2(\varphi_1-\theta),\\
S_{1112}(\theta)&=r_0\sin 4(\varphi_0-\theta)+2r_1\sin 2(\varphi_1-\theta),\\
S_{1122}(\theta)&=-t_0+2t_1-r_0\cos 4(\varphi_0-\theta),\\
S_{1212}(\theta)&=t_0-r_0\cos 4(\varphi_0-\theta),\\
S_{1222}(\theta)&=-r_0\sin 4(\varphi_0-\theta)+2r_1\sin 2(\varphi_1-\theta),\\
S_{2222}(\theta)&=t_0+2t_1+r_0\cos 4(\varphi_0-\theta)-4r_1\cos 2(\varphi_1-\theta).
\end{aligned} \tag{4.73}$$

Then, we use the definitions of the technical constants, Sect. 2.3, where we inject the above expressions for the S_{ijkl}. We get hence:

- Young's moduli:

$$\begin{aligned}
E_1(\theta)&=\frac{1}{S_{1111}(\theta)}=\frac{1}{t_0+2t_1+r_0\cos 4(\varphi_0-\theta)+4r_1\cos 2(\varphi_1-\theta)};\\
E_2(\theta)&=\frac{1}{S_{2222}(\theta)}=\frac{1}{t_0+2t_1+r_0\cos 4(\varphi_0-\theta)-4r_1\cos 2(\varphi_1-\theta)};
\end{aligned} \tag{4.74}$$

- shear modulus:

$$G_{12}(\theta)=\frac{1}{4S_{1212}(\theta)}=\frac{1}{4[t_0-r_0\cos 4(\varphi_0-\theta)]}; \tag{4.75}$$

- Poisson's coefficient:

$$\nu_{12}(\theta)=-\frac{S_{1122}(\theta)}{S_{1111}(\theta)}=\frac{t_0-2t_1+r_0\cos 4(\varphi_0-\theta)}{t_0+2t_1+r_0\cos 4(\varphi_0-\theta)+4r_1\cos 2(\varphi_1-\theta)}; \tag{4.76}$$

- coefficients of mutual influence of the first type:

$$\begin{aligned}\eta_{1,12}(\theta) &= \frac{S_{1112}(\theta)}{2S_{1212}(\theta)} = \frac{r_0 \sin 4\,(\varphi_0-\theta) + 2r_1 \sin 2\,(\varphi_1-\theta)}{2\,[t_0 - r_0 \cos 4\,(\varphi_0-\theta)]},\\ \eta_{2,12}(\theta) &= \frac{S_{1222}(\theta)}{2S_{1212}(\theta)} = \frac{-r_0 \sin 4\,(\varphi_0-\theta) + 2r_1 \sin 2\,(\varphi_1-\theta)}{2\,[t_0 - r_0 \cos 4\,(\varphi_0-\theta)]};\end{aligned} \tag{4.77}$$

- coefficients of mutual influence of the second type:

$$\begin{aligned}\eta_{12,1}(\theta) &= 2\frac{S_{1112}(\theta)}{S_{1111}(\theta)} = 2\frac{r_0 \sin 4\,(\varphi_0-\theta) + 2r_1 \sin 2\,(\varphi_1-\theta)}{t_0 + 2t_1 + r_0 \cos 4\,(\varphi_0-\theta) + 4r_1 \cos 2\,(\varphi_1-\theta)},\\ \eta_{12,2}(\theta) &= 2\frac{S_{1222}(\theta)}{2S_{2222}(\theta)} = 2\frac{-r_0 \sin 4\,(\varphi_0-\theta) + 2r_1 \sin 2\,(\varphi_1-\theta)}{t_0 + 2t_1 + r_0 \cos 4\,(\varphi_0-\theta) - 4r_1 \cos 2\,(\varphi_1-\theta)}.\end{aligned} \tag{4.78}$$

Using Eq. (4.68) it is also possible to express the technical constants as functions of the stiffness polar invariants; in the most general case, this leads to very long expressions, that we omit here. Nevertheless, it is interesting to consider the case of isotropic materials; for such a situation, Eq. (4.68) reduce to

$$t_0 = \frac{1}{4T_0}, \quad t_1 = \frac{1}{16T_1}, \quad r_0 = 0, \quad r_1 = 0, \tag{4.79}$$

so we get

- Young's modulus:

$$E = \frac{1}{t_0 + 2t_1} = \frac{8T_0T_1}{T_0 + 2T_1}; \tag{4.80}$$

- Shear modulus:

$$G = \frac{1}{4t_0} = T_0; \tag{4.81}$$

- Poisson's coefficient:

$$\nu = \frac{t_0 - 2t_1}{t_0 + 2t_1} = \frac{2T_1 - T_0}{2T_1 + T_0}. \tag{4.82}$$

The remaining coefficients are of course null for isotropic materials.

Another modulus is usually introduced for isotropic materials: the *bulk modulus* κ:

$$\forall\, \boldsymbol{\sigma} = p\mathbf{I}, \quad \kappa := \frac{p}{\operatorname{tr}\boldsymbol{\varepsilon}}. \tag{4.83}$$

Applying this definition to the plane anisotropic case[9] gives

$$\kappa = \frac{1}{S_{1111}(\theta) + 2S_{1122}(\theta) + S_{2222}(\theta)} = \frac{1}{8t_1}, \tag{4.84}$$

which, for a material at least square symmetric ($R_1 = r_1 = 0$), gives also

$$\kappa = 2T_1. \tag{4.85}$$

We have hence a physical meaning for the polar invariants of isotropy: t_0 and T_0 are linked to the shear modulus, while t_1 and T_1 to the bulk one. We will see in Sect. 4.10 that the existence of these two different parts of the isotropic phase corresponds to the physical fact that for classical elastic materials the whole of the strain energy can be split, under some conditions, into two different parts, a spherical and a deviatoric one, the first linked to volume changes, and ruled by the bulk modulus, i.e. by T_1, the second related to shape changes and ruled by the shear modulus, i.e. by T_0 (for the isotropic case).

The relations between the Lamé's constants and the polar invariants can also be given:

$$\kappa = \lambda + \mu, \; G = \mu \;\Rightarrow\; \lambda = 2T_1 - T_0, \; \mu = T_0. \tag{4.86}$$

Finally, looking at Eqs. (4.81), (4.84) and (4.85), we see that the classical constants G and κ corresponds to the polar invariants t_0 and t_1, that generalizes them also to the anisotropic case (i.e. t_0 and t_1, or equivalently T_0 and T_1 reduce to G and κ when the material becomes isotropic).

4.10 Polar Decomposition of the Strain Energy

Let us consider a layer subjected to some stresses $\boldsymbol{\sigma}$, whose polar components are T, R and Φ, that produce the strain $\boldsymbol{\varepsilon}$, described by its polar components t, r and φ. Then, Eq. (4.63), the strain energy V is

$$V = \frac{1}{2}\boldsymbol{\sigma} \cdot \boldsymbol{\varepsilon} = T\,t + R\,r\cos 2(\Phi - \varphi). \tag{4.87}$$

[9] The definition of κ is the same for isotropic or anisotropic materials, and Eq. (4.83) is valid in any case; nevertheless, the mechanical idea in the definition of κ is to measure the mechanical hydrostatic pressure to be done on the material to obtain a unitary change of volume, $\mathrm{tr}\boldsymbol{\varepsilon} = 1$; it is somewhat understood, in doing this, that all the mechanical stress produce uniquely a change of volume, not of shape, i.e. that the deformation itself is spherical. This fact is always true not only for isotropic materials, but also for a square symmetric material, Sect. 2.5, hence when $r_1 = 0$; in fact, in such a case $S_{1111}(\theta) = S_{2222}(\theta)$ and $S_{1112}(\theta) + S_{1222}(\theta) = 0\ \forall\theta$. That is why, though Eq. (4.83) has a general validity, the same notion of κ is usually restricted to the use with isotropic materials.

For what said in Sect. 4.5.1, it is[10]

$$\begin{aligned} V_s &:= \frac{1}{2}\boldsymbol{\varepsilon}_{sph}\cdot\boldsymbol{\sigma}_{sph} = T\,t,\\ V_d &:= \frac{1}{2}\boldsymbol{\varepsilon}_{dev}\cdot\boldsymbol{\sigma}_{dev} = R\,r\cos 2(\Phi-\varphi). \end{aligned} \tag{4.88}$$

We introduce now the material behavior, injecting Eq. (4.67) into the Hooke's law (T_0, T_1, R_0, R_1 and $\Phi_0 - \Phi_1$ are intended here to be the polar components of $\mathbb{E}$). After some standard manipulations, we get

$$\begin{aligned} V = \frac{1}{2}\boldsymbol{\varepsilon}\cdot\mathbb{E}\boldsymbol{\varepsilon} &= 4T_1t^2 + 8R_1\cos 2(\Phi_1-\varphi)r\,t+\\ &+2\left[T_0 + R_0\cos 4(\Phi_0-\varphi)\right]r^2. \end{aligned} \tag{4.89}$$

The change δV caused by a variation $\delta\boldsymbol{\varepsilon}$ of the stain state is, Eq. (2.2),

$$\delta V = \boldsymbol{\sigma}\cdot\delta\boldsymbol{\varepsilon} = 2T\delta t + 2R\cos 2(\Phi-\varphi)\delta r + 4R\,r\sin 2(\Phi-\varphi)\delta\varphi, \tag{4.90}$$

and hence the spherical and deviatoric parts of $\boldsymbol{\sigma}$ are

$$\begin{aligned} T &= \frac{1}{2}\frac{\partial V}{\partial t},\\ Re^{2i\Phi} &= \frac{1}{2}\left(\frac{\partial V}{\partial r} + \frac{i}{2r}\frac{\partial V}{\partial\varphi}\right)e^{2i\varphi}. \end{aligned} \tag{4.91}$$

Injecting Eq. (4.90) in (4.91) gives

$$\begin{aligned} T &= 4T_1t + 4R_1r\cos 2(\Phi_1-\varphi),\\ Re^{2i\Phi} &= 2T_0re^{2i\varphi} + 2R_0re^{2i(2\Phi_0-\varphi)} + 4R_1te^{2i\Phi_1}. \end{aligned} \tag{4.92}$$

The above relations show a fact already discussed in Sect. 2.5: for an anisotropic material, also in $\mathbb{R}^2$, in the most general case the spherical and deviatoric parts of $\boldsymbol{\sigma}$ depend on both the spherical and deviatoric parts of $\boldsymbol{\varepsilon}$.

Using these relations in the expressions of V_s and V_d gives

$$\begin{aligned} V_s &= 4T_1t^2 + 4R_1r\,t\cos 2(\Phi_1-\varphi),\\ V_d &= 2r^2\left[T_0 + R_0\cos 4(\Phi_0-\varphi)\right] + 4R_1r\,t\cos 2(\Phi_1-\varphi). \end{aligned} \tag{4.93}$$

We can then observe the role played by the different polar invariants of $\mathbb{E}$ in the decomposition of the strain energy: T_1 affects only V_s, T_0 and R_0 only V_d while R_1

[10] The reader should remark that $V_s \neq V_{sph}$ and $V_d \neq V_{dev}$, cf. Sect. 2.5.

couples V_s with V_d. For materials with $R_1 = 0$, this case is analyzed in Sect. 4.12.4, the two parts are uncoupled. It is then clear, and simple to be checked, that for materials with $R_1 = 0$, Eq. (2.151) is satisfied and consequently that

$$\begin{aligned} V_s = V_{sph} &\rightarrow \frac{1}{2}\boldsymbol{\varepsilon}_{sph}\cdot\boldsymbol{\sigma}_{sph} = \frac{1}{2}\boldsymbol{\varepsilon}_{sph}\cdot\mathbb{E}\boldsymbol{\varepsilon}_{sph},\\ V_d = V_{dev} &\rightarrow \frac{1}{2}\boldsymbol{\varepsilon}_{dev}\cdot\boldsymbol{\sigma}_{dev} = \frac{1}{2}\boldsymbol{\varepsilon}_{dev}\cdot\mathbb{E}\boldsymbol{\varepsilon}_{dev}, \end{aligned} \tag{4.94}$$

which implies

$$V = V_{sph} + V_{dev} = V_s + V_d. \tag{4.95}$$

Finally, *the minimal requirement, in* $\mathbb{R}^2$, *for decomposing the strain energy in a spherical and deviatoric part is that* $R_1 = 0$.

4.11 Bounds on the Polar Invariants

The positiveness of the strain energy V gives the bounds on the components of $\mathbb{E}$, so also on its polar invariants. V is a quadratic form of r and t, Eq. (4.89), that can be written as

$$V = \{r, t\}\cdot\begin{bmatrix} 2\left[T_0 + R_0\cos 4(\Phi_0 - \varphi)\right] & 4R_1\cos 2(\Phi_1 - \varphi)\\ 4R_1\cos 2(\Phi_1 - \varphi) & 4T_1 \end{bmatrix}\begin{Bmatrix} r\\ t \end{Bmatrix}. \tag{4.96}$$

$V > 0\ \forall\{r, t\}$ if and only if the matrix in the previous equation is positive definite. This happens if and only if, Theorem 2,

$$\left.\begin{aligned} &T_0 + R_0\cos 4(\Phi_0 - \varphi) > 0,\\ &T_1\left[T_0 + R_0\cos 4(\Phi_0 - \varphi)\right] > 2R_1^2\cos^2 2(\Phi_1 - \varphi), \end{aligned}\right. \quad \forall\varphi. \tag{4.97}$$

To be noticed that, because the term at the second member of Eq. $(4.97)_2$ is a square, hence not negative, Eq. $(4.97)_1$ necessarily implies

$$T_1 > 0. \tag{4.98}$$

Conditions (4.97) can be transformed to obtain relations for the only polar invariants. This can be done in the following way (Vannucci and Desmorat 2015): first, we transform Eq. $(4.97)_2$ introducing the angle

$$\alpha = \Phi_1 - \varphi \tag{4.99}$$

which implies

$$\Phi_0 - \varphi = \Delta\Phi + \alpha, \tag{4.100}$$

where

$$\Delta\Phi = \Phi_0 - \Phi_1. \tag{4.101}$$

Equation $(4.97)_2$ becomes hence

$$T_1\left[T_0 + R_0\cos 4(\Delta\Phi + \alpha)\right] > 2R_1^2\cos^2 2\alpha \ \ \forall\alpha, \tag{4.102}$$

that can be transformed, using standard trigonometric identities, first to

$$T_0T_1 - R_1^2 + \left[\left(T_1R_0\cos 4\Delta\Phi - R_1^2\right)\cos 4\alpha - T_1R_0\sin 4\Delta\Phi\sin 4\alpha\right] > 0 \ \ \forall\alpha, \tag{4.103}$$

then to

$$T_0T_1 - R_1^2 > \sqrt{(T_1R_0\cos 4\Delta\Phi - R_1^2)^2 + T_1^2R_0^2\sin^2 4\Delta\Phi}\cos 4(\alpha - \varpi) \ \ \forall\alpha, \tag{4.104}$$

where

$$\varpi = \frac{1}{4}\arctan\frac{T_1R_0\sin 4\Delta\Phi}{R_1^2 - T_1R_0\cos 4\Delta\Phi}, \tag{4.105}$$

a function of only invariants of $\mathbb{E}$. Being the quantity under the sign of square root in (4.104) strictly positive, conditions $(4.97)_1$ and (4.104) to be true $\forall\varphi$ resume, with some simple manipulations, to

$$\begin{aligned} &T_0 - R_0 > 0,\\ &T_0T_1 - R_1^2 > 0,\\ &T_1(T_0^2 - R_0^2) - 2R_1^2\left[T_0 - R_0\cos 4\Delta\Phi\right] > 0. \end{aligned} \tag{4.106}$$

Condition $(4.106)_2$ is less restrictive than condition $(4.106)_3$, and can be discarded. To show this, let us transform Eq. (4.106) to a dimensionless form upon introduction of the ratios

$$\xi = \frac{T_0T_1}{R_1^2}, \ \eta = \frac{R_0}{T_0}. \tag{4.107}$$

To remark that by Eqs. (4.55), (4.98) and $(4.106)_1$, ξ and η cannot be negative quantities. Introducing Eq. (4.107) into (4.106) gives

$$\eta < 1, \quad \xi > 1, \quad \xi > 2\frac{1 - \eta\cos 4\Delta\Phi}{1 - \eta^2}. \tag{4.108}$$

Then, condition $(4.108)_3$ is more restrictive than condition $(4.108)_2$ if

$$2\frac{1 - \eta\cos 4\Delta\Phi}{1 - \eta^2} \geq 1, \tag{4.109}$$

thanks to (4.108_1) equivalent to

$$\eta^2 - 2\eta\cos 4\Delta\Phi + 1 \geq 0, \tag{4.110}$$

which is always true, as it is easily checked. Finally, condition $(4.106)_2$ can be discarded because less restrictive than condition $(4.106)_3$ and the only invariant conditions for positive definiteness of $\mathbb{E}$ are Eq. $(4.106)_{1,3}$, along with the two conditions (4.55), intrinsic to the polar method:

$$\begin{aligned} &T_0 - R_0 > 0,\\ &T_1(T_0^2 - R_0^2) - 2R_1^2\left[T_0 - R_0\cos 4(\Phi_0 - \Phi_1)\right] > 0,\\ &R_0 \geq 0,\\ &R_1 \geq 0. \end{aligned} \tag{4.111}$$

To remark also that conditions (4.98) and (4.111) imply that the isotropic part of $\mathbb{E}$ is strictly positive:

$$T_0 > 0, \quad T_1 > 0. \tag{4.112}$$

The above *four intrinsic conditions* (4.111) *are valid for a completely anisotropic planar material*; their reduction due to the existence of an elastic symmetry are considered in Sect. 4.12.

Finally, we notice that Eq. $(4.111)_2$ is equivalent to state that Δ, Eq. (4.71), is necessarily a positive quantity.

4.12 Symmetries

We consider in this section the consequences of the existence of elastic symmetries on tensor $\mathbb{T}$. The analysis of the elastic symmetries can be done in different ways; we follow here the original approach of Verchery, but other approaches are possible, cf. Vincenti (2002), p. 45.

Quantities L_1, L_2, Q_1, Q_2, C_1 and C_2 are tensor invariants under the action of a frame rotation. Nevertheless, a mirror symmetry with respect to an axis inclined of the angle α on the axis of x_1 does not leave unchanged all of these quantities. This can be seen in the following way: such a symmetry is described by the complex variable transformation

$$z'' = s^2\bar{z}, \quad s = e^{i\alpha}; \tag{4.113}$$

applying the Verchery's transformation (4.1) we get

$$\begin{aligned} \mathsf{X}^{1''} &= \frac{1}{\sqrt{2}}\bar{k}\,z'' = \frac{1}{\sqrt{2}}\bar{k}\,s^2\bar{z} = -i\,s^2\mathsf{X}^2,\\ \mathsf{X}^{2''} &= \frac{1}{\sqrt{2}}k\,\bar{z}'' = \frac{1}{\sqrt{2}}k\,\bar{s}^2\,z = i\,\bar{s}^2\mathsf{X}^1, \end{aligned} \tag{4.114}$$

that in matrix form reads like

$$\mathsf{X}^{cont''} = \mathbf{S}_1 \mathsf{X}^{cont} \rightarrow \begin{Bmatrix} \mathsf{X}^{1''} \\ \mathsf{X}^{2''} \end{Bmatrix} = \begin{bmatrix} 0 & -i\,s^2 \\ i\,\bar{s}^2 & 0 \end{bmatrix} \begin{Bmatrix} \mathsf{X}^1 \\ \mathsf{X}^2 \end{Bmatrix}. \tag{4.115}$$

This result shows that $\mathsf{X}^1\mathsf{X}^2$ is still the only invariant for a vector: a mirror symmetry does not affect the norm of a vector. The *symmetry matrix* has a typical structure, given by the Verchery's transformation: it is *anti-diagonal*. This is true for the symmetry matrices of any rank tensors, that can be constructed using the same procedure of matrices $\mathbf{m}_j$. We obtain hence, for rank-two tensors (null components are not indicated),

$$\mathsf{L}^{cont''} = \mathbf{S}_2 \mathsf{L}^{cont} \rightarrow \begin{Bmatrix} \mathsf{L}^{11''} \\ \mathsf{L}^{12''} \\ \mathsf{L}^{21''} \\ \mathsf{L}^{22''} \end{Bmatrix} = \begin{bmatrix} & & & -s^4 \\ & & 1 & \\ & 1 & & \\ -\bar{s}^4 & & & \end{bmatrix} \begin{Bmatrix} \mathsf{L}^{11} \\ \mathsf{L}^{12} \\ \mathsf{L}^{21} \\ \mathsf{L}^{22} \end{Bmatrix}, \tag{4.116}$$

which shows that a symmetry does not add any more information: L^{12}, L^{21} and $\mathsf{L}^{11}\mathsf{L}^{22}$ are still tensor invariants also under a mirror symmetry. In other words, mirror symmetries have no effects on plane rank-two tensors.

In the case of fourth rank tensors, we obtain

$$\mathsf{T}^{cont''} = \mathbf{S}_4 \mathsf{T}^{cont} \rightarrow$$

$$\begin{Bmatrix} \mathsf{T}^{1111''} \\ \mathsf{T}^{1112''} \\ \mathsf{T}^{1121''} \\ \mathsf{T}^{1122''} \\ \mathsf{T}^{1211''} \\ \mathsf{T}^{1212''} \\ \mathsf{T}^{1221''} \\ \mathsf{T}^{1222''} \\ \mathsf{T}^{2111''} \\ \mathsf{T}^{2112''} \\ \mathsf{T}^{2121''} \\ \mathsf{T}^{2122''} \\ \mathsf{T}^{2211''} \\ \mathsf{T}^{2212''} \\ \mathsf{T}^{2221''} \\ \mathsf{T}^{2222''} \end{Bmatrix} = \begin{bmatrix} &&&&&&&&&&&&&&& s^8 \\ &&&&&&&&&&&&&& -s^4 & \\ &&&&&&&&&&&&& -s^4 && \\ &&&&&&&&&&&& 1 &&& \\ &&&&&&&&&&& -s^4 &&&& \\ &&&&&&&&&& 1 &&&&& \\ &&&&&&&&& 1 &&&&&& \\ &&&&&&&& -\bar{s}^4 &&&&&&& \\ &&&&&&& -s^4 &&&&&&&& \\ &&&&&& 1 &&&&&&&&& \\ &&&&& 1 &&&&&&&&&& \\ &&&& -\bar{s}^4 &&&&&&&&&&& \\ &&& 1 &&&&&&&&&&&& \\ && -\bar{s}^4 &&&&&&&&&&&&& \\ & -\bar{s}^4 &&&&&&&&&&&&&& \\ \bar{s}^8 &&&&&&&&&&&&&&& \end{bmatrix} \begin{Bmatrix} \mathsf{T}^{1111} \\ \mathsf{T}^{1112} \\ \mathsf{T}^{1121} \\ \mathsf{T}^{1122} \\ \mathsf{T}^{1211} \\ \mathsf{T}^{1212} \\ \mathsf{T}^{1221} \\ \mathsf{T}^{1222} \\ \mathsf{T}^{2111} \\ \mathsf{T}^{2112} \\ \mathsf{T}^{2121} \\ \mathsf{T}^{2122} \\ \mathsf{T}^{2211} \\ \mathsf{T}^{2212} \\ \mathsf{T}^{2221} \\ \mathsf{T}^{2222} \end{Bmatrix}, \tag{4.117}$$

that for an elasticity tensor becomes

$$
\mathsf{T}^{cont''} = \mathbf{S}_4 \mathsf{T}^{cont} \rightarrow
$$
$$
\left\{\begin{array}{c} \mathsf{T}^{1111''} \\ \mathsf{T}^{1112''} \\ \mathsf{T}^{1122''} \\ \mathsf{T}^{1212''} \\ \mathsf{T}^{1222''} \\ \mathsf{T}^{2222''} \end{array}\right\} = \begin{bmatrix} 0 & 0 & 0 & 0 & 0 & s^8 \\ 0 & 0 & 0 & 0 & -s^4 & 0 \\ 0 & 0 & 1 & 0 & 0 & 0 \\ 0 & 0 & 0 & 1 & 0 & 0 \\ 0 & -\overline{s}^4 & 0 & 0 & 0 & 0 \\ \overline{s}^8 & 0 & 0 & 0 & 0 & 0 \end{bmatrix} \left\{\begin{array}{c} \mathsf{T}^{1111} \\ \mathsf{T}^{1112} \\ \mathsf{T}^{1122} \\ \mathsf{T}^{1212} \\ \mathsf{T}^{1222} \\ \mathsf{T}^{2222} \end{array}\right\}, \tag{4.118}
$$

where the anti-diagonal structure is only apparently lost, due to the removed components.

A scrutiny of Eq. (4.118) shows immediately that L_1, L_2, Q_1 and Q_2 are still invariants also under the action of a mirror symmetry, but not C_1 and C_2 because

$$
\begin{aligned}
C_1'' + iC_2'' &= \mathsf{T}^{1111''}\left(\mathsf{T}^{1222''}\right)^2 = s^8\mathsf{T}^{2222}\left(\overline{s}^4\mathsf{T}^{1112}\right)^2 = \\
&= \overline{\mathsf{T}}^{1111}\left(\overline{\mathsf{T}}^{1222}\right)^2 = C_1 - iC_2 :
\end{aligned} \tag{4.119}
$$

C_2 is antisymmetric as effect of the mirror symmetry. To study the consequences of the mirror symmetry, we operate a rotation of axes, choosing the new frame so that the bisector of the first quadrant coincide with the axes of mirror symmetry. For such a choice, in Eq. (4.25) it must be

$$
\theta = \alpha - \frac{\pi}{4} \quad \Rightarrow \quad r = k\,\overline{s}. \tag{4.120}
$$

Then, Eq. (4.121) gives

$$
\left\{\begin{array}{c} \mathsf{T}^{1111'} \\ \mathsf{T}^{1112'} \\ \mathsf{T}^{1122'} \\ \mathsf{T}^{1212'} \\ \mathsf{T}^{1222'} \\ \mathsf{T}^{2222'} \end{array}\right\} = \begin{bmatrix} -e^{-4i\alpha} & & & & & \\ & ie^{-2i\alpha} & & & & \\ & & 1 & & & \\ & & & 1 & & \\ & & & & -ie^{2i\alpha} & \\ & & & & & -e^{4i\alpha} \end{bmatrix} \left\{\begin{array}{c} \mathsf{T}^{1111} \\ \mathsf{T}^{1112} \\ \mathsf{T}^{1122} \\ \mathsf{T}^{1212} \\ \mathsf{T}^{1222} \\ \mathsf{T}^{2222} \end{array}\right\}. \tag{4.121}
$$

For the same choice of the new frame, the axes of x_1' and x_2' are equivalent with respect to the mirror symmetry, which implies, Eq. (4.21),

$$
\begin{aligned}
\mathsf{T}^{1111'} &= \mathsf{T}^{2222'} = \overline{\mathsf{T}}^{1111'}, \\
\mathsf{T}^{1112'} &= \mathsf{T}^{1222'} = \overline{\mathsf{T}}^{1112'},
\end{aligned} \tag{4.122}
$$

and hence that[11]

$$\mathsf{T}^{1111'} = -e^{-4i\alpha}\mathsf{T}^{1111} \in \mathbb{R}$$
$$\mathsf{T}^{1222'} = -ie^{2i\alpha}\mathsf{T}^{1222} \in \mathbb{R}; \tag{4.123}$$

by consequence, for the cubic invariants we get

$$C_1 + iC_2 = \mathsf{T}^{1111}\left(\mathsf{T}^{1222}\right)^2 = \mathsf{T}^{1111'}\left(\mathsf{T}^{1222'}\right)^2 \in \mathbb{R} \;\Rightarrow\; C_2 = 0. \tag{4.124}$$

This result opens the way to examine the algebraic characterization of elastic symmetries in $\mathbb{R}^2$. First of all, we remark that if α is the direction of an axis of symmetry, then $\beta = \alpha + \pi/2$ is also the direction of an axis of symmetry. In fact, if the direction of β becomes the bisector of a new frame $\{x_1'', x_2''\}$, then $x_1'' = x_2', x_2'' = -x_1'$: the axes x_1'' and x_2'' are, of course, still equivalent with respect to a mirror symmetry, that can be only that of β, their bisector. This fact just shows that in $\mathbb{R}^2$ the monoclinic syngony cannot exist, the minimal symmetry condition being that of the orthorhombic syngony, i.e. of orthotropic tensors $\mathbb{T}$.

The direction of the mirror can be obtained considering that the imaginary part of $\mathsf{T}^{1222'}$ must be null:

$$\operatorname{Im}\left(\mathsf{T}^{1222'}\right) = \operatorname{Im}\left(-ie^{2i\alpha}\mathsf{T}^{1222}\right) = 0 \;\Rightarrow$$
$$\tan 2\alpha = \frac{\operatorname{Re}\left(\mathsf{T}^{1222}\right)}{\operatorname{Im}\left(\mathsf{T}^{1222}\right)} = \frac{2(T_{1112} + T_{1222})}{T_{1111} - T_{2222}}. \tag{4.125}$$

The general condition for the existence of a mirror symmetry and hence, for what said above, for the tensor $\mathbb{T}$ to be orthotropic, is Eq. (4.124): $C_2 = 0$. The syzygy (4.45) becomes then

$$C_1^2 = Q_1 Q_2^2 \;\Rightarrow\; Q_1 = \left(\frac{C_1}{Q_2}\right)^2. \tag{4.126}$$

so that in case of orthotropy, there are only four independent nonzero invariants: L_1, L_2, Q_2 and C_1.[12] The above equation let us obtain the *general algebraic relation characterizing all the types of elastic symmetry in* $\mathbb{R}^2$: using Eq. (4.56) in (4.126) we get the condition

$$R_0 R_1^2 \sin 4(\Phi_0 - \Phi_1) = 0. \tag{4.127}$$

Such condition depends upon three invariants, $R_0, R_1, \Phi_0 - \Phi_1$, and can be satisfied when these invariants take some special values. To each value of one of the above

[11]Remembering Eq. (4.54)$_{3,4}$, we remark that $\mathsf{T}^{cont} \in \mathbb{R}$ for a material having an axis of mirror symmetry tilted of $\pi/4$ on the axis of x_1.

[12]It is important to preserve, in the set of the independent invariants, the invariant of the highest degree, that is why we keep C_1 in the list.

three invariants root of Eq. (4.127) corresponds a different case of elastic symmetry in $\mathbb{R}^2$. To remark that condition (4.127) is an *intrinsic characterization of elastic symmetries in* $\mathbb{R}^2$, because it makes use of only tensor invariants. So, all the following special cases are also intrinsic conditions of orthotropy and so on. Let us consider all of them separately.

4.12.1 Ordinary Orthotropy

The first solution to (4.127) that we consider is

$$\begin{aligned}
&\sin 4(\Phi_0-\Phi_1)=0 \;\Rightarrow\; \Phi_0-\Phi_1=K\frac{\pi}{4},\; K\in\{0,1\} \;\Rightarrow\; C_2=0 \;\Rightarrow\\
&(T_{1112}-T_{1222})\left[(T_{1111}-T_{2222})^2-4\,(T_{1112}+T_{1222})^2\right]-\\
&(T_{1112}+T_{1222})\,(T_{1111}-T_{2222})\,(T_{1111}-2T_{1122}-4T_{1212}+T_{2222})=0.
\end{aligned} \tag{4.128}$$

Condition (4.128) depends upon a cubic invariant; it is the first invariant characterization of orthotropy in $\mathbb{R}^2$ and was explicitly given by Verchery and Vong (1986). It characterizes intrinsically *ordinary orthotropy*, i.e. common orthotropy, as the particular anisotropic situation where *the shift angle between the two anisotropy phases is a multiple of* $\pi/4$; clearly, due to the periodicity of the functions, only two cases are meaningful: 0 or $\pi/4$.

This result shows that, generally speaking, for the same set of invariants T_0, T_1, R_0 and R_1 *two possible and distinct orthotropic materials can exist*: one with $K=0$ and the other one with $K=1$. This fact is interesting *per se* and because it shows that an algebraic analysis of symmetries, based upon the study of the invariants, gives more information than a mere geometric study of the symmetries.

If for an ordinarily orthotropic material a frame rotation of Φ_1 is operated, Eq. (4.67) can be written as

$$\begin{aligned}
T_{1111}(\theta)&=T_0+2T_1+(-1)^K R_0\cos 4\theta+4R_1\cos 2\theta,\\
T_{1112}(\theta)&=-\,(-1)^K R_0\sin 4\theta-2R_1\sin 2\theta,\\
T_{1122}(\theta)&=-T_0+2T_1-(-1)^K R_0\cos 4\theta,\\
T_{1212}(\theta)&=T_0-(-1)^K R_0\cos 4\theta,\\
T_{1222}(\theta)&=(-1)^K R_0\sin 4\theta-2R_1\sin 2\theta,\\
T_{2222}(\theta)&=T_0+2T_1+(-1)^K R_0\cos 4\theta-4R_1\cos 2\theta.
\end{aligned} \tag{4.129}$$

This is the form of the polar representation normally used for orthotropic layers; of course, it corresponds to choose the frame where $\Phi_1=0$.

The parameter K, that is an invariant, characterizes ordinary orthotropy; its importance has been observed in different studies (Vannucci 2009). In particular K plays a fundamental role in several optimization problems: an optimal solution to a given problem becomes the *anti-optimal*, i.e. the worst one, when K switches from 0 to 1 and vice-versa: if a solution is optimal for a material with $K = 0$ (or $K = 1$) it is *anti-optimal*, i.e. the worst one, for a material with $K = 1$ ($K = 0$). To have an idea of the influence of parameter K, i.e. of the type of ordinary orthotropy, let us consider two examples.

The first one concerns the variation of the normal stiffness, i.e. of the component $T_{1111}(\theta)$, Eq. (4.129)$_1$ (for the normal compliance, the results are analogous). We want to know of which type is its variation with θ: how many are its stationary points, where they are located etc. The derivatives of $T_{1111}(\theta)$ are

$$\begin{aligned}\frac{dT_{1111}}{d\theta} &= -8R_1\left[(-1)^K\rho\cos 2\theta + 1\right]\sin 2\theta,\\ \frac{d^2T_{1111}}{d\theta^2} &= -16R_1\left[(-1)^K\rho\cos 4\theta + \cos 2\theta\right],\end{aligned} \tag{4.130}$$

where

$$\rho = \frac{R_0}{R_1} \tag{4.131}$$

is a dimensionless parameter called the *anisotropy ratio* which measures the relative importance of the two anisotropy phases. From Eq. (4.130)$_1$ we find that possible stationary points are

$$\theta_1 = 0, \quad \theta_2 = \frac{1}{2}\arccos\frac{(-1)^{K+1}}{\rho}, \quad \theta_3 = \frac{\pi}{2}, \tag{4.132}$$

with the solution θ_2 that exists if and only if $\rho > 1$. For these θ, T_{1111} takes the values

$$\begin{aligned}T_{1111}(\theta_1) &= T_0 + 2T_1 + (-1)^K R_0 + 4R_1,\\ T_{1111}(\theta_2) &= T_0 + 2T_1 - (-1)^K\left(R_0 + 2\frac{R_1}{\rho}\right),\\ T_{1111}(\theta_3) &= T_0 + 2T_1 + (-1)^K R_0 - 4R_1,\end{aligned} \tag{4.133}$$

We remark also that for $K = 0$, $\theta_2 \in [\pi/4, \pi/2)$, while for $K = 1$, $\theta_2 \in (0, \pi/4[$. The study of the quality of the stationary points can be done using Eq. (4.130)$_2$; we get

$$\begin{aligned}\left.\frac{d^2T_{1111}}{d\theta^2}\right|_{\theta_1} &= -16R_1\left[(-1)^K\rho + 1\right],\\ \left.\frac{d^2T_{1111}}{d\theta^2}\right|_{\theta_2} &= -16R_1(-1)^K\frac{1-\rho^2}{\rho},\\ \left.\frac{d^2T_{1111}}{d\theta^2}\right|_{\theta_3} &= -16R_1\left[(-1)^K\rho - 1\right].\end{aligned} \tag{4.134}$$

Table 4.1 Stationary points of $T_{1111}(\theta)$ for ordinary orthotropy in $\mathbb{R}^2$

$K=0$			
$\rho \le 1$	θ_1	Global max:	$T_{1111}=T_0+2T_1+R_0+4R_1$
	θ_3	Global min:	$T_{1111}=T_0+2T_1+R_0-4R_1$
$\rho > 1$	θ_1	Global max:	$T_{1111}=T_0+2T_1+R_0+4R_1$
	θ_2	Global min:	$T_{1111}=T_0+2T_1-R_0-2\frac{R_1}{\rho}$
	θ_3	Local max:	$T_{1111}=T_0+2T_1+R_0-4R_1$
$K=1$			
$\rho \le 1$	θ_1	Global max:	$T_{1111}=T_0+2T_1-R_0+4R_1$
	θ_3	Global min:	$T_{1111}=T_0+2T_1-R_0-4R_1$
$\rho > 1$	θ_1	Local max:	$T_{1111}=T_0+2T_1-R_0+4R_1$
	θ_2	Global max:	$T_{1111}=T_0+2T_1+R_0+2\frac{R_1}{\rho}$
	θ_3	Global min:	$T_{1111}=T_0+2T_1-R_0-4R_1$

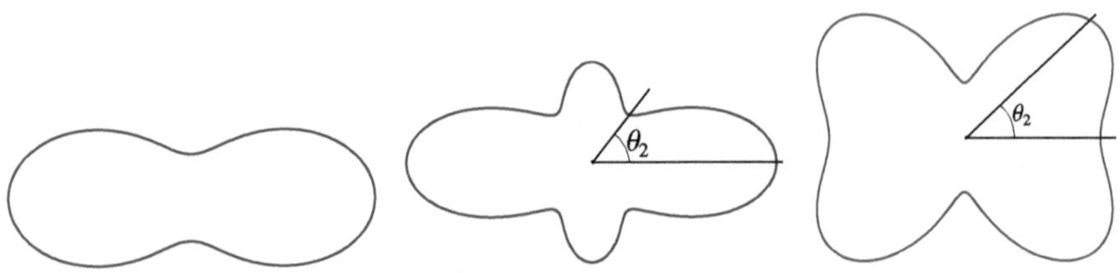

(a) $\rho < 1, K = 0, 1$ (b) $\rho > 1, K = 0$ (c) $\rho > 1, K = 1$

Fig. 4.2 Different cases of $T_{1111}(\theta)$ for ordinary orthotropy in $\mathbb{R}^2$

The results are summarized in Table 4.1 and represented in Fig. 4.2. It can be remarked that the intermediary stationary point changes from a global minimum to a global maximum when K changes from 0 to 1.

The second example, concerns a simple optimization problem: a plate is formed by bonding together two identical orthotropic layers. The problem consists in finding the orientation angles $\delta_1 \neq \delta_2$ of the two layers that maximize the in-plane shear stiffness G_{12}.

G_{12} is simply the average of the moduli T_{1212} of the two layers, to be written in the same common frame:

$$G_{12} = \frac{1}{2}\left[T_{1212}(\delta_1) + T_{1212}(\delta_2)\right], \tag{4.135}$$

that with the polar formalism becomes

$$G_{12} = T_0 - (-1)^K R_0 \eta, \quad \eta = \frac{\cos 4\delta_1 + \cos 4\delta_2}{2}, \quad -1 \le \eta \le 1. \tag{4.136}$$

It is immediate to see that the maximum of G_{12} is get for $\eta = -1$ if $K = 0$, but for $\eta = 1$ if $K = 1$; in the two cases, $G_{12}^{max} = T_0 + R_0$. Because it must be $\delta_1 \neq \delta_2$, the solution for the case $K = 0$ is $\delta_1 = \pm\pi/4$, $\delta_2 = -\delta_1$, while for the case $K = 1$ it is $\delta_1 = 0$, $\delta_2 = \pi/2$ (or indifferently $\delta_1 = \pi/2$, $\delta_2 = 0$; actually, it can be easily proved that in this case the number of plies in the two directions does not matter: the unicity of the solution is lost, with respect to the case $K = 0$, Vannucci 2009). It can be also remarked what already said about the effect of K: in both the cases, the optimal solution for a value of K is the anti-optimal one for the other K: $G_{12}^{min} = T_0 - R_0$, obtained for $\eta = 1$ when $K = 0$ and for $\eta = -1$ when $K = 1$.

The two cases of $K = 0$ or $K = 1$ corresponds to what Pedersen (1989, 1990b), names *high* ($K = 1$) *or low* ($K = 0$) *shear modulus materials*; the above example shows the reason of such a denomination, but the former example as well as the results of other studies on K, reveals that its importance is far greater than that of a mere distinction of orthotropic layers based upon the value of their shear modulus.

Two questions concern $\mathbb{S}$, the inverse of $\mathbb{T}$: how is it oriented the orthotropy of $\mathbb{S}$ and of which type is it? To this purpose, we use Eq. (4.68) $_{3,4}$, that after a rotation of Φ_1 become

$$\begin{aligned} r_0 e^{4i(\varphi_0-\Phi_1)} &= \frac{2}{\Delta}\left[R_1^2 - T_1 R_0 e^{4i(\Phi_0-\Phi_1)}\right], \\ r_1 e^{2i(\varphi_1-\Phi_1)} &= -\frac{R_1}{\Delta}\left[T_0 - R_0 e^{4i(\Phi_0-\Phi_1)}\right], \end{aligned} \tag{4.137}$$

and, because $\mathbb{T}$ is orthotropic, Eq. (4.128),

$$\begin{aligned} r_0 e^{4i(\varphi_0-\Phi_1)} &= \frac{2}{\Delta}\left[R_1^2 - (-1)^K T_1 R_0\right], \\ r_1 e^{2i(\varphi_1-\Phi_1)} &= -\frac{1}{\Delta} R_1\left[T_0 - (-1)^K R_0\right]. \end{aligned} \tag{4.138}$$

Both the right-hand terms in Eq. (4.138) are real valued, which implies

$$\begin{aligned} \sin 4(\varphi_0 - \Phi_1) = 0 \ &\Rightarrow\ \varphi_0 = \Phi_1 + \beta_0\frac{\pi}{4}, \\ \sin 2(\varphi_1 - \Phi_1) = 0 \ &\Rightarrow\ \varphi_1 = \Phi_1 + \beta_1\frac{\pi}{2}, \end{aligned} \quad \beta_0, \beta_1 \in \{0, 1\}. \tag{4.139}$$

Let us consider first φ_1: the real part of Eq. (4.138)$_2$ is

$$r_1 \cos 2(\varphi_1 - \Phi_1) = (-1)^{\beta_1} r_1 = -\frac{1}{\Delta} R_1\left[T_0 - (-1)^K R_0\right]. \tag{4.140}$$

In the above equation, it is

$$T_0 - (-1)^K R_0 > 0, \quad \Delta > 0, \quad R_1 > 0, \quad r_1 > 0, \tag{4.141}$$

because of conditions (4.55) and (4.97). Then, it is necessarily

$$\beta_1 = 1 \;\Rightarrow\; \varphi_1 = \Phi_1 + \frac{\pi}{2}. \tag{4.142}$$

This result states that $\mathbb{S}$ is *always turned of* $\pi/2$ *with respect to* $\mathbb{T}$.

We pass now to analyze φ_0: the real part of Eq. (4.138)$_1$ is

$$\begin{aligned} &r_0 \cos 4(\varphi_0 - \Phi_1) = \frac{2}{\Delta}\left[R_1^2 - (-1)^K T_1 R_0\right] \;\Rightarrow \\ &(-1)^{\beta_0} = \frac{2}{r_0\Delta}\left[R_1^2 - (-1)^K T_1 R_0\right]. \end{aligned} \tag{4.143}$$

Because of Eqs. (4.55) and (4.97), both the quantities Δ and r_0 are positive, so:

$$\beta_0 = 0 \iff R_1^2 - (-1)^K T_1 R_0 > 0 \rightarrow \begin{cases} K = 0: \; R_1^2 - T_1 R_0 > 0, \\ K = 1: \; R_1^2 + T_1 R_0 > 0 \text{ always.} \end{cases} \tag{4.144}$$

By consequence

$$\begin{aligned} &\beta_0 = 0 \Rightarrow \varphi_0 = \Phi_1 \;\text{ when }\; \begin{cases} K = 0 \;\text{ and }\; R_1^2 > T_1 R_0, \\ \text{or} \\ K = 1, \end{cases} \\ &\beta_0 = 1 \Rightarrow \varphi_0 = \Phi_1 + \frac{\pi}{4} \;\text{ when }\; K = 0 \;\text{ and }\; R_1^2 < T_1 R_0. \end{aligned} \tag{4.145}$$

Then, the difference between the two polar angles of $\mathbb{S}$ can be only

$$\varphi_0 - \varphi_1 = (\beta_0 - 2)\frac{\pi}{4}, \tag{4.146}$$

which, being a multiple of $\pi/4$, shows that if $\mathbb{T}$ is ordinarily orthotropic, so also is its inverse $\mathbb{S}$. Hence, putting, as already done for $\mathbb{T}$,

$$\varphi_0 - \varphi_1 = k\frac{\pi}{4}, \quad k = \beta - 2, \tag{4.147}$$

we get that

$$\begin{aligned} &\left.\begin{aligned} &K = 0 \;\text{ and }\; R_1^2 > T_1 R_0 \\ &\text{or} \\ &K = 1 \end{aligned}\right\} \;\Rightarrow\; k = 0, \\ &K = 0 \;\text{ and }\; R_1^2 < T_1 R_0 \;\Rightarrow\; k = 1. \end{aligned} \tag{4.148}$$

Finally, an elasticity tensor and its inverse, when ordinarily orthotropic, *can be of a different type*; in particular, the possible combinations are three: $(K = 0, k = 0)$, $(K = 0, k = 1)$, $(K = 1, k = 0)$.

The bounds on polar invariants (4.111) in the case of ordinarily orthotropic materials become

$$\begin{aligned} &T_0 > R_0, \\ &T_1\left[T_0 + (-1)^K R_0\right] > 2R_1^2, \\ &R_0 \geq 0, \\ &R_1 \geq 0. \end{aligned} \tag{4.149}$$

Equation $(4.149)_2$ suggests a graphical representation: the level lines of the surface

$$\mathscr{S} = \frac{2R_1^2}{T_1} \tag{4.150}$$

are the intersection with the planes

$$T_0 + (-1)^K R_0 = \gamma. \tag{4.151}$$

For the same T_0 and R_0, the constant γ takes the values

$$\begin{aligned} \gamma_0 &= T_0 + R_0 \ \text{ for } \ K = 0, \\ \gamma_1 &= T_0 - R_0 \ \text{ for } \ K = 1, \end{aligned} \tag{4.152}$$

with of course $\gamma_0 > \gamma_1$. So, the two planes intersect the surface $\mathscr{S}$ through two different level curves, the one corresponding to $K = 0$ higher than that of $K = 1$, see Fig. 4.3. As a consequence, if for a couple T_1, R_1 condition $(4.149)_2$ is satisfied for $K = 0$, it is possible that the same is not true when $K = 1$. In this sense, materials

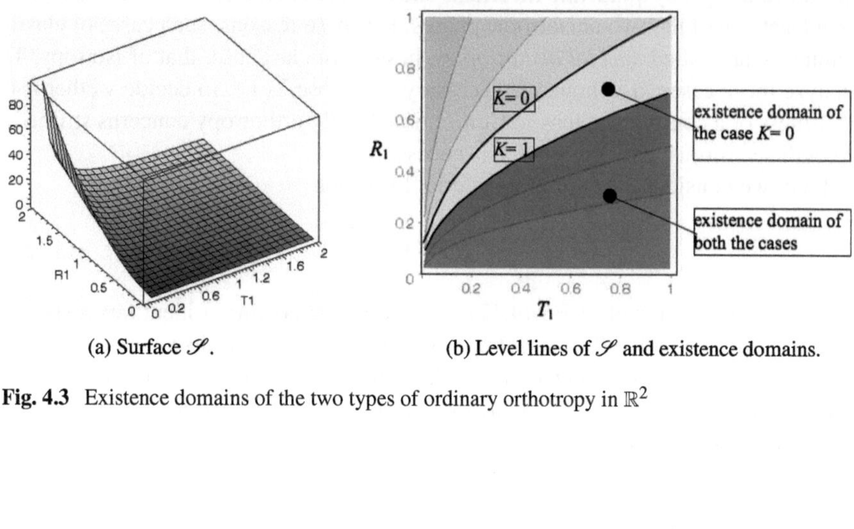

(a) Surface $\mathscr{S}$.

(b) Level lines of $\mathscr{S}$ and existence domains.

Fig. 4.3 Existence domains of the two types of ordinary orthotropy in $\mathbb{R}^2$

Table 4.2 Characteristics of the different types of elastic symmetries in $\mathbb{R}^2$

Symmetry type	Polar condition	Independent invariants	Nonzero invariants
Ordinary orthotropy $K=0$	$\Phi_0-\Phi_1=0$	4	$L_1, L_2, Q_1=\left(\frac{C_1}{Q_2}\right)^2, Q_2, C_1$
Ordinary orthotropy $K=1$	$\Phi_0-\Phi_1=\frac{\pi}{4}$	4	$L_1, L_2, Q_1=\left(\frac{C_1}{Q_2}\right)^2, Q_2, C_1$
R_0-orthotropy	$R_0=0$	3	L_1, L_2, Q_2
r_0-orthotropy	$r_0=0$	3	L_1, L_2, Q_2
Square symmetry	$R_1=0$	3	L_1, L_2, Q_1
Isotropy	$R_0=0, R_1=0$	2	L_1, L_2

with $K=1$ are less probable than materials with $K=0$, nonetheless they can exist, cf. Table 4.2.

Finally, we have seen that what is commonly considered the ordinary orthotropy in $\mathbb{R}^2$ is actually composed by two distinct cases, that have quite different mechanical properties. This type of symmetry is identified by a cubic invariant, that in the end can be represented by a simple integer, K, which can get only two values, 0 and 1. It is possible that for a same material, the stiffness and the compliance tensors are ordinarily orthotropic of different types.

4.12.2 R_0-Orthotropy

The general equation of elastic symmetries in $\mathbb{R}^2$, Eq. (4.127), can be satisfied also by other conditions than root (4.128). Algebraically speaking, unlike in the case of ordinary orthotropy, detected by a cubic invariant, all the other solutions are linked to special values get by *quadratic* invariants and they are characterized by the vanishing of at least one of the two anisotropic phases. For these reasons, such cases of elastic symmetry are called *special orthotropies*, besides the last case, that of isotropy. To analyze these cases, it is however necessary to choose $\mathbb{T}$, i.e. to decide wether it is a stiffness or a compliance tensor. Conventionally, R_0-orthotropy concerns stiffness. We will see further why this choice is necessary.

First, we consider the case of a material for which

$$R_0=0. \tag{4.153}$$

Of course, this is a root of Eq. (4.127), so the above condition identifies a special orthotropy, the so-called R_0*−orthotropy* (Vannucci 2002b). The discovery of this type of special orthotropy has been done thanks to the polar formalism and it constitutes a rather strange case of elastic behavior, whose existence has been later discovered also in $\mathbb{R}^3$ (Forte 2005).

It is easily recognized that

$$R_0 = 0 \;\Rightarrow\; \begin{cases} Q_1 = C_1 = 0, \; \mathsf{T}^{1111} = \mathsf{T}^{2222} = 0, \\ (T_{1111} - 2T_{1122} - 4T_{1212} + T_{2222})^2 + 16(T_{1112} - T_{1222})^2 = 0. \end{cases} \tag{4.154}$$

Though this case of elastic symmetry presents two orthogonal axes of mirror symmetry, just like in ordinary orthotropy, it has some peculiar characteristics. First of all, the Cartesian components of a R_0-orthotropic material are (we have put $\Phi_1 = 0$ for fixing the frame)

$$\begin{aligned} T_{1111}(\theta) &= T_0 + 2T_1 + 4R_1 \cos 2\theta, \\ T_{1112}(\theta) &= -\,2R_1 \sin 2\theta, \\ T_{1122}(\theta) &= -T_0 + 2T_1, \\ T_{1212}(\theta) &= T_0, \\ T_{1222}(\theta) &= -\,2R_1 \sin 2\theta, \\ T_{2222}(\theta) &= T_0 + 2T_1 - 4R_1 \cos 2\theta. \end{aligned} \tag{4.155}$$

For a material R_0-orthotropic, the anisotropic phase depending on R_0 is absent. By consequence, some of the components, T_{1122} and T_{1212}, are *isotropic*, while the other ones, depending upon the circular functions of 2θ, *change like the components of a second-rank tensor*. We are hence faced to a very strange case, that of a fourth-rank tensor whose components do not vary according to the tensor law and, in addition, with some of them frame independent. Moreover, unlike what happens in all the other cases of anisotropy, $T_{1112}(\theta) = T_{1222}(\theta)\ \forall\theta$.

Only three invariants are nonzero: L_1, L_2 and Q_2; the polar angle Φ_0 is now meaningless. This case of orthotropy is not characterized by a special value of the phase angle between the two anisotropic phases, but by the *absence* of one of them.

Let us now consider what happens for the compliance tensor $\mathbb{T}^{-1}$, that we still indicate by $\mathbb{S}$: when $R_0 = 0$, Eq. (4.68) becomes

$$\begin{aligned} t_0 &= \frac{T_0 T_1 - R_1^2}{4T_0(T_0 T_1 - 2R_1^2)}, \\ t_1 &= \frac{T_0}{16(T_0 T_1 - 2R_1^2)}, \\ r_0 e^{4i\varphi_0} &= \frac{R_1^2 e^{4i\Phi_1}}{4T_0(T_0 T_1 - 2R_1^2)}, \\ r_1 e^{2i\varphi_1} &= -\frac{R_1 e^{2i\Phi_1}}{8(T_0 T_1 - 2R_1^2)}, \end{aligned} \tag{4.156}$$

and by consequence

$$r_0 = \frac{R_1^2}{4T_0(T_0T_1 - 2R_1^2)}, \qquad \varphi_0 = \Phi_1,$$
$$r_1 = \frac{R_1}{8(T_0T_1 - 2R_1^2)}, \qquad \varphi_1 = \Phi_1 + \frac{\pi}{2}. \tag{4.157}$$

In obtaining the results in Eq. (4.157), it has been considered that the denominator of both the terms is positive, cf. further in this section.

As remarked in Eq. (4.72), $R_0 = 0 \nRightarrow r0 = 0$: the compliance tensor $\mathbb{S}$ depends on both the anisotropic phases, that is, its components preserve a higher degree of symmetry than those of $\mathbb{T}$. This is a rather unusual case, where stiffness and compliance of the same material do not have the same kind of variation, the same morphology. In addition, tensor $\mathbb{S}$ has always $k = 0$ (in fact, k is even, so it can be assumed equal to zero, because $\varphi_0 - \varphi_1 = -\pi/2 = -2\pi/4$).

Nevertheless, just like $\mathbb{T}$, also $\mathbb{S}$ depends upon only three independent nonzero invariants, because it is easily recognized from Eq. (4.156) that

$$r_0 = \frac{r_1^2}{t_1}. \tag{4.158}$$

Hence, for a chosen frame, i.e. for a fixed Φ_1, φ_0 and φ_1 are fixed too, and the only polar moduli t_0, t_1 and r_1 are sufficient to completely determine $\mathbb{S}$. If we put $\Phi_1 = 0$, we obtain

$$\begin{aligned}
S_{1111} &= t_0 + 2t_1 + \frac{r_1^2}{t_1}\cos 4\theta - 4r_1\cos 2\theta,\\
S_{1112} &= -\frac{r_1^2}{t_1}\sin 4\theta + 2r_1\sin 2\theta,\\
S_{1122} &= -t_0 + 2t_1 - \frac{r_1^2}{t_1}\cos 4\theta,\\
S_{1212} &= t_0 - \frac{r_1^2}{t_1}\cos 4\theta,\\
S_{1222} &= \frac{r_1^2}{t_1}\sin 4\theta + 2r_1\sin 2\theta,\\
S_{2222} &= t_0 + 2t_1 + \frac{r_1^2}{t_1}\cos 4\theta + 4r_1\cos 2\theta,
\end{aligned} \tag{4.159}$$

or, injecting Eq. (4.156) into the previous equation,

$$\begin{aligned}
S_{1111} &= \frac{1}{8(T_0T_1-2R_1^2)}\left[T_0+2T_1+2\frac{R_1^2}{T_0}(\cos 4\theta-1)-4R_1\cos 2\theta\right],\\
S_{1112} &= \frac{R_1}{4(T_0T_1-2R_1^2)}\left(-\frac{R_1}{T_0}\sin 4\theta+\sin 2\theta\right),\\
S_{1122} &= \frac{1}{8(T_0T_1-2R_1^2)}\left[T_0-2T_1-2\frac{R_1^2}{T_0}(\cos 4\theta-1)\right],\\
S_{1212} &= \frac{1}{4(T_0T_1-2R_1^2)}\left[T_1-\frac{R_1^2}{T_0}(\cos 4\theta+1)\right],\\
S_{1222} &= \frac{R_1}{4(T_0T_1-2R_1^2)}\left(\frac{R_1}{T_0}\sin 4\theta+\sin 2\theta\right),\\
S_{2222} &= \frac{1}{8(T_0T_1-2R_1^2)}\left[T_0+2T_1+2\frac{R_1^2}{T_0}(\cos 4\theta-1)+4R_1\cos 2\theta\right].
\end{aligned} \tag{4.160}$$

Contrarily to what happens for $\mathbb{T}$, S_{1122} and S_{1212} are not isotropic and $S_{1112} \neq S_{1222}$; nevertheless, just as for any common orthotropic layer, both them are null in correspondence of the two symmetry axes.

The general bounds (4.111) become, for R_0-orthotropy,

$$T_0 > \frac{2R_1^2}{T_1}, \quad R_1 > 0, \tag{4.161}$$

hence only two intrinsic bounds are sufficient.

For a R_0-orthotropic material, an analysis of the variation of T_{1111} like that done in the previous section for an ordinarily orthotropic material is meaningless, because the anisotropic phase varying with 4θ is lacking. Nevertheless, we can perform a similar analysis on the Young's modulus

$$E_1(\theta) = \frac{1}{S_{1111}(\theta)}, \tag{4.162}$$

studying the variation of $S_{1111}(\theta)$. First of all, we remark that

$$\begin{aligned}
&\frac{dS_{1111}}{d\theta} = 8S_{1112} \Rightarrow \frac{dS_{1111}}{d\theta} = 0 \text{ for}\\
&\theta_1 = 0,\\
&\theta_2 = \frac{1}{2}\arccos\frac{t_1}{r_1} = \frac{1}{2}\arccos\frac{T_0}{2R_1}, \quad \omega \in \left[0, \frac{\pi}{4}\right],\\
&\theta_3 = \frac{\pi}{2}.
\end{aligned} \tag{4.163}$$

Of course, the root θ_2 exists if and only if

$$t_1 < r_1 \text{ or } T_0 < 2R_1. \tag{4.164}$$

Because

$$\begin{aligned}
\frac{d^2 S_{1111}}{d\theta^2} &= -4r_1\left[2\frac{r_1}{t_1}\cos^2 2\theta - \cos 2\theta - \frac{r_1}{t_1}\right] \Rightarrow\\
\left.\frac{d^2 S_{1111}}{d\theta^2}\right|_{\theta_1} &= 4r_1\frac{t_1 - r_1}{t_1},\\
\left.\frac{d^2 S_{1111}}{d\theta^2}\right|_{\theta_2} &= 4(t_1 + r_1)\frac{t_1 - r_1}{t_1},\\
\left.\frac{d^2 S_{1111}}{d\theta^2}\right|_{\theta_3} &= -4r_1\frac{t_1 + r_1}{t_1},
\end{aligned} \tag{4.165}$$

and

$$\begin{aligned}
S_{1111}(\theta_1) &= t_0 + 2t_1 + \frac{r_1^2}{t_1} - 4r_1 = \frac{T_0 + 2T_1 - 4R_1}{8(T_0T_1 - 2R_1^2)},\\
S_{1111}(\theta_2) &= \frac{t_0t_1 - r_1^2}{t_1} = \frac{1}{4T_0},\\
S_{1111}(\theta_3) &= t_0 + 2t_1 + \frac{r_1^2}{t_1} + 4r_1 = \frac{T_0 + 2T_1 + 4R_1}{8(T_0T_1 - 2R_1^2)},
\end{aligned} \tag{4.166}$$

two cases are possible:

1. $t_1 > r_1$ ($T_0 > 2R_1$): $S_{1111}(\theta)$ has a global minimum for $\theta = 0$ and a global maximum for $\theta = \pi/2$; hence, $E_1(\theta)$ is maximum for $\theta = 0$ and minimum for $\theta = \pi/2$;
2. $t_1 < r_1$ ($T_0 < 2R_1$): $S_{1111}(\theta)$ has a local minimum for $\theta = 0$, an absolute minimum for $\theta = \theta_2$ and an absolute maximum for $\theta = \pi/2$; hence, $E_1(\theta)$ is maximum for $\theta = \theta_2$ and minimum for $\theta = \pi/2$, while for $\theta = 0$ it has only a local maximum.

The two possibilities are sketched in Fig. 4.4a. They concern the case of two materials having the same T_0 and T_1 but different R_1: $T_0 = 1, T_1 = 0.8, R_1 = 0.1$ when $T_0 > 2R_1$ and $R_1 = 0.6$ when $T_0 < 2R_1$, respectively represented by points A and B in Fig. 4.4b.

Condition (4.164) can be considered to determine which is the part of the domain of existence of the material, in the plane (T_1, R_1) and for a given positive value T_0 of T_0, where the above cases 1 and 2 are possible. The intersection of the plane given by condition (4.164) and the surface corresponding to the $(4.161)_1$, which is the surface $\mathscr{S}$ in Fig. 4.3, is a straight line whose projection on the plane (T_1, R_1) is the straight line $R_1 = T_1$. Let P be the point of the plane (T_1, R_1) representative of a given material, with $T_0 = \mathsf{T}_0$. So, Fig. 4.4b, three cases are possible:

- $P \in \Omega_1 : T_0 > 2R_1^2/T_1 > 2R_1 \rightarrow$ only case 1 is possible ($T_0 > 2R_1; T_1 < R_1$);
- $P \in \Omega_2 : T_0 > 2R_1 > 2R_1^2/T_1 \rightarrow$ again, only case 1 is possible ($T_0 > 2R_1; T_1 > R_1$); this is the case of point A in Fig. 4.4b;

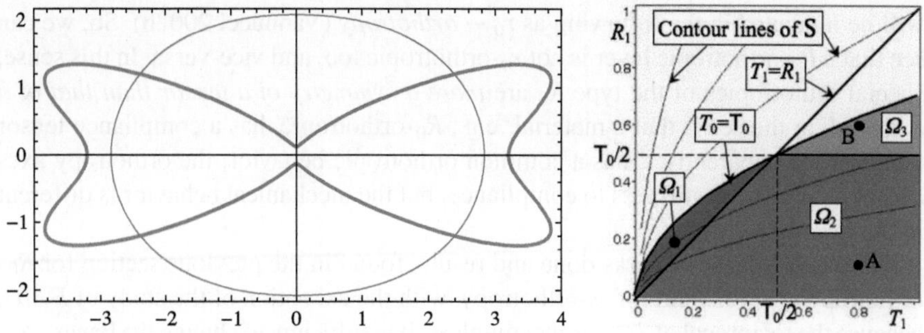

(a) Directional diagrams of $E_1(\theta)$; $T_0 > 2R_1$, thin line, $T_0 < 2R_1$, thick line.

(b) Existence domains.

Fig. 4.4 R_0-orthotropic materials

- $P \in \Omega_3 : 2R_1 > T_0 > 2R_1^2/T_1 \ \rightarrow$ only case 2 is possible ($T_0 < 2R_1$; $T_1 > R_1$); this is the case of point B in Fig. 4.4b.

It can be seen that if $T_1 < T_0/2$, only case 1 is possible, while if $R_1 > T_0/2$ only case 2 can exist.

Finally, one can wonder if R_0-orthotropic materials do really exist. Actually, they do; in fact, it is rather simple, using the polar formalism and the classical lamination theory, to see that a R_0-orthotropic lamina can be fabricated reinforcing an isotropic matrix by unidirectional fibers arranged in equal quantity along two directions tilted of 45 (Vannucci 2002b).

A special property of R_0-orthotropic layers, is linked to the sensitivity of a laminate to layers' orientation defects. It has been shown, Vannucci (2001), Vincenti et al. (2003), that the influence of such defects on the uncoupling and quasi-homogeneity of a laminate[13] depends on the anisotropy ratio ρ, Eq. (4.131). In particular, the sensitivity to uncoupling or quasi-homogeneity is minimal when $\rho = 0$, i.e. when the laminate is composed by R_0-orthotropic layers, cf. Sect. 5.3.6.

4.12.3 r_0-Orthotropy

It has already been noticed that relations (4.68) are perfectly symmetric, i.e., they can be rewritten swapping the polar compliance constants with the polar stiffness constants, i.e., putting upper-case letters at the left-hand side and lower-case letters at the right-hand side of relations (4.68). This circumstance, together with Eq. (4.72), i.e. the fact that whenever $R_0 = 0$, then $r_0 \neq 0$, implies the existence of another special orthotropy, an analog of R_0-orthotropy, but concerning compliance, not stiffness: it

[13] A laminate is said to be *quasi-homogeneous* if the bending and extension response are uncoupled and equal (Vannucci and Verchery 2001), see Sect. 5.2.8.

will be indicated in the following as $r_0 - orthotropy$ (Vannucci 2002b). So, we can see that a R_0-orthotropic layer is not r_0-orthotropic too, and vice-versa. In this sense, special orthotropies of the type R_0 are *more a symmetry of a tensor than that of a material*, in the sense that a material, e.g., R_0-orthotropic, has a compliance tensor that, at least apparently,[14] has a common orthotropic behavior: the orthotropy axes do not change from stiffness to compliance, but the mechanical behavior is different in the two cases.

Of course, all the remarks done and results found in the previous section for R_0-orthotropy are still valid for r_0-orthotropy, with the exception of the study of $E_1(\theta)$, because the reciprocal of T_{1111} is meaningless, it is sufficient to change the lower-case letters with capital letters to all the polar components and the word *stiffness* with the word *compliance*.

Something different can be said about the technical constants; in fact, putting $\varphi_1 = 0$, the compliance tensor $\mathbb{S}$ looks like

$$\begin{aligned} S_{1111}(\theta) &= t_0 + 2t_1 + 4r_1 \cos 2\theta, \\ S_{1112}(\theta) &= -2r_1 \sin 2\theta, \\ S_{1122}(\theta) &= -t_0 + 2t_1, \\ S_{1212}(\theta) &= t_0, \\ S_{1222}(\theta) &= -2r_1 \sin 2\theta, \\ S_{2222}(\theta) &= t_0 + 2t_1 - 4r_1 \cos 2\theta, \end{aligned} \tag{4.167}$$

which gives

$$\begin{aligned} E_1(\theta) &= \frac{1}{S_{1111}(\theta)} = \frac{1}{t_0 + 2t_1 + 4r_1 \cos 2\theta}, \\ G_{12}(\theta) &= \frac{1}{4S_{1212}(\theta)} = \frac{1}{4t_0}, \\ \nu_{12}(\theta) &= -\frac{S_{1122}(\theta)}{S_{1111}(\theta)} = \frac{t_0 - 2t_1}{t_0 + 2t_1 + 4r_1 \cos 2\theta}, \\ \eta_{1,12}(\theta) = \eta_{2,12}(\theta) &= \frac{S_{1222}(\theta)}{S_{1212}(\theta)} = -2\frac{r_1 \sin 2\theta}{t_0}. \end{aligned} \tag{4.168}$$

We can hence remark that $E_1(\theta)$, $\nu_{12}(\theta)$ and $\eta_{1,12}(\theta)$ vary with 2θ, while the shear modulus $G_{12}(\theta)$ is isotropic. This is a basic characteristic of r_0-orthotropic materials. It was observed experimentally since the fifties that paper has this characteristic (Hono and Onogi 1951; Campbell 1961), but only lately an explanation of this fact in the framework of classical elasticity has been done, thanks to the polar formalism (Vannucci 2010).

[14] "Apparently" because if one makes experimental tests on the components of $\mathbb{S}$ or traces the directional diagrams of its components, they look like those of an ordinarily orthotropic material with $k = 0$, the difference is in the special value get by r_0, Eq. (4.158).

Just like for R_0-orthotropy, only three nonzero independent invariants are sufficient to completely determine $\mathbb{S}$: t_0, t_1, r_1. From Eq. (4.168) we get also

$$t_0 = \frac{1}{4G_{12}},\\ t_1 = \frac{1}{2}\left(\frac{1}{4G_{12}} - \frac{\nu_{12}}{E_1}\right),\\ r_1 = \frac{1}{4}\left(\frac{1+\nu_{12}}{E_1} - \frac{1}{2G_{12}}\right). \tag{4.169}$$

The general bounds (4.111) become, for r_0-orthotropy,

$$t_0 > \frac{2r_1^2}{t_1}, \quad r_1 > 0, \tag{4.170}$$

like in the case of R_0-orthotropy, so also in this case, of course, only two intrinsic bounds are sufficient. Using Eq. (4.169), the above bounds can be rewritten also in terms of technical constants:

$$\frac{1+\nu_{12}}{E_1} - \frac{1}{2G_{12}} > 0,\\ E_1 > G_{12}(1+\nu_{12})^2. \tag{4.171}$$

Finally, just like for the previous case of R_0-orthotropic materials, it is easy to see that for the stiffness tensor it is

$$R_0 = \frac{R_1^2}{T_1}, \quad K = 0. \tag{4.172}$$

4.12.4 Square Symmetry

Another root of the general equation of elastic symmetries in $\mathbb{R}^2$, Eq. (4.127), is

$$R_1 = 0. \tag{4.173}$$

Just like the case of R_0-orthotropy, also in this case an anisotropy phase, the one varying with 2θ, vanishes, so it is a special orthotropy, determined once more by a quadratic invariant:

$$R_1 = 0 \ \Rightarrow \ \begin{cases} Q_2 = C_1 = 0, \ \mathsf{T}^{1112} = \mathsf{T}^{1222} = 0, \\ (T_{1111} - T_{2222})^2 + 4(T_{1112} + T_{1222})^2 = 0. \end{cases} \tag{4.174}$$

The only nonzero invariants are L_1, L_2 and Q_1. In this case, the polar angle Φ_1 is meaningless, so the frame can be fixed only fixing a value for Φ_0. Choosing $\Phi_0 = 0$, the Cartesian components of $\mathbb{T}$ are

$$\begin{aligned}
T_{1111}(\theta)&=T_0+2T_1+R_0\cos 4\theta,\\
T_{1112}(\theta)&=-R_0\sin 4\theta,\\
T_{1122}(\theta)&=-T_0+2T_1-R_0\cos 4\theta,\\
T_{1212}(\theta)&=T_0-R_0\cos 4\theta,\\
T_{1222}(\theta)&=R_0\sin 4\theta,\\
T_{2222}(\theta)&=T_0+2T_1+R_0\cos 4\theta.
\end{aligned} \tag{4.175}$$

We can remark that all the components are periodic of $\pi/2$:

$$T_{ijkl}\left(\theta+\frac{\pi}{2}\right)=T_{ijkl}(\theta)\ \ \forall\theta. \tag{4.176}$$

For this reason, this special orthotropy is known in the literature as *square symmetry* and actually, it is the corresponding, in $\mathbb{R}^2$, of the cubic syngony. This fact can be immediately appreciated looking at the directional diagram of its components, a typical example is in Fig. 4.32. Such materials can be fabricated reinforcing an isotropic matrix with a balanced fabric, i.e. by a fabric having the same amount of fibers in warp and weft. We remark also that components T_{1122} and T_{1212} are the same of the case of ordinary orthotropy with $K = 0$ and that

$$T_{1111}(\theta)=T_{2222}(\theta),\quad T_{1112}(\theta)=-T_{1222}(\theta)\ \ \forall\theta. \tag{4.177}$$

Because everything is periodic of $\pi/2$, there is another couple of mirror symmetry axes, tilted of $\pi/4$ with respect to the directions Φ_0, $\Phi_0+\pi/2$. In fact, Eq. (4.123), the direction α of the mirror symmetry is given by

$$\begin{aligned}
&\mathrm{Im}\left(\mathsf{T}^{1111'}\right)=\mathrm{Im}\left(-ie^{2i\alpha}\mathsf{T}^{1111}\right)=0\ \Rightarrow\\
&\tan 4\alpha=\tan 4\left(\alpha+\frac{\pi}{4}\right)=\frac{\mathrm{Re}\left(\mathsf{T}^{1111}\right)}{\mathrm{Im}\left(\mathsf{T}^{1111}\right)}=\frac{T_{1111}-2T_{1122}-4T_{1212}+T_{2222}}{4(T_{1112}-T_{1222})}.
\end{aligned} \tag{4.178}$$

Unlike the case of R_0-orthotropy, for Eq. $(4.72)_1$ when a material has $R_1 = 0$ it has also $r_1 = 0$: square symmetry is a property of both the stiffness and the compliance tensors. Also, for square symmetric materials, tensors $\mathbb{T}$ and $\mathbb{S}$ preserve the typical variation with the orientation: their components vary with 4θ.

The general bounds for the polar invariants (4.111) become now

$$\begin{aligned}
&T_1(T_0-R_0)>0,\\
&R_0\geq 0.
\end{aligned} \tag{4.179}$$

4.12.5 Isotropy

The last possible syngony for a planar material is isotropy; in this case, every angle α must determine the direction of a mirror symmetry. This means that α must be, at the same time, the solution of Eq. (4.125) and of Eq. (4.178), which gives the condition

$$\begin{aligned} &\mathrm{T}^{1111}=\mathrm{T}^{1112}=0 \ \Rightarrow\ Q_1=Q_2=C_1=0, \ \Rightarrow\ R_0=R_1=0 \ \Rightarrow \\ &T_{1112}=T_{1222}=0,\ \ T_{2222}=T_{1111},\ \ T_{1111}=T_{1122}+2T_{1212}. \end{aligned} \tag{4.180}$$

Algebraically, isotropy is hence characterized by the fact that the two anisotropy phases are null; it can be remarked also that a material is isotropic if and only if the conditions for the two special orthotropies are satisfied at the same time: algebraically, isotropy is determined by the vanishing of two quadratic invariants. Alternatively, isotropy can be determined by a unique condition in place of the two polar relations $R_0=R_1=0$,

$$\begin{aligned} &R_0^2+R_1^2=0 \Rightarrow \\ &\left[(T_{1111}-2T_{1122}-4T_{1212}+T_{2222})^2+16(T_{1112}-T_{1222})^2\right]^2+ \\ &\left[(T_{1111}-T_{2222})^2+4(T_{1112}+T_{1222})^2\right]^2=0 \end{aligned} \tag{4.181}$$

which makes use of a fourth degree invariant.

Finally, using Eq. (4.86), we can give the polar expression of the Lamé's equations (2.139):

$$\boldsymbol{\sigma}=2T_0\boldsymbol{\varepsilon}+(2T_1-T_0)\mathrm{tr}\boldsymbol{\varepsilon}\ \mathbf{I}, \tag{4.182}$$

and of their reversal:

$$\boldsymbol{\varepsilon}=\frac{1}{2T_0}\left(\boldsymbol{\sigma}-\frac{2T_1-T_0}{6T_1-T_0}\mathrm{tr}\boldsymbol{\sigma}\ \mathbf{I}\right). \tag{4.183}$$

4.12.6 Final Considerations About Elastic Symmetries in $\mathbb{R}^2$

The results found in the previous sections, deserve some commentary:

- from a purely *geometric* point of view, i.e. merely considering the elastic symmetries, nothing differentiates ordinary orthotropy from the special orthotropy $R_0=0$: both of them have only a couple of mutually orthogonal symmetry axes. From the algebraic point of view, they are different: they depend upon a different number of independent nonzero invariants and they are determined by invariant conditions concerning invariants of a different order. They also are interpreted differently: ordinary orthotropy corresponds to a precise value taken by the phase angle between the two anisotropic phases, R_0-orthotropy to the vanishing of the anisotropic phase varying with 4θ. Also, while ordinary orthotropy preserves the same morphology also for the inverse tensor, though it is possible a change of type, from $K = 0$ to $k = 1$, R_0-orthotropy does not preserve the same morphology for the compliance tensor, whose components depend upon the two anisotropic phases. From a mechanical point of view, R_0-orthotropic materials have a behavior somewhat different from ordinary orthotropy, e.g. the components vary like those of a second-rank tensor or are isotropic.

- Square symmetric materials share some of the remarks done for R_0-orthotropy, but geometrically speaking they are different from them and from ordinary orthotropy because they have two couples of mutually orthogonal symmetry axes tilted of $\pi/4$. This gives a periodicity of $\pi/2$ to all of the components.
- It can be seen that special orthotropies have some other interesting mechanical properties that are not possessed by ordinarily orthotropic materials. In particular (Vannucci and Verchery 2001), everybody knows that superposing n orthotropic plies does not give an orthotropic laminate, unless the layers have appropriate mutual orientations: ordinary orthotropy is not preserved by the homogenization process that determines the final mechanical characteristics of a laminate, for all of its behaviors: extension, coupling, bending. This is not the case for special orthotropies: a laminate composed by R_0-orthotropic layers will always be R_0-orthotropic, in extension, coupling and bending, no matter of the layer orientations. This is true also for laminates composed by square symmetric layers. In other words, special orthotropies are preserved throughout the homogenization process describing the properties of a laminate. In this sense, quadratic invariants are *more invariants* than the cubic one.
- In the end, there are *six possible cases of algebraically distinct elastic symmetries in* $\mathbb{R}^2$: ordinary orthotropy with $K = 0$ or $K = 1$, R_0-orthotropy, r_0-orthotropy, square symmetry and isotropy.

For ending this part about elastic symmetries in $\mathbb{R}^2$, we resume in Table 4.2 the properties of the different cases seen above.

4.13 The Polar Formulae with the Kelvin's Notation

All the relations given in the previous sections for the polar formalism make use of the tensor notation, using four indexes. We give here also their expression with the Kelvin's notation.

- Cartesian components

$$
\begin{aligned}
T_{11}(\theta)&=T_0+2T_1+R_0\cos 4\,(\Phi_0-\theta)+4R_1\cos 2\,(\Phi_1-\theta),\\
T_{16}(\theta)&=\sqrt{2}\,[R_0\sin 4\,(\Phi_0-\theta)+2R_1\sin 2\,(\Phi_1-\theta)],\\
T_{12}(\theta)&=-T_0+2T_1-R_0\cos 4\,(\Phi_0-\theta),\\
T_{66}(\theta)&=2\,[T_0-R_0\cos 4\,(\Phi_0-\theta)],\\
T_{26}(\theta)&=\sqrt{2}\,[-R_0\sin 4\,(\Phi_0-\theta)+2R_1\sin 2\,(\Phi_1-\theta)],\\
T_{22}(\theta)&=T_0+2T_1+R_0\cos 4\,(\Phi_0-\theta)-4R_1\cos 2\,(\Phi_1-\theta).
\end{aligned}
\tag{4.184}
$$

- Polar parameters

$$
\begin{aligned}
T_0 &= \frac{1}{8}(T_{11}-2T_{12}+2T_{66}+T_{22}),\\
T_1 &= \frac{1}{8}(T_{11}+2T_{12}+T_{22}),\\
R_0 &= \frac{1}{8}\sqrt{(T_{11}-2T_{12}-2T_{66}+T_{22})^2+8(T_{16}-T_{26})^2},\\
R_1 &= \frac{1}{8}\sqrt{(T_{11}-T_{22})^2+2(T_{16}+T_{26})^2},\\
\tan 4\Phi_0 &= \frac{2\sqrt{2}(T_{16}-T_{26})}{T_{11}-2T_{12}-2T_{66}+T_{22}},\\
\tan 2\Phi_1 &= \frac{\sqrt{2}\,(T_{16}+T_{26})}{T_{11}-T_{22}}.
\end{aligned}
\tag{4.185}
$$

4.14 Comparison with the Tsai and Pagano Parameters

The polar formalism and the Tsai and Pagano parameters are methods for the analysis of plane anisotropy. By consequence, there must be some analogies and differences between these two approaches.

Injecting Eq. (4.184) into (3.29) we get

$$
\begin{aligned}
U_1 &= T_0+2T_1,\\
U_2 &= 4R_1\cos 2\Phi_1,\\
U_3 &= R_0\cos 4\Phi_0,\\
U_4 &= -T_0+2T_1,\\
U_5 &= T_0,\\
U_6 &= 2R_1\sin 2\Phi_1,\\
U_7 &= R_0\sin 4\Phi_0,
\end{aligned}
\tag{4.186}
$$

and conversely

$$
\begin{aligned}
T_0 &= U_5,\\
T_1 &= \frac{U_1 - U_5}{2} = \frac{U_4 + U_5}{2},\\
R_0 &= \sqrt{U_3^2 + U_7^2},\\
R_1 &= \frac{1}{4}\sqrt{U_2^2 + 4U_6^2},\\
\tan 4\Phi_0 &= \frac{U_7}{U_3},\\
\tan 2\Phi_1 &= 2\frac{U_6}{U_2},
\end{aligned}
\tag{4.187}
$$

that used in Eq. (4.56) give also

$$
\begin{aligned}
L_1 &= U_1 - U_4 = 2U_5,\\
L_2 &= \frac{U_1 + U_4}{2} = U_1 - U_5,\\
Q_1 &= 4\left(U_3^2 + U_7^2\right),\\
Q_2 &= \frac{U_2^2 + 4U_6^2}{4},\\
C_1 &= \frac{1}{2}\left[U_3\left(U_2^2 - 4U_6^2\right) + 4U_2U_6U_7\right],\\
C_2 &= \frac{1}{2}\left[U_7\left(U_2^2 - 4U_6^2\right) - 4U_2U_3U_6\right].
\end{aligned}
\tag{4.188}
$$

We can hence easily see that only U_1, U_4 and U_5 are tensor invariants; all of them are linear invariants. As we know that only two independent linear invariants can exist for a planar elastic tensor, see Sect. 4.4, this explains why only two among U_1, U_4 and U_5 are independent, Eq. (3.30). Neglecting the use of higher degree invariants, the method of Tsai and Pagano cannot be a tensor representation where some properties appear explicitly, like the third-order invariant expressing the phase angle between the two anisotropic phases, $\Phi_0 - \Phi_1$. For the same reason, it is not immediate to obtain with the Tsai and Pagano parameters the direction of the mirror symmetries when U_6 and U_7 are nonzero, nor to give a decomposition of the strain energy like that given in Sect. 4.10.

Equation $(4.188)_6$ gives the equivalent of the intrinsic condition for ordinary orthotropy, Eq. (4.128):

$$
C_2 = 0 \;\Rightarrow\; U_7\left(U_2^2 - 4U_6^2\right) - 4U_2U_3U_6 = 0. \tag{4.189}
$$

This condition was first given by Pedersen (1990a); however, it is not completely equivalent to (4.128), because to distinguish between the two cases of $K = 0$ and $K = 1$, another condition must be given:

$$K = 0 \rightarrow U_3 > 0, \\ K = 1 \rightarrow U_3 < 0. \tag{4.190}$$

As said in Sect. 4.12.1, the two cases are named respectively low and high shear modulus. Two parameters must vanish for a R_0-orthotropic material

$$U_3 = U_7 = 0, \tag{4.191}$$

and two also for the square symmetry,

$$U_2 = U_6 = 0, \tag{4.192}$$

while isotropy corresponds to four conditions:

$$U_2 = U_3 = U_6 = U_7 = 0. \tag{4.193}$$

We see hence that the polar formalism can express in a more synthetic way all the elastic symmetries in $\mathbb{R}^2$, using invariant parameters that have a clear physical meaning.

4.15 Special Plane Elastic Anisotropic Materials

The analysis of plane anisotropy made so far is tacitly based upon the assumption of *classical elastic body*. The mechanical response of such a body is described by an elastic tensor $\mathbb{E}$ characterized by having the minor and major symmetries, Eqs. (2.13) and (2.11). A large part of existing materials belongs to such a category, namely the most part of materials used for structural purposes, like metallic alloys, wood, composite materials, concrete and so on.

Nevertheless, materials with different tensor symmetries[15] can exist and we briefly consider them in this section. On one side, going towards an ancient, celebrated scientific diatribe in elasticity, we first consider the so-called *rari-constant materials*, having supplementary tensor symmetries adding to the minor and major ones of classical materials. Then, now looking at the most recent researches in mechanics of materials, we shortly analyze *complex materials*, calling with this name all the elastic materials that do not possess all of the minor or major symmetries.

There is a characteristic fact in all these cases: *the number of tensor symmetries is linked to the number of tensor invariants*. In particular, we will see that to any increase of the number of tensor symmetries corresponds a decrease of the number of tensor invariants and vice-versa, when the tensor symmetries decrease, the number of tensor invariants increases.

---

[15] We remember, see Sect. 2.1, that we call *tensor* or *index symmetry* any equivalence of the positions of an index for two or more components of the elastic tensor.

Because the tensor invariants are linked to the material symmetries, it is to be expected that also the pattern of material symmetries changes with that of the tensor symmetries. Actually, this is not automatic: it is so for complex materials, where the number and types of special orthotropies is radically changed with respect to the case of classical materials, but it is not so for rari-constant materials, whose anisotropic part is not affected by the presence of supplementary tensor symmetries. This result puts in evidence an important fact: *there is a link between the possible material symmetries and the tensor symmetries*, i.e. the type of the mechanical response of the material. This fact shows once more that a mere analysis of anisotropy based exclusively upon considerations of *geometric* symmetries of the matter, cannot be exhaustive.

In all the cases, however, the study is greatly facilitated by the use of the polar formalism: the different conditions of symmetry of the elastic response emerge directly and simply as purely algebraic conditions offered by the analysis of the polar invariants, while an analysis based upon considerations of symmetry of the matter should be rather cumbersome.

4.15.1 Rari-Constant Materials

4.15.1.1 A Brief Historical Background

Elastic materials whose behavior is described by a smaller number of parameters have been widely studied in the past and their existence has been the subject of one of the most famous diatribes in the theory of elasticity: that between what Pearson (Todhunter and Pearson 1886, p. 496) named *multi-constant* and *rari-constant* materials (Benvenuto 1991, v. 1, p. 227; Dugas 1950, p. 398; Love 1944, p. 6, p. 13).

The idea of rari-constant materials stems from the early works of Navier (1827) and his model of matter, known as *molecular theory*, first presented at *Académie des Sciences* on May 14, 1821. Basically, the model proposed by Navier aims at explaining the behavior of elastic solids as that of a lattice of particles (*molecules*) interacting together via central forces proportional to their mutual distance. This is not a new idea: it has its last foundation in the works of Newton (1687). For what concerns the mechanics of solids, the true initiator of the molecular theory is considered to be Boscovich (1758); other works on this topic, before the *mémoire* of Navier, are those of Poisson (1811, 1814–1816) on the equilibrium of bent plates, while subsequent fundamental contributions are due to Cauchy (1828a, b), still Poisson (1829) and Saint-Venant (1844a).

The basic idea of the classical molecular approach of Navier and Cauchy, the continuum as a limit of a discrete lattice of particles interacting together via central forces, has a direct consequence (Stackgold 1950; Doyle and Ericksen 1956): 15 moduli describe the behavior of a completely anisotropic body in 3D, and only one modulus suffices to determine it for an isotropic material. These results was not confirmed by experimental tests, so doubts existed about its validity, until the molecular approach was completely by-passed by the theory proposed in 1837 by Green (1839): no underlying microscopic structure of the matter, considered as a continuum, is assumed, and the basic property defining the elastic behavior is *energetic*: in non dissipative processes the internal forces derive from a quadratic potential, see Sect. 2.1.

The consequences of such an assumption lead to the multi-constant model: 21 independent moduli are necessary to describe the elastic response of a completely anisotropic body in 3D, which reduce to only 2 for an isotropic material. The results of the Green's theory were confirmed by experience which, together with its much simpler theoretical background, ensured the success of the multi-constant theory. Nonetheless, the diatribe between the molecular, rari-constant, and continuum, multi-constant, theories lasted a long period: which is the right number of elastic constants and the correct model of elastic continuum?

The further developments of the molecular model by Poincaré (1892), and Voigt (1910), are refined models that, enriching in different ways the original model of Navier, obtain multi-constant theories starting from a molecular model, see Ostoja-Starzewski (2007), Capecchi et al. (2011). More recently, ideas inspired by the Navier–Cauchy approach has produced molecular dynamics models or models for explicating the behavior of complex bodies.

As an effect of this diatribe, the two models are usually considered as opposing and somewhat irreconcilable, though different researchers has made attempts to show that this is not the case, cf. Born (1915), Love (1944), Note B, p. 616, Muskhelishvili (1953), p. 55.

The results presented below, taken from Vannucci and Desmorat (2016), concern the planar case and show some new results for an old problem: there exist two dual types of rari-constant materials and the classical Cauchy–Poisson conditions are not sufficient to characterize such a material: the only true necessary and sufficient condition is the number of independent linear tensor invariants, that must be of one.

4.15.1.2 The Polar Approach to the Study of Planar Rari-Constant Materials

Within the classical paradigm of elasticity, see Sect. 2.1, $\mathbb{E}$ possesses the minor, Eq. (2.13), and major, Eq. (2.11), symmetries of the indexes, and it describes a so-called *multi-constant material*: we know that for the complete anisotropic case, a whole of 21 independent components (18 tensor invariants and 3 frame dependent parameters) determine the material behavior; they reduce to only two for the isotropic case. In the plane case, there are 5 invariants plus a quantity taking into account for the frame orientation.

Let us ponder the consequences of the existence of 6 supplementary index symmetries, the so-called *Cauchy–Poisson symmetries*:

$$E_{ijkl} = E_{ikjl}, \tag{4.194}$$

that for the plane case reduce to the only supplementary condition

$$E_{1122} = E_{1212}. \tag{4.195}$$

It is immediate to recognize that in such a case the behavior is described by only 12 tensor invariants plus 3 quantities fixing the frame, for a whole of 15 independent components. In the plane case, we have 5 independent components, 4 of which are invariants, and isotropy is always described by a unique invariant quantity. Finally, the existence of supplementary index symmetries decreases the number of the material parameters needed to describe the material behavior; that is why, materials of such a type are called *rari-constant*.

Let us concentrate on the planar case; from now on, we identify *rari-constant tensors* with those satisfying the Cauchy–Poisson conditions, and we show that identifying *rari-constant materials* is not so simple, because there are two possible and dual rari-constant materials, at least in $\mathbb{R}^2$.

We can easily state now the algebraic conditions for the *elastic tensor* $\mathbb{E}$ in $\mathbb{R}^2$ to be rari-constant:

Theorem 10 *$\mathbb{E}$ is a rari-constant elastic tensor in $\mathbb{R}^2$* $\iff$ $T_0 = T_1$.

Proof The proof is immediate: if $\mathbb{E}$ is a rari-constant tensor, then $E_{1212}(\theta) = E_{1122}(\theta)\ \forall\theta$, and Eq. (4.67)$_{3,4}$ give $T_0 = T_1$. Conversely, if $T_0 = T_1$, then Eq. (4.66)$_{1,2}$ give $E_{1212}(\theta) = E_{1122}(\theta)\ \forall\theta$.

Let us consider all the consequences of such a result:

- the number of independent tensor invariants is linked to the number of index symmetries; in particular, a supplementary index symmetry corresponds to the identity of two invariants, so that the number of independent invariants is decreased by one;
- the rari-constant condition affects only the isotropic part of $\mathbb{E}$, i.e. only its linear invariants: the anisotropic part is not touched by the Cauchy–Poisson conditions, so that multi- and rari-constant materials share all the same types of elastic symmetries;
- the bounds on the polar parameters, Eq. (4.111), do not exclude the existence of the case $T_0 = T_1$: in the classical frame of continuum elastic bodies, materials with a rari-constant tensor $\mathbb{E}$ are *possible*;
- the existence of multi-constant materials with $T_0 = T_1$ *is not allowed*; this point is essential: apparently, just because Eq. (4.111) do not exclude the case $T_0 = T_1$ for multi-constant materials, then such materials could exist; nevertheless, this is not possible, because of Theorem 10; physically, this means that whenever $T_0 = T_1$, then tensor $\mathbb{E}$ is *necessarily* rari-constant: $E_{1212}(\theta) = E_{1122}(\theta)\ \forall\theta$: a particular value of the tensor invariants determine a change of the algebraic properties of the elastic tensor;

A fundamental remark can now be done: *all what has been said for* $\mathbb{E}$ *is equally valid for* $\mathbb{S}$: we can define a *dual class* of rari-constant materials, where the Cauchy–Poisson conditions are valid for the compliance tensor $\mathbb{S}$. We name in the following *direct-* and *inverse-* rari-constant materials those for which the Cauchy–Poisson condition (4.195) holds respectively for $\mathbb{E}$ or for $\mathbb{S}$. These two classes are *necessarily distinct*, i.e. it cannot exist a material being at the same time direct- and inverse-rari-constant: the Cauchy–Poisson conditions cannot be satisfied at the same time by $\mathbb{E}$ and $\mathbb{S}$. That is why the name rari-constant has been used not only to denote a class of materials, but also a type of elastic tensor: this distinction is necessary in the following.

For proving why a material cannot be at the same time direct- and rari-constant, we need first a preliminary result:

Theorem 11 *The value*

$$T_0 = \frac{4R_1^2 - R_0^2}{3} \tag{4.196}$$

is incompatible with the elastic bounds (4.111) on the polar invariants for direct-rari-constant materials, i.e. when $T_1 = T_0$.

Proof Replacing Eq. (4.196) into $(4.111)_1$ and taking into account for Eqs. $(4.111)_{3,4}$ gives

$$R_1 > R_0 > 0. \tag{4.197}$$

Now, injecting Eq. (4.196) into $(4.111)_2$ we get, after posing

$$\rho = \frac{R_0}{R_1}, \quad C = \cos 4(\Phi_0 - \Phi_1), \quad 0 \leq \rho < 1, \ -1 \leq C \leq 1, \tag{4.198}$$

$$\sqrt{\frac{4-\rho^2}{3}} < \frac{3\rho\, C}{1+2\rho^2}, \tag{4.199}$$

a condition that is satisfied if and only if

$$\begin{cases} \dfrac{4-\rho^2}{3} \geq 0, \\ \dfrac{3\rho\, C}{1+2\rho^2} \geq 0, \\ \dfrac{4-\rho^2}{3} < \dfrac{9\rho^2 C^2}{(1+2\rho^2)^2}. \end{cases} \tag{4.200}$$

Condition $(4.200)_1$ gives $\rho \leq 2$, which is redundant because of Eq. $(4.198)_3$, condition $(4.200)_2$ limits Eq. $(4.198)_4$ to $0 \leq C \leq 1$ while condition $(4.200)_3$ can be rewritten as

$$f = \frac{(4-\rho^2)(1+2\rho^2)^2}{27\rho^2} < C^2, \tag{4.201}$$

which is never satisfied because $f > 1 = \max C^2$ for $0 \leq \rho < 1$, as it can be easily recognized.

The isotropic case is trivial, for Eq. (4.196) should give $T_0 = 0$ which corresponds to a material with a null stiffness, hence it is impossible.

The two cases of special orthotropies are also impossible; in fact, the case of square symmetry, $R_1 = 0$, should imply a negative value for T_0^2, Eq. (4.196), while that of R_0-orthotropy, $R_0 = 0 \Rightarrow \rho = 0$, gives $f \to \infty$.

Theorem 12 *The Cauchy–Poisson condition (4.195) cannot be satisfied at the same time by* $\mathbb{E}$ *and* $\mathbb{S}$.

Proof Be $\mathbb{E}$ rari-constant, i.e. $E_{1122} = E_{1212}$; then $T_0 = T_1$ by Theorem 10. The polar invariants of $\mathbb{S}$ can then be calculated through Eqs. (4.68) and (4.71) that in this case become:

$$\begin{aligned}
t_0 &= \frac{2}{\Delta}\left(T_0^2 - R_1^2\right),\\
t_1 &= \frac{1}{2\Delta}\left(T_0^2 - R_0^2\right),\\
r_0 e^{4i\varphi_0} &= \frac{2}{\Delta}\left(R_1^2 e^{4i\Phi_1} - T_0 R_0 e^{4i\Phi_0}\right),\\
r_1 e^{2i\varphi_1} &= -\frac{R_1 e^{2i\Phi_1}}{\Delta}\left[T_0 - R_0 e^{4i(\Phi_0-\Phi_1)}\right].
\end{aligned} \tag{4.202}$$

with

$$\Delta = 8T_0\left(T_0^2 - R_0^2\right) - 16R_1^2\left[T_0 - R_0\cos 4\left(\Phi_0 - \Phi_1\right)\right]. \tag{4.203}$$

It is then apparent that

$$t_0 = t_1 \iff T_0^2 = \frac{4R_1^2 - R_0^2}{3}. \tag{4.204}$$

This value of T_0 is incompatible with the elastic bounds (4.111), as shown in Theorem 11, and hence, $t_0 \neq t_1$ when $T_0 = T_1$, so by Theorem 10 applied to $\mathbb{S}$, $S_{1212} \neq S_{1122}$.

The consequence is immediate: it is not correct to identify *automatically* rari-constant materials in $\mathbb{R}^2$ with the Cauchy–Poisson condition, because this concerns only one of the two elastic tensors of the material.

So, if $\mathbb{E}$ is rari-constant, it has only 5 distinct Cartesian components, but its inverse, $\mathbb{S}$ has 6 different components. Conversely, if $\mathbb{S}$ is rari-constant, it has 5 distinct Cartesian components, but they are 6 for $\mathbb{E}$. Nevertheless, in both the cases the number of independent tensor invariants is 4. In fact, if $\mathbb{E}$ is rari-constant, then $T_0 = T_1$ and by Eq. (4.202) we get

$$t_1 = \frac{T_0^2 - R_0^2}{4(T_0^2 - R_1^2)}t_0. \tag{4.205}$$

Hence, though $t_1 \neq t_0$, it is proportional to t_0. Of course, a similar relation exists for the dual case of $\mathbb{S}$ rari-constant, it is sufficient to swap lower- and upper-case letters.

Finally, there are two dual families of rari-constant materials:

- the *direct* rari-constant materials:

$$\begin{aligned} &E_{1212}(\theta) = E_{1122}(\theta)\ \forall\theta,\\ &T_0 = T_1,\\ &S_{1212}(\theta) \neq S_{1122}(\theta),\\ &t_1 = \frac{T_0^2 - R_0^2}{4(T_0^2 - R_1^2)} t_0, \end{aligned} \tag{4.206}$$

and

$$\begin{aligned} &E_{1111}(\theta) = 3T_0 + R_0\cos 4(\Phi_0 - \theta) + 4R_1\cos 2(\Phi_1 - \theta),\\ &E_{1112}(\theta) = R_0\sin 4(\Phi_0 - \theta) + 2R_1\sin 2(\Phi_1 - \theta),\\ &E_{1122}(\theta) = E_{1212}(\theta) = T_0 - R_0\cos 4(\Phi_0 - \theta),\\ &E_{1222}(\theta) = -R_0\sin 4(\Phi_0 - \theta) + 2R_1\sin 2(\Phi_1 - \theta),\\ &E_{2222}(\theta) = 3T_0 + R_0\cos 4(\Phi_0 - \theta) - 4R_1\cos 2(\Phi_1 - \theta). \end{aligned} \tag{4.207}$$

- the *inverse* rari-constant materials:

$$\begin{aligned} &S_{1212}(\theta) = S_{1122}(\theta)\ \forall\theta,\\ &t_0 = t_1,\\ &E_{1212}(\theta) \neq E_{1122}(\theta)\ \forall\theta,\\ &T_1 = \frac{t_0^2 - r_0^2}{4(t_0^2 - r_1^2)} T_0, \end{aligned} \tag{4.208}$$

and

$$\begin{aligned} &S_{1111}(\theta) = 3t_0 + r_0\cos 4(\varphi_0 - \theta) + 4r_1\cos 2(\varphi_1 - \theta),\\ &S_{1112}(\theta) = r_0\sin 4(\varphi_0 - \theta) + 2r_1\sin 2(\varphi_1 - \theta),\\ &S_{1122}(\theta) = S_{1212}(\theta) = t_0 - r_0\cos 4(\varphi_0 - \theta),\\ &S_{1222}(\theta) = -r_0\sin 4(\varphi_0 - \theta) + 2r_1\sin 2(\varphi_1 - \theta),\\ &S_{2222}(\theta) = 3t_0 + r_0\cos 4(\varphi_0 - \theta) - 4r_1\cos 2(\varphi_1 - \theta). \end{aligned} \tag{4.209}$$

A further consideration concerns strain energy, V: the general expression (4.89) becomes

$$V = 4T_0 t^2 + 8R_1\cos 2(\Phi_1 - \varphi) r\, t + 2\,[T_0 + R_0\cos 4(\Phi_0 - \varphi)]\, r^2, \tag{4.210}$$

for direct- rari-constant materials, while for the inverse- ones we get

$$V = 4\alpha T_0 t^2 + 8R_1\cos 2(\Phi_1 - \varphi) r\, t + 2\,[T_0 + R_0\cos 4(\Phi_0 - \varphi)]\, r^2, \tag{4.211}$$

where

$$\alpha = \frac{t_0^2 - r_0^2}{4(t_0^2 - r_1^2)}. \tag{4.212}$$

Finally, if we consider that special orthotropies are characterized by the vanishing of a tensor invariant, i.e. $R_0 = 0$ for the case of R_0-orthotropy, while $R_1 = 0$ for square-symmetry, or by being an invariant a function of the other ones, for the case of r_0-orthotropic materials, then it is clear that the only necessary and sufficient condition for identifying a rari-constant material, regardless of its type, i.e. independently of the number of distinct Cartesian components for $\mathbb{E}$ or $\mathbb{S}$, is that the number of *independent linear tensor invariants must be one*.

Two last remarks: first, while rari-constant materials can actually exist, multi-constant materials with $T_0 = T_1$ or $t_0 = t_1$ are not allowed. Then, condition $T_0 = T_1$ clearly indicates that the anisotropic part of a plane elastic tensor is necessarily rari-constant; in other words, in $\mathbb{R}^2$ only the isotropic part is responsible for the behavior to be multi- or rari-constant.

4.15.1.3 How to Fabricate Rari-Constant Anisotropic Layers

We show now that it is actually possible to fabricate both the cases of direct- and inverse- rari-constant layers. This can be done using appropriate volume fractions of unidirectional fibers to reinforce an isotropic matrix. Just as an example, rather simple and having the only purpose to show that these materials can be designed, we use the classical technical laws of homogenization (Jones 1999) (E indicates a Young's modulus, G a shear modulus and ν a Poisson's coefficient):

$$\begin{aligned} E_1 &= v_f E_f + (1 - v_f) E_m, \\ E_2 &= \frac{E_f E_m}{(1 - v_f) E_f + v_f E_m}, \\ G_{12} &= \frac{G_f G_m}{(1 - v_f) G_f + v_f G_m}, \\ \nu_{12} &= v_f \nu_f + (1 - v_f) \nu_m. \end{aligned} \tag{4.213}$$

The above relations give the technical constants for an orthotropic layer reinforced by unidirectional fibers; the subscript f denotes a constant of the fibers and m of the matrix; v_f is the volume fraction of the fibers, $0 \leq v_f \leq 1$.

We pose

$$E_f = m\, E_m, \quad \nu_f = n\, \nu_m, \tag{4.214}$$

and remark that usually $m \gg 1$ while $n \sim 1$. Then, remembering that for the isotropic case (so are assumed to be the fibers and the matrix)

$$G = \frac{E}{2(1 + \nu)}, \tag{4.215}$$

we obtain

$$\begin{aligned}
E_1 &= [1+(m-1)v_f]E_m,\\
E_2 &= \frac{m\,E_m}{m+v_f(1-m)},\\
G_{12} &= \frac{m\,E_m}{2[m(1-v_f)(1+\nu_m)+v_f(1+n\,\nu_m)]},\\
\nu_{12} &= [1+(n-1)v_f]\nu_m.
\end{aligned} \tag{4.216}$$

Then, for an orthotropic layer in a plane stress state it is, cf. Eq. (3.70),

$$\begin{aligned}
E_{1111} &= \frac{E_1}{1-\nu_{12}\nu_{21}},\\
E_{2222} &= \frac{E_2}{1-\nu_{12}\nu_{21}},\\
E_{1122} &= \frac{\nu_{12}E_2}{1-\nu_{12}\nu_{21}},\\
E_{1212} &= G_{12},\\
E_{1112} &= E_{1222} = 0,
\end{aligned} \tag{4.217}$$

and, see Eq. (2.95),

$$\nu_{21} = \nu_{12}\frac{E_2}{E_1}. \tag{4.218}$$

Finally, one should remember Eq. $(4.66)_{1,2}$:

$$\begin{aligned}
8T_0 &= E_{1111} - 2E_{1122} + 4E_{1212} + E_{2222},\\
8T_1 &= E_{1111} + 2E_{1122} + E_{2222}.
\end{aligned} \tag{4.219}$$

Then, let us now consider the direct- rari-constant case: the condition to be satisfied is

$$T_0 = T_1, \tag{4.220}$$

completely equivalent to the Cauchy–Poisson condition (4.195). Injecting successively Eq. (4.216) into (4.217) and then into Eq. (4.219), after some standard passages condition (4.220) becomes

$$\begin{aligned}
&[1+(m-1)v_f][m+v_f(1-m)] - m\,\nu_m^2[1+(n-1)v_f]^2 -\\
&2\nu_m[1+(m-1)v_f][1+(n-1)v_f][m(1-v_f)(1+\nu_m)+v_f(1+n\,\nu_m)] = 0.
\end{aligned} \tag{4.221}$$

The only material parameter entering the above condition is ν_m; this condition depends upon m, n and v_f. The solution of such equation, for different values of ν_m, is plotted in the left diagram of Fig. 4.5. For a given couple of materials, fibers

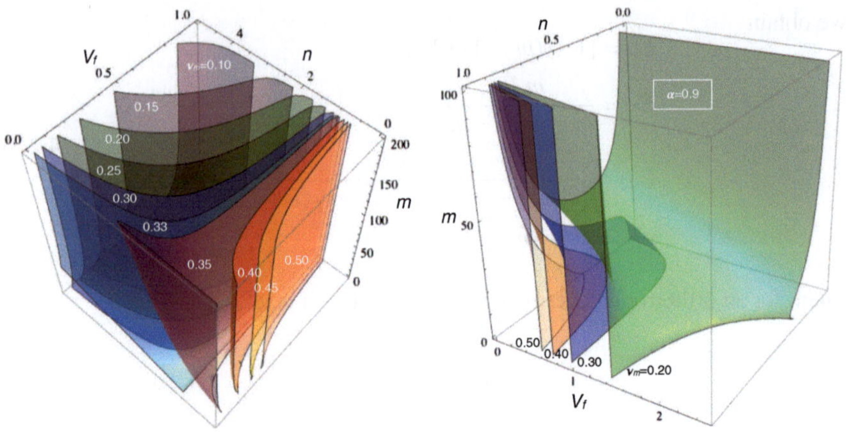

(a) Direct rari-constant materials (b) Inverse rari-constant materials

Fig. 4.5 Solutions for rari-constant anisotropic layers

and matrix, m and n are fixed and Eq. (4.221) allows for finding a value of the volume fraction of fibers v_f resulting in a direct- rari-constant layer.

The inverse- rari-constant case is characterized by

$$t_0 = t_1, \tag{4.222}$$

which gives, for the stiffness constant, the condition

$$T_1 = \alpha T_0, \tag{4.223}$$

with α given by Eq. (4.212). The passage to the stiffness components is necessary because the homogenization formulae (4.213) are given as functions of these ones. Unlike the previous case of direct rari-constant layers, now condition (4.223) is not equivalent to the Cauchy–Poisson's one, but to, Eq. (4.219),

$$(E_{1111} + E_{2222})(1-\alpha) + 2E_{1122}(1+\alpha) - 4\alpha E_{1212} = 0, \tag{4.224}$$

which becomes, after introducing Eqs. (4.216), (4.217) and some standard passages,

$$\begin{aligned}
&[1+(m-1)v_f]\{[(m-1)^2 v_f^2 - (m-1)^2 v_f - 2m](\alpha-1)+\\
&2m(\alpha+1)[1+v_f(n-1)]\nu_m\}[v_f(1+n\,\nu_m) + m(1-v_f)(1+\nu_m)]+\\
&2m\alpha\{(v_f-1)v_f + m^2(v_f-1)v_f+\\
&m[[1+v_f(n-1)]^2\nu_m^2] - 2v_f^2 + 2v_f - 1\} = 0.
\end{aligned} \tag{4.225}$$

In this case, the solution depends also upon the parameter α, i.e. upon the compliance invariants, those of $\mathbb{S}$, of the final layer. These can be calculated only inverting the stiffness tensor $\mathbb{E}$, so the problem is nonlinear and the computation can be done iterating. Anyway, like in the case of the direct- rari-consant layers, also in this case it is possible to find from Eq. (4.225) a value of the volume fraction of the fibers v_f to obtain an inverse- rari-constant layer; Eq. (4.225) is of the fourth degree in v_f, hence a numerical procedure is needed for the resolution. The solution of Eq. (4.225) is plotted in the right diagram of Fig. 4.5.

A question concerns α: can α take any value? Actually not, and this can be seen in the following way: introduce the two dimensionless parameters

$$\eta_0 = \frac{r_0}{t_0},\ \eta_1 = \frac{r_1}{t_0}. \tag{4.226}$$

Then,

$$\alpha = \frac{1-\eta_0^2}{4(1-\eta_1^2)}, \tag{4.227}$$

and we can study its variation as a function of η_0 and η_1, considering that the bounds (4.111) applied to an orthotropic $\mathbb{S}$ and for $t_0 = t_1$ give

$$\begin{aligned} &0 \leq \eta_0 < 1,\\ &0 \leq \eta_1 < \sqrt{\frac{1+(-1)^k\eta_0}{2}}. \end{aligned} \tag{4.228}$$

Function $\alpha(\eta_0, \eta_1)$ is plotted in Fig. 4.6 and it is evident that $0 < \alpha < 1$. Incidentally, the fact that for inverse- rari-constant layers it cannot be $\alpha = 1$ confirms that a material cannot be at the same time direct- and inverse- rari-constant.

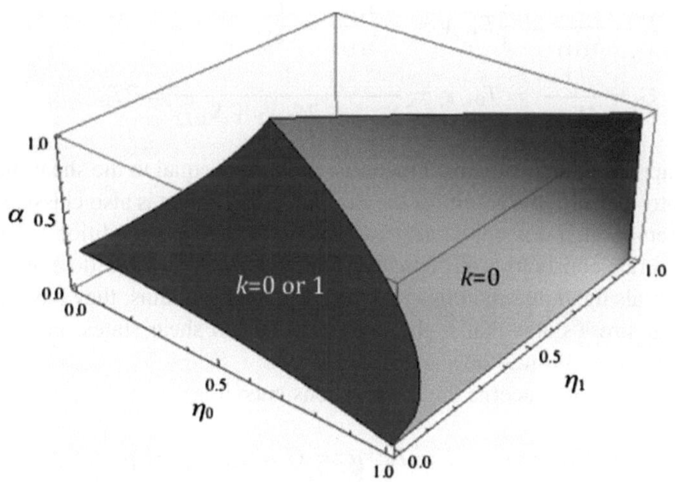

Fig. 4.6 Function $\alpha(\eta_0, \eta_1)$

4.15.1.4 The Isotropic Case

Two isotropic rari-constant materials can exist, the direct- and the inverse- one. Let us briefly consider their properties.

The direct case first: isotropy is characterized by the vanishing of the anisotropic part, i.e. by

$$R_0 = R_1 = 0 \Rightarrow r_0 = r_1 = 0. \tag{4.229}$$

The stiffness behavior is uniquely determined by T_0:

$$E_{1111}(\theta) = E_{2222}(\theta) = 3T_0,\ E_{1122}(\theta) = T_0,\ E_{1112}(\theta) = E_{1222}(\theta) = 0\ \forall\theta. \tag{4.230}$$

For $\mathbb{S}$, it is

$$t_0 = \frac{1}{4T_0},\ t_1 = \frac{1}{16T_0} \Rightarrow t_0 = 4t_1 \tag{4.231}$$

and, $\forall\theta$,

$$\begin{aligned}
&S_{1111}(\theta) = S_{2222}(\theta) = t_0 + 2t_1 = \frac{3}{2}t_0 = \frac{3}{8T_0},\\
&S_{1122}(\theta) = -t_0 + 2t_1 = -\frac{t_0}{2} = -\frac{1}{8T_0},\\
&S_{1212}(\theta) = t_0 = \frac{1}{4T_0} \Rightarrow S_{1212}(\theta) = -2S_{1122}(\theta),\\
&S_{1112}(\theta) = S_{1222}(\theta) = 0.
\end{aligned} \tag{4.232}$$

We can also introduce the classic technical constants:

$$\begin{aligned}
E &= \frac{1}{S_{1111}} = \frac{8}{3}T_0,\ \nu = -\frac{S_{1122}}{S_{1111}} = \frac{1}{3},\\
G &= \frac{1}{4S_{1212}} = T_0,\ \kappa = \frac{1}{S_{1111} + 2S_{1122} + S_{2222}} = 2T_0.
\end{aligned} \tag{4.233}$$

It is then apparent the mechanical meaning of T_0: it is equal to the shear modulus G for the isotropic case; the result for the Poisson's coefficient is also classical, but it is worth to remark that it is only a *necessary* but *not sufficient* condition for a material to be direct rari-constant: multi-constant materials with $\nu = 1/3$ do exist. Also, for these materials the bulk modulus κ is twice the shear modulus: they have a stiffness to spherical stress states that is the double of that to shear states. For the normal stiffness, this is $8/3$ times the shear one.

Finally, for what concerns the Lamé's constants, it is

$$\lambda := \kappa - G = T_0,\ \mu := G = T_0 \Rightarrow \lambda = \mu, \tag{4.234}$$

a classical result.

Let us now turn the attention to inverse- rari-constant materials; now, t_0 uniquely determines all the distinct components of $\mathbb{S}$:

$$S_{1111}(\theta) = S_{2222}(\theta) = 3t_0,\ S_{1122}(\theta) = t_0,\ S_{1112}(\theta) = S_{1222}(\theta) = 0\ \forall\theta. \quad (4.235)$$

For tensor $\mathbb{E}$, we get

$$T_0 = \frac{1}{4t_0},\ T_1 = \frac{1}{16t_0} \Rightarrow T_0 = 4T_1 \quad (4.236)$$

and, $\forall\theta$,

$$\begin{aligned} E_{1111}(\theta) &= E_{2222}(\theta) = T_0 + 2T_1 = \frac{3}{2}T_0 = \frac{3}{8t_0},\\ E_{1122}(\theta) &= -T_0 + 2T_1 = -\frac{T_0}{2} = -\frac{1}{8t_0},\\ E_{1212}(\theta) &= T_0 = \frac{1}{4t_0} \Rightarrow E_{1212}(\theta) = -2E_{1122}(\theta),\\ E_{1112}(\theta) &= E_{1222}(\theta) = 0. \end{aligned} \quad (4.237)$$

Now, the technical constants are

$$E = \frac{4}{3}T_0,\ \nu = -\frac{1}{3},\ G = T_0,\ \kappa = \frac{T_0}{2}. \quad (4.238)$$

Inverse- rari-constant materials are hence necessarily materials with a negative Poisson's coefficient, whose value is exactly the opposite of the direct case; nevertheless, they can exist. Also, their normal stiffness is just half and their bulk modulus a fourth of the corresponding direct case ones. Now, the spherical stiffness is smaller than the shear one.

The Lamé's constants now are

$$\lambda = -\frac{T_0}{2},\ \mu = T_0 \Rightarrow \lambda = -\frac{\mu}{2}, \quad (4.239)$$

i.e. λ is negative; nevertheless, thanks to Eq. (4.111), the bounds on the values of the Lamé's constants in $\mathbb{R}^2$ are satisfied:

$$\mu = T_0 > 0, \lambda + \mu = \frac{T_0}{2} > 0. \quad (4.240)$$

Concerning the strain energy, Eq. (4.210) becomes

$$V = 2T_0(2t^2 + r^2), \quad (4.241)$$

while Eq. (4.211), being for the isotropic case $\alpha = 1/4$, becomes

$$V = T_0(t^2 + 2r^2). \tag{4.242}$$

It is interesting to notice, as easily seen from Eq. (4.93), that in the isotropic case V_s is proportional to t^2 and V_d to r^2, so that $V_s \sim 2V_d$ for the direct case while $V_s \sim V_d/2$ for the inverse one. In other words, direct rari-constant materials stock better the elastic energy under the form of energy linked to volume changes than to shape changes, while for the inverse ones it is exactly the opposite.

4.15.2 Complex Materials

The case of complex materials, indicating here bodies whose elastic tensor has only a part of the minor and/or major symmetries, has been addressed in a theoretical work by Vannucci and Verchery (2010). In particular, two cases have been examined: the first one, when $\mathbb{E}$ does not have the minor symmetries, and the second one when it has not the major ones. In both the cases, the way to handle the problem is that typical of the polar approach, of course modified by the different number of index symmetries. This is actually the key point: the influence that the index symmetries have on the anisotropic behavior. We give here only some details on both the cases considered in the cited paper, referring the reader to the same article for the theoretical developments, that are rather technical.

Let us first consider the case of a material whose elastic tensor does not have the minor symmetries; in such a case the elastic tensor has 10 independent Cartesian components in the planar case, represented by an equal number of polar parameters, 9 of them being tensor invariants:

$$\begin{aligned}
E_{1111} &= T_0 + T_1 + T_2 + R_0 \cos 4\Phi_0 + 2R_1 \cos 2\Phi_1 + 2R_2 \cos 2\Phi_2,\\
E_{1112} &= -T_3 + R_0 \sin 4\Phi_0 + 2R_2 \sin 2\Phi_2,\\
E_{1121} &= T_3 + R_0 \sin 4\Phi_0 + 2R_1 \sin 2\Phi_1,\\
E_{1122} &= -T_0 + T_1 + T_2 - R_0 \cos 4\Phi_0,\\
E_{1212} &= T_0 + T_1 - T_2 - R_0 \cos 4\Phi_0 + 2R_1 \cos 2\Phi_1 - 2R_2 \cos 2\Phi_2,\\
E_{1221} &= T_0 - T_1 + T_2 - R_0 \cos 4\Phi_0,\\
E_{1222} &= -T_3 - R_0 \sin 4\Phi_0 + 2R_1 \sin 2\Phi_1,\\
E_{2121} &= T_0 + T_1 - T_2 - R_0 \cos 4\Phi_0 - 2R_1 \cos 2\Phi_1 + 2R_2 \cos 2\Phi_2,\\
E_{1112} &= T_3 - R_0 \sin 4\Phi_0 + 2R_2 \sin 2\Phi_2,\\
E_{2222} &= T_0 + T_1 + T_2 + R_0 \cos 4\Phi_0 - 2R_1 \cos 2\Phi_1 - 2R_2 \cos 2\Phi_2.
\end{aligned} \tag{4.243}$$

The invariants are all the polar moduli T_0, T_1 etc. and the differences of the polar angles $\Phi_0 - \Phi_2$ and $\Phi_1 - \Phi_2$. For this case, it can be proved that ordinary orthotropy corresponds to the conditions

$$\Phi_0 - \Phi_1 = K_{01}\frac{\pi}{4}, \quad \Phi_0 - \Phi_2 = K_{02}\frac{\pi}{4}, \quad \Phi_1 - \Phi_2 = K_{12}\frac{\pi}{2}. \tag{4.244}$$

As a consequence, there are four possible different ordinary orthotropic materials sharing the same polar moduli and determined by the combinations $K_{02} = K_{12} = 0$, $K_{02} = 1$ and $K_{12} = 0$, $K_{02} = 0$ and $K_{12} = 1$, $K_{02} = K_{12} = 1$.

Besides these four ordinary cases, there are six different special orthotropies, characterized by the following conditions:

$$\begin{aligned} R_0 &= 0,\ K_{12} = 0, \quad R_0 = 0,\ K_{12} = 1,\\ R_1 &= 0,\ K_{02} = 0, \quad R_1 = 0,\ K_{02} = 1,\\ R_2 &= 0,\ K_{01} = 0, \quad R_2 = 0,\ K_{01} = 1. \end{aligned} \tag{4.245}$$

It is interesting also to remark that for these materials, isotropy is given by the conditions:

$$T_3 = R_0 = R_1 = R_2 = 0; \tag{4.246}$$

there is hence a condition on a linear invariant, T_3, needed to ensure the invariance of the material response under a mirror symmetry about an axis. The relations between the Cartesian and polar components in this case are

$$\begin{aligned} E_{1111} &= E_{2222} = T_0 + T_1 + T_2,\\ E_{1122} &= -T_0 + T_1 + T_2,\\ E_{1212} &= E_{2121} = T_0 + T_1 - T_2,\\ E_{1221} &= T_0 - T_1 + T_2, \end{aligned} \tag{4.247}$$

the remaining components being null. Isotropy is hence determined by three independent moduli, not by two as for classical materials.

For the second case, a tensor without the major symmetries, there are 9 independent components and it is:

$$\begin{aligned} E_{1111} &= T_0 + 2T_1 + R_0 \cos 4\Phi_0 + 2R_1 \cos 2\Phi_1 + 2R_2 \cos 2\Phi_2,\\ E_{1112} &= -T_3 + R_0 \sin 4\Phi_0 + 2R_2 \sin 2\Phi_2,\\ E_{1122} &= -T_0 + 2T_1 - R_0 \cos 4\Phi_0 + 2R_1 \cos 2\Phi_1 - 2R_2 \cos 2\Phi_2,\\ E_{1211} &= T_3 + R_0 \sin 4\Phi_0 + 2R_1 \sin 2\Phi_1,\\ E_{1212} &= T_0 - R_0 \cos 4\Phi_0,\\ E_{1222} &= -T_3 - R_0 \sin 4\Phi_0 + 2R_1 \sin 2\Phi_1,\\ E_{2211} &= -T_0 + 2T_1 - R_0 \cos 4\Phi_0 - 2R_1 \cos 2\Phi_1 + 2R_2 \cos 2\Phi_2,\\ E_{1121} &= T_3 - R_0 \sin 4\Phi_0 + 2R_2 \sin 2\Phi_2,\\ E_{2222} &= T_0 + 2T_1 + R_0 \cos 4\Phi_0 - 2R_1 \cos 2\Phi_1 - 2R_2 \cos 2\Phi_2. \end{aligned} \tag{4.248}$$

The elastic behaviors in the two cases, in term of invariants, differ only for a term of the isotropic part, this implying as additional result that the whole discussion of anisotropy does not change with respect to the previous case. In particular, the number and types of orthotropies are quite the same of the previous case. In particular, it is easily seen that isotropy in this case perfectly coincides with that of classical materials.

The two examples briefly considered above, clearly show that there is an influence of the tensor symmetries, i.e. of the algebraic structure of the elastic tensor, on the elastic symmetries.

Finally, the following considerations can be done:

- the number of independent invariants, and hence of parameters determining intrinsically the behavior of the material, depends upon the number and type of index symmetries;
- the number and types of algebraically distinct types of orthotropy depends upon the index symmetries of the elastic tensor;
- in some cases of special orthotropy, some Cartesian components are null, or constant or vary with the orientation angle like a second-rank tensor component;
- the number of independent elastic constants in isotropy is two only if the elastic tensor has the minor indicial symmetries, otherwise the constants are three;
- it is easily recognized that classical hyperelastic materials can be recovered from more complex elastic materials as a particular case.

4.16 Special Topics of the Polar Formalism

4.16.1 *Polar Projectors*

We introduce now a new formalism for plane elasticity (Desmorat and Vannucci 2014), inspired by the Kelvin decomposition[16] but based upon the polar formalism. The way we find it is rather easy: Eq. (4.67) shows that the matrix representing tensor $\mathbb{E}$ in the Kelvin formalism, Eq. (4.184), can be written as

[16]The Kelvin decomposition (Kelvin 1856, 1878), basically consists in the diagonalization of the elasticity matrix as defined in Eq. (2.24); some simple algebraic passages show that the elasticity tensor $\mathbb{E}$ can be represented as

$$\mathbb{E} = \lambda^i \mathbb{P}^i, \quad i = 1, ..., 6,$$

where the *Kelvin projectors* $\mathbb{P}^i$ are fourth-rank dimensionless tensors defined as (no summation over i)

$$\mathbb{P}^i = \mathbf{E}^i \otimes \mathbf{E}^i.$$

The scalars λ^i and the second-rank tensors $\mathbf{E}^i$ are the couples eigenvalue-eigenvector of the equation (no summation over i)

$$\mathbb{E}\mathbf{E}^i = \lambda^i \mathbf{E}^i.$$

The λ^i are homogeneous to a modulus, and are called the *Kelvin moduli*, while the $\mathbf{E}^i$ are homogeneous to a strain and are called the *Kelvin modes*.

$$\mathbb{E}(\theta) = T_0\ \mathbb{T}_0 + 2T_1\ \mathbb{T}_1 + R_{0c}\ \mathbb{R}_{0c} + R_{0s}\ \mathbb{R}_{0s} + 2R_1\ \mathbb{R}_1, \tag{4.249}$$

where, for the sake of simplicity and without loss of generality, we have fixed a frame choosing $\Phi_1 = 0$, and put

$$R_0 \cos 4\Phi_0 = R_{0c}, \quad R_0 \sin 4\Phi_0 = R_{0s}. \tag{4.250}$$

It is immediately recognized that R_{0c} and R_{0s} are tensor invariants too. If the material is ordinarily orthotropic, then, see Eq. (4.128)$_2$ and recall that we have put $\Phi_1 = 0$,

$$\Phi_0 - \Phi_1 (= \Phi_0) = K\frac{\pi}{4}, \ K = 0, 1 \ \rightarrow \ R_{0c} = (-1)^K R_0, \ \ R_{0s} = 0, \tag{4.251}$$

while the invariant R_{0c} vanishes whenever

$$\Phi_0 - \Phi_1 (= \Phi_0) = \frac{\pi}{8} + \kappa\frac{\pi}{4}, \ \ \kappa \in \mathbb{N} \ \rightarrow \ R_{0c} = 0. \tag{4.252}$$

Finally, the particular case of r_0-orthotropy gives, see Eqs. (4.172) and (4.251),

$$R_{0c} = -\frac{R_1^2}{T_1}, \ \ R_{0s} = 0. \tag{4.253}$$

In Eq. (4.249), $\mathbb{T}_0$, $\mathbb{T}_1$, $\mathbb{R}_{0c}$, $\mathbb{R}_{0s}$ and $\mathbb{R}_1$ are fourth-rank tensors of $\mathbb{R}^2$ of the type of elasticity, defined by (the set $\{\mathbf{e}_1, \mathbf{e}_2\}$ is an orthonormal basis of $\mathbb{R}^2$)

(Footnote 16 continued)

Calling *Kelvin strains* and *Kelvin stresses* respectively each one of the tensors

$$\boldsymbol{\varepsilon}^i = \mathbb{P}^i \boldsymbol{\varepsilon}, \ \ \boldsymbol{\sigma}^i = \mathbb{P}^i \boldsymbol{\sigma},$$

by their same construction, Kelvin strains and stresses are mutually orthogonal:

$$\boldsymbol{\varepsilon}^i \cdot \boldsymbol{\varepsilon}^j = 0, \ \ \boldsymbol{\sigma}^i \cdot \boldsymbol{\sigma}^j = 0.$$

Let us consider now the strain energy V stored in an elastic body:

$$V = \frac{1}{2}\boldsymbol{\varepsilon} \cdot \mathbb{E}\boldsymbol{\varepsilon} = \frac{1}{2}\lambda^i \boldsymbol{\varepsilon}^i \cdot \mathbb{P}^i \boldsymbol{\varepsilon} = \frac{1}{2}\lambda^i \boldsymbol{\varepsilon}^i \cdot \boldsymbol{\varepsilon}^i.$$

V can hence be decomposed into three terms V^i,

$$V^i = \frac{1}{2}\lambda^{\underline{i}} \boldsymbol{\varepsilon}^{\underline{i}} \cdot \boldsymbol{\varepsilon}^{\underline{i}} \ \ \forall i \in \{\mathrm{I}, \mathrm{II}, \mathrm{III}\},$$

each one of these three terms being associated to the corresponding Kelvin mode. For this reason, we will denote them as *Kelvin modal energies*.

For more details about the Kelvin decomposition, see Rychlewski (1984), Mehrabadi and Cowin (1990), François (1995, 2012), Desmorat and Marull (2011), de Saxcé and Vallée (2013), Desmorat and Vannucci (2014).

$$
\begin{aligned}
\mathbb{T}_0 &= (\mathbf{e}_1 \otimes \mathbf{e}_1 - \mathbf{e}_2 \otimes \mathbf{e}_2) \otimes (\mathbf{e}_1 \otimes \mathbf{e}_1 - \mathbf{e}_2 \otimes \mathbf{e}_2) + \\
&\quad (\mathbf{e}_1 \otimes \mathbf{e}_2 + \mathbf{e}_2 \otimes \mathbf{e}_1) \otimes (\mathbf{e}_1 \otimes \mathbf{e}_2 + \mathbf{e}_2 \otimes \mathbf{e}_1), \\
\mathbb{T}_1 &= (\mathbf{e}_1 \otimes \mathbf{e}_1 + \mathbf{e}_2 \otimes \mathbf{e}_2) \otimes (\mathbf{e}_1 \otimes \mathbf{e}_1 + \mathbf{e}_2 \otimes \mathbf{e}_2), \\
\mathbb{R}_{0c} &= c_4[(\mathbf{e}_1 \otimes \mathbf{e}_1 - \mathbf{e}_2 \otimes \mathbf{e}_2) \otimes (\mathbf{e}_1 \otimes \mathbf{e}_1 - \mathbf{e}_2 \otimes \mathbf{e}_2) - \\
&\quad (\mathbf{e}_1 \otimes \mathbf{e}_2 + \mathbf{e}_2 \otimes \mathbf{e}_1) \otimes (\mathbf{e}_1 \otimes \mathbf{e}_2 + \mathbf{e}_2 \otimes \mathbf{e}_1)] - \\
&\quad s_4[(\mathbf{e}_1 \otimes \mathbf{e}_1 - \mathbf{e}_2 \otimes \mathbf{e}_2) \otimes (\mathbf{e}_1 \otimes \mathbf{e}_2 + \mathbf{e}_2 \otimes \mathbf{e}_1) + \\
&\quad (\mathbf{e}_1 \otimes \mathbf{e}_2 + \mathbf{e}_2 \otimes \mathbf{e}_1) \otimes (\mathbf{e}_1 \otimes \mathbf{e}_1 - \mathbf{e}_2 \otimes \mathbf{e}_2)], \\
\mathbb{R}_{0s} &= s_4[(\mathbf{e}_1 \otimes \mathbf{e}_1 - \mathbf{e}_2 \otimes \mathbf{e}_2) \otimes (\mathbf{e}_1 \otimes \mathbf{e}_1 - \mathbf{e}_2 \otimes \mathbf{e}_2) - \\
&\quad (\mathbf{e}_1 \otimes \mathbf{e}_2 + \mathbf{e}_2 \otimes \mathbf{e}_1) \otimes (\mathbf{e}_1 \otimes \mathbf{e}_2 + \mathbf{e}_2 \otimes \mathbf{e}_1)] + \\
&\quad c_4[(\mathbf{e}_1 \otimes \mathbf{e}_1 - \mathbf{e}_2 \otimes \mathbf{e}_2) \otimes (\mathbf{e}_1 \otimes \mathbf{e}_2 + \mathbf{e}_2 \otimes \mathbf{e}_1) + \\
&\quad (\mathbf{e}_1 \otimes \mathbf{e}_2 + \mathbf{e}_2 \otimes \mathbf{e}_1) \otimes (\mathbf{e}_1 \otimes \mathbf{e}_1 - \mathbf{e}_2 \otimes \mathbf{e}_2)], \\
\mathbb{R}_1 &= 2\, c_2[(\mathbf{e}_1 \otimes \mathbf{e}_1) \otimes (\mathbf{e}_1 \otimes \mathbf{e}_1) - (\mathbf{e}_2 \otimes \mathbf{e}_2) \otimes (\mathbf{e}_2 \otimes \mathbf{e}_2)] + \\
&\quad s_2[(\mathbf{e}_1 \otimes \mathbf{e}_1 + \mathbf{e}_2 \otimes \mathbf{e}_2) \otimes (\mathbf{e}_1 \otimes \mathbf{e}_2 + \mathbf{e}_2 \otimes \mathbf{e}_1) + \\
&\quad (\mathbf{e}_1 \otimes \mathbf{e}_2 + \mathbf{e}_2 \otimes \mathbf{e}_1) \otimes (\mathbf{e}_1 \otimes \mathbf{e}_1 + \mathbf{e}_2 \otimes \mathbf{e}_2)],
\end{aligned} \tag{4.254}
$$

where, for the sake of conciseness,

$$
\begin{aligned}
c_2 &= \cos 2\theta, \quad s_2 = \sin 2\theta, \\
c_4 &= \cos 4\theta, \quad s_4 = \sin 4\theta.
\end{aligned} \tag{4.255}
$$

Tensors $\mathbb{T}_0, \mathbb{T}_1, \mathbb{R}_{0c}, \mathbb{R}_{0s}$ and $\mathbb{R}_1$ play a role similar to that of the Kelvin projectors, in the sense that they give a decomposition of the elasticity tensor $\mathbb{C}$. For this reason we will call them *polar projectors* and denote them in a general way by $\mathbb{P}^i_{pol}$, i.e.

$$
\mathbb{P}^i_{pol} \in \{\mathbb{T}_1, \mathbb{T}_0, \mathbb{R}_{0c}, \mathbb{R}_{0s}, \mathbb{R}_1\}. \tag{4.256}
$$

Their matrix form is quite simple:

$$
\begin{aligned}
&\mathbb{T}_0 = \begin{bmatrix} 1 & -1 & 0 \\ -1 & 1 & 0 \\ 0 & 0 & 2 \end{bmatrix}, \quad \mathbb{T}_1 = \begin{bmatrix} 1 & 1 & 0 \\ 1 & 1 & 0 \\ 0 & 0 & 0 \end{bmatrix}, \quad \mathbb{R}_{0c} = \begin{bmatrix} c_4 & -c_4 & -\sqrt{2}\, s_4 \\ -c_4 & c_4 & \sqrt{2}\, s_4 \\ -\sqrt{2}\, s_4 & \sqrt{2}\, s_4 & -2\, c_4 \end{bmatrix}, \\
&\mathbb{R}_{0s} = \begin{bmatrix} s_4 & -s_4 & \sqrt{2}\, c_4 \\ -s_4 & s_4 & -\sqrt{2}\, c_4 \\ \sqrt{2}\, c_4 & -\sqrt{2}\, c_4 & -2\, s_4 \end{bmatrix}, \quad \mathbb{R}_1 = \begin{bmatrix} 2\, c_2 & 0 & \sqrt{2}\, s_2 \\ 0 & -2\, c_2 & \sqrt{2}\, s_2 \\ \sqrt{2}\, s_2 & \sqrt{2}\, s_2 & 0 \end{bmatrix}.
\end{aligned} \tag{4.257}
$$

A relation similar to Eq. (4.249) holds also for tensor $\mathbb{S}$, it is sufficient to replace the polar stiffness moduli by the corresponding compliance ones, in practice, they must be written with lowercase letters, while the projectors are exactly the same.

The polar projectors have some algebraic properties, listed below:

$$\begin{aligned}
&\|\mathbb{T}_1\| = 2, \ \ \|\mathbb{T}_0\| = \|\mathbb{R}_{0c}\| = \|\mathbb{R}_{0s}\| = \|\mathbb{R}_1\| = 2\sqrt{2},\\
&\det\left(\mathbb{P}^i_{pol}\right) = 0,\\
&\mathbb{P}^i_{pol}\cdot\mathbb{P}^j_{pol} = 0, \ \ \forall\, \mathbb{P}^i_{pol} \neq \mathbb{P}^j_{pol} \in \{\mathbb{T}_1, \mathbb{T}_0, \mathbb{R}_{0c}, \mathbb{R}_{0s}, \mathbb{R}_1\}.\\
&\textstyle\sum_i \mathbb{P}^i_{pol} \neq \mathbb{I},
\end{aligned} \tag{4.258}$$

So, the polar projectors are singular orthogonal tensors, whose norm is not the unity and whose sum is not the identity. Actually, they constitute another way to decompose an elasticity tensor, that has some particular features and links with the material symmetries, through the polar invariants. It is worth recalling that the coefficients of the linear combination giving the elastic tensor, Eq. (4.249), are tensor invariants and that all the dependence upon the orientation θ is in the three projectors $\mathbb{R}_{0c}$, $\mathbb{R}_{0s}$ and $\mathbb{R}_1$, linked to the anisotropic part, while the two projectors $\mathbb{T}_0$ and $\mathbb{T}_1$, linked to the isotropic part, are insensible to the orientation. Fixing the orientation to zero gives

$$\theta = 0 \rightarrow \quad \mathbb{R}_{0c} = \begin{bmatrix} 1 & -1 & 0\\ -1 & 1 & 0\\ 0 & 0 & -2\end{bmatrix}, \quad \mathbb{R}_{0s} = \begin{bmatrix} 0 & 0 & \sqrt{2}\\ 0 & 0 & -\sqrt{2}\\ \sqrt{2} & -\sqrt{2} & 0\end{bmatrix},$$
$$\mathbb{R}_1 = \begin{bmatrix} 2 & 0 & 0\\ 0 & -2 & 0\\ 0 & 0 & 0\end{bmatrix}. \tag{4.259}$$

We already know, see Sect. 4.12, that some material symmetries, i.e., algebraically speaking, some particular values of the polar invariants, make one or more anisotropic polar invariants vanish, so eliminating the corresponding polar projectors from the sum of tensors in Eq. (4.249). Namely, as already recalled in Eq. (4.251), ordinary orthotropy eliminates $\mathbb{R}_{0s}$, while condition (4.252) eliminates $\mathbb{R}_{0c}$. Also, R_0-orthotropy, Eq. (4.153), eliminates both $\mathbb{R}_{0c}$ and $\mathbb{R}_{0s}$, while square symmetry, Eq. (4.173), $\mathbb{R}_1$ and finally, isotropy eliminates all of them.

Let us now introduce the tensors

$$\boldsymbol{\varepsilon}^i_{pol} = \mathbb{P}^i_{pol}\boldsymbol{\varepsilon}, \ \ \boldsymbol{\sigma}^i_{pol} = \mathbb{P}^i_{pol}\boldsymbol{\sigma} \ \ \forall i \in \{\text{I}, ..., \text{V}\}, \tag{4.260}$$

that we will call *polar strains* and *polar stresses* respectively. Applying Eq. (4.63) to the strain tensor $\boldsymbol{\varepsilon}$, one obtains

$$
\begin{aligned}
&\boldsymbol{\varepsilon}^{\mathrm{I}}_{pol}=\mathbb{T}_0\boldsymbol{\varepsilon}=\left\{\begin{array}{c}2r\cos 2(\varphi-\theta)\\-2r\cos 2(\varphi-\theta)\\2\sqrt{2}r\sin 2(\varphi-\theta)\end{array}\right\},\\
&\boldsymbol{\varepsilon}^{\mathrm{II}}_{pol}=\mathbb{T}_1\boldsymbol{\varepsilon}=\left\{\begin{array}{c}2t\\2t\\0\end{array}\right\},\\
&\boldsymbol{\varepsilon}^{\mathrm{III}}_{pol}=\mathbb{R}_{0c}\boldsymbol{\varepsilon}=\left\{\begin{array}{c}2r\cos 2(\varphi+\theta)\\-2r\cos 2(\varphi+\theta)\\-2\sqrt{2}r\sin 2(\varphi+\theta)\end{array}\right\},\\
&\boldsymbol{\varepsilon}^{\mathrm{IV}}_{pol}=\mathbb{R}_{0s}\boldsymbol{\varepsilon}=\left\{\begin{array}{c}2r\sin 2(\varphi+\theta)\\-2r\sin 2(\varphi+\theta)\\2\sqrt{2}r\cos 2(\varphi+\theta)\end{array}\right\},\\
&\boldsymbol{\varepsilon}^{\mathrm{V}}_{pol}=\mathbb{R}_1\boldsymbol{\varepsilon}=\left\{\begin{array}{c}2t\cos 2\theta+2r\cos 2(\varphi-2\theta)\\-2t\cos 2\theta+2r\cos 2(\varphi-2\theta)\\2\sqrt{2}t\sin 2\theta\end{array}\right\},
\end{aligned}
\tag{4.261}
$$

If one fixes the direction $\theta=0$, then the above expressions simplify to

$$
\theta=0\ \rightarrow\ \left\{
\begin{aligned}
&\boldsymbol{\varepsilon}^{\mathrm{I}}_{pol}=\mathbb{T}_0\boldsymbol{\varepsilon}=\left\{\begin{array}{c}\varepsilon_1-\varepsilon_2\\-\varepsilon_1+\varepsilon_2\\2\varepsilon_6\end{array}\right\}=\left\{\begin{array}{c}2r\cos 2\varphi\\-2r\cos 2\varphi\\2\sqrt{2}r\sin 2\varphi\end{array}\right\},\\
&\boldsymbol{\varepsilon}^{\mathrm{II}}_{pol}=\mathbb{T}_1\boldsymbol{\varepsilon}=\left\{\begin{array}{c}\varepsilon_1+\varepsilon_2\\\varepsilon_1+\varepsilon_2\\0\end{array}\right\}=\left\{\begin{array}{c}2t\\2t\\0\end{array}\right\},\\
&\boldsymbol{\varepsilon}^{\mathrm{III}}_{pol}=\mathbb{R}_{0c}\boldsymbol{\varepsilon}=\left\{\begin{array}{c}\varepsilon_1-\varepsilon_2\\-\varepsilon_1+\varepsilon_2\\-2\varepsilon_6\end{array}\right\}=\left\{\begin{array}{c}2r\cos 2\varphi\\-2r\cos 2\varphi\\-2\sqrt{2}r\sin 2\varphi\end{array}\right\},\\
&\boldsymbol{\varepsilon}^{\mathrm{IV}}_{pol}=\mathbb{R}_{0s}\boldsymbol{\varepsilon}=\left\{\begin{array}{c}\sqrt{2}\varepsilon_6\\-\sqrt{2}\varepsilon_6\\\sqrt{2}\,(\varepsilon_1-\varepsilon_2)\end{array}\right\}=\left\{\begin{array}{c}2r\sin 2\varphi\\-2r\sin 2\varphi\\2\sqrt{2}r\cos 2\varphi\end{array}\right\},\\
&\boldsymbol{\varepsilon}^{\mathrm{V}}_{pol}=\mathbb{R}_1\boldsymbol{\varepsilon}=\left\{\begin{array}{c}2\varepsilon_1\\-2\varepsilon_2\\0\end{array}\right\}=\left\{\begin{array}{c}2t+2r\cos 2\varphi\\-2t+2r\cos 2\varphi\\0\end{array}\right\}.
\end{aligned}\right.
\tag{4.262}
$$

Similar expressions hold also for the polar stresses $\boldsymbol{\sigma}^i_{pol}$, it is sufficient to replace t, r and φ by T, R and Φ, respectively. It is worth noting that the polar strains and stresses are not necessarily orthogonal and they do not decompose strain and stress:

$$
\sum_{i=\mathrm{I}}^{\mathrm{V}}\boldsymbol{\varepsilon}^i\neq\boldsymbol{\varepsilon},\ \ \sum_{i=\mathrm{I}}^{\mathrm{V}}\boldsymbol{\sigma}^i\neq\boldsymbol{\sigma}.
\tag{4.263}
$$

On the other hand, noticing that

$$\boldsymbol{\varepsilon}^{\text{I}}_{pol} = 2\,\boldsymbol{\varepsilon}^{dev},\quad \boldsymbol{\varepsilon}^{\text{II}}_{pol} = 2\,\boldsymbol{\varepsilon}^{sph}, \tag{4.264}$$

it is possible to decompose strains or stresses with the first two polar strains or stresses:

$$\begin{aligned} \boldsymbol{\varepsilon} &= \boldsymbol{\varepsilon}^{sph} + \boldsymbol{\varepsilon}^{dev} = \frac{1}{2}\left(\boldsymbol{\varepsilon}^{\text{I}}_{pol} + \boldsymbol{\varepsilon}^{\text{II}}_{pol}\right), \\ \boldsymbol{\sigma} &= \boldsymbol{\sigma}^{sph} + \boldsymbol{\sigma}^{dev} = \frac{1}{2}\left(\boldsymbol{\sigma}^{\text{I}}_{pol} + \boldsymbol{\sigma}^{\text{II}}_{pol}\right). \end{aligned} \tag{4.265}$$

Unlike the case of $\boldsymbol{\sigma}$ and $\boldsymbol{\varepsilon}$, the stored elastic energy can be perfectly decomposed by the polar stresses or strains. Let us consider, for instance, the strain energy (similar results exist of course also for the stress energy, it is sufficient to change $\boldsymbol{\varepsilon}$ with $\boldsymbol{\sigma}$ and switch all the polar components denoted by lowercase letters to capital letters, and those denoted by capital letters to lowercase letters):

$$V = \frac{1}{2}\boldsymbol{\varepsilon}\cdot\mathbb{E}\boldsymbol{\varepsilon} = \frac{1}{2}\boldsymbol{\varepsilon}\cdot\left[T_0\,\mathbb{T}_0 + 2T_1\,\mathbb{T}_1 + R_{0c}\,\mathbb{R}_{0c} + R_{0s}\,\mathbb{R}_{0s} + 2(-1)^h R_1\,\mathbb{R}_1\right]\boldsymbol{\varepsilon}. \tag{4.266}$$

Using Eq. (4.260) we can get:

$$V = \sum_{i=\text{I}}^{\text{V}} V^i_{pol}, \tag{4.267}$$

with

$$\begin{aligned} V^i_{pol} &= \tfrac{1}{2}\boldsymbol{\varepsilon}\cdot\beta^i_{pol}\boldsymbol{\varepsilon}^i_{pol} \quad \forall \in \{I, ..., V\} \quad \rightarrow \\ V^{\text{I}}_{pol} &= \tfrac{1}{2}\boldsymbol{\varepsilon}\cdot T_0\boldsymbol{\varepsilon}^{\text{I}}_{pol} = \tfrac{1}{2}T_0\left[(\varepsilon_1-\varepsilon_2)^2 + 2\varepsilon_6^2\right] = 2T_0\,r^2, \\ V^{\text{II}}_{pol} &= \tfrac{1}{2}\boldsymbol{\varepsilon}\cdot 2\,T_1\boldsymbol{\varepsilon}^{\text{II}}_{pol} = T_1\,(\varepsilon_1+\varepsilon_2)^2 = 4T_1\,t^2, \\ V^{\text{III}}_{pol} &= \tfrac{1}{2}\boldsymbol{\varepsilon}\cdot R_{0c}\boldsymbol{\varepsilon}^{\text{III}}_{pol} = \tfrac{1}{2}R_{0c}\left[(\varepsilon_1-\varepsilon_2)^2 - 2\varepsilon_6^2\right] = 2R_{0c}\,r^2\cos 4\varphi, \\ V^{\text{IV}}_{pol} &= \tfrac{1}{2}\boldsymbol{\varepsilon}\cdot R_{0s}\boldsymbol{\varepsilon}^{\text{IV}}_{pol} = \sqrt{2}R_{0s}\,(\varepsilon_1-\varepsilon_2)\,\varepsilon_6 = 2R_{0s}\,r^2\sin 4\varphi, \\ V^{\text{V}}_{pol} &= \tfrac{1}{2}\boldsymbol{\varepsilon}\cdot 2\,R_1\boldsymbol{\varepsilon}^{\text{V}}_{pol} = 2R_1\left(\varepsilon_1^2-\varepsilon_2^2\right) = 8R_1 t\,r\cos 2\varphi, \end{aligned} \tag{4.268}$$

where we have indicated by β^i_{pol} any one of the polar invariants; the values of the β^i_{pol} are apparent from Eq. (4.268). The different terms V^i_{pol} are the *polar strain energies*; each one of them is linked to a polar modulus and to a part of the strain tensor: V^{I}_{pol}, V^{III}_{pol} and V^{IV}_{pol} depends upon the deviatoric part of $\boldsymbol{\varepsilon}$, V^{II}_{pol} upon its spherical part and V^{V}_{pol} on both of them. The sum of the terms in (4.268) is of course equal to the expression of V given in (4.89), how it can be easily checked.

The above polar energies can be further decomposed. Recalling (4.265), we get

$$V^i_{pol} = \frac{1}{2}(\boldsymbol{\varepsilon}^{sph} + \boldsymbol{\varepsilon}^{dev})\cdot\beta^i\boldsymbol{\varepsilon}^i_{pol} = V^{sph\ i}_{pol} + V^{dev\ i}_{pol} \quad \forall \in \{I, ..., V\}, \tag{4.269}$$

where

$$V_{pol}^{sph\ i}=\frac{1}{2}\boldsymbol{\varepsilon}^{sph}\cdot\beta^{\underline{i}}\boldsymbol{\varepsilon}_{pol}^{\underline{i}}=\frac{1}{4}\boldsymbol{\varepsilon}^{\mathrm{II}}\cdot\beta^{\underline{i}}\boldsymbol{\varepsilon}_{pol}^{\underline{i}},\quad V_{pol}^{dev\ i}=\frac{1}{2}\boldsymbol{\varepsilon}^{dev}\cdot\beta^{\underline{i}}\boldsymbol{\varepsilon}_{pol}^{\underline{i}}=\frac{1}{4}\boldsymbol{\varepsilon}^{\mathrm{I}}\cdot\beta^{\underline{i}}\boldsymbol{\varepsilon}_{pol}^{\underline{i}}. \tag{4.270}$$

The total number of non null $V_{pol}^{sph\ i}$ and $V_{pol}^{dev\ i}$ is not of ten, as one could expect, but only of six; this happens because, as can be easily checked using Eq. (4.261), it is

$$\boldsymbol{\varepsilon}_{pol}^{\mathrm{II}}\cdot\boldsymbol{\varepsilon}_{pol}^{i}=0\ \ \forall i\in\{\mathrm{I,III,IV}\},\quad \boldsymbol{\varepsilon}_{pol}^{\mathrm{III}}\cdot\boldsymbol{\varepsilon}_{pol}^{\mathrm{IV}}=0. \tag{4.271}$$

We hence find

$$\begin{aligned}
&V_{pol}^{sph\ \mathrm{I}}=0,\quad V_{pol}^{dev\ \mathrm{I}}=V_{pol}^{\mathrm{I}}=2\,T_0r^2,\\
&V_{pol}^{sph\ \mathrm{II}}=V_{pol}^{\mathrm{II}}=4\,T_1t^2,\quad V_{pol}^{dev\ \mathrm{II}}=0,\\
&V_{pol}^{sph\ \mathrm{III}}=0,\quad V_{pol}^{dev\ \mathrm{III}}=V_{pol}^{\mathrm{III}}=2\,R_{0c}r^2\cos 4\varphi,\\
&V_{pol}^{sph\ \mathrm{IV}}=0,\quad V_{pol}^{dev\ \mathrm{IV}}=V_{pol}^{\mathrm{IV}}=2\,R_{0s}r^2\sin 4\varphi,\\
&V_{pol}^{sph\ \mathrm{V}}=V_{pol}^{dev\ \mathrm{V}}=\tfrac{1}{2}V_{pol}^{\mathrm{V}}=4\,R_1t\,r\cos 2\varphi.
\end{aligned} \tag{4.272}$$

The only polar energy to have both the spherical and deviatoric part non null, and identical, is V_{pol}^{V}. This term is more properly a mixt energy because always depending on both the spherical and deviatoric parts of $\boldsymbol{\varepsilon}$. Also,

$$V_{pol}^{\mathrm{I}}>V_{pol}^{\mathrm{III}},\quad V_{pol}^{\mathrm{I}}>V_{pol}^{\mathrm{IV}}\quad\forall\{r,\ \varphi\}, \tag{4.273}$$

because it is always $T_0>R_0$, Eq. (4.111)$_1$, and hence, *a fortiori*, $T_0>R_{0c}$ and $T_0>R_{0s}$, Eq. (4.250).

The main properties, with respect to the Kelvin decomposition, of the tensorial decomposition based on the polar projectors, are:

- similarly to the Kelvin decomposition, the polar decomposition introduces identical polar projectors when considering the stiffness tensor $\mathbb{E}$ or the compliance tensor $\mathbb{S}$;
- *the polar decomposition has a direct interpretation in terms of material symmetry*, which is not the case for the Kelvin decomposition outside the isotropic and square symmetry cases;
- *the projectors are intrinsinc in the polar decomposition for any type of symmetry considered.* More precisely, the polar projectors do not depend on any modulus for every type of symmetry. This is true in the case of the Kelvin decomposition only for isotropy and square symmetry;
- the dependence of the polar projectors with any global frame considered is *explicitly* given by the angle θ;
- *two polar projectors are found to be constant with respect to any material behaviour*, i.e. to any polar moduli and to the global frame considered;

- *in the isotropic case,* the Kelvin and polar decomposition are identical and the projectors are found to be constant;
- *in the square symmetry case,* only one Kelvin and polar projector are identical. For the Kelvin decomposition, the associated Kelvin stresses and strains are the spherical part, the diagonal of the deviatoric part and the out of diagonal of the devitoric part. For the polar decomposition, the associated polar stresses and strains are the spherical part, the deviatoric part and the deviatoric part with an opposite sign for the out of diagonal terms;
- for any considered symmetry, two of the polar strains (or stresses) are equal to twice the spherical and deviatoric parts, which results in the fact that *the strain (or stress) is equal to the sum of the first two polar strains (or stresses).*

We can introduce another decomposition of $\boldsymbol{\sigma}$, that can be obtained using Eq. (4.249) and the Hooke's law (2.12):

$$\boldsymbol{\sigma} = \mu^{\text{I}}_{pol} + 2\,\mu^{\text{II}}_{pol} + \mu^{\text{III}}_{pol} + \mu^{\text{IV}}_{pol} + 2\,\mu^{\text{V}}_{pol}, \tag{4.274}$$

where

$$\begin{gathered} \mu^{\text{I}}_{pol} = T_0\,\boldsymbol{\varepsilon}^{\text{I}}_{pol},\;\; \mu^{\text{II}}_{pol} = T_1\,\boldsymbol{\varepsilon}^{\text{II}}_{pol},\;\; \mu^{\text{III}}_{pol} = R_{0c}\,\boldsymbol{\varepsilon}^{\text{III}}_{pol},\\ \mu^{\text{IV}}_{pol} = R_{0s}\,\boldsymbol{\varepsilon}^{\text{IV}}_{pol},\;\; \mu^{\text{V}}_{pol} = R_1\,\boldsymbol{\varepsilon}^{\text{V}}_{pol}. \end{gathered} \tag{4.275}$$

It follows immediately that *a polar stress can never be proportional to its corresponding polar strain*; in fact, if one considers for instance the case of $\boldsymbol{\sigma}^{\text{I}}_{pol}$ and $\boldsymbol{\varepsilon}^{\text{I}}_{pol}$, it is

$$\boldsymbol{\sigma}^{\text{I}}_{pol} = \mu^{\text{I}}_{pol} = T_0\,\boldsymbol{\varepsilon}^{\text{I}}_{pol} \quad\Longleftrightarrow\quad \boldsymbol{\sigma} = T_0\,\boldsymbol{\varepsilon}, \tag{4.276}$$

which can never be true, also in the isotropic case, by virtue of (4.249), because the polar isotropic moduli T_0 and T_1 are strictly positive, Eqs. (4.98) and (4.111). This explains also, mechanically, why *the polar stresses cannot decompose* $\boldsymbol{\sigma}$, Eq. (4.263). A similar result is valid, of course, also for the polar strains.

4.16.2 Interaction of Geometry and Anisotropy

Anisotropy is a mere fact of material properties and, as such, normally its effects should be the same no matter of the problem at hand, i.e. the anisotropic behaviour should not be altered by other factors. Actually, this is not the case. We show in this section that, in the case of flat plates at least, the geometry and boundary conditions of the plate interact with the anisotropic properties of the plate, so modifying its overall elastic response to, e.g., buckling, or vibrations (Vannucci 2012).

The following considerations arise from a research concerning the influence of anisotropy on the flexural response of laminates (Vannucci 2009). Here, the scope is to show that, in some sense, geometry *filters* the anisotropy of the plate. To better understand, let us consider again Eq. (4.67); as already remarked in Sect. 4.7, each

Cartesian component is just the superposition of a maximum of three contributions: a constant term, the isotropic part, and two oscillating terms, two waves with period respectively $\pi/2$ and $\pi/4$, representing together the anisotropic part.

Well, geometry is able, in some cases, to *make one or both of this two waves disappear in the elastic response of a plate*. In some way, *geometry acts on material properties just like, in signal analysis, a filter acts upon the harmonics of a signal: a part or all of the oscillating terms describing the anisotropy of the material disappear from the elastic response of the plate*. That is why we can talk, in such cases, of *filtering anisotropy*, an unusual expression in composite mechanics for a curious phenomenon.

To show this fact, we consider here the flexural behavior of an anisotropic rectangular laminate composed by identical layers and bending-extension uncoupled, with sides length a and b, respectively along the axes x and y. Along its boundary, the plate is simply supported and its mass per unit area is μ. A constraint is imposed on the anisotropy of the plate: the bending tensor $\mathbb{D}$ is orthotropic and the axes of orthotropy are aligned with the axes of the plate. This assumption, along with that of uncoupling, is needed for having exact solutions for flexural problems, see for instance (Jones 1999). The question of obtaining uncoupled laminates orthotropic in bending has been addressed in very few works, and the reader can namely refer to Valot and Vannucci (2005) for further details on the matter. Here, we bound ourselves to remark that it is possible to find uncoupled laminates with $\mathbb{D}$ orthotropic.[17]

In the above assumptions, it can be shown (Vannucci 2009), that (the polar parameters in the following equations are those of the basic layer):

- the compliance J of the plate, that is a measure of its bending stiffness,[18] when loaded by a sinusoidal load orthogonal to its mean surface, is

$$J=\frac{\gamma_{pq}}{p^4h^3\left(1+\chi^2\right)^2\sqrt{R_0^2+R_1^2}}\frac{1}{\varphi\left(\xi_{13},\xi_{15}\right)}; \tag{4.277}$$

- the buckling load multiplier λ_{pq} for the mode pq when the plate is loaded by in-plane forces N_x and N_y is given by

[17] In this section, some concepts introduced extensively in Chap. 5 are used; the reader is hence addressed to this chapter to have details, concepts and nomenclature on laminates.

[18] The *compliance* of a structure is defined as the overall work of the external forces acting upon the structure; from the Clapeyron's Theorem, we have that

$$J=\int_\Omega \mathbf{f}\cdot\mathbf{u}\, d\omega=\int_\Omega \boldsymbol{\sigma}\cdot\boldsymbol{\varepsilon}\, d\omega.$$

It is evident that, for a given set of applied forces $\mathbf{f}$, the less the compliance, the less the displacements $\mathbf{u}$, i.e. the highest the stiffness. That is why the minimization of the compliance J is often used, in structural optimization, as a standard formulation for the problems of stiffness maximization. This has also some mathematical nice properties: unlike the minimization of the displacement of some specific points of the structure, that are hold by local functionals, J is a global functional, which has important, positive consequences in variational calculus, see for instance (Banichuk 1983).

$$\lambda_{pq}=\frac{\pi^2p^2h^3}{12a^2}\frac{\left(1+\chi^2\right)^2\sqrt{R_0^2+R_1^2}}{N_x+N_y\chi^2}\varphi\left(\xi_{13},\xi_{15}\right); \tag{4.278}$$

- the frequency of transversal vibrations ω_{pq} for the mode pq is expressed by

$$\omega_{pq}^2=\frac{\pi^4p^4h^3}{12\mu a^4}\left(1+\chi^2\right)^2\sqrt{R_0^2+R_1^2}\varphi\left(\xi_{13},\xi_{15}\right). \tag{4.279}$$

In the above equations, h is the laminate's thickness, γ_{pq} is a coefficient depending on the geometry of the plate and on the loading, p and q are the number of half-waves in the directions x and y respectively, while the dimensionless parameter χ is the ratio of the wavelengths in the two directions, i.e.

$$\chi=\frac{a}{b}\frac{q}{p}. \tag{4.280}$$

Finally, $\varphi(\xi_{13},\xi_{15})$ is

$$\varphi\left(\xi_{13},\xi_{15}\right)=\tau+\frac{1}{\sqrt{1+\rho^2}}\left[(-1)^k\rho\xi_{13}\frac{\chi^4-6\chi^2+1}{\left(1+\chi^2\right)^2}+4\xi_{15}\frac{1-\chi^2}{1+\chi^2}\right], \tag{4.281}$$

and it is a dimensionless function of dimensionless parameters and variables:

$$\tau=\frac{T_0+2T_1}{\sqrt{R_0^2+R_1^2}} \tag{4.282}$$

is the *isotropy-to-anisotropy ratio*; ρ is the anisotropy ratio already introduced in Eq. (4.131) while

$$\begin{cases}\xi_{13}=\sum_{j=1}^{n}d_j\cos 4\delta_j,\\ \xi_{15}=\sum_{j=1}^{n}d_j\cos 2\delta_j,\end{cases}\quad d_j=\frac{1}{n^3}[12j(j-n-1)+4+3n(n+2)]. \tag{4.283}$$

are the *bending lamination parameters* (Jones 1999), with n the number of layers and δ_j the orientation angles of the plies.

Without rephrasing here all the theory of lamination parameters, introduced in detail in Sect. 5.2.12, it is worth noticing that the final bending properties are completely determined by the choice of the material and of a *lamination point*, i.e. of a couple (ξ_{13},ξ_{15}) (cf. Sect. 6.4.2). For orthotropic uncoupled laminates, the set of lamination points defines in the plane ξ_{13}–ξ_{15} a feasible domain Ω in the form of a parabolic sector, bounded by the conditions (Jones 1999; Miki 1982)

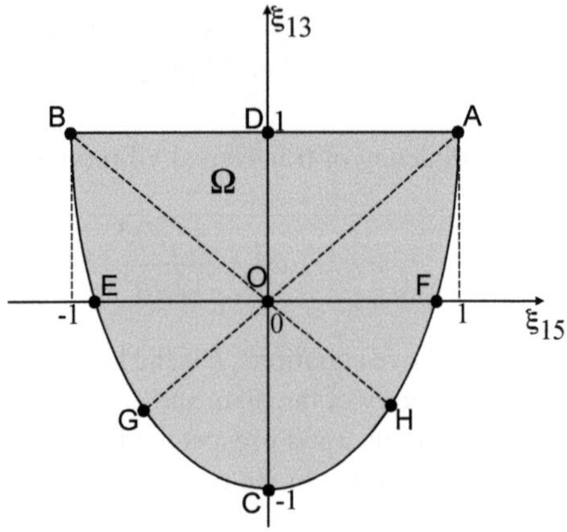

Fig. 4.7 Domain of the lamination parameters

$$\begin{aligned} 2\xi_{15}^2-1\leq\xi_{13}\leq 1,\\ -1\leq\xi_{15}\leq 1. \end{aligned} \tag{4.284}$$

On this domain, some points and lines correspond to particular stacking sequences and in particular, with reference to Fig. 4.7

- all the *cross-ply laminates*, i.e. having all the layers at 0 or $\pi/2$, belong to the straight line AB;
- all the *angle-ply laminates*, i.e. having half the layers at the orientation δ and the other half at $-\delta$, belong to the parabolic arc ACB;
- all the *unidirectional laminates*, with all the layers aligned at 0, are represented by the lamination point A $= (1, 1)$;
- all the *unidirectional laminates*, with all the layers aligned at $\pi/2$, are represented by the lamination point B $= (-1, 1)$;
- all the *balanced cross-ply laminates*, i.e. having the same number of layers at 0 and at $\pi/2$, are represented by the lamination point D$= (0, 1)$;
- all the angle-ply laminates with $\alpha = \pi/4$ are represented by the lamination point C $= (0, -1)$;
- all the angle-ply laminates with $\alpha = \pi/8$ are represented by the lamination point F $= (1/\sqrt{2}, 0)$;
- all the angle-ply laminates with $\alpha = 3\pi/8$ are represented by the lamination point E $= (-1/\sqrt{2}, 0)$;
- all the angle-ply laminates with $\alpha = \pi/6$ are represented by the lamination point H $= (1/2, -1/2)$;
- all the angle-ply laminates with $\alpha = \pi/3$ are represented by the lamination point G $= (-1/2, -1/2)$;
- all the isotropic laminates are represented by the lamination point O $= (0, 0)$.

Nevertheless, there is not a bijective correspondence between lamination points and stacking sequences: a given bending behavior is uniquely determined by one and only one lamination point, but several different stacking sequences can correspond to the same lamination point, and hence be mechanically equivalent. In other words, different laminates can have the same bending behavior. For instance, there is not a unique sequence to obtain isotropy, see for instance (Vannucci and Verchery 2002), and on the parabolic boundary of the feasible domain one can find also sequences that do not belong to the angle-ply set.

From the above Eqs. (4.277)–(4.281), it is apparent that in the three cases considered here and regarding all the possible situations concerning the flexural response of the plate, this response is always a function of $\varphi(\xi_{13}, \xi_{15})$. Hence, for minimizing J or for maximizing λ_{pq} or ω_{pq}, the problem is always reduced to

$$\max_{\xi_{13},\xi_{15}} \varphi\left(\xi_{13},\xi_{15}\right). \tag{4.285}$$

It is worth noticing that in the dimensionless function $\varphi(\xi_{13}, \xi_{15})$, thanks to the polar formalism, the isotropic part is well separated from the anisotropic one, this last being the only one to interact with geometry. So, in some special cases, $\varphi(\xi_{13}, \xi_{15})$ can be reduced to its only isotropic part:

$$\varphi\left(\xi_{13},\xi_{15}\right)=\tau. \tag{4.286}$$

In these circumstances, the flexural behavior is no more affected by the stacking sequence, nor by the anisotropy of the material. This annihilation of the anisotropic part of the elastic response happens for particular values of the anisotropy and/or for particular geometries and in such cases, the *laminate behaves like an isotropic plate, though it is anisotropic*. Let us consider now the possible conditions leading to such a strange situation. First of all, looking at Eq. (4.281), there are two cases independent from the lamination point $\varphi(\xi_{13}, \xi_{15})$, and given simply by special condition on anisotropy and geometry:

1. $\rho = 0$ and $\chi = 1$; this is the case of R_0-orthotropic materials, for which $R_0 = 0$, and of plates having the same wavelength along x and y:

$$\chi=\frac{a}{b}\frac{q}{p}=1 \iff \frac{a}{b}=\frac{p}{q}. \tag{4.287}$$

This happens for instance for square plates when diagonal modes, i.e. having $p = q$, are considered, but also for other situations. In such cases,

$$\begin{aligned}
&\varphi\left(\xi_{13},\xi_{15}\right)=\tau=\frac{T_0+2T_1}{R_1},\\
&J=\frac{\gamma_{pq}}{4\,p^4h^3\left(T_0+2T_1\right)},\\
&\lambda_{pq}=\frac{\pi^2p^2h^3}{3\,a^2}\frac{\sqrt{1+\nu^2}}{1+\nu}\frac{T_0+2T_1}{\sqrt{N_x^2+N_y^2}},\\
&\omega_{pq}^2=\frac{\pi^4p^4h^3}{3\,\mu\,a^4}\left(T_0+2T_1\right).
\end{aligned} \tag{4.288}$$

So, there is no trace of anisotropy in the expressions of J, λ_{pq} and ω_{pq} in Eq. (4.288): all the contributions given by the anisotropic part of the material composing the basic layers have disappeared and, despite the fact that the material and the plate are anisotropic, the responses in Eq. (4.288) depends only upon the isotropic part of the material and they are exactly the same that belong to a plate having the same geometry and composed by an isotropic material whose polar parameters T_0 and T_1 are identical to those of the material actually composing the anisotropic plate. This result is independent of the stack and of the anisotropy of the laminate, i.e. of $\mathbb{D}$. Namely, it is not needed that the laminate be R_0-orthotropic in bending.

2. $\rho = \infty$ and $\chi = \sqrt{2} \pm 1$; this is the case of laminates composed of square-symmetric layers, for which $R_1 = 0$, and having

$$\frac{a}{b}=\frac{p}{q}\left(\sqrt{2}\pm 1\right). \tag{4.289}$$

In this case, the observations made for the previous case can be repeated *verbatim*, in particular the laminate does not need to be square-symmetric in bending, and now

$$\begin{aligned}
&\varphi\left(\xi_{13},\xi_{15}\right)=\tau=\frac{T_0+2T_1}{R_0},\\
&J=\frac{\gamma_{pq}}{8\left(3\pm2\sqrt{2}\right)\,p^4h^3\left(T_0+2T_1\right)},\\
&\lambda_{pq}=\frac{2\left(3\pm2\sqrt{2}\right)\pi^2p^2h^3}{3\,a^2}\frac{\sqrt{1+\nu^2}}{1+\nu\left(3\pm2\sqrt{2}\right)}\frac{T_0+2T_1}{\sqrt{N_x^2+N_y^2}},\\
&\omega_{pq}^2=\frac{2\left(3\pm2\sqrt{2}\right)\pi^4p^4h^3}{3\,\mu\,a^4}\left(T_0+2T_1\right).
\end{aligned} \tag{4.290}$$

There are other sufficient conditions determining an isotropic-like flexural response of the plate: they are all those that render $\varphi(\xi_{13}, \xi_{15}) = \tau$, but they depends upon special values of the lamination point, i.e. they are stack-dependent. Some of these sufficient conditions are:

3. $\rho = 0$ and $\xi_{15} = 0$; this is the case of laminates composed of R_0-orthotropic layers and with a lamination point belonging to the straight line CD; on this line, lie all the combinations of layers with orientations 0, $\pi/2$ and $\pm\pi/4$. Such sequences are called *generalized quasi-isotropic*, and are used extensively in aeronautical composite structures.
4. $\chi = \sqrt{2} \pm 1$ and $\xi_{15} = 0$; the lamination points are the same as in the previous case, but now the layers do not have to be R_0-orthotropic, the essential condition concerns now geometry, cfr. case 2.
5. $\rho = \infty$ and $\xi_{13} = 0$; it is the case of laminates composed of square symmetric layers and with the lamination point belonging to the line EF. On this line there are the laminates composed by combinations of two angle-ply laminates, one with $\alpha = \pi/8$ and the other one with $\alpha = 3\pi/8$.
6. $\chi = 1$ and $\xi_{13} = 0$; the lamination points are the same of the previous case, but now the material properties of the basic layer have no importance, the essential condition concerns now geometry, cfr. case 1.
7. Generally speaking, for a not specially orthotropic material, i.e. for $\rho \neq 0$ and $\rho \neq \infty$, the condition determining $\varphi(\xi_{13}, \xi_{15}) = \tau$ is simply

$$\xi_{13} = \frac{4}{(-1)^K \rho} \frac{\chi^4 - 1}{\chi^4 - 6\chi^2 + 1} \xi_{15}. \tag{4.291}$$

The above equation constitutes, for a given material and geometry, the relation to be satisfied by the lamination parameters in order to obtain a flexural isotropic response. On the domain Ω, this relation corresponds to a straight line, always passing through the origin O (isotropic point).

To end this section, it is interesting to remark that the geometry of the stack and/or of the plate can filter also partially the anisotropy of the material. For instance, Eqs. (4.280) and (4.131) show immediately that all the laminates whose lamination point lies on the line EF ($\xi_{13} = 0$) and/or having $\chi = \sqrt{2} \pm 1$, cancel the contribution of R_0 to the flexural response of the plate. In this case, the response is just like that of a laminate composed by R_0 -orthotropic plies: the geometry and/or the stack act in such a case as a filter on the anisotropy of the material. A similar effect happens, and this case is more interesting for applications, when the lamination point lies on the line CD ($\xi_{15} = 0$) and/or $\chi = 1$. Now, it is the component R_1 to be filtered: the laminate behaves just as one composed of square symmetric layers, i.e. as if it was $R_1 = 0$, without necessarily being square orthotropic.

All the above cases show clearly that in elasticity geometry interacts with the anisotropy of the material or, in other words, that the anisotropy of the same structure can vary according to the cases (for instance varying the ratio p/q). Of course, the cases presented in this section are particularly simple, due to the geometry of the plate; nevertheless, effects similar to those described above are likely to happen also for other geometries, but analytical solutions cannot be found and a similar analysis is much more difficult to be done.

4.16.3 Wrinkling of Anisotropic Membranes

Anisotropic membranes are widely used in several domains, spanning from civil to aerospace engineering: large-span roofs, sails, vehicle airbags and inflatable deployable space structures are just a few examples of possible applications.

To obtain high stiffness and strength, membranes are often reinforced by fibers, so obtaining an anisotropic continuum. The mechanical response of such membranes is always nonlinear, for two main reasons: the amplitude of the displacements, usually non negligible in regard to the membrane dimensions, and local buckling, due to the very low bending stiffness, resulting in the wrinkling of some parts of the membrane.

The phenomenon of wrinkling enables a membrane to resist applied loads, whenever possible, without transmitting relevant compressive stresses (a textile membrane is unable to transmit compressive forces). Wrinkling is frequently observed, in particular whenever the membrane is not sufficiently pretensioned. In case of wrinkling, the stress field is mainly assessed via two different numerical solutions strategies, see e.g. Wong and Pellegrino (2006):

- in the first one, the equilibrium problem is solved by means of three-dimensional geometrically nonlinear analysis, with the membrane modeled as a thin shell;
- in the second approach, a two-dimensional nonlinear analysis is performed using a membrane model where constitutive relations are suitably modified in some way in order to take into account for wrinkling.

The 3D nonlinear approach, though very interesting because it can give a rather complete information about the phenomenon of wrinkling, has different numerical drawbacks: due to the extremely low bending stiffness, very refined meshes or high-order elements must be used to obtain reasonable results, while a more serious problem is the dependence of the solution on the actual distribution of the membrane's thicknesses. This is seldom known within a reasonable degree of approximation and the high accuracy of the results obtainable via this method may sometimes be illusory.

Two-dimensional nonlinear analysis methods are based on the so-called *tension field theory* (Reissner 1938), an approximate model which completely disregards the compressive stresses. The numerical solution procedures are much simpler than in the 3D approach: stresses and in-plane displacements, that are determined in the tension field theory, represent *local average values*. About wrinkles, no information on their amplitude or number can be obtained but, once a wrinkle criterion is chosen, the regions of the membrane where they appear can be identified.

Within this approach, the mechanical response of anisotropic wrinkled membranes has been studied either by introducing an inelastic strain term known as *wrinkle strain* (Wu 1974; Barsotti and Ligarò 2002), or modifying the constitutive law through the definition of a suitable relaxed strain energy density, cf. Pipkin (1993, 1994), Epstein (1999).

In the case of anisotropic wrinkled membranes, it is interesting to know how the anisotropy of the material influences the response of the membrane. Such a study has been conducted in Barsotti and Vannucci (2013) using the polar formalism. The use of tensor invariants is particularly interesting in this case of nonlinear response, where geometry, and hence orientation, affects the response of the structure. We give hereafter a brief account of the principal results.

4.16.3.1 The Mechanical Problem

We consider the problem of a plane membrane subjected to given loads. The solution to such a problem, also in the linear setting, is very difficult to be found, because the membrane cannot transmit compressive stresses. As soon as the first small compressive stresses appear, the membrane wrinkles: transversal displacements, giving rise to the formation of small amplitude foldings, appear, at least on some parts of its surface, see Fig. 4.8. We search for an approximate solution by making use of a simple two-dimensional asymptotic model in which a compression-free stress field is assumed by hypothesis through the introduction of a suitable material nonlinearity. The problem can be reduced to the plane one shown in Fig. 4.9, where Ω is the region of the plane occupied by the membrane in its initial configuration. The load distributed on the surface, Ω, and that acting on the unconstrained part of the border, $\partial\Omega_f$, are indicated by $\mathbf{b}$ and $\mathbf{f}$, respectively. The equilibrium problem can be stated as:

$$\text{on } \Omega : \begin{cases} div\boldsymbol{\sigma} + \mathbf{b} = \mathbf{O}, \\ \boldsymbol{\sigma}\mathbf{v} \cdot \mathbf{v} \geq 0 \,\forall \mathbf{v} : \|\mathbf{v}\| = 1, \end{cases} \tag{4.292}$$
$$\text{on } \partial\Omega_f : \boldsymbol{\sigma}\mathbf{n} = \mathbf{f}.$$

Following an approach originally proposed by Wu (1974), a stress distribution approximating the actual one may be obtained considering the so-called *pseudo*

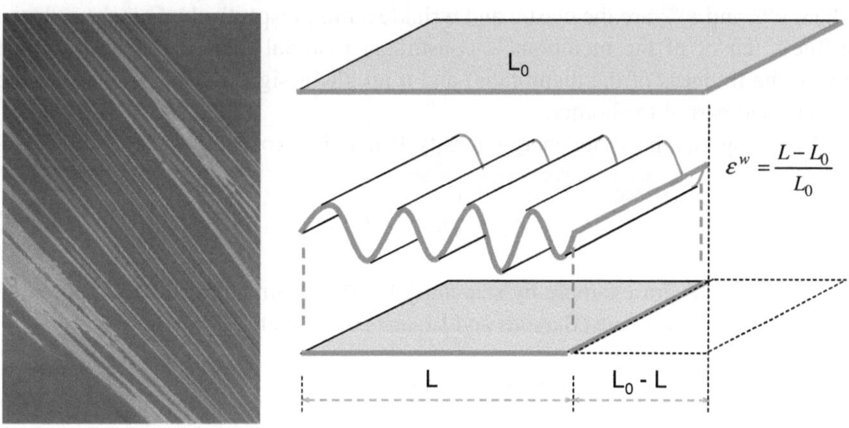

Fig. 4.8 The phenomenon of wrinkling; **a** wrinkles on a polyethylene film, **b** geometric interpretation of wrinkle strain

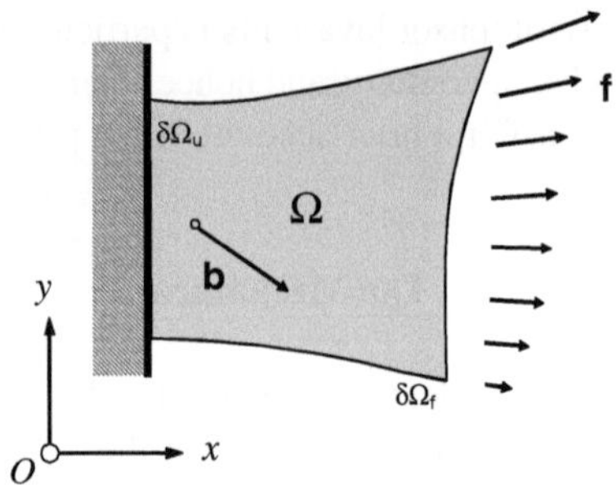

Fig. 4.9 The scheme of the problem for an initially plane anisotropic membrane

deformed surface of the membrane, which can be thought of as the projection of the true deformed surface on the plane containing the initial surface, see Fig. 4.8b. The *wrinkle strain* is defined as the additional negative semi-definite inelastic strain related to the out-of-plane component of the displacements and in the framework of small strain, the total strain is assumed to be the algebraic sum of the elastic and wrinkle strains.

Indicating with $\boldsymbol{\varepsilon}^{(w)}$ the negative semi-definite wrinkle strain tensor, the equilibrium problem can be stated as:

$$\begin{aligned}
&\text{on } \Omega: \begin{cases} \boldsymbol{\varepsilon} = \boldsymbol{\varepsilon}^{(e)} + \boldsymbol{\varepsilon}^{(w)}, \\ \boldsymbol{\sigma} = \mathbb{Q}\boldsymbol{\varepsilon}^{(e)}, \\ \div \boldsymbol{\sigma} + \mathbf{b} = \mathbf{O}, \\ \boldsymbol{\varepsilon}^{(w)}\mathbf{v}\cdot\mathbf{v} \leq 0\, \forall \mathbf{v} : ||\mathbf{v}|| = 1\,, \end{cases}\\
&\text{on } \partial\Omega_f : \boldsymbol{\sigma}\mathbf{n} = \mathbf{f},\\
&\text{on } \partial\Omega_u : \mathbf{u} = \hat{\mathbf{u}},
\end{aligned} \tag{4.293}$$

where $\boldsymbol{\varepsilon}^{(e)}$ and $\boldsymbol{\varepsilon}^{(w)}$ are the elastic and wrinkle strain, respectively, $\mathbb{Q}$ is the reduced stiffness tensor of the membrane's constituent material (plane stress is assumed due to the thinness of the membrane) and $\hat{\mathbf{u}}$ are the assigned displacements on the constrained part of the border.

Assuming that the strain energy density V may be expressed as:

$$V = \frac{1}{2}\mathbb{Q}\boldsymbol{\varepsilon}^{(e)} : \boldsymbol{\varepsilon}^{(e)}, \tag{4.294}$$

it can be proved, for example by searching for the minimum of the total potential energy for the membrane (Barsotti and Ligarò 2002), that:

$$\begin{aligned} \boldsymbol{\sigma}\mathbf{v}\cdot\mathbf{v} \geq 0,\ \forall \mathbf{v} : ||\mathbf{v}|| = 1,\\ \boldsymbol{\sigma}\cdot\boldsymbol{\varepsilon}^{(w)} = 0, \end{aligned} \tag{4.295}$$

that is, the Cauchy stress tensor is positive semi-definite and coaxial with the wrinkle strain tensor. As a consequence, only one of the following three stress regimes may take place at any point in Ω:

1. *taut*, both principal stresses are positive ($\boldsymbol{\sigma} > \mathbf{O}$, $\boldsymbol{\varepsilon}^{(w)} = \mathbf{O}$);
2. *wrinkled*, only one principal stress is positive, the other one is zero ($\boldsymbol{\sigma} \geq \mathbf{O}$, $\boldsymbol{\varepsilon}^{(w)} \leq \mathbf{O}$);
3. *inactive*, both principal stresses are zero ($\boldsymbol{\sigma} = \mathbf{O}$, $\boldsymbol{\varepsilon}^{(w)} = \boldsymbol{\varepsilon} < \mathbf{O}$).

These three stress regimes are characterized by completely different elastic responses. The first one coincides with the *standard* elastic behavior of the material, while in the wrinkled regime the membrane behaves as a sort of orthotropic material having one elasticity modulus equal to zero and whose directions of elastic symmetry change from point to point according to that of the principal stress. Finally, in the inactive regime the membrane is stress-free. The overall strain $\boldsymbol{\varepsilon}$ uniquely determines the stress regime that will take place. Inactive and taut stress regimes respectively correspond to the inequalities:

$$\begin{aligned} &\mathrm{tr}\boldsymbol{\varepsilon} < 0\,, \quad &&\det \boldsymbol{\varepsilon} > 0\,, \quad &&\text{(inactive state)},\\ &\mathrm{tr}(\mathbb{Q}\boldsymbol{\varepsilon}) > 0\,, &&\det(\mathbb{Q}\boldsymbol{\varepsilon}) > 0\,, &&\text{(taut state)}. \end{aligned} \tag{4.296}$$

When neither of the two sets of inequalities is verified, the membrane is wrinkled.

Determining the state of stress for assigned strains is straightforward for both taut and inactive regimes; in case of wrinkled regime, the constitutive relation $(4.293)_2$ and the *orthogonality condition* (4.295) allow for determining the three unknowns quantities (one principal value for stress and wrinkle strain and the inclination angle of the principal stress direction).

4.16.3.2 A Polar Analysis of the Effects of Anisotropy on Wrinkling

For an assigned strain state, be ε_1 and ε_2 the principal strain values and φ the angle formed between the directions of the principal strains and that of the material elastic symmetry (the membrane is assumed to be orthotropic). We want to determine which is the stress regime in the membrane, among the three above, in function of the elastic constants. The partitioning of the plane of principal strains $(\varepsilon_1, \varepsilon_2)$ into taut, wrinkled and inactive states is shown in Fig. 4.10. The separating lines between wrinkled and inactive stress states coincide with the negative part of the ε_1 and ε_2 axes. The separating lines between wrinkled and taut stress states, instead, are the positive part of the pair of straight lines solution of the quadratic equation:

$$\sigma_{11}\sigma_{22} - \sigma_{12}^2 = 0\,, \tag{4.297}$$

where the $\sigma_{\alpha\beta}$ are the stress components with respect to the principal strain axes x_1 and x_2, cf. Fig. 4.10.

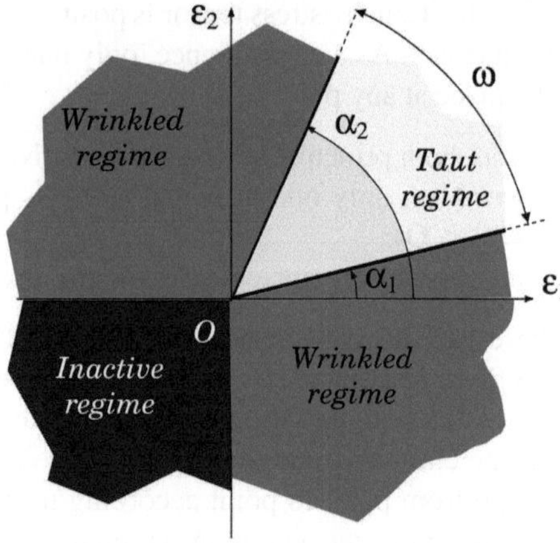

Fig. 4.10 Taut, wrinkled and inactive regions of the $\varepsilon_1, \varepsilon_2$ plane

Once the directions of principal strain are fixed, the inclination angles, α_1 and α_2, of the two straight lines separating taut from wrinkled stress regimes depend only on the values of the material elastic constants. The angular difference $\omega = \alpha_2 - \alpha_1$ gives a measure of the width of the region of the plane of principal strains corresponding to taut membrane stress regimes. A large ω means that a greater number of strain states will be characterized by the absence of wrinkles, while, on the contrary, when ω is small, wrinkles will appear more frequently. That is why we can take ω as a measure of the material's susceptibility to wrinkling.

Angles α_1 and α_2 depend on the angle φ formed between the directions of elastic symmetry and those of the principal strains, see Fig. 4.10. Injecting Eq. (4.293)$_2$ into Eq. (4.297), setting $\varepsilon_{11} = \varepsilon_1$, $\varepsilon_{22} = \varepsilon_2$, $\varepsilon_{12} = 0$, and expressing the components of the elasticity matrix $\mathbb{Q}$ in terms of the polar parameters, Eq. (4.67), we obtain, after some standard passages,

$$\begin{aligned} \alpha_1 &= \arctan\left(\frac{A_1 + A_2\cos 4\varphi + 2\sqrt{A_3 + A_4\cos 4\varphi}}{A_5 + A_6\cos 2\varphi + A_2\cos 4\varphi}\right), \\ \alpha_2 &= \frac{\pi}{2} - \arctan\left(\frac{A_5 + A_6\cos 2\varphi + A_2\cos 4\varphi}{A_1 + A_2\cos 4\varphi - 2\sqrt{A_3 + A_4\cos 4\varphi}}\right) \end{aligned} \tag{4.298}$$

and

$$\omega = \alpha_2 - \alpha_1 = \frac{\pi}{2} - \arctan\left(\frac{A_5 + A_2\cos 4\varphi}{4\sqrt{A_3 + A_4\cos 4\varphi}}\right), \tag{4.299}$$

where

$$
\begin{aligned}
A_1 &= -T_0^2 - 4T_1^2 - R_0^2 + 6R_1^2\,,\\
A_2 &= 2R_1^2 - 2(-1)^K R_0 T_0\,,\\
A_3 &= 4T_1^2(T_0^2 + R_0^2) - 2R_1^2(4T_0T_1 + T_0^2 - 4R_1^2 + R_0^2) -\\
&\quad 4(-1)^K R_1^2 R_0(2T_1 - T_0)\,,\\
A_4 &= 8T_0T_1^2R_0 - 4R_0R_1^2(T_0 + 2T_1) +\\
&\quad 2(-1)^K R_1^2(T_0^2 - 4T_0T_1 + 4R_1^2 + R_0^2)\,,\\
A_5 &= -T_0^2 + 4T_1^2 - R_0^2 - 2R_1^2\,,\\
A_6 &= 4R_1(T_0 - 2T_1 + (-1)^K R_0)\,.
\end{aligned}
\tag{4.300}
$$

The above relations clearly show that ω is stationary for $\varphi = \pi/4$. As can be easily verified, there are special values of R_0 and R_1 that make ω a constant. In other words, the angular width ω turns out to be independent of the inclination φ of the principal strain axes not only in the obvious case of an isotropic material. For instance, if $K = 0$, putting $R_0 = 2T_1 - T_0$ and $R_1 = \sqrt{R_0 T_0}$, Eq. (4.299) reduces to:

$$
\omega = \frac{\pi}{2} - \arctan\left(\frac{A_5}{4\sqrt{A_3}}\right)\,. \tag{4.301}
$$

which is independent from φ. The variation of α_1 and α_2 as functions of φ is plotted in Fig. 4.12 for two orthotropic materials with $K = 0$ and $K = 1$. The values of the polar parameters are: $T_0 = 1.3$, $T_1 = 0.8$, $R_0 = 0.7$, $R_1 = 0.3$ and $\Phi_1 = 0$ (Fig. 4.11).

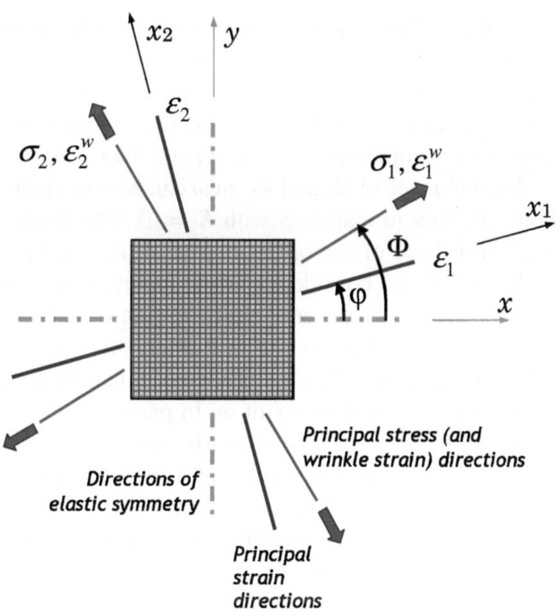

Fig. 4.11 Orientation of principal and elastic symmetry directions

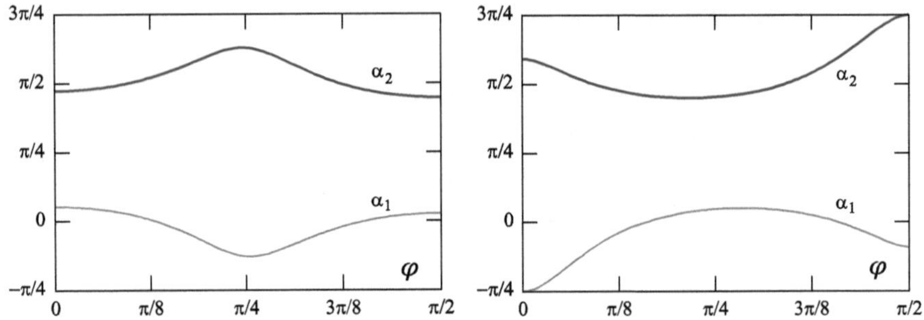

Fig. 4.12 Angles α_1 and α_2 as functions of φ: *left* $K = 0$, *right* $K = 1$

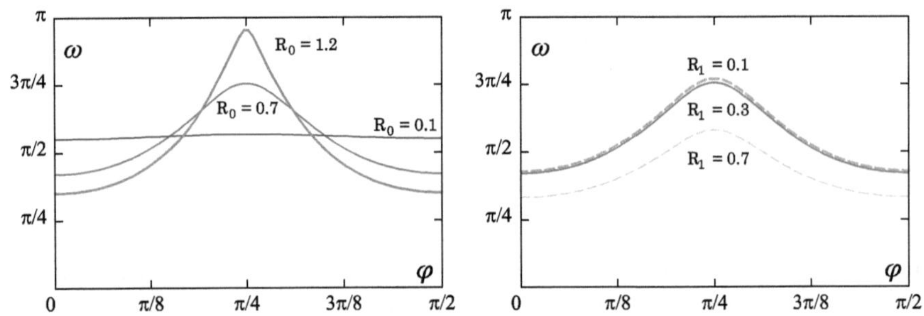

Fig. 4.13 Amplitude ω of the taut region as function of φ ($T_0 = 1.3$, $T_1 = 0.8$, $K = 0$); *left*, $R_1 = 0.3$; *right*, $R_0 = 0.7$

Figure 4.12 clearly illustrates the opposing behaviors of the ordinary orthotropies: for $K = 0$, the angular difference $\alpha_2 - \alpha_1$ is minimum when the principal strain directions coincide with those of elastic symmetry, i.e. for $\varphi = 0$ and $\varphi = \pi/2$, while it is maximum for $\varphi = \pi/4$. On the contrary, for $K = 1$, the same angular difference becomes a maximum for $\varphi = 0$ or $\varphi = \pi/2$ and a minimum for $\varphi = \pi/4$.

The influence of R_0 and R_1 on ω can be seen in the diagrams plotted in Fig. 4.13, both referring to materials with $K = 0$. The figure shows a plot of the values of ω against those of the orientation φ of the principal strain axes. Different values of R_0 and R_1 are considered, while the other two polar parameters describing the isotropic part of the material elastic response are kept constant and equal to $T_0 = 1.3$ and $T_1 = 0.8$, respectively. In the left diagram R_1 is kept constant, while three different values for R_0 are considered. It can be seen that small values of R_0 yield more uniform distributions of ω. In particular, for $R_0 = 0.1$ the width of the taut region in the ε_1, ε_2 plane is nearly constant and seems to be independent of the orientation angle φ of the principal strain axes. In the right diagram, on the contrary, R_0 is kept fixed while three different values are assigned to parameter R_1. As it can be seen, when R_1 increases the diagram is uniformly shifted downward. The diagrams plotted in Fig. 4.13 highlight the remarkably different roles played by the two polar parameters. Whereas R_0 seems to be associated with higher or lower uniformity of

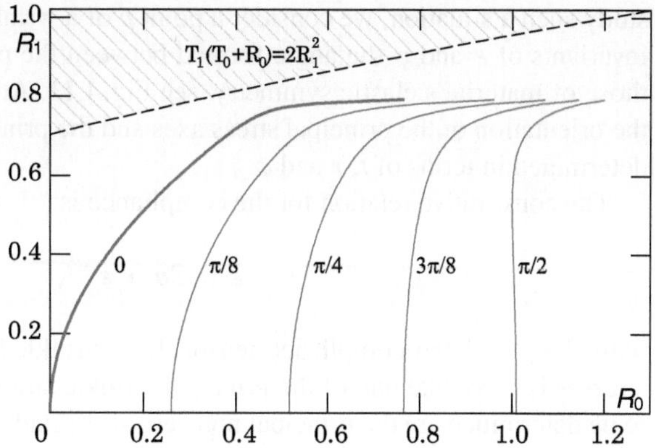

Fig. 4.14 Level curves of the difference $\omega(\pi/4) - \omega(0)$ in the (R_0, R_1) plane ($T_0 = 1.3$, $T_1 = 0.8$, $K = 0$)

the distribution of the ω values, a change in R_1 instead seems to affect all the values of ω, regardless of the direction of the principal strain axes. These observations can be useful for the design of orthotropic membranes, especially in all those boundary values problems where the displacements are assigned. With the aim of providing a more detailed picture of the influence on wrinkling of the two polar parameters R_0 and R_1, we examine the difference $\omega(\pi/4) - \omega(0)$, which can be considered a measure of the degree of uniformity of the ω diagram. Figure 4.14 shows some level curves of this difference, each one corresponding to different, but equally uniform, width distributions of the taut region ω.

From the diagrams, it is evident that the level curves are inclined towards the R_1 axis for much of their lengths. This confirms the observation made above regarding the role of R_0: a change in R_0 is likely to produce a significant change in the shape of the ω diagram, which becomes less uniform as R_0 increases. On the contrary, a change in R_1 would be expected to cause only small variations in the shape of the ω diagram.

A final remark concerns the level curve corresponding to $\omega(\pi/4) - \omega(0) = 0$, the thick curve on the left of the diagram. The points of this curve represent orthotropic materials that, as far as the extension of taut states in the plane of the principal strains is concerned, behave exactly like isotropic ones: their anisotropic behavior is hidden. This is another case, besides those considered in Sect. 4.16.2 for laminates, where anisotropy is, so to speak, *filtered* by the structure. The level curves plotted in Fig. 4.14 may be a useful tool for the design of orthotropic membranes.

4.16.3.3 A Polar Analysis of the Stress State in a Wrinkled Anisotropic Membrane

When a membrane is wrinkled, a one-dimensional purely tensile stress state exists like if the membrane was made of a sort of orthotropic material with the axes aligned with the principal directions of stress, so, generally speaking, varying pointwise. To study such a situation, we consider a point p of a wrinkled region and be t and r the invariants of $\boldsymbol{\varepsilon}$ and φ the angle formed between the principal strain directions and those of material's elastic symmetry, see Fig. 4.11. At p the principal tensile stress, the orientation of the principal stress axes and the principal wrinkle strain can all be determined in terms of t, r and φ.

The constitutive relation for the compliance is:

$$\boldsymbol{\varepsilon} = \mathbb{S}\boldsymbol{\sigma} + \boldsymbol{\varepsilon}^{(w)}, \tag{4.302}$$

with $\mathbb{S} = \mathbb{Q}^{-1}$ the compliance tensor. In a wrinkled region, one of the principal stresses is zero, like one of the principal wrinkle strains. Thus, the three unknowns to be determined are the principal non zero stress and wrinkle strain, σ and $\varepsilon^{(w)}$, and the orientation Φ of the non zero principal stress direction. Equation (4.302) gives a set of three equations that enable determining these unknowns in terms of the overall strain $\boldsymbol{\varepsilon}$.

Since, Eq. (4.295), $\boldsymbol{\sigma}$ and $\boldsymbol{\varepsilon}$ are coaxial, the stress and wrinkle strain can be written as $\boldsymbol{\sigma} = (\sigma, 0, 0)$ and $\boldsymbol{\varepsilon}^{(w)} = (0, \varepsilon^{(w)}, 0)$, with respect to the principal stress axes. Injecting these expressions into Eq. (4.302) and expressing the components of $\boldsymbol{\varepsilon} = (\varepsilon_{11}, \varepsilon_{22}, \varepsilon_{12})$ in terms of t and r,

$$\varepsilon_{11} = t + r\cos 2(\varphi - \Phi),\ \varepsilon_{22} = t - r\cos 2(\varphi - \Phi),\ \varepsilon_{12} = r\sin 2(\varphi - \Phi), \tag{4.303}$$

simple passages allow for obtaining

$$\begin{aligned} &r(2t_1 + t_0)\sin 2(\Phi - \varphi) - 2t\, r_1 \sin 2\Phi - (-1)^k t\, r_0 \sin 4\Phi + \\ &+ r\, r_1 \sin 2(2\Phi - \varphi) - 3r\, r_1 \sin 2\varphi - (-1)^k r\, r_0 \sin 2(\Phi + 2\varphi) = 0. \end{aligned} \tag{4.304}$$

This equation allows for determining the angle Φ formed between the principal stress directions and the orthotropy axes, once the other angle φ known, or viceversa.

Once the orientation Φ of the principal stress axes has been determined, the principal tensile stress σ and the principal wrinkle strain $\varepsilon^{(w)}$ can be found using the two following relations, both derived from (4.302):

$$\sigma = \frac{t + r\cos 2(\varphi - \Phi)}{2t_1 + t_0 + 4r_1 \cos 2\Phi + (-1)^k r_0 \cos 4\Phi}, \tag{4.305}$$

and

$$\varepsilon^{(w)} = t - r\cos 2(\varphi - \Phi) - \left[2t_1 - t_0 - (-1)^k r_0 \cos 4\Phi\right]\sigma, \tag{4.306}$$

where t_0, t_1, r_0, r_1 and k are the polar parameters of $\mathbb{S}$. Equations (4.304), (4.305) and (4.306) allow for completely determining the stress state at any point within a wrinkled membrane region as functions of the strain state.

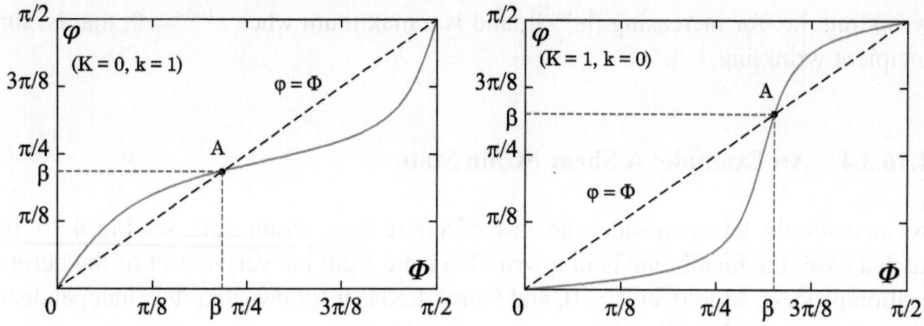

Fig. 4.15 Inclination of principal strain and stress directions for incipient wrinkling for two sets of polar invariants of $\mathbb{S}$

For incipient wrinkling conditions ($\varepsilon^{(w)} = 0$), the relation between Φ and φ is independent from the strain invariants r e t according to the following relation:

$$(t_0+2r_1 \cos 2\Phi+(-1)^k r_0 \cos 4\Phi) \tan 2(\varphi-\Phi)+2r_1 \sin 2\Phi+(-1)^k r_0 \sin 4\Phi = 0\,. \tag{4.307}$$

In Fig. 4.15 we plot two examples of the curve implicitly defined by the above relation. The polar parameters of $\mathbb{S}$ have been chosen so as to correspond to the same two orthotropic materials considered in the previous section. The left diagram is obtained putting $t_0 = 1.9$, $t_1 = 0.6$, $r_0 = 0.94$, $r_1 = 0.18$, $\varphi_0 = \pi/4$ and $\varphi_1 = 0$, i.e. $k = 1$, while the polar parameters for the right diagram are $t_0 = 1.9$, $t_1 = 0.6$, $r_0 = 1.3$, $r_1 = 0.6$, $\varphi_0 = 0$ and $\varphi_1 = 0$, i.e. $k = 0$. The first case corresponds to a material having $\Phi_0 = \Phi_1 = 0$, i.e. $K = 0$, while for the second one $\Phi_0 = \pi/4$ and $\Phi_1 = 0$, i.e. $K = 1$. We plot in the same diagrams also the line $\Phi = \varphi$ to highlight the phase shift between the principal stress and strain directions.

It is worth noting that for $\varphi = \Phi = \beta$ (point A in Fig. 4.15), stress and strain are coaxial, regardless of the values of the wrinkle strain and principal strains. In other words, when the principal strain, or stress, axes are rotated of the angle β with respect to the axes of elastic symmetry, the wrinkled membrane behaves as if it was isotropic. Expression (4.307) yields the following expression for β:

$$\beta = \frac{1}{2} \arccos\left[-(-1)^k \frac{r_1}{r_0}\right]. \tag{4.308}$$

This equation clearly shows that the inclination angle β associated to coaxial stress and strain states may exist only if $r_1 < r_0$. In this case, β is contained within the range $(0, \pi/4)$ or $(\pi/4, \pi/2)$, depending on whether $k = 1$ or $k = 0$. If, on the contrary, $r_1 \geq r_0$, stress and strain are coaxial only in the obvious cases of $\Phi = 0$ or $\Phi = \pi/2$.

Finally, by recalling (4.302) it is an easy task to deduce that, if the stress state at any given point on the membrane belonging to a wrinkled region is kept fixed, the phase shift between the principal stress and strain directions decreases with increasing wrinkling, i.e. for increasing $||\varepsilon^{(w)}||$, and is a maximum when $\varepsilon^{(w)} = 0$, that is, for incipient wrinkling.

4.16.3.4 An Example: A Shear Strain State

As an example, let us consider the case of a pure shear strain state, see Fig. 4.16. In such a case, the membrane is in a wrinkled state from the very outset of the deformation process. Moreover, $t = 0$, and from (4.304) it follows that Φ is independent from the magnitude of the shear strain u/h and is thus a function of φ only.

The two quantities that completely describe the stress state in the wrinkled membrane are the orientation of the principal tensile stress direction and the magnitude of the principal stress. Concerning the angle Φ formed between the principal stress directions and those of elastic symmetry of the material, the diagram of the function $\Phi(\varphi)$, obtained putting $t = 0$ in Eq. (4.304), is shown in Fig. 4.17. In order to highlight the influence of r_0, three different values of r_0 are considered for the two cases, corresponding to $k = 0$ and $k = 1$.

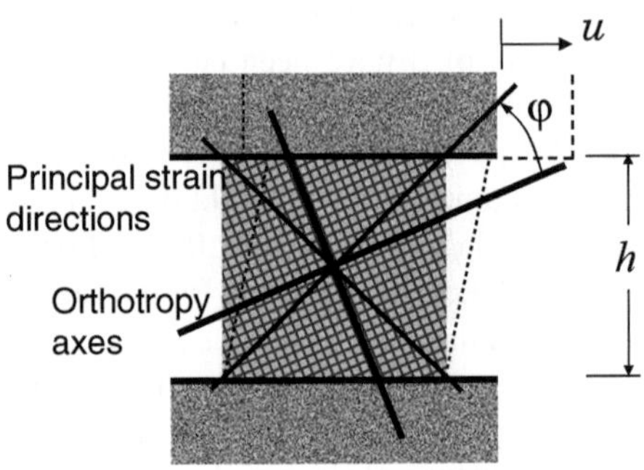

Fig. 4.16 Orthotropic membrane under pure shear

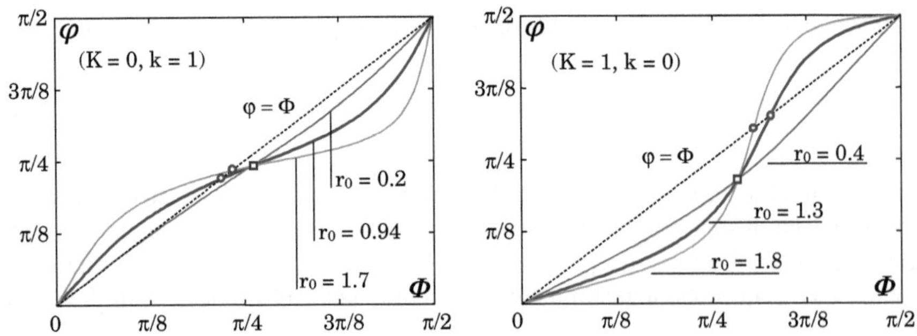

Fig. 4.17 Orientation angle Φ of the principal tensile stress direction for pure shear strain states for two sets of polar invariants of $\mathbb{S}$

The left diagram corresponds to the case in which $t_0 = 1.9$, $t_1 = 0.6$, $r_1 = 0.18$, $\varphi_0 = \pi/4$, $\varphi_1 = 0$, i.e. $k = 1$, and r_0 takes the values $r_0 = 0.2$, 0.94, 1.7. In the right diagram the polar parameters are instead $t_0 = 1.9$, $t_1 = 0.6$, $r_1 = 0.6$, $\varphi_0 = 0$ and $\varphi_1 = 0$, so that $k = 0$, while $r_0 = 0.4$, 1.3, 1.8. In order to highlight the phase shift between the principal stress and strain directions, the same figure also shows a plot of the straight line $\Phi = \varphi$.

Note that the maximum phase shift between the principal stress and strain directions decreases for decreasing r_0, both for $k = 0$ and $k = 1$. This is consistent with the results of the previous section, see Fig. 4.14, where an analogous effect of R_0 was described. An interesting finding is the presence of a particular orientation of the principal strain axes (see the points indicated with small squares in the figure), to which correspond stress states whose orientation Φ is independent from r_0. Moreover, as already observed with reference to Fig. 4.15, there is a particular orientation of the principal strain direction which yields a null phase shift for two of the three values of r_0 (see the points indicated with small circles in the figure).

The principal tensile stress values σ may be obtained from Eqs. (4.304) and (4.305). The expression for σ as a function of the principal stress direction Φ is:

$$\sigma = \frac{r}{\sqrt{(2t_1 + t_0 + 4r_1\cos 2\Phi + (-1)^k r_0 \cos 4\Phi)^2 + (2r_1 \sin 2\Phi + (-1)^k r_0 \sin 4\Phi)^2}}. \tag{4.309}$$

As an example, the diagram of the principal stress σ is shown in Fig. 4.18 for the same values of the polar parameters as above. The ordinates are the dimensionless ratio values σ/σ_0, obtained by dividing the principal tensile stress by $\sigma_0 = \sigma(\Phi = 0, r_0 = 0.94)$ for $k = 1$ and by $\sigma_0 = \sigma(\Phi = 0, r_0 = 1.3)$ for $k = 0$. The tensile stress varies widely with the orientation Φ of the principal stress axes for all three r_0 values, thus highlighting the substantial differences in elastic response to shear of orthotropic materials having different values of r_0 and r_1. Recalling that a one-to-one increasing monotonic correspondence exists between the principal stress and strain

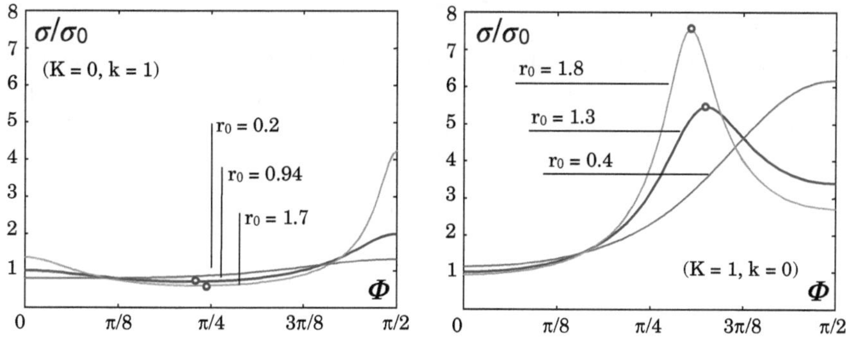

Fig. 4.18 Principal tensile stress σ as function of principal stress directions for two sets of polar invariants of $\mathbb{S}$

directions, i.e. between angles Φ and φ, cf. Fig. 4.17, the diagrams of the principal stress as a function of φ may be easily built.

Figure 4.18 confirms once again that the two classes of orthotropic materials corresponding to $k = 0$ and $k = 1$ exhibit opposing behaviors. For $k = 0$, a maximum point is present within the interval $[0, \pi/2]$ in two of the three cases examined, while for $k = 1$ a minimum point appears. Differentiating Eq. (4.309), it is a simple matter to verify that the local maximum (or minimum) points for σ (the small circles in the figure) correspond to an inclination of the principal stress direction $\Phi = \beta$, i.e. the same inclination that makes the stress and strain coaxial, according to (4.308).

The results shown in Figs. 4.17 and 4.18 indicate that r_0 and r_1 heavily influence the stress state. Moreover, as can be seen from Fig. 4.15, even two orthotropic materials sharing the same values of the stiffness polar parameters, except K, may give completely different compliance responses.

4.17 Applications of the Polar Formalism to Other Fields

The polar formalism is not a mathematical technique for representing exclusively elasticity tensors, but in general any type of planar phenomenon described by a tensor of any rank. The technique for adapting the polar formalism to a tensor of a given rank is just the same described in this chapter for the elastic case, namely the transformation matrices can be constructed using the rules given in Sects. 4.2, 4.3 and 4.12.

The polar formalism has been used for representing also physical phenomena different from elasticity, namely it has been applied to plane piezoelectricity, to a particular problem of damage and to formulate a classical strength criteria for anisotropic materials by tensor invariants. A brief account of all these topics is given below.

4.17.1 Plane Piezoelectricity

In $\mathbb{R}^2$ the third-rank tensor of piezoelectric moduli $\mathscr{D}$ introduced in Sect. 1.3.4, Eqs. (1.14) and (1.16), can be represented by the polar formalism (Vannucci 2007); the index symmetries (1.17) reduce in such a case to only two: $D_{112} = D_{121}$, $D_{212} = D_{221}$. The polar analysis leads to the following result:

$$\begin{aligned}
D_{111}(\theta) =& A_1[\cos 3(\alpha_1 - \theta) + \sin 3(\alpha_1 - \theta)] - A_2[\cos(\alpha_2 - \theta) + \\
& \sin(\alpha_2 - \theta)] + A_3[\cos(\alpha_3 - \theta) - \sin(\alpha_3 - \theta)], \\
D_{112}(\theta) =& A_1[-\cos 3(\alpha_1 - \theta) + \sin 3(\alpha_1 - \theta)] + \\
& A_3[\cos(\alpha_3 - \theta) + \sin(\alpha_3 - \theta)], \\
D_{122}(\theta) =& -A_1[\cos 3(\alpha_1 - \theta) + \sin 3(\alpha_1 - \theta)] - A_2[\cos(\alpha_2 - \theta) + \\
& \sin(\alpha_2 - \theta)] - A_3[\cos(\alpha_3 - \theta) - \sin(\alpha_3 - \theta)],
\end{aligned} \tag{4.310}$$

$$\begin{aligned}D_{211}(\theta) =&A_1[-\cos 3(\alpha_1-\theta)+\sin 3(\alpha_1-\theta)]+A_2[\cos(\alpha_2-\theta)-\\&\sin(\alpha_2-\theta)]-A_3[\cos(\alpha_3-\theta)+\sin(\alpha_3-\theta)],\\D_{221}(\theta) =&-A_1[\cos 3(\alpha_1-\theta)+\sin 3(\alpha_1-\theta)]+\\&A_3[\cos(\alpha_3-\theta)-\sin(\alpha_3-\theta)],\\D_{222}(\theta) =&A_1[\cos 3(\alpha_1-\theta)-\sin 3(\alpha_1-\theta)]+A_2[\cos(\alpha_2-\theta)-\\&\sin(\alpha_2-\theta)]+A_3[\cos(\alpha_3-\theta)+\sin(\alpha_3-\theta)].\end{aligned}$$

One can see that, unlike for even-rank tensors, where the functions sine and cosine appear separately in the expressions of the Cartesian components, for the case of a third-rank tensor, and more generally this is true for all the odd-rank tensors, these two functions appear jointly in each component and they are affected by the same modulus. Also, it is to be noticed that $\mathscr{D}$ has not an isotropic part: then, all of its components oscillates around a null mean and all of them take necessarily negative and positive values in $[0, 2\pi]$.

The invariants are the three moduli A_1, A_2 and A_3 along with the two polar angles differences $\alpha_1-\alpha_3,\ \alpha_2-\alpha_3$. As usual, fixing one among the three polar angles corresponds to fix a frame.

There are several possible conditions of material symmetry for such a tensor:

- *ordinary piezoelectric symmetry*: it is characterized by the following simultaneous values of the angular invariants

$$\begin{cases}\alpha_1-\alpha_3=k_1\dfrac{\pi}{3},\\ \alpha_2-\alpha_3=\dfrac{\pi}{2}+k_2\pi,\end{cases}\quad k_1, k_2\in\{0,1\};\tag{4.311}$$

  the different cases of ordinary piezoelectric symmetry are hence 4, corresponding to the possible combinations of the values of k_1 and k_2
- *first special piezoelectric symmetry* (we indicate by special symmetries, as in the case of the elastic tensor, the symmetries hold by the vanishing of at least one modulus):

$$A_3=0;\tag{4.312}$$

- *second special piezoelectric symmetry*:

$$A_1=A_2=0;\tag{4.313}$$

  in this case it is $D_{111}=D_{221}=-D_{122},\ D_{222}=D_{112}=-D_{211}\ \forall\theta$;
- *third special piezoelectric symmetry*:

$$A_1=A_3=0;\tag{4.314}$$

  in this case it is $D_{112}=D_{221}=0,\ D_{111}=D_{122},\ D_{222}=D_{211}\ \forall\theta$;

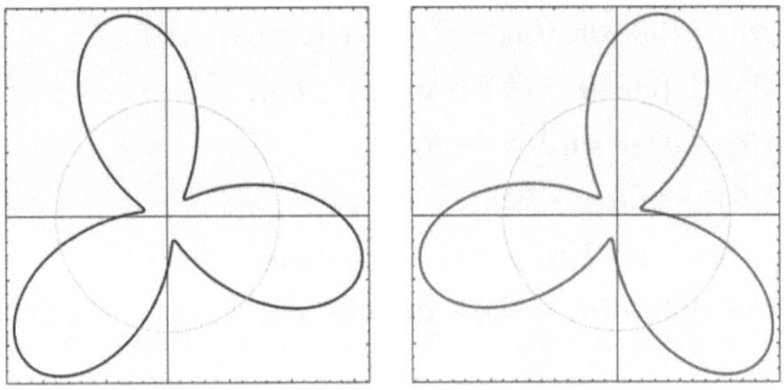

(a) D_{111} (b) D_{112} (c) D_{122}

Fig. 4.19 Qualitative directional diagrams of three components of a piezoelectric anisotropic tensor $\mathscr{D}$ with $A_2 = A_3 = 0$

- *fourth special piezoelectric symmetry*:

$$A_2 = A_3 = 0; \tag{4.315}$$

in this case it is $D_{122} = D_{221} = -D_{111}$, $D_{112} = D_{211} = -D_{222}\ \forall\theta$.

In Fig. 4.19 we show the directional diagram of three Cartesian components of an anisotropic piezoelectric tensor $\mathscr{D}$ with $A_2 = A_3 = 0$. The grey circle traced in the diagrams indicates the zero: when the diagram is inside the circle, the component is negative.

4.17.2 Anisotropic Damage of Isotropic Layers

We consider the anisotropy induced by damage on an initially isotropic layer, cf. Vannucci and Desmorat (2015), and in particular the following questions:

- if the elastic tensor of the virgin material is $\mathbb{E}$, which is the final tensor $\widetilde{\mathbb{E}}$?
- what are the bounds for the elastic moduli of $\widetilde{\mathbb{E}}$?
- and those for the characteristics of the damage tensor $\mathbb{D}$?

To this purpose, we define the damage tensor $\mathbb{D}$ as a fourth-rank tensor with minor and major tensor symmetries, such that the elastic tensor $\widetilde{\mathbb{E}}$ of the damaged material linearly depends upon $\mathbb{E}$ and $\mathbb{D}$, cf. Chaboche (1982), Leckie and Onat (1980), Sidoroff (1980), Chow (1987), Lemaitre et al. (2009):

$$\widetilde{\mathbb{E}} = \mathbb{E} - \widehat{\mathbb{E}}, \quad \text{with} \quad \widehat{\mathbb{E}} = \frac{\mathbb{E}\mathbb{D} + \mathbb{D}\mathbb{E}}{2}. \tag{4.316}$$

The elastic tensor $\mathbb{E}$ of the intact material and the damaged elastic tensor $\widetilde{\mathbb{E}}$ must be positive definite, as a consequence of the positiveness of the elastic potential. In a thermodynamical framework, the positive semi-definiteness of the loss of stiffness tensor $\widehat{\mathbb{E}}$ is equivalent to a positive intrinsic dissipation due to linear elasticity-damage coupling.[19] The damage tensor $\mathbb{D}$ is assumed to be positive semi-definite.

The above questions can be effectively investigated using the polar formalism: $\mathbb{E}$ and $\mathbb{D}$ can be represented by the classical polar representation for elasticity-like tensors and the bounds for their positiveness are known explicitly, Eq. (4.111).

We then obtain the polar invariants of $\widetilde{\mathbb{E}}$ as functions of those of $\mathbb{E}$ and $\mathbb{D}$; while we assume that the initial material is isotropic, we consider all the possible transformations for the damaged material, leading to a final elastic behavior that can be completely anisotropic, orthotropic, specially orthotropic or also isotropic.

If the damage tensor $\mathbb{D}$ is represented within the polar formalism as

$$\begin{aligned}
D_{1111}(\theta) &= D_0+2D_1+S_0\cos 4(\Psi_0-\theta)+4S_1\cos 2(\Psi_1-\theta),\\
D_{1112}(\theta) &= S_0\sin 4(\Psi_0-\theta)+2S_1\sin 2(\Psi_1-\theta),\\
D_{1122}(\theta) &= -D_0+2D_1-S_0\cos 4(\Psi_0-\theta),\\
D_{1212}(\theta) &= D_0-S_0\cos 4(\Psi_0-\theta),\\
D_{1222}(\theta) &= -S_0\sin 4(\Psi_0-\theta)+2S_1\sin 2(\Psi_1-\theta),\\
D_{2222}(\theta) &= D_0+2D_1+S_0\cos 4(\Psi_0-\theta)-4S_1\cos 2(\Psi_1-\theta).
\end{aligned} \tag{4.317}$$

then the damaged stiffness tensor $\widetilde{\mathbb{E}}$ is given by

$$\begin{aligned}
\widetilde{E}_{1111}(\theta) &= T_0(1-2D_0)+2T_1(1-4D_1)-2T_0S_0\cos 4(\Psi_0-\theta)-\\
&\qquad -4(T_0+2T_1)S_1\cos 2(\Psi_1-\theta),\\
\widetilde{E}_{1112}(\theta) &= -2T_0S_0\sin 4(\Psi_0-\theta)-2(T_0+2T_1)S_1\sin 2(\Psi_1-\theta),\\
\widetilde{E}_{1122}(\theta) &= -T_0(1-2D_0)+2T_1(1-4D_1)+2T_0S_0\cos 4(\Psi_0-\theta),\\
\widetilde{E}_{1212}(\theta) &= T_0(1-2D_0)+2T_0S_0\cos 4(\Psi_0-\theta),\\
\widetilde{E}_{1222}(\theta) &= 2T_0S_0\sin 4(\Psi_0-\theta)-2(T_0+2T_1)S_1\sin 2(\Psi_1-\theta),\\
\widetilde{E}_{2222}(\theta) &= T_0(1-2D_0)+2T_1(1-4D_1)-2T_0S_0\cos 4(\Psi_0-\theta)+\\
&\qquad +4(T_0+2T_1)S_1\cos 2(\Psi_1-\theta),
\end{aligned} \tag{4.318}$$

with T_0 and T_1 the unique two polar components of the undamaged isotropic stiffness tensor $\mathbb{E}$. We find the polar parameters of $\widetilde{\mathbb{E}}$, indicated in the following by a $\sim$, comparing the above equations with the usual polar expressions of an elastic tensor, Eq. (4.67):

$$\begin{aligned}
\widetilde{T}_0 &= T_0(1-2D_0),\\
\widetilde{T}_1 &= T_1(1-4D_1),\\
\widetilde{R}_0 &= 2T_0S_0,\\
\widetilde{R}_1 &= (T_0+2T_1)S_1,\\
\widetilde{\Phi}_0 &= \Psi_0+\tfrac{\pi}{4},\\
\widetilde{\Phi}_1 &= \Psi_1+\tfrac{\pi}{2}.
\end{aligned} \tag{4.319}$$

[19]For a proof of this statement, see (Vannucci and Desmorat, 2015), Sect. 4 and Appendix.

Some remarks about these results:

- an advantage of the polar formalism, apparent from the developments above, is the uncoupling of the expressions of the parameters of $\widetilde{\mathbb{E}}$ as functions of those of $\mathbb{D}$: each one of the polar parameters of $\widetilde{\mathbb{E}}$ depends exclusively upon the corresponding polar parameter of $\mathbb{D}$;
- Equation (4.319) show that the damage symmetries, i.e. the corresponding for $\mathbb{D}$ of the ordinary and special orthotropies of $\mathbb{E}$, determine, each one, exactly the same elastic symmetry of the same type for the damaged elastic tensor $\widetilde{\mathbb{E}}$ and inversely;
- Equations $(4.318)_{1,6}$ and $(4.319)_6$ show that the axis of the strongest component of $\widetilde{\mathbb{E}}$, i.e. $\widetilde{E}_{2222}$, is turned of $\pi/2$ with respect to the direction of the strongest component of $\mathbb{D}$, D_{1111}. This is quite natural, because the material is more severely damaged along the direction of D_{1111}, so that, finally, $\widetilde{E}_{1111}(\theta=0) < \widetilde{E}_{2222}(\theta=0)$. Also the harmonic depending upon 4θ is turned of $\pi/4$, Eq. $(4.319)_5$, which gives for the angular invariant of $\widetilde{\mathbb{E}}$

$$\widetilde{\Phi}_0-\widetilde{\Phi}_1=\Psi_0-\Psi_1-\frac{\pi}{4}; \tag{4.320}$$

- the last result shows a rather surprising fact: the damaged elasticity tensor $\widetilde{\mathbb{E}}$ cannot have the same form of ordinary orthotropy of the damage tensor $\mathbb{D}$. In fact, for $\mathbb{D}$ orthotropic with

$$\Psi_0-\Psi_1=L\frac{\pi}{4},\ L=\{0,1\}, \tag{4.321}$$

  $\widetilde{\mathbb{E}}$ is orthotropic with

$$\widetilde{\Phi}_0-\widetilde{\Phi}_1=\widetilde{K}\frac{\pi}{4},\ \widetilde{K}=L-1. \tag{4.322}$$

  So, for $L=1$, $\widetilde{K}=0$ and for $L=0$, $\widetilde{K}=1$ (the sign does not matter).

The conditions of positive semi-definiteness for $\mathbb{D}$ and $\widehat{\mathbb{E}}$ and positive definiteness for $\widetilde{\mathbb{E}}$ provide the conditions to determine the bounds on the values of their moduli, once those on $\mathbb{E}$ known. It can be proved that the positive semi-definiteness of $\widehat{\mathbb{E}}$ always implies the positive semi-definiteness of $\mathbb{D}$, and is even equivalent in some particular cases related to the anisotropy induced by damage.

Using the polar formalism, it is possible to give an explicit expression for the bounds on the polar invariants of $\mathbb{D}$ and $\widehat{\mathbb{E}}$. Starting from the simpler case, that of an isotropic tensor $\widetilde{\mathbb{E}}$, we consider all the possible cases of elastic symmetries for $\widetilde{\mathbb{E}}$, until the most general case of a completely anisotropic $\widetilde{\mathbb{E}}$, and show that the admissible domain for the moduli is convex in all the cases; in some of them a graphical representation is also possible. The results in the most general case are summarized in Table 4.3, where the following ratios have been introduced:

$$\tau_1=\frac{2T_1}{T_0},\ \widetilde{\tau}_0=\frac{\widetilde{T}_0}{T_0},\ \widetilde{\tau}_1=\frac{2\widetilde{T}_1}{T_0},\ \widetilde{\rho}_0=\frac{\widetilde{R}_0}{T_0},\ \widetilde{\rho}_1=\frac{\widetilde{R}_1}{T_0}. \tag{4.323}$$

Table 4.3 Minimal set of dimensionless polar bounds in the completely anisotropic case

	Polar bounds for $\mathbb{D}$	Polar bounds for $\widetilde{\mathbb{E}}$
B5	$2(D_0+S_0)<1$	$\widetilde{\tau}_0>\widetilde{\rho}_0$
B6	$\frac{\tau_1}{4(1+\tau_1)^2}(1-4D_1)[(1-2D_0)^2-$ $-4S_0^2]>S_1^2[1-2D_0+$ $2S_0\cos 4(\Psi_0-\Psi_1)]$	$\widetilde{\tau}_1(\widetilde{\tau}_0^2-\widetilde{\rho}_0^2)>$ $4\widetilde{\rho}_1^2\left[\widetilde{\tau}_0-\widetilde{\rho}_0\cos 4(\widetilde{\Phi}_0-\widetilde{\Phi}_1)\right]$
B7	$S_0\geq 0$	$\widetilde{\rho}_0\geq 0$
B8	$S_1\geq 0$	$\widetilde{\rho}_1\geq 0$
B9	$D_0\geq S_0$	$\widetilde{\tau}_0+\widetilde{\rho}_0\leq 1$
B10	$D_1(D_0^2-S_0^2)\geq\frac{(1+\tau_1)^2}{2\tau_1}S_1^2[D_0-$ $-S_0\cos 4(\Psi_0-\Psi_1)]$	$(\tau_1-\widetilde{\tau}_1)[(1-\widetilde{\tau}_0)^2-\widetilde{\rho}_0^2]\geq$ $4\widetilde{\rho}_1^2[1-\widetilde{\tau}_0+\widetilde{\rho}_0\cos 4(\widetilde{\Phi}_0-\widetilde{\Phi}_1)]$

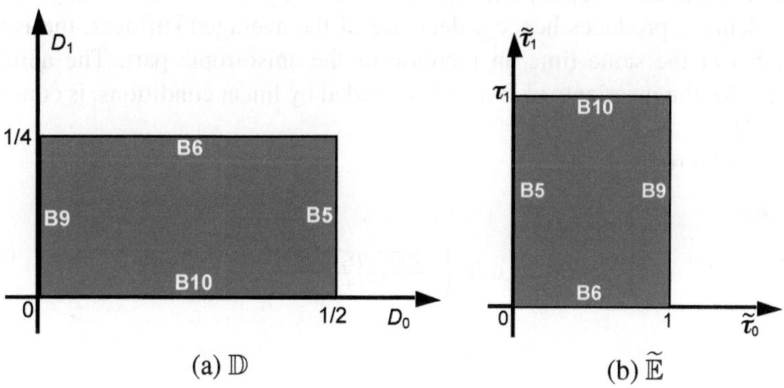

Fig. 4.20 Admissible domain for the case of isotropic damaged material

The ratio $\tau_1>0$ will hence be the only term representing the mechanical characteristics of the undamaged material.

The general results presented in Table 4.3 can be easily specialized to the different cases of material symmetry of the final behavior:

- $\widetilde{\mathbb{E}}$ isotropic:

$$\begin{cases} 0\leq D_0<\frac{1}{2}, \\ 0\leq D_1<\frac{1}{4}, \end{cases} \tag{4.324}$$

$$\begin{cases} 0<\widetilde{\tau}_0\leq 1, \\ 0<\widetilde{\tau}_1\leq \tau_1, \end{cases} \rightarrow \begin{cases} 0<\widetilde{T}_0\leq T_0, \\ 0<\widetilde{T}_1\leq T_1. \end{cases} \tag{4.325}$$

The isotropic part of $\mathbb{E}$ is hence diminished by damage; the admissible domain is clearly convex, see Fig. 4.20

- $\widetilde{\mathbb{E}}$ square symmetric:

$$\begin{cases} 0 \leq D_1 < \frac{1}{4}, \\ 0 \leq S_0 < \min\left\{D_0; \frac{1}{2} - D_0\right\}, \end{cases} \tag{4.326}$$

$$\begin{cases} 0 < \widetilde{\tau}_1 \leq \tau_1, \\ 0 \leq \widetilde{\rho}_0 < \min\left\{\widetilde{\tau}_0; 1 - \widetilde{\tau}_0\right\}, \end{cases} \rightarrow \begin{cases} 0 < \widetilde{T}_0 \leq T_0, \\ 0 < \widetilde{T}_1 \leq T_1, \\ 0 \leq \widetilde{R}_0 < \min\left\{\widetilde{T}_0; T_0 - \widetilde{T}_0\right\}. \end{cases} \tag{4.327}$$

Also for square symmetry, the isotropic part of $\mathbb{E}$ is decreased by damage, but at the same time the anisotropic part, here represented by the only term $\widetilde{R}_0$, grows from zero: damage produces hence a decrease of the averaged stiffness, the isotropic part, but at the same time an increase of the anisotropic part. The admissible domain for the invariants of $\mathbb{D}$ and $\widetilde{\mathbb{E}}$, bounded by linear conditions, is convex, see Fig. 4.21.

- $\widetilde{\mathbb{E}}$ R_0−orthotropic:

$$\begin{cases} 0 \leq D_0 < \frac{1}{2}, \\ 0 \leq S_1 < \min\left\{\frac{\sqrt{\tau_1(1-2D_0)(1-4D_1)}}{2(1+\tau_1)}; \frac{\sqrt{2\tau_1 D_0 D_1}}{1+\tau_1}\right\}, \end{cases} \tag{4.328}$$

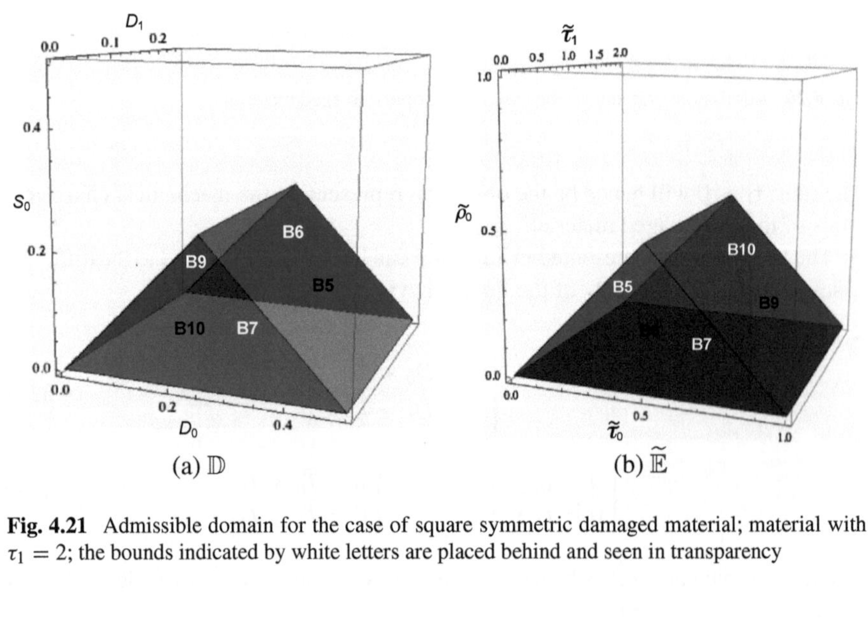

(a) $\mathbb{D}$ (b) $\widetilde{\mathbb{E}}$

Fig. 4.21 Admissible domain for the case of square symmetric damaged material; material with $\tau_1 = 2$; the bounds indicated by white letters are placed behind and seen in transparency

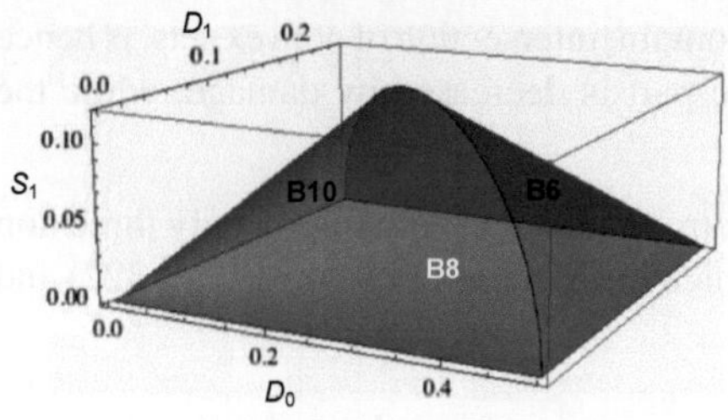

(a) $\mathbb{D}$

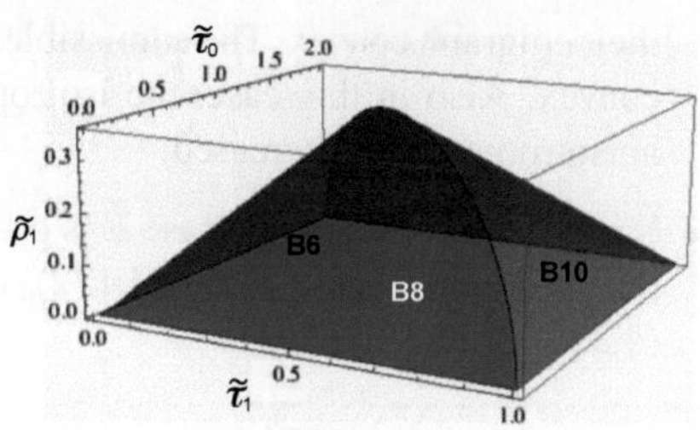

(b) $\widetilde{\mathbb{E}}$

Fig. 4.22 Admissible domain for the case of R_0−orthotropic damaged material; material with $\tau_1 = 2$; the bounds indicated by white letters are placed behind and seen in transparency

$$\begin{cases} 0 < \widetilde{\tau}_0 \leq 1, \\ 0 \leq \widetilde{\rho}_1 < \min\left\{\frac{\sqrt{\widetilde{\tau}_0\widetilde{\tau}_1}}{2}; \frac{\sqrt{(1-\widetilde{\tau}_0)(\tau_1-\widetilde{\tau}_1)}}{2}\right\}, \end{cases} \rightarrow$$
$$\begin{cases} 0 < \widetilde{T}_0 \leq T_0, \\ 0 < \widetilde{T}_1 \leq T_1, \\ 0 \leq \widetilde{R}_1 < \min\left\{\sqrt{\frac{\widetilde{T}_0\widetilde{T}_1}{2}}; \sqrt{\frac{(T_0-\widetilde{T}_0)(T_1-\widetilde{T}_1)}{2}}\right\}. \end{cases} \tag{4.329}$$

In this case too the isotropic part of $\mathbb{E}$ is decreased by damage and the anisotropic part, the term $\widetilde{R}_1$ alone, grows from zero. The above bounds define a convex admissible domain for the invariants of $\mathbb{D}$ and $\widetilde{\mathbb{E}}$, see Fig. 4.22.

- $\widetilde{\mathbb{E}}$ ordinarily orthotropic:

$$\begin{cases} 0 \leq S_0 < \min\left\{D_0; \frac{1}{2} - D_0\right\}, \\ 0 \leq S_1 < \min\left\{\frac{\sqrt{2\tau_1 D_1[D_0+(-1)^L S_0]}}{1+\tau_1}; \frac{\sqrt{\tau_1(1-4D_1)[1-2D_0-(-1)^L S_0]}}{2(1+\tau_1)}\right\}, \end{cases} \tag{4.330}$$

$$\begin{cases} 0 \leq \widetilde{\rho}_0 < \min\left\{\widetilde{\tau}_0; 1-\widetilde{\tau}_0\right\}, \\ 0 \leq \widetilde{\rho}_1 < \min\left\{\frac{\sqrt{\widetilde{\tau}_1[\widetilde{\tau}_0+(-1)^{\widetilde{K}}\widetilde{\rho}_0]}}{2}; \frac{\sqrt{[1-\widetilde{\tau}_0-(-1)^{\widetilde{K}}\widetilde{\rho}_0](\tau_1-\widetilde{\tau}_1)}}{2}\right\}, \end{cases} \rightarrow$$
$$\begin{cases} 0 < \widetilde{T}_0 \leq T_0, \\ 0 < \widetilde{T}_1 \leq T_1, \\ 0 \leq \widetilde{R}_0 < \min\left\{\widetilde{T}_0; T_0-\widetilde{T}_0\right\}, \\ 0 \leq \widetilde{R}_1 < \min\left\{\sqrt{\frac{\widetilde{T}_1[\widetilde{T}_0+(-1)^{\widetilde{K}}\widetilde{R}_0]}{2}}; \sqrt{\frac{[T_0-\widetilde{T}_0-(-1)^{\widetilde{K}}\widetilde{R}_0](T_1-\widetilde{T}_1)}{2}}\right\}. \end{cases} \tag{4.331}$$

Unfortunately, it is not possible to give a graphical representation of the domains defined by the above bounds, because they are functions of four independent quantities. Nevertheless, these domains are convex; in fact, the conditions in Eq. (4.330) or (4.331) are either linear or with a Hessian matrix whose eigenvalues are either null or negative, as it can be easily checked. So, such functions are concave and their epigraph convex. The admissible domain, intersection of convex sets, is hence convex. Also in this case, the isotropic part is decreased by damage, while the anisotropic one is increased.

- $\widetilde{\mathbb{S}}$ r_0−orthotropic: in this case $\widetilde{\mathbb{E}}$ is orthotropic but depending upon only three non null invariants thanks to Eq. (4.172), which inserted into Eqs. (4.319), (4.322) and (4.323) gives

$$S_0 = \frac{(1+\tau_1)^2}{\tau_1}\frac{S_1^2}{1-4D_1}, \quad \widetilde{\rho}_0 = 2\frac{\widetilde{\rho}_1^2}{\widetilde{\tau}_1} = 2\frac{(1+\tau_1)^2}{\tau_1}\frac{S_1^2}{1-4D_1}, \quad L = 1. \tag{4.332}$$

If an isotropic layer is damaged according to the previous conditions, it is transformed into a material whose behavior is of the same type of that of a sheet of paper (Vannucci 2010). The bounds in this case are

$$\begin{cases} 0 \leq D_0 < \frac{1}{2}, \\ 0 \leq D_1 < \frac{1}{4}, \\ 0 \leq S_1 < \min\left\{\frac{1}{1+\tau_1}\sqrt{\frac{\tau_1(1-2D_0)(1-4D_1)}{3}}; \frac{1}{1+\tau_1}\sqrt{\frac{2\tau_1 D_0 D_1(1-4D_1)}{1-2D_1}}\right\}. \end{cases} \tag{4.333}$$

$$\begin{aligned} &\begin{cases} 0 < \widetilde{\tau}_0 \leq 1, \\ 0 < \widetilde{\tau}_1 \leq \tau_1, \\ 0 \leq \widetilde{\rho}_1 < \min\left\{\sqrt{\frac{\widetilde{\tau}_0\widetilde{\tau}_1}{2}}; \sqrt{\frac{\widetilde{\tau}_1(1-\widetilde{\tau}_0)(\tau_1-\widetilde{\tau}_1)}{2(1+\tau_1)}}\right\}, \end{cases} \rightarrow \\ &\begin{cases} 0 < \widetilde{T}_0 \leq T_0, \\ 0 < \widetilde{T}_1 \leq T_1, \\ 0 \leq \widetilde{R}_1 < \min\left\{\sqrt{\widetilde{T}_0\widetilde{T}_1}; \sqrt{\frac{2\widetilde{T}_1(T_0-\widetilde{T}_0)(T_1-\widetilde{T}_1)}{T_0+2T_1}}\right\}. \end{cases} \end{aligned} \tag{4.334}$$

We remark that Fig. 4.23 clearly confirms that the admissible domain is convex, as it has to be, being this one a particular case of ordinary orthotropy. Once more, the isotropic part decreases and the anisotropic one increases as a consequence of damage.

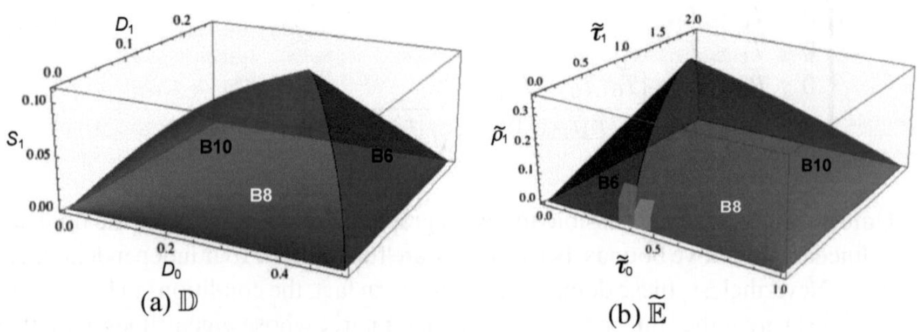

Fig. 4.23 Admissible domain for the case of r_0−orthotropic damaged material; material with $\tau_1 = 2$; the bounds indicated by white letters are placed behind and seen in transparency

4.17.3 Tensor Strength Criteria for Anisotropic Layers

The strength of anisotropic bodies cannot be checked using the same criteria of the isotropic ones, for different reasons; namely, anisotropy concerns also strength, hence it cannot be represented by a unique quantity, the Von Mises criterion cannot be used because it limits uniquely the deviatoric part of the strain energy, and we have seen in Sect. 2.5 that generally speaking it is not possible to isolate such a part of the total strain energy for an anisotropic material.

Strength criteria specially conceived for anisotropic bodies have been proposed in the past, see Jones (1999). We focus here on the *tensor criteria*, so called because all of them limit the value of a quadratic form of a given tensor. In the planar case, such a tensor can be represented by the polar formalism, so obtaining an intrinsic expression of the strength criterion, cf. Catapano et al. (2012). It is worth noting that all the anisotropic strength criteria are formulated for orthotropic materials: none of such criteria can be adapted to a completely anisotropic material. In the most general case, a tensor strength criterion can be put in the form

$$\boldsymbol{\sigma}\cdot\mathbb{F}\boldsymbol{\sigma}+\boldsymbol{\sigma}\cdot\mathbf{F}\leq 1, \tag{4.335}$$

where $\mathbb{F}$ is a fourth-rank tensor, while $\mathbf{F}$ a second-rank one. This last is introduced to take account of a possible difference of the strength in tension and compression. In all the criteria, it is tacitly understood that $\mathbb{F}$ has the same tensor symmetries of an elasticity tensor, i.e. the minor and major ones, Eqs. (2.13) and (2.11), while $\mathbf{F}$ is symmetric. Hence, they can be represented with the polar formalism in the classical way.

The Cartesian components of $\mathbb{F}$ and $\mathbf{F}$ are given in Table 4.4 for the criteria considered in the following. The terms appearing in the table are:

- X_t: tension strength in the direction x_1;
- X_c: compression strength in the direction x_1;
- Y_t: tension strength in the direction x_2;
- Y_c: compression strength in the direction x_2;
- S: shear strength;
- F_{12}^*: dimensionless term to be determined upon biaxial tests and depending on the material and the stress field.

The directions x_1 and x_2 are the orthotropy directions of the layer.

As apparent from Table 4.4, the components of $\mathbf{F}$ and $\mathbb{F}$ have the dimensions of the reciprocal of a strength or of a square of such a quantity, respectively. So, they describe, in some sense, the inverse of a strength of the material; that is why we can call them the *weakness tensors*.

We give below the polar expression for three of the most common anisotropic strength criteria in $\mathbb{R}^2$.

Table 4.4 Cartesian components of $\mathbb{F}$ and $\mathbf{F}$ for three strength criteria

	Tsai-Hill	Hoffman	Tsai–Wu
F_{11}	$\frac{1}{Y^2}$	$\frac{1}{Y_tY_c}$	$\frac{1}{Y_tY_c}$
F_{12}	$-\frac{1}{2X^2}$	$-\frac{1}{2X_tX_c}$	$\frac{F_{12}^*}{\sqrt{X_tX_cY_tY_c}}$
F_{22}	$\frac{1}{X^2}$	$\frac{1}{X_tX_c}$	$\frac{1}{X_tX_c}$
F_{66}	$\frac{1}{4S^2}$	$\frac{1}{4S^2}$	$\frac{1}{4S^2}$
F_1	0	$\frac{Y_c-Y_t}{Y_tY_c}$	$\frac{Y_c-Y_t}{Y_tY_c}$
F_2	0	$\frac{X_c-X_t}{X_tX_c}$	$\frac{X_c-X_t}{X_tX_c}$
F_6	0	0	0

4.17.3.1 The Tsai-Hill Criterion

This criterion (Tsai 1965, 1968), is an adaptation of the yielding criterion of Hill (1948) originally proposed for metallic alloys and as such appropriate for materials having the same strength in tension and compression. The polar expression of the criterion is

$$\begin{aligned} F_{Hill} =& 4R^2\gamma_0 + 8T^2\gamma_1 + (-1)^l 4\lambda_0 R^2 \cos 4(\omega_1 - \Phi) + \\ & 16TR\lambda_1 \cos 2(\omega_1 - \Phi) \leq 1. \end{aligned} \tag{4.336}$$

In the above equation, T, R and Φ are the polar parameters of the stress tensor $\boldsymbol{\sigma}$, γ_0, γ_1, λ_0, λ_1, ω_0 and ω_1 are the polar parameters of $\mathbb{F}$, while $\mathbf{F} = \mathbf{O}$, because no difference exists between the tension and compression strengths, as said above; being the criterion suited for orthotropic materials, we have posed, cf. Sect. 4.12.1,

$$\omega_0 = \omega_1 + l\frac{\pi}{4}, \tag{4.337}$$

as will be done also for the next criteria.

4.17.3.2 The Hoffmann Criterion

The Hoffman criterion (Hoffman 1967), can take into account for a difference of the strength in tension and in compression; its polar formulation is

$$\begin{aligned} F_{Hoff} =& 4R^2\gamma_0 + 8T^2\gamma_1 + (-1)^l 4\lambda_0 R^2 \cos 4(\omega_1 - \Phi) + \\ & 16TR\lambda_1 \cos 2(\omega_1 - \Phi) + 2T\gamma + 2\lambda R \cos 2(\omega - \Phi) \leq 1, \end{aligned} \tag{4.338}$$

with $\gamma,\ \lambda$ and ω the polar parameters of $\mathbf{F}$.

4.17.3.3 The Tsai–Wu Criterion

The Tsai–Wu criterion (Tsai and Wu 1971), like the previous one, can take into account for a different strength in tension and compression and its general expression is similar to that of the Hoffmann criterion while it can be proved that its polar form is completely analogous:

$$\begin{aligned} F_{TW} =& 4R^2\gamma_0 + 8T^2\gamma_1 + (-1)^l 4\lambda_0 R^2 \cos 4(\omega_1 - \Phi) + \\ & 16TR\lambda_1 \cos 2(\omega_1 - \Phi) + 2T\gamma + 2\lambda R \cos 2(\omega - \Phi) \leq 1. \end{aligned} \tag{4.339}$$

4.17.3.4 Polar Representation of the Strength Moduli

The polar components of $\mathbb{F}$, for an orthotropic material, in terms of strength properties can be easily obtained introducing in the Cartesian components of Table 4.4 their polar expressions, Eqs. (4.52) and (4.57):

$$\begin{aligned} 2\gamma &= \frac{Y_c - Y_t}{Y_t Y_c} + \frac{X_c - X_t}{X_t X_c}, \\ 2\lambda e^{2i\omega} &= \frac{Y_c - Y_t}{Y_t Y_c} - \frac{X_c - X_t}{X_t X_c}; \end{aligned} \tag{4.340}$$

$$\begin{aligned} 8\gamma_0 &= \frac{1}{X_t X_c} + \frac{1}{Y_t Y_c} + \frac{1}{S^2} - 2\frac{F_{12}^*}{\sqrt{X_t X_c Y_t Y_c}}, \\ 8\gamma_1 &= \frac{1}{X_t X_c} + \frac{1}{Y_t Y_c} + 2\frac{F_{12}^*}{\sqrt{X_t X_c Y_t Y_c}}, \\ 8(-1)^l \lambda_0 e^{4i\omega_1} &= \frac{1}{X_t X_c} + \frac{1}{Y_t Y_c} - \frac{1}{S^2} - 2\frac{F_{12}^*}{\sqrt{X_t X_c Y_t Y_c}}, \\ 8\lambda_1 e^{2i\omega_1} &= \frac{1}{Y_t Y_c} - \frac{1}{X_t X_c}. \end{aligned} \tag{4.341}$$

Concerning the anisotropic terms of $\mathbb{F}$, we assume that λ_0 and λ_1 are positives. This corresponds to the following conditions:

$$\begin{aligned}
&\frac{2}{X_tX_c}+\frac{1}{Y_tY_c}-2\frac{F_{12}^*}{\sqrt{X_tX_cY_tY_c}}>\frac{1}{S^2} \text{ if } l=0,\\
&\frac{2}{X_tX_c}+\frac{1}{Y_tY_c}-2\frac{F_{12}^*}{\sqrt{X_tX_cY_tY_c}}<\frac{1}{S^2} \text{ if } l=1.\\
&Y_tY_c<X_tX_c.
\end{aligned} \tag{4.342}$$

Therefore, the two types of orthotropy are directly linked to this two inequalities.

4.17.3.5 Two Practical Cases

The value of the polar invariants for two orthotropic layers, see Table 4.5 are reported in Tables 4.6, 4.7 and 4.8. In particular, four cases are presented for an E-glass epoxy lamina for the criterion of Tsai-Hill, with two different values of the strength, one in tension and the other in compression. These ones, when used in a criterion that does not consider this possibility, give rise to two different sets of invariants, one for the tension and the other for the compression case. With the use of the Hoffman's failure criterion polar invariants are univocal, see Table 4.7. Concerning the Tsai–Wu's criterion, some polar parameters are linear function of the term $F_{12}^* \in [-1; 1]$:

$$\begin{cases}
8\gamma_0=\dfrac{1}{X_tX_c}+\dfrac{1}{Y_tY_c}+\dfrac{1}{S^2}-2\dfrac{F_{12}^*}{\sqrt{X_tX_cY_tY_c}},\\
8\gamma_1=\dfrac{1}{X_tX_c}+\dfrac{1}{Y_tY_c}+2\dfrac{F_{12}^*}{\sqrt{X_tX_cY_tY_c}},\\
8(-1)^l\lambda_0e^{4i\omega_1}=\dfrac{1}{X_tX_c}+\dfrac{1}{Y_tY_c}-\dfrac{1}{S^2}-2\dfrac{F_{12}^*}{\sqrt{X_tX_cY_tY_c}}.
\end{cases} \tag{4.343}$$

Table 4.5 Mechanical strength properties of orthotropic materials, [MPa]

Material	X_t	X_c	Y_t	Y_c	S
E-glass epoxy	1080	620	39	128	89
Carbon/Epoxy	1447	1447	51.7	206	93

Table 4.6 Weakness polar components of the generic E-glass epoxy, criterion of Tsai-Hill

X	Y	$\times 10^{-5}$	$\times 10^{-6}$	$\times 10^{-6}$	$\times 10^{-6}$	l	ω_1
		γ_0 [MPa^{-2}]	γ_1 [MPa^{-2}]	λ_0 [MPa^{-2}]	λ_1 [MPa^{-2}]		
X_T	Y_T	9.807	82.29	66.51	82.08	0	0
X_C	Y_C	2.373	7.955	7.826	7.304	1	0
X_T	Y_C	2.352	7.737	8.044	7.522	1	0
X_C	Y_T	9.829	82.51	66.73	81.86	0	0

Table 4.7 Weakness polar components for orthotropic materials, criterion of Hoffman

Material	$\times 10^{-5}$	$\times 10^{-5}$	$\times 10^{-6}$	$\times 10^{-5}$	l	ω_1	$\times 10^{-3}$	$\times 10^{-3}$	ω
	γ_0 [MPa^{-2}]	γ_1 [MPa^{-2}]	λ_0 [MPa^{-2}]	λ_1 [MPa^{-2}]			γ	λ [MPa^{-1}]	
E-glass epoxy	4.12	2.50	9.63	2.49	0	0	8.57	9.26	0
Carbon/Epoxy	2.63	1.17	2.60	1.17	1	0	7.24	7.24	0

Table 4.8 Weakness polar components for orthotropic materials, criterion of Tsai–Wu

Material	$\times 10^{-5}$	$\times 10^{-5}$	$\times 10^{-6}$	$\times 10^{-5}$	l	ω_1	$\times 10^{-3}$	$\times 10^{-3}$	ω
	γ_0 [MPa^{-2}]	γ_1 [MPa^{-2}]	λ_0 [MPa^{-2}]	λ_1 [MPa^{-2}]			γ [MPa^{-1}]	λ [MPa^{-1}]	
E-glass epoxy	[4.53:3.67]	[2.09:2.96]	[13.8:5.12]	2.49	0	0	8.57	9.26	0
Carbon/ Epoxy	[2.79:2.46]	[1.01:1.35]	[0.98:4.33]	1.17	1	0	7.24	7.24	0

Their numerical values are presented in Table 4.8. A fixed F_{12}^* cannot be given because its value actually depends upon the experimental test. Nevertheless, thanks to the polar formalism, we can see that the term F_{12}^* is independent from the stress field in two cases:

$$\begin{aligned} &\lambda_0 = 0 : F_{12}^* = \frac{\sqrt{X_t X_c Y_t Y_c}}{2}\left(\frac{1}{X_t X_c} + \frac{1}{Y_t Y_c} - \frac{1}{S^2}\right), \\ &\text{Isotropy, i.e. } \lambda_0 = \lambda_1 = 0 : F_{12}^* = 1 - \frac{X_t X_c}{2S^2}. \end{aligned} \tag{4.344}$$

The first one represents a material with the special orthotropy $\lambda_0 = 0$ for which the characterization of the orthotropic weakness tensor $\mathbb{F}$ is given by three, and not four, polar invariants: γ_0, γ_1 and λ_1. As shown in Eq. (4.344), this phenomenon has a consequence on the description of strength: the term F_{12}^* of Tsai–Wu tensor $\mathbb{F}$ has an exact value that depends only upon the known properties of normal and shear strength. The second case of Eq. (4.344) concerns isotropy. Of course, in this case the further reduction of independent components *a fortiori* leads to an explicit formulation of F_{12}^*.

The other polar terms of the Tsai–Wu tensors, independent from F_{12}^*, are equivalents to those of the Hoffman's criterion.

4.18 Some Examples of Planar Anisotropic Materials

We end this chapter giving some examples of planar anisotropic materials, i.e. of materials in the form of thin layers. For the most part of them, they are commercial composite materials for structural use. For all of them, the directional diagrams of

Table 4.9 Examples of planar anisotropic materials

	Wood[a]	Carbon/ Epoxy[b]	Boron/ Epoxy[c]	S-Glass/ Epoxy[d]	Kevlar/ Epoxy[e]	Ice[f]	TiB_2 [g]	Carbon/ Epoxy[h]	Glass/ Epoxy[i]	BR45a[j]	BR60[k]
E_1	10.04	181.00	205.00	47.66	86.90	11.75	253.81	54.00	29.70	40.40	30.90
E_2	0.42	10.30	18.50	13.31	5.52	9.61	387.60	54.00	29.70	19.60	42.60
G_{12}	0.75	7.17	5.59	4.75	2.14	3.00	250.00	4.00	5.30	25.00	14.00
ν_{12}	0.23	0.28	0.23	0.27	0.34	0.27	0.29	0.04	0.17	0.75	0.34
E_{max}	10.04	181.00	205.00	47.66	86.90	11.75	510.51	54.00	29.70	57.82	42.60
$\theta_{E_{max}}$ ()	0	0	0	0	0	0	53.19	0, 90	0, 90	34.44	90
E_{min}	0.42	10.30	15.62	12.36	5.26	8.33	253.81	14.02	16.35	19.60	30.90
$\theta_{E_{min}}$ ()	90	90	57.35	60.8	64.73	50.60	0	45	45	90	0
Q_{11}	10.06	181.81	206.00	48.65	87.54	12.51	291.76	54.11	30.58	55.56	36.76
Q_{22}	0.42	10.35	18.59	13.59	5.56	10.22	445.56	54.11	30.58	26.96	50.68
Q_{66}	1.50	14.34	11.18	9.50	4.28	6.00	500.00	8.00	10.60	50.00	28.00
Q_{12}	0.10	2.89	4.27	3.67	1.89	2.78	130.05	2.43	5.20	20.22	17.23
T_0	1.66	26.88	29.80	92.38	12.23	3.65	184.65	14.92	8.99	17.76	13.62
T_1	1.34	24.74	29.14	86.97	12.11	3.54	124.71	14.14	8.94	15.37	15.24

(continued)

Table 4.9 (continued)

	Wood[a]	Carbon/Epoxy[b]	Boron/Epoxy[c]	S-Glass/Epoxy[d]	Kevlar/Epoxy[e]	Ice[f]	TiB_2 [g]	Carbon/Epoxy[h]	Glass/Epoxy[i]	BR45a[j]	BR60[k]
R_0	0.91	19.71	24.21	44.86	10.09	0.65	65.35	10.92	3.70	7.24	0.38
R_1	1.20	21.43	23.42	43.82	10.25	0.28	19.23	0	0	3.57	1.74
Φ_0	0	0	0	0	0	0	$\pi/4$	0	0	$\pi/4$	$\pi/4$
Φ_1	0	0	0	0	0	0	$\pi/2$	0	0	0	$\pi/2$
K	0	0	0	0	0	0	1	–	–	1	1
ρ	0.76	0.92	1.03	1.02	0.98	2.32	3.40	∞	∞	2.03	0.22
τ	2.88	2.62	2.61	4.25	2.53	15.16	6.37	3.96	7.26	6.01	24.76

[a] Pine wood, *source* (Lekhnitskii 1950)
[b] Carbon/Epoxy T300/5208, *source* (Tsai and Hahn 1980)
[c] Boron/Epoxy B(4)-55054, *source* (Tsai and Hahn 1980)
[d] S-Glass/Epoxy S2-449/SP 381, *source* (AAVV 2002)
[e] Kevlar/Epoxy 149, *source* (Daniel and Ishai 1994)
[f] Ice of the Mendenhall Glacier, 270°K, *source* (Landolt and Börnstein 1992)
[g] Titanium Boride, *source* (Landolt and Börnstein 1992)
[h] Carbon/Epoxy balanced fabric, *source* (Gay 2014)
[i] Glass/Epoxy balanced fabric 7781/5245C, *source* (Daniel and Ishai 1994)
[j] Braided Carbon/Epoxy BR45a, *source* (Falzon and Herszberg 1998)
[k] Braided Carbon/Epoxy BR60, *source* (Falzon and Herszberg 1998)
Moduli are in GPa, the Q_{ij}s are in the Kelvin's notation

E_1, G_{12}, ν_{12}, $\eta_{12,1}$ and $\eta_{12,2}$ are traced, normalized with respect to their highest value. A grey circle is traced in the diagrams of the coefficients, indicating the zero: when the diagram is inside the circle, the coefficient is negative. The values of the relevant moduli are given in Table 4.9. Two dimensionless polar parameters are also indicated: the anisotropy ratio ρ, Eq. (4.131), and the isotropy to anisotropy ratio τ, Eq. (4.282).

The calculations have been done using the polar method and the values of the moduli have been deduced, if the case, by the corresponding 3D values transforming the stiffness matrix $[C]$ to the reduced stiffness matrix $[Q]$ valid in the assumption of plane stress, according to Eq. (3.62).

It can be noticed the extreme anisotropy of materials like Pine Wood, Carbon/Epoxy T300/5208, Boron/Epoxy B(4)-55054, Kevlar/Epoxy 149. Their common characteristic is to be reinforced by unidirectional fibers, that for the Pine Wood are the natural wood-fibers, while for the others, the aligned fibers of Carbon, Boron or Kevlar immersed into the isotropic matrix of Epoxy resin for reinforcing it. Balanced fabrics are composites where the reinforcement is given by a fabric where there is the same amount of fibers in warp and weft. The directional diagrams of such materials clearly show that, contrarily to what often believed, such materials are not isotropic, but square symmetric. Braided composites are tissues where warp and weft are not perpendicular, but tilted of another angle: for BR45a, this angle is 45°, for BR60 it is 60°.

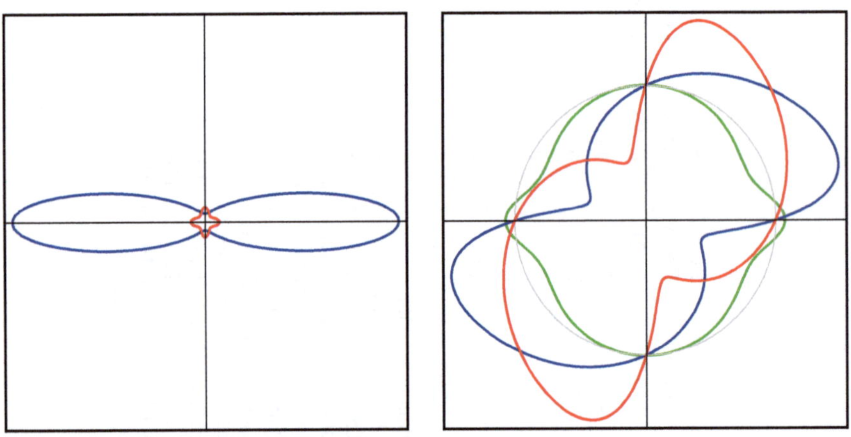

(a) E_1 (blue) and G_{12} (red). (b) ν_{12} (green), $\eta_{12,1}$ (blue) and $\eta_{12,2}$ (red).

Fig. 4.24 Directional 2D diagrams of Pine Wood

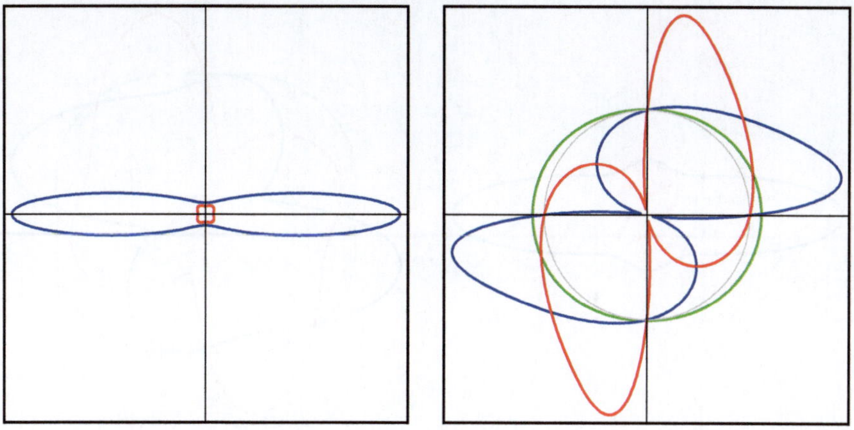

(a) E_1 (blue) and G_{12} (red).

(b) ν_{12} (green), $\eta_{12,1}$ (blue) and $\eta_{12,2}$ (red).

Fig. 4.25 Directional 2D diagrams of Carbon/Epoxy T300/5208

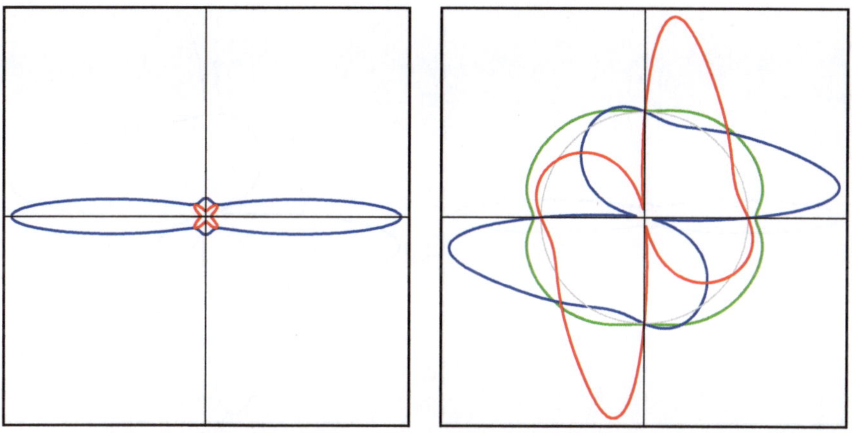

(a) E_1 (blue) and G_{12} (red).

(b) ν_{12} (green), $\eta_{12,1}$ (blue) and $\eta_{12,2}$ (red).

Fig. 4.26 Directional 2D diagrams of Boron/Epoxy B(4)-55054

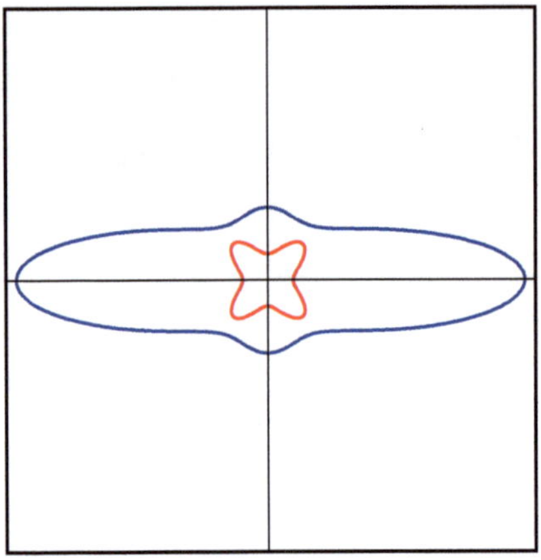

(a) E_1 (blue) and G_{12} (red).

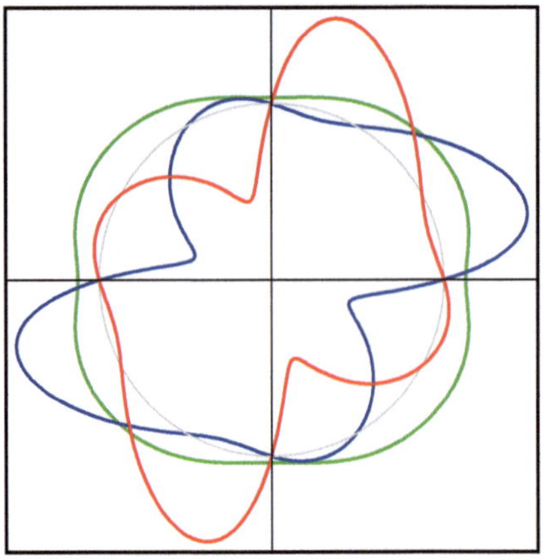

(b) ν_{12} (green), $\eta_{12,1}$ (blue) and $\eta_{12,2}$ (red).

Fig. 4.27 Directional 2D diagrams of S-Glass/Epoxy S2-449/SP 381

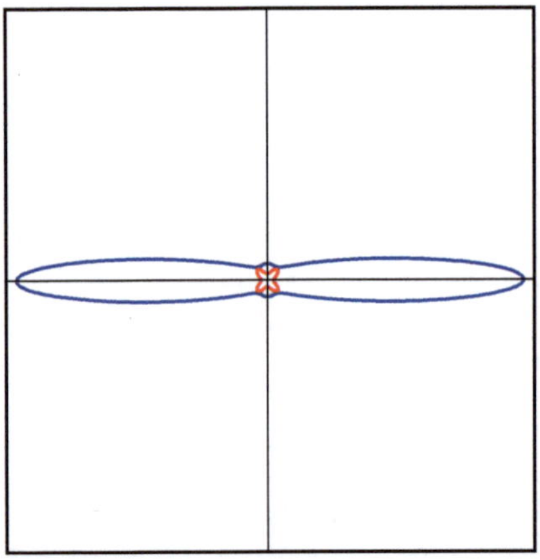

(a) E_1 (blue) and G_{12} (red).

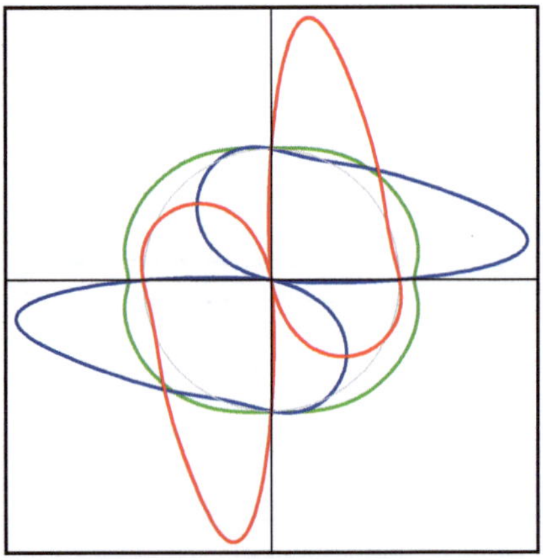

(b) ν_{12} (green), $\eta_{12,1}$ (blue) and $\eta_{12,2}$ (red).

Fig. 4.28 Directional 2D diagrams of Kevlar/Epoxy 149

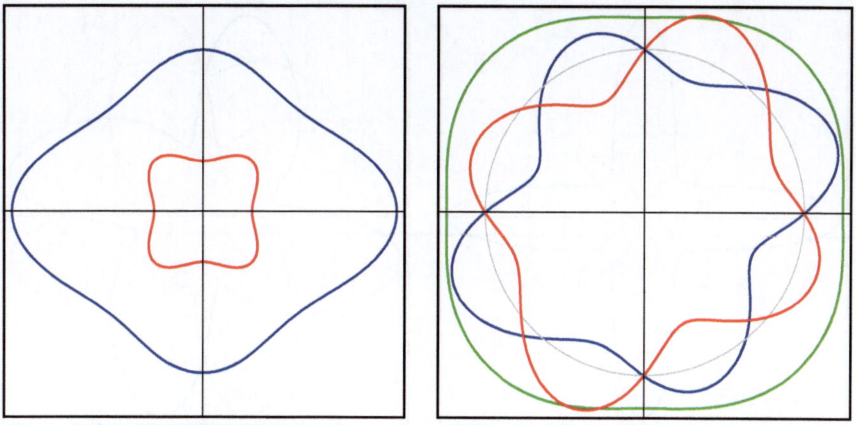

(a) E_1 (blue) and G_{12} (red). (b) ν_{12} (green), $\eta_{12,1}$ (blue) and $\eta_{12,2}$ (red).

Fig. 4.29 Directional 2D diagrams of Ice of the Mendenhall Glacier, 270°K

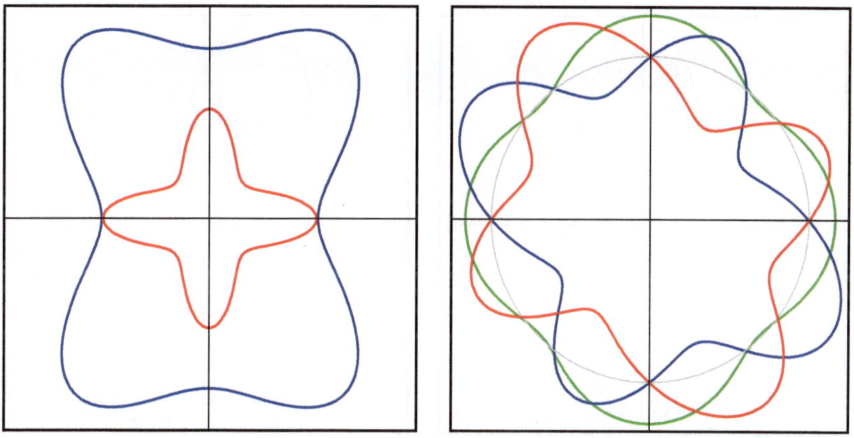

(a) E_1 (blue) and G_{12} (red). (b) ν_{12} (green), $\eta_{12,1}$ (blue) and $\eta_{12,2}$ (red).

Fig. 4.30 Directional 2D diagrams of Titanium Boride (TiB_2)

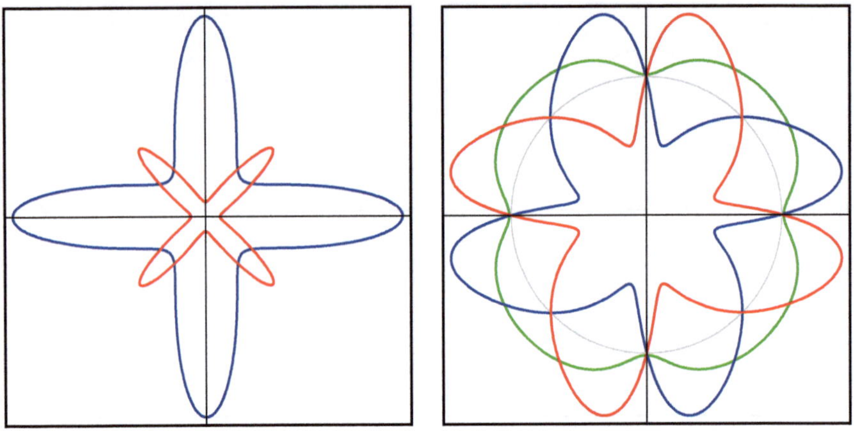

(a) E_1 (blue) and G_{12} (red). (b) ν_{12} (green), $\eta_{12,1}$ (blue) and $\eta_{12,2}$ (red).

Fig. 4.31 Directional 2D diagrams of Carbon/Epoxy balanced fabric

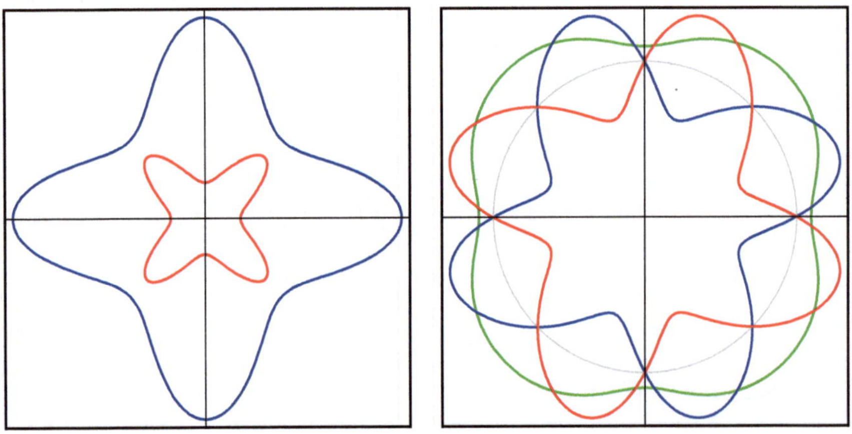

(a) E_1 (blue) and G_{12} (red). (b) ν_{12} (green), $\eta_{12,1}$ (blue) and $\eta_{12,2}$ (red).

Fig. 4.32 Directional 2D diagrams of Glass/Epoxy 7781/5245C balanced fabric

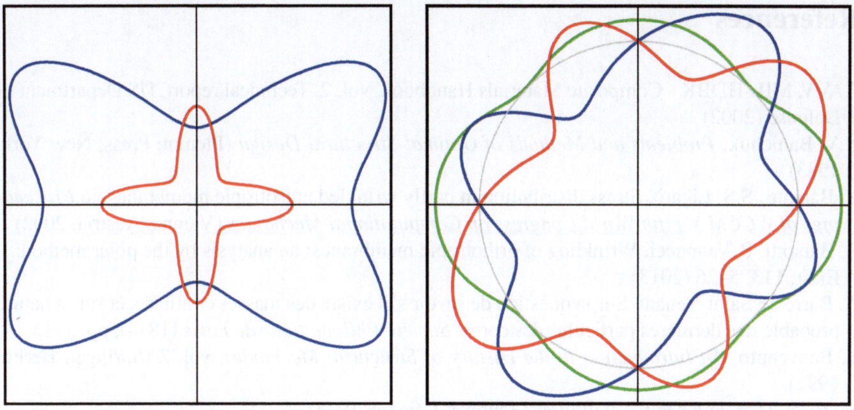

(a) E_1 (blue) and G_{12} (red). (b) ν_{12} (green), $\eta_{12,1}$ (blue) and $\eta_{12,2}$ (red).

Fig. 4.33 Directional 2D diagrams of Braided Carbon/Epoxy BR45a

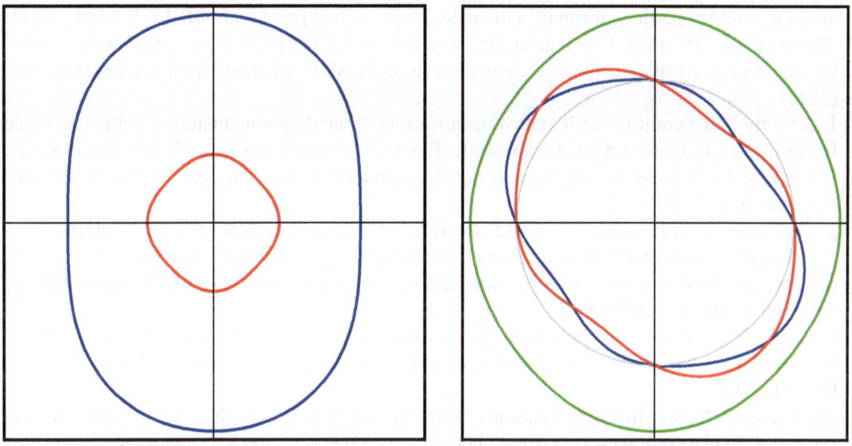

(a) E_1 (blue) and G_{12} (red). (b) ν_{12} (green), $\eta_{12,1}$ (blue) and $\eta_{12,2}$ (red).

Fig. 4.34 Directional 2D diagrams of Braided Carbon/Epoxy BR60

References

AAVV. MIL-HDBK - Composite Materials Handbook, vol. 2. Technical report, US Department of Defense (2002)

N.V. Banichuk, *Problems and Methods of Optimal Structural Design* (Plenum Press, New York, 1983)

R. Barsotti, S.S. Ligarò, Stress distribution in partly wrinkled anisotropic membranes, in *Proceedings of WCCM V Fifth World Congress on Computational Mechanics* (Vienna, Austria, 2002)

R. Barsotti, P. Vannucci, Wrinkling of orthotropic membranes: an analysis by the polar method. J. Elast. **113**, 5–26 (2013)

A. Barré de Saint-Venant, Sur la question de savoir s'il existe des masses continues et sur la nature probable des dernières particules des corps. *Société Philomatique de Paris* (1844), pp. 3–15

E. Benvenuto, *An Introduction to the History of Structural Mechanics*, vol. 2 (Springer, Berlin, 1991)

M. Born, *Dynamik der Krystallgitter* (Teubner, Leipzig, 1915)

R.G. Boscovich,*Theoria philosophiae naturalis redacta ad unicam legem virium in natura existentium*. Officina Libraria Kaliwodiana (Vienna, Austria, 1758)

J.G. Campbell, The in-plane elastic constants of paper. Aust. J. Appl. Sci. **12**, 356–357 (1961)

D. Capecchi, G. Ruta, P. Trovalusci, Voigt and Poincaré's mechanistic-energetic approaches to linear elasticity and suggestions for multiscale modelling. Arch. Appl. Mech. **81**, 1573–1584 (2011)

A. Catapano, B. Desmorat, P. Vannucci, Invariant formulation of phenomenological failure criteria for orthotropic sheets and optimisation of their strength. Math. Methods Appl.Sci. **35**, 1842–1858 (2012)

A.L. Cauchy, Sur l'équilibre et le mouvement d'un système de points matériels sollicités par des forces d'attraction ou de repulsion mutuelle. Exercices de mathématiques **3**, 188–212 (1828a)

A.L. Cauchy, De la pression ou tension dans un système de points matériels. Exercices de mathématiques **3**, 213–236 (1828b)

J.L. Chaboche. Le concept de contrainte effective appliqué à l'elasticité et à la viscoplasticité en présence d'un endommagement anisotrope, in *Proceedings of Colloque Euromech 115 (Villard-de-Lans, 1979): Comportement mécanique des matériaux anisotropes* (Paris, France, 1982), pp. 737–760. Editions du CNRS

C.L. Chow, On evolution laws of anisotropic damage. Eng. Fract. Mech. **34**, 679–701 (1987)

I.M. Daniel, O. Ishai, *Engineering Mechanics of Composite Materials* (Oxford University Press, Oxford, 1994)

G. de Saxcé, C. Vallée, Invariant measures of the lack of symmetry with respect to the symmetry groups of 2d elasticity tensors. J. Elast. **111**, 21–39 (2013)

R. Desmorat, R. Marull, Non-quadratic Kelvin modes based plasticity criteria for anisotropic materials. Int. J. Plast. **27**, 328–351 (2011)

B. Desmorat, P. Vannucci, An alternative to the Kelvin decomposition for plane anisotropic elasticity. Math. Methods Appl. Sci. **38**, 164–175 (2014)

T.C. Doyle, J.L. Ericksen, Nonlinear elasticity. Adv. Appl. Mech. **4**, 53–115 (1956)

R. Dugas, *Histoire de la mécanique* (Editions du Griffon, Neuchâtel, Switzerland, 1950)

M. Epstein, On the wrinkling of anisotropic elastic membranes. J. Elast. **55**, 99–109 (1999)

P.J. Falzon, I. Herszberg, Mechanical performance of 2-d braided carbon/epoxy composites. Compos. Sci. Technol. **58**, 253–265 (1998)

S. Forte. Classi di simmetria in elasticità piana, in *Proceedings of AIMETA XVII - 17th Congress of Theoretical and Applied Mechanics* (Florence, Italy, 2005)

M. François, *Idéntification des symétries matérielles de matériaux anisotropes*. Ph.D. thesis, University Pierre et Marie Curie, Paris, France (1995)

M. François, A damage model based on Kelvin eigentensors and Curie principle. Mech. Mater. **44**, 23–34 (2012)

D. Gay, *Composite Materials Design and Applications*, 3rd edn. (CRC Press, Boca Raton, 2014)

A. Green, W. Zerna, *Theoretical Elasticity* (Clarendon, Oxford, 1954)

G. Green, On the laws of refelxion and refraction of light at the common surface of two non-crystallized media. Camb. Philos. Soc. Trans. **7**, 245–269 (1839)

R. Hill, A theory of the yielding and plastic flow of anisotropic metals. Proc. R. Soc. A **193**, 281–297 (1948)

O. Hoffman, The brittle strength of orthotropic materials. J. Compos. Mater. **1**, 200–206 (1967)

M. Hono, S. Onogi, Dynamic measurements of physical properties of pulp and paper by audiofrequency sound. J. Appl. Phys. **22**, 971–977 (1951)

R.M. Jones, *Mech. Compos. Mater.*, 2nd edn. (Taylor & Francis, Philadelphia, 1999)

H.H. Landolt, R. Börnstein, *Numerical Data and Functional Relationships in Science and Technology, III/29/a*, Second and Higher Order Elastic Constants (Springer, Berlin, 1992)

F.A. Leckie, E.T. Onat, Tensorial nature of damage measuring internal variables, in *In Proceedings of IUTAM Colloquium Physical Non-linearities in Structural Mechanics*, ed. by J. Hult, J. Lemaitre (France, Senlis, 1980), pp. 140–155

S.G. Lekhnitskii, Theory of Elasticity of an Anisotropic Elastic Body. English translation (1963), ed. by P. Fern (Holden-Day, San Francisco, CA, 1950)

J. Lemaitre, J.L. Chaboche, A. Benallal, R. Desmorat, *Mécanique des matériaux solides* (Dunod, Paris, 2009)

A.E.H. Love, *A Treatise on the Mathematical Theory of Elasticity* (Dover, New York, 1944)

M.M. Mehrabadi, S.C. Cowin, Eigentensors of linear anisotropic elastic materials. Q. J. Mech. Appl. Math. **43**, 15–41 (1990)

M. Miki. Material design of composite laminates with required in-plane elastic properties, in *Proceedings of ICCM 4 - Fourth International Conference on Composite Materials* (Tokio, Japan, 1982), pp. 1725–1731

L.M. Milne-Thomson, *Plane Elastic Systems* (Springer, Berlin, 1960)

N.I. Muskhelishvili, *Some Basic Problems of the Mathematical Theory of Elasticity* (P. Noordhoff Ltd, Gröningen, 1953)

L. Navier, Mémoire sur les lois de l'équilibre et du mouvement des solides élastiques. Mémoires de l'Académie Royale des Sciences de l'Institut National **7**, 375–393 (1827)

I. Newton. *Philosophiae Naturalis Principia Mathematica.* J. Streater (London, UK, 1687)

M. Ostoja-Starzewski, *Microstructural Randomness and Scaling in Mechanics of Materials* (Chapmann and Hall/CRC, New York, 2007)

P. Pedersen, On optimal orientation of orthotropic materials. Struct. Optim. **1**, 101–106 (1989)

P. Pedersen, Combining material and element rotation in one formula. Commun. Appl. Numer. Methods **6**, 549–555 (1990a)

P. Pedersen, Bounds on elastic energy in solids of orthotropic materials. Struct. Optim. **2**, 55–62 (1990b)

A.C. Pipkin, The relaxed eneergydensity for isotropic elastic membranes. IMA J. Appl. Math. **50**, 225–237 (1993)

A.C. Pipkin, Relaxed energy densities for large deformations of membranes. IMA J. Appl. Math. **52**, 297–308 (1994)

H. Poincaré, *Leçons sur la théorie de l'élasticité* (Carré, Paris, 1892)

S.D. Poisson, *Traité de mécanique* (Courcier, Paris, 1811)

S.D. Poisson, Mémoire sur les surfaces élastiques. Mémoires de l'Académie Royale des Sciences de l'Institut National **9**, 167–225 (1814–1816)

S.D. Poisson, Mémoire sur l'équilibre et le mouvement des corps élastiques. Mémoires de l'Académie Royale des Sciences de l'Institut National **8**, 357–570 (1829)

E. Reissner. On tension field theory, in *Proceedings of 5th International congress on applied Mechanics* (1938), pp. 88–92

J. Rychlewski, On Hooke's law. Prikl. Matem. Mekhan. **48**, 420–435 (1984)

F. Sidoroff, Description of anisotropic damage. Application to elasticity, in *Proceedings of IUTAM Colloquium Physical Non-linearities in Structural Mechanics*, ed. By J. Hult, J. Lemaitre (Senlis, France, 1980), pp. 237–244

I. Stackgold, The cauchy relations in a molecular theory of elasticity. Q. Appl. Math. **8**, 169–186 (1950)
I. Todhunter, K. Pearson, *History of the Theory of Elasticity*, vol. 1 (Cambridge University Press, Cambridge, 1886)
W. Thomson Lord Kelvin, Elements of a mathematical theory of elasticity. Philos. Trans. R. Soc. **146**, 481–498 (1856)
W. Thomson Lord Kelvin, Mathematical theory of elasticity. Encycl. Br. **7**, 819–825 (1878)
S.W. Tsai, Strength characteristics of composite materials. Contractor report cr-224, NASA (1965)
S.W. Tsai, Strength Theories of Filament Structures, in *Fundamental Aspects of Fiber Reinforced Plastic Composites*, ed. by R. Schwartz, H. Schwartz (Wiley Interscience, New York, 1968), pp. 3–11
S.W. Tsai, T. Hahn, *Introduction to Composite Materials* (Technomic, Stamford, 1980)
S.W. Tsai, E.M. Wu, A general theory of strength for anisotropic materials. J. Compos. Mater. **5**, 58–80 (1971)
E. Valot, P. Vannucci, Some exact solutions for fully orthotropic laminates. Compos. Struct. **69**, 157–166 (2005)
P. Vannucci, On bending-tension coupling of laminates. J. Elast. **64**, 13–28 (2001)
P. Vannucci. HDR thesis. University of Burgundy (2002a), http://tel.archives-ouvertes.fr/tel-00625958/fr/
P. Vannucci, A special planar orthotropic material. J. Elast. **67**, 81–96 (2002b)
P. Vannucci, Plane anisotropy by the polar method. Meccanica **40**, 437–454 (2005)
P. Vannucci, The polar analysis of a third order piezoelectricity-like plane tensor. Int. J. Solids Struct. **44**, 7803–7815 (2007)
P. Vannucci, Influence of invariant material parameters on the flexural optimal design of thin anisotropic laminates. Int. J. Mech. Sci. **51**, 192–203 (2009)
P. Vannucci, On special orthotropy of paper. J. Elast. **99**, 75–83 (2010)
P. Vannucci, Strange laminates. Math. Methods Appl. Sci. **35**, 1532–1546 (2012)
P. Vannucci, B. Desmorat, Analytical bounds for damage induced planar anisotropy. Int. J. Solids Struct. **60–61**, 96–106 (2015)
P. Vannucci, B. Desmorat, Plane anisotropic rari-constant materials. Math. Methods Appl. Sci. **39**, 3271–3281 (2016)
P. Vannucci, G. Verchery, Stiffness design of laminates using the polar method. Int. J. Solids Struct. **38**, 9281–9294 (2001)
P. Vannucci, G. Verchery, A new method for generating fully isotropic laminates. Compos. Struct. **58**, 75–82 (2002)
P. Vannucci, G. Verchery, Anisotropy of plane complex elastic bodies. Int. J. Solids Struct. **47**, 1154–1166 (2010)
G. Verchery. Les invariants des tenseurs d'ordre 4 du type de l'élasticité, in *Proceedings of Colloque Euromech 115 (Villard-de-Lans, 1979): Comportement mécanique des matériaux anisotropes* (Editions du CNRS, Paris, 1982), pp. 93–104
G. Verchery, T.S. Vong, Une méthode d'aide graphique à la conception des séquences d'empilement dans les stratifiés, in *Proceedings of JNC5 - Cinquièmes Journées Nationales sur les Composites* (Paris, France, 1986), pp. 267–280
A. Vincenti, *Conception et optimisation des composites stratifiés par méthode polaire et algorithme génétique*. Ph.D. thesis, ISAT - University of Burgundy, Nevers, France (2002)
A. Vincenti, P. Vannucci, G. Verchery, Influence of orientation errors on quasi-homogeneity of composite laminates. Compos. Sci. Technol. **63**, 739–749 (2003)
W. Voigt, *Lehrbuch der Kristallphysik* (B.G. Teubner, Leipzig, 1910)
Y.W. Wong, S. Pellegrino, Wrinkled membranes - Part III: numerical simulations. J. Mech. Mater. Struct. **1**, 61–93 (2006)
C.H. Wu, Plane linear wrinkle elasticity without body force. Technical report, Departement of Materials Engineering - University of Illinois, Chicago, IL (1974)

Chapter 5
Anisotropic Laminates

Abstract This chapter concerns the mechanical properties of anisotropic laminates. After an introduction to the Classical Laminated Plates Theory (CLPT), different topics, mainly developed with the aid of the polar formalism, are treated: coupling, heterogeneity of the elastic behavior, quasi-homogeneity, the case of identical layers, the lamination parameters and the geometrical bounds. Successively, the conditions for a laminate to have some special general properties are investigated: bending-extension uncoupling, quasi-homogeneity, ordinary and special orthotropies, isotropy. The category of quasi-trivial solutions is introduced and several examples of sequences of this type are given. A theoretical and numerical study on the effects of orientation errors on coupling and quasi-homogeneity is presented, and the extension of the CLPT to thermal and hygral properties is detailed. Finally, a brief account of higher-order theories for laminates is given.

5.1 Introduction

Anisotropic structures are composed, for the most part of times, by plates and shells, also beams are usually obtained assembling plates. Such structural members are normally made bonding together anisotropic layers to realize laminated plates or shells. We are hence concerned in this chapter with the mechanics of laminates. This subject is a classical topic in mechanics and the literature on it is extremely vast. However, such literature regards more the mechanical model of laminates than the description of their mechanical properties. Because in this text we are in fact concerned with anisotropy, we will focus essentially on the consequences of anisotropy on the constitutive lax of laminates. We will see that, because the mechanics of laminates is not ruled by a unique elastic tensor, the matter is rather, let us say, articulated.

We will bound ourselves essentially to the classical laminated plates theory; this is sufficient for our scopes, and it is worth noting since now that a change of mechanical model will *not* alter the elasticity tensors describing extension, coupling and bending of laminates, but the definition of the curvature tensor. That is why the anisotropy of a laminate is not model-dependent and the use of the classical theory is sufficient to

P. Vannucci, *Anisotropic Elasticity*, Lecture Notes in Applied and Computational Mechanics 85, DOI 10.1007/978-981-10-5439-6_5

analyze the relevant anisotropy properties of the laminate. The case of the through-the-thickness properties, appearing in higher order theories, is however considered too.

A special attention is devoted to the use of the polar formalism for the description of the anisotropy of laminates and to the description of particular classes of laminates, often used in practical applications. Finally, thermal and hygral properties are also introduced.

5.2 Fundamentals of the Classical Laminated Plates Theory

5.2.1 The Assumptions of the Classical Model

The Classical Laminated Plates Theory, in short CLPT, is based upon the kinematical model of Kirchhoff, see for instance (Villaggio 1997), applied here to the case of plates composed by plies bonded together. The fundamental assumptions are:

- plies perfectly bonded together: no slip between adjacent plies is allowed;
- linearly elastic anisotropic plies;
- small displacements, rotations and strains;
- small thickness of the plate compared to a characteristic in-plane dimension;
- the Kirchhoff kinematical model is axiomatically assumed for the description of the displacement fields.

Besides the usual assumptions of the classical theory of elasticity, that of non slipping contact between plies is also done, while we recall that the Kirchhoff kinematical model is actually composed by three distinct assumptions: any material straight segment originally orthogonal to the mid-plane remains

- a straight segment;
- normal to the deformed mid surface;
- a segment of the same original length.

5.2.2 The Kinematical Consequences of the Kirchhoff Hypotheses

For the first and third assumptions of Kirchhoff, the displacement $\mathbf{u}$ of a point $P(x_1, x_2, x_3)$ along x_1 (along x_2 the result is similar) is, see Fig. 5.1:

$$u(x_1, x_2, x_3) = u_0(x_1, x_2) - x_3 \sin\beta, \tag{5.1}$$

while along x_3 it is

$$w(x_1, x_2, x_3) = w_0(x_1, x_2) + x_3(\cos\beta - 1) \tag{5.2}$$

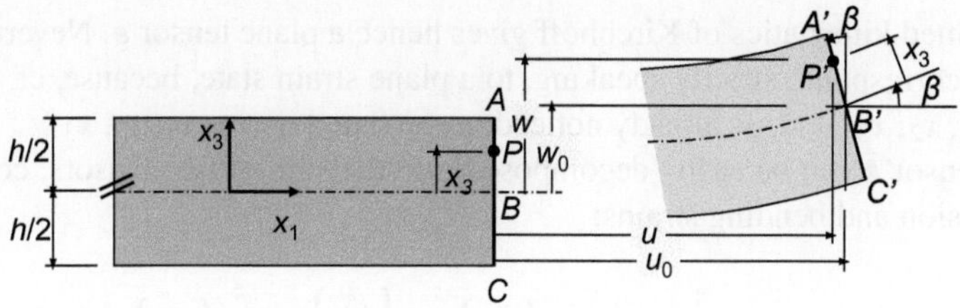

Fig. 5.1 The Kirchhoff kinematical model: displacement of a material segment, orthogonal to the mid-plane, in the plane $x - z$

where with the subscript 0 we denote a quantity referred to a point of the mid-plane. For the second Kirchhoff assumption and for the hypotheses of small displacements and rotations:

$$\beta \simeq \sin\beta \simeq \tan\beta = \frac{\partial w_0}{\partial x_1}, \quad \cos\beta \simeq 1. \tag{5.3}$$

As a consequence

$$\begin{aligned} u(x_1, x_2, x_3) &= u_0(x_1, x_2) - x_3 \frac{\partial w_0(x_1, x_2)}{\partial x_1}, \\ v(x_1, x_2, x_3) &= v_0(x_1, x_2) - x_3 \frac{\partial w_0(x_1, x_2)}{\partial x_2}, \\ w(x_1, x_2) &= w_0(x_1, x_2). \end{aligned} \tag{5.4}$$

Hence, the displacement field **u** is neither a plane vector nor a plane field; nevertheless, its third component w is a plane field.

5.2.3 *The Strain and Stress Tensors*

The kinematics of Kirchhoff is hence linear with respect to x_3. Applying now Eq. (1.25), we get the components of the strain tensor $\boldsymbol{\varepsilon}$:

$$\begin{aligned} \varepsilon_{11} &= \frac{\partial u_0(x_1, x_2)}{\partial x_1} - x_3 \frac{\partial^2 w_0(x_1, x_2)}{\partial x_1^2}, \\ \varepsilon_{12} &= \frac{1}{2}\left[\frac{\partial u_0(x_1, x_2)}{\partial x_2} + \frac{\partial v_0(x_1, x_2)}{\partial x_1}\right] - x_3 \frac{\partial^2 w_0(x_1, x_2)}{\partial x_1 \partial x_2}, \\ \varepsilon_{22} &= \frac{\partial v_0(x_1, x_2)}{\partial x_2} - x_3 \frac{\partial^2 w_0(x_1, x_2)}{\partial x_2^2}, \\ \varepsilon_{13} &= \varepsilon_{23} = \varepsilon_{33} = 0. \end{aligned} \tag{5.5}$$

The assumed kinematics of Kirchhoff gives hence a plane tensor $\boldsymbol{\varepsilon}$. Nevertheless, it does not correspond, strictly speaking, to a plane strain state, because, cf. Sect. 3.4, $\boldsymbol{\varepsilon} = \boldsymbol{\varepsilon}(x_1, x_2, x_3)$ and, as already noticed, $u_3 \neq 0$ and $\mathbf{u} = \mathbf{u}(x_1, x_2, x_3)$.

The tensor $\boldsymbol{\varepsilon}$ can be easily decomposed into the sum of two tensors, concerning the extension and bending strains:

$$\boldsymbol{\varepsilon} = \boldsymbol{\varepsilon}^0 + x_3 \boldsymbol{\kappa} \quad \rightarrow \quad \begin{Bmatrix} \varepsilon_1 \\ \varepsilon_2 \\ \varepsilon_6 \end{Bmatrix} = \begin{Bmatrix} \varepsilon_1^0 \\ \varepsilon_2^0 \\ \varepsilon_6^0 \end{Bmatrix} + x_3 \begin{Bmatrix} \kappa_1 \\ \kappa_2 \\ \kappa_6 \end{Bmatrix}, \tag{5.6}$$

where we have used the Kelvin's notation and introduced the two tensors $\boldsymbol{\varepsilon}^0$, *the mid-plane extension strain tensor*

$$\boldsymbol{\varepsilon}^0 = \begin{Bmatrix} \varepsilon_1^0 \\ \varepsilon_2^0 \\ \varepsilon_6^0 \end{Bmatrix} = \begin{Bmatrix} \frac{\partial u_0(x_1,x_2)}{\partial x_1} \\ \frac{\partial v_0(x_1,x_2)}{\partial x_2} \\ \frac{1}{\sqrt{2}} \left[\frac{\partial u_0(x_1,x_2)}{\partial x_2} + \frac{\partial v_0(x_1,x_2)}{\partial x_1} \right] \end{Bmatrix}, \tag{5.7}$$

and $\boldsymbol{\kappa}$, the *curvature tensor*

$$\boldsymbol{\kappa} = \begin{Bmatrix} \kappa_1 \\ \kappa_2 \\ \kappa_6 \end{Bmatrix} = - \begin{Bmatrix} \frac{\partial^2 w_0(x_1,x_2)}{\partial x_1^2} \\ \frac{\partial^2 w_0(x_1,x_2)}{\partial x_2^2} \\ \sqrt{2} \frac{\partial^2 w_0(x_1,x_2)}{\partial x_1 \partial x_2} \end{Bmatrix}. \tag{5.8}$$

The stress field $\boldsymbol{\sigma}$ is found using the Hooke's law, Eq. (2.12); the basic assumption is that of layers that are transversely isotropic with the axis x_1, see Fig. 3.1, as symmetry axis (for a discussion on the importance of this assumption, that actually reflects the practical reality of layers, whose fibers are always parallel to the faces, see Sect. 3.7). Then, see Sect. 2.2.8,

$$[C] = \begin{bmatrix} C_{11} & C_{12} & C_{12} & 0 & 0 & 0 \\ C_{12} & C_{22} & C_{23} & 0 & 0 & 0 \\ C_{12} & C_{23} & C_{22} & 0 & 0 & 0 \\ 0 & 0 & 0 & C_{22} - C_{23} & 0 & 0 \\ 0 & 0 & 0 & 0 & C_{66} & 0 \\ 0 & 0 & 0 & 0 & 0 & C_{66} \end{bmatrix} \tag{5.9}$$

It is easily seen that in a frame rotated through an angle θ it is

$$\begin{aligned}
C'_{13} &= C_{12}\cos^2\theta + C_{23}\sin^2\theta,\\
C'_{23} &= C_{12}\sin^2\theta + C_{23}\cos^2\theta,\\
C'_{44} &= (C_{22} - C_{23})\cos^2\theta + C_{55}\sin^2\theta,\\
C'_{45} &= (C_{22} - C_{23} - C_{55})\sin\theta\cos\theta,\\
C'_{55} &= (C_{22} - C_{23})\sin^2\theta + C_{55}\cos^2\theta,\\
C'_{63} &= \sqrt{2}(C_{23} - C_{12})\sin\theta\ \cos\theta,\\
C'_{33} &= C_{22},\ \ C'_{14} = C'_{24} = C'_{34} = C'_{64} = C'_{15} = C'_{25} = C'_{35} = C'_{65} = 0.
\end{aligned} \tag{5.10}$$

Then

$$\begin{aligned}
&\sigma'_4 = \sigma'_5 = 0,\\
&\sigma'_3 = (C_{12}\cos^2\theta + C_{23}\sin^2\theta)\varepsilon'_1 + (C_{12}\sin^2\theta + C_{23}\cos^2\theta)\varepsilon'_2 +\\
&\sqrt{2}(C_{23} - C_{12})\sin\theta\ \cos\theta\,\varepsilon'_6.
\end{aligned} \tag{5.11}$$

Generally speaking, $\sigma'_3 \neq 0 \rightarrow$ the stress state is *not plane*, though transverse shear stresses are null. What is commonly admitted, on a heuristic base, is that $\sigma_3 = 0$. This in the end corresponds to admit that the *stress tensor is plane too*. Nevertheless, one should notice that unlike in the case of a true plane stress field, both the stress and strain fields are *not plane fields*:

$$\boldsymbol{\varepsilon} = \boldsymbol{\varepsilon}^p(x_1, x_2, x_3), \quad \boldsymbol{\sigma} = \boldsymbol{\sigma}^p(x_1, x_2, x_3) \tag{5.12}$$

So, this case does *not corresponds* to any of the cases seen in Chap. 3: plane strain or stress, generalized plane stress or strain, plane deformation. In particular, contrarily to what is commonly said, it is *not* a generalized plane stress state.

Finally, the constitutive law is that of a plane stress state, with the stiffness matrix $[C]$ that is replaced by the reduced stiffness matrix $[Q]$. In the material (orthotropy) frame,

$$\{\sigma\} = [Q]\{\varepsilon\} \rightarrow \{\sigma\} = \begin{Bmatrix} \sigma_1 \\ \sigma_2 \\ \sigma_6 \end{Bmatrix} = \begin{bmatrix} Q_{11} & Q_{12} & 0 \\ Q_{12} & Q_{22} & 0 \\ 0 & 0 & Q_{66} \end{bmatrix} \begin{Bmatrix} \varepsilon_1 \\ \varepsilon_2 \\ \varepsilon_6 \end{Bmatrix} \tag{5.13}$$

where

$$Q_{ij} = C_{ij} - \frac{C_{i3}C_{j3}}{C_{33}},\ i, j = 1, 2, 6 \rightarrow [Q] = \begin{bmatrix} C_{11} - \frac{C_{13}^2}{C_{33}} & C_{12} - \frac{C_{13}C_{23}}{C_{33}} & 0 \\ & C_{22} - \frac{C_{23}^2}{C_{33}} & 0 \\ sym & & C_{66} \end{bmatrix}. \tag{5.14}$$

To remark that in a general frame $[Q]$ is full. The result of the constitutive law for the special case considered and of the Kirchhoff kinematics is that $\sigma_4 = \sigma_5 = 0$ in any frame.

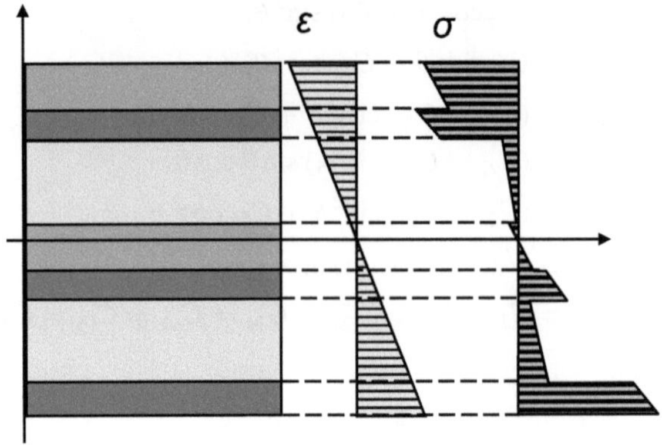

Fig. 5.2 Sketch of the distribution of the strain and stress components through the plate's thickness according to the Kirchhoff model

A consequence of this fact is that equilibrium is possible only for extension and pure bending, but *not for shearing actions* (i.e. loads normal to the mid-plane cannot be equilibrated). This inconsistency, typical of the Kirchhoff theory, is solved in the case of single layer plates using the equilibrium equations. This is not possible in general for laminates. The use of higher order theories only can solve this problem for laminates, see Sect. 5.5.

To notice that, unlike the strain field $\boldsymbol{\varepsilon}$, the stress field $\boldsymbol{\sigma}$ is, generally speaking, *not continuous at the interfaces*, see Fig. 5.2, due to the change of constitutive law from one ply to another one.

5.2.4 Internal Actions

With reference to Fig. 5.3, we define the tensors of *in-plane actions* $\mathbf{N}$ and of *bending moments* $\mathbf{M}$ as follows:

$$\mathbf{N} = \int_{-\frac{h}{2}}^{\frac{h}{2}} \boldsymbol{\sigma}\, dx_3, \ \mathbf{M} = \int_{-\frac{h}{2}}^{\frac{h}{2}} x_3 \boldsymbol{\sigma}\, dx_3 \ \rightarrow$$

$$\begin{Bmatrix} N_1 \\ N_2 \\ N_6 \end{Bmatrix} = \begin{Bmatrix} \int_{-\frac{h}{2}}^{\frac{h}{2}} \sigma_1\, dx_3 \\ \int_{-\frac{h}{2}}^{\frac{h}{2}} \sigma_2\, dx_3 \\ \int_{-\frac{h}{2}}^{\frac{h}{2}} \sigma_6\, dx_3 \end{Bmatrix}, \ \begin{Bmatrix} M_1 \\ M_2 \\ M_6 \end{Bmatrix} = \begin{Bmatrix} \int_{-\frac{h}{2}}^{\frac{h}{2}} \sigma_1 x_3\, dx_3 \\ \int_{-\frac{h}{2}}^{\frac{h}{2}} \sigma_2 x_3\, dx_3 \\ \int_{-\frac{h}{2}}^{\frac{h}{2}} \sigma_6 x_3\, dx_3 \end{Bmatrix}. \tag{5.15}$$

In injecting the constitutive law into the above equations, we need to consider that, being such a law different layer by layer, either because layers are different or because they are identical but differently oriented, the integrals must be split into the sum of n terms. In the end we get[1] (δ_k is the orientation of the k-th ply)

[1]With a slight abuse of notation, and for the sake of simplicity, we indicate the reduced stiffness matrix $[Q]$ by $\mathbb{Q}$.

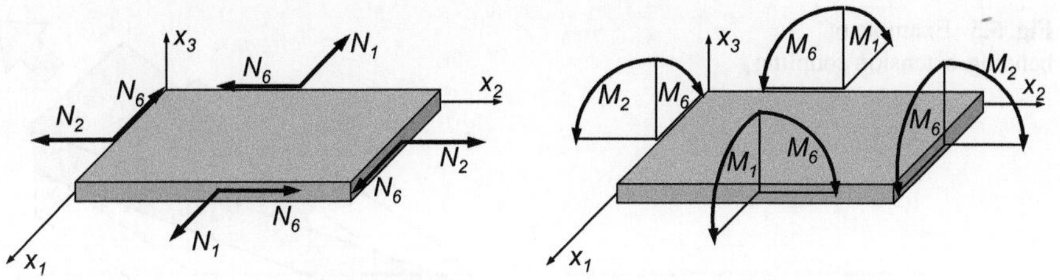

Fig. 5.3 Scheme of the internal actions for the Kirchhoff model

Fig. 5.4 Layers' numbering scheme

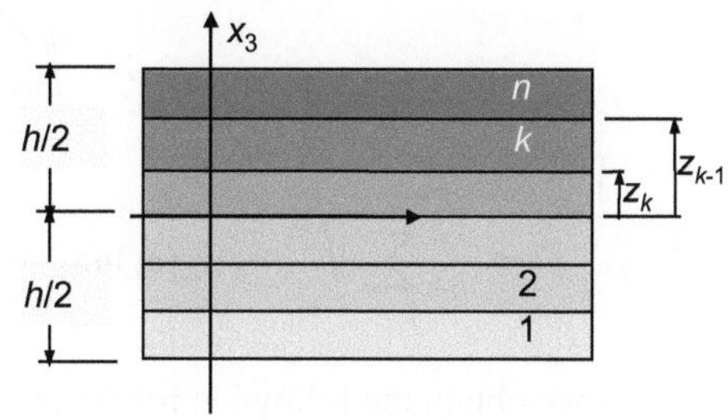

$$\mathbf{N}=\sum_{k=1}^{n}\int_{z_k-1}^{z_k}\boldsymbol{\sigma}_k dx_3=\sum_{k=1}^{n}\int_{z_k-1}^{z_k}\mathbb{Q}_k(\delta_k)(\boldsymbol{\varepsilon}_0+x_3\boldsymbol{\kappa})dx_3 \tag{5.16}$$

$$\mathbf{M}=\sum_{k=1}^{n}\int_{z_k-1}^{z_k}x_3\boldsymbol{\sigma}_k dx_3=\sum_{k=1}^{n}\int_{z_k-1}^{z_k}x_3\mathbb{Q}_k(\delta_k)(\boldsymbol{\varepsilon}_0+x_3\boldsymbol{\kappa})dx_3 \tag{5.17}$$

5.2.5 *The Laminates' Fundamental Law*

After numbering the layers and interfaces like in Fig. 5.4, we now introduce the tensors

- $\mathbb{A}$, tensor of the *extension behavior*

$$\mathbb{A}=\frac{1}{h}\sum_{k=1}^{n}\int_{z_{k-1}}^{z_k}\mathbb{Q}_k(\delta_k)dx_3=\frac{1}{h}\sum_{k=1}^{n}(z_k-z_{k-1})\mathbb{Q}_k(\delta_k) \tag{5.18}$$

- $\mathbb{B}$, tensor of the *coupling behavior*

$$\mathbb{B}=\frac{2}{h^2}\sum_{k=1}^{n}\int_{z_{k-1}}^{z_k}x_3\mathbb{Q}_k(\delta_k)dx_3=\frac{1}{h^2}\sum_{k=1}^{n}(z_k^2-z_{k-1}^2)\mathbb{Q}_k(\delta_k) \tag{5.19}$$

- $\mathbb{D}$, tensor of the *bending behavior*

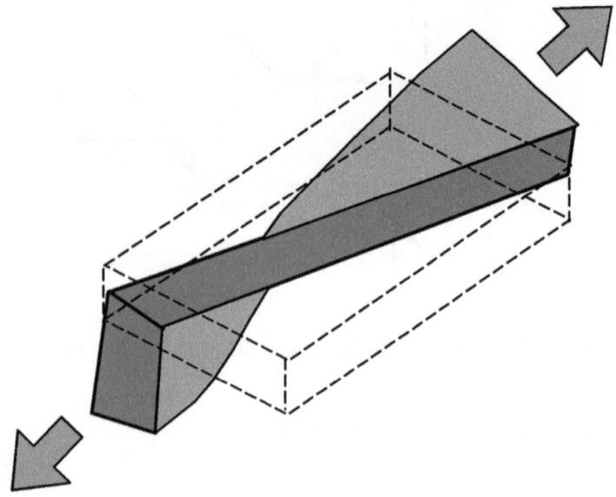

Fig. 5.5 Example of bending-extension coupling

$$\mathbb{D}=\frac{12}{h^3}\sum_{k=1}^{n}\int_{z_{k-1}}^{z_k}x_3^2\mathbb{Q}_k(\delta_k)dx_3=\frac{4}{h^3}\sum_{k=1}^{n}(z_k^3-z_{k-1}^3)\mathbb{Q}_k(\delta_k) \tag{5.20}$$

Then we obtain the following relations

$$\left\{\frac{\mathbf{N}}{\mathbf{M}}\right\}=\left[\begin{array}{c|c} h\mathbb{A} & \frac{h^2}{2}\mathbb{B} \\ \hline \frac{h^2}{2}\mathbb{B} & \frac{h^3}{12}\mathbb{D}\end{array}\right]\left\{\frac{\boldsymbol{\varepsilon}^0}{\boldsymbol{\kappa}}\right\} \tag{5.21}$$

This is the *fundamental law of laminates*, which actually is the constitutive law for laminates, linking the internal actions to the laminate's deformation. By construction,

$$\mathbb{A}=\mathbb{A}^\top,\ \mathbb{B}=\mathbb{B}^\top,\ \mathbb{D}=\mathbb{D}^\top. \tag{5.22}$$

In this way, the mechanics of a multi-layer plate is reduced to that of an *equivalent single layer* plate, whose elastic tensor is $\mathbb{A}$ for its extension behavior, $\mathbb{D}$ for its bending behavior, while $\mathbb{B}$ couples these last ones, see Sect. 5.2.6. Nevertheless, there are some special differences with respect to a single layer plate; let us see all of them.

5.2.6 Bending-Extension Coupling

A first difference is the existence of a *coupling* between extension and bending. In fact, generally speaking,

$$\mathbb{B}\neq\mathbb{O}. \tag{5.23}$$

So, an extension produces also a curvature of the plate, see Fig. 5.6, while a bending moment stretches the mid-plane. A laminate with $\mathbb{B}=\mathbb{O}$ is said to be *uncoupled*; in this case, it is simply

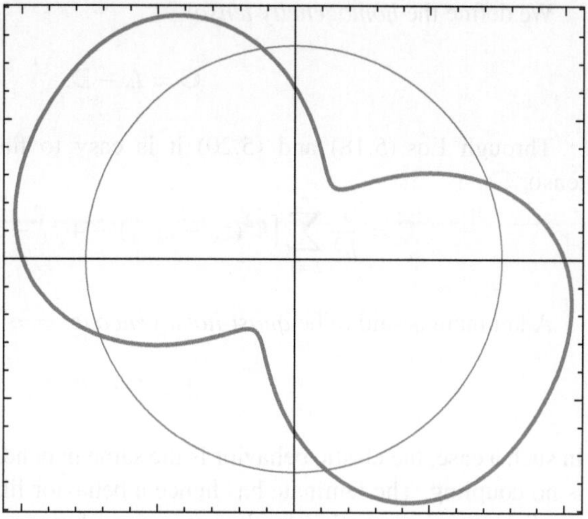

Fig. 5.6 Heterogeneity of the laminates' behavior: *thin line* A_{1111}, *thick line* D_{1111}

$$\mathbf{N} = h\mathbb{A}\boldsymbol{\varepsilon}^0, \quad \mathbf{M} = \frac{h^3}{12}\mathbb{D}\boldsymbol{\kappa}. \tag{5.24}$$

5.2.7 *Heterogeneity of the Elastic Behavior*

A second difference is the *heterogeneity of the elastic behavior*: generally speaking,

$$\mathbb{A} \neq \mathbb{D}. \tag{5.25}$$

The laminate has a *different* elastic response, at any direction, in extension and in bending, like if it were composed by two different materials. In Fig. 5.6, an example with the components $\mathbb{A}_{1111}$ and $\mathbb{D}_{1111}$ for a same laminate; the first one is isotropic, the second one orthotropic.

5.2.8 *Quasi-homogeneous Laminates*

The concept of quasi-homogeneous laminates has been introduced by G. Verchery in 1988, (Kandil and Verchery 1988).[2]

[2]In their paper of 1988, Verchery and Kandil introduce the concept of quasi-homogeneity with regards to the vanishing of $\mathbb{C}$ only, so without imposing the same for $\mathbb{B}$. Anyway, in following papers, the definition (5.28) has been used, see Vannucci et al. (1998), Vannucci and Verchery (2001a), Vannucci and Verchery (2001b), Vannucci and Verchery (2002), Vincenti et al. (2003), Valot and Vannucci (2005), Verchery (2007), Verchery (2011). Other authors have considered the

We define the *homogeneity tensor*

$$\mathbb{C} = \mathbb{A} - \mathbb{D}. \tag{5.26}$$

Through Eqs. (5.18) and (5.20) it is easy to find the general expression of tensor $\mathbb{C}$:

$$\mathbb{C} = \frac{1}{h^3}\sum_{k=1}^{n}\left[h^2(z_k - z_{k-1}) - 4(z_k^3 - z_{k-1}^3)\right]\mathbb{Q}_k(\delta_k). \tag{5.27}$$

A laminate is said to be *quasi-homogeneous* $\Longleftrightarrow$

$$\mathbb{B} = \mathbb{C} = \mathbb{O}. \tag{5.28}$$

In such a case, the elastic behavior is the same in bending and in extension and there is no coupling. The laminate has hence a behavior like that of a homogeneous, i.e. single layer, plate, that's why the name quasi-homogeneous.

5.2.9 Inverting the Fundamental Law of Laminates

From the fundamental law of laminates, Eq. (5.21), we get

$$\mathbf{M} = \frac{h^2}{2}\mathbb{B}\boldsymbol{\varepsilon}^0 + \frac{h^3}{12}\mathbb{D}\boldsymbol{\kappa} \rightarrow \boldsymbol{\kappa} = \frac{12}{h^3}\mathbb{D}^{-1}(\mathbf{M} - \frac{h^2}{2}\mathbb{B}\boldsymbol{\varepsilon}^0), \tag{5.29}$$

and injecting this in the equation for $\mathbf{N}$ we obtain

$$\mathbf{N} = h\mathbb{A}\boldsymbol{\varepsilon}^0 + \frac{h^2}{2}\mathbb{B}\boldsymbol{\kappa} \rightarrow \mathbf{N} = h\mathbb{A}\boldsymbol{\varepsilon}^0 + \frac{h^2}{2}\mathbb{B}\frac{12}{h^3}\mathbb{D}^{-1}(\mathbf{M} - \frac{h^2}{2}\mathbb{B}\boldsymbol{\varepsilon}^0). \tag{5.30}$$

Hence, resolving the last equation with respect to $\boldsymbol{\varepsilon}^0$,

$$\boldsymbol{\varepsilon}^0 = \frac{1}{h}\mathscr{A}\mathbf{N} + \frac{2}{h^2}\mathscr{B}_1\mathbf{M} \tag{5.31}$$

and in a similar way

$$\boldsymbol{\kappa} = \frac{2}{h^2}\mathscr{B}_2\mathbf{N} + \frac{12}{h^3}\mathscr{D}\mathbf{M}, \tag{5.32}$$

(Footnote 2 continued)
existence of such a kind of laminates, namely (Wu and Avery 1992), in a paper of 1992, where they announce the existence of these laminates, although without giving any example of them, for patenting such sequences. M. Grédiac, in some papers on a numerical method for the search of some types of laminates, considers also the search for quasi-homogeneous laminates, (Grédiac 1998, 1999a, b, 2001).

where

$$\begin{aligned}\mathscr{A}&=(\mathbb{A}-3\mathbb{B}\mathbb{D}^{-1}\mathbb{B})^{-1}, \quad \mathscr{B}_1=-3\mathscr{A}\mathbb{B}\mathbb{D}^{-1},\\ \mathscr{D}&=(\mathbb{D}-3\mathbb{B}\mathbb{A}^{-1}\mathbb{B})^{-1}, \quad \mathscr{B}_2=-3\mathscr{D}\mathbb{B}\mathbb{A}^{-1}.\end{aligned} \tag{5.33}$$

$\mathscr{A}$ and $\mathscr{D}$ are the compliance extension and bending tensors, respectively, while $\mathscr{B}_1$ and $\mathscr{B}_2$ are the compliance coupling tensors. We find now another *strange behavior* of laminates: $\mathbb{B}$ has, apparently, not an unique inverse, but two. More specifically, if we write the inverse of the fundamental law (5.21) in a symbolic matrix form,

$$\left\{\begin{array}{c}\boldsymbol{\varepsilon}^0\\ \hline \boldsymbol{\kappa}\end{array}\right\}=\left[\begin{array}{c|c}\frac{1}{h}\mathscr{A} & \frac{2}{h^2}\mathscr{B}_1\\ \hline \frac{2}{h^2}\mathscr{B}_2 & \frac{12}{h^3}\mathscr{D}\end{array}\right]\left\{\begin{array}{c}\mathbf{N}\\ \hline \mathbf{M}\end{array}\right\}, \tag{5.34}$$

we do not find at the upper right position and at the lower left one the same tensor. Nevertheless, one should remark that the symbolic matrix in Eq. (5.34) is exactly the inverse of matrix in Eq. (5.21): the uniqueness of the inverse of the global matrix is preserved.

Moreover, because the transpose of the inverse is equal to the inverse of the transpose and because tensors $\mathbb{A}$, $\mathbb{B}$ and $\mathbb{D}$ are symmetric, it is easy to check that

$$\mathscr{A}^\top=\left[(\mathbb{A}-3\mathbb{B}\mathbb{D}^{-1}\mathbb{B})^\top\right]^{-1}=(\mathbb{A}-3\mathbb{B}\mathbb{D}^{-1}\mathbb{B})^{-1}=\mathscr{A} \tag{5.35}$$

and similarly

$$\mathscr{D}^\top=\left[(\mathbb{D}-3\mathbb{B}\mathbb{A}^{-1}\mathbb{B})\top\right]^{-1}=(\mathbb{D}-3\mathbb{B}\mathbb{A}^{-1}\mathbb{B})^{-1}=\mathscr{D}, \tag{5.36}$$

but

$$\mathscr{B}_1\neq\mathscr{B}_2. \tag{5.37}$$

In fact

$$\begin{aligned}\mathscr{B}_1^\top&=(-3\mathscr{A}\mathbb{B}\mathbb{D}^{-1})^\top=-3\mathbb{D}^{-1}\mathbb{B}\mathscr{A}\neq\mathscr{B}_1,\\ \mathscr{B}_2^\top&=(-3\mathscr{D}\mathbb{B}\mathbb{A}^{-1})^\top=-3\mathbb{A}^{-1}\mathbb{B}\mathscr{D}\neq\mathscr{B}_2,\end{aligned} \tag{5.38}$$

Hence, $\mathscr{B}_1$ and $\mathscr{B}_2$ are not only different, but *un-symmetric too* → $\mathbb{B}$ is *not a classical elastic tensor*, because its compliance corresponding is not unique and without the major symmetries. In addition, $\mathbb{B}$ is *not definite* and no bounds can be given to its components.

Anyway, $\mathscr{B}_1$ and $\mathscr{B}_2$ are not independent and the global symmetry of matrix in Eq. (5.34) is preserved; in fact

$$\begin{aligned}(\mathscr{B}_1^\top)^{-1}&=-\frac{1}{3}\mathscr{A}^{-1}\mathbb{B}^{-1}\mathbb{D}=-\frac{1}{3}\mathbb{A}\mathbb{B}^{-1}\mathbb{D}+\mathbb{B},\\ \mathscr{B}_2^{-1}&=-\frac{1}{3}\mathbb{A}\mathbb{B}^{-1}\mathscr{D}^{-1}=-\frac{1}{3}\mathbb{A}\mathbb{B}^{-1}\mathbb{D}+\mathbb{B}=(\mathscr{B}_1^\top)^{-1}.\end{aligned} \tag{5.39}$$

Hence, for the uniqueness of the inverse,

$$\mathscr{B}_2^\top = \mathscr{B}_1 := \mathscr{B}, \tag{5.40}$$

so finally we can write the inverse of the fundamental law (5.34) in the form

$$\left\{\begin{array}{c}\boldsymbol{\varepsilon}^0\\ \hline \boldsymbol{\kappa}\end{array}\right\} = \left[\begin{array}{c|c}\frac{1}{h}\mathscr{A} & \frac{2}{h^2}\mathscr{B}\\ \hline \frac{2}{h^2}\mathscr{B}^\top & \frac{12}{h^3}\mathscr{D}\end{array}\right]\left\{\begin{array}{c}\mathbf{N}\\ \hline \mathbf{M}\end{array}\right\}. \tag{5.41}$$

We remark that for uncoupled laminates

$$\mathbb{B} = \mathbb{O} \iff \mathscr{B} = \mathbb{O} \Rightarrow \mathscr{A} = \mathbb{A}^{-1}, \ \ \mathscr{D} = \mathbb{D}^{-1}. \tag{5.42}$$

Exclusively in this case, the symmetries of $\mathbb{A}$ or of $\mathbb{D}$ are the *same in stiffness and in compliance*. In all the other cases, i.e. when a laminate is coupled, the stiffness and compliance tensors of the extension and bending behaviors have, generally speaking, *different symmetries*. For the sake of an example, let us imagine that for an uncoupled laminate $\mathbb{A}$ is orthotropic: then $\mathscr{A}$ is orthotropic too and with the same axes, which in general is false if $\mathbb{B} \neq \mathbb{O}$, cf. Sect. 5.3.4, Note 12.

This is another strange fact of laminates with respect to single layer plates and corroborates, once more, the idea that the symmetry of the elastic behavior is first of all a *matter of algebraic properties of an elastic tensor*, more than a material symmetry of the plate, that in the case of coupled laminates cannot be clearly defined.

5.2.10 Laminates Made of Identical Plies

If the n plies composing a laminate of thickness h are all identical, i.e. if they are made of the *same material* and have the *same thickness*, then the upper face of the k–th layer is at the position

$$z_k = \frac{2k - n}{2n} h \tag{5.43}$$

and the equations giving $\mathbb{A}$, $\mathbb{B}$, $\mathbb{C}$ and $\mathbb{D}$ become

$$\begin{aligned}\mathbb{A} &= \sum_{k=1}^{n} a_k \mathbb{Q}(\delta_k), \ \ \mathbb{B} = \sum_{k=1}^{n} b_k \mathbb{Q}(\delta_k),\\ \mathbb{C} &= \sum_{k=1}^{n} c_k \mathbb{Q}(\delta_k), \ \ \mathbb{D} = \sum_{k=1}^{n} d_k \mathbb{Q}(\delta_k),\end{aligned} \tag{5.44}$$

where

$$
\begin{aligned}
a_k &= \frac{1}{n},\ b_k = \frac{1}{n^2}(2k-n-1),\\
c_k &= a_k - d_k = \frac{1}{n^3}[-2(6k^2+n^2)+6n(2k-1)+4(3k-1)] \\
d_k &= \frac{1}{n^3}\left[12k(k-n-1)+4+3n(n+2)\right],
\end{aligned}
\tag{5.45}
$$

Because $a_k = 1/n\ \forall k$, then the *stacking sequence does not affect the extension behavior*, which is determined only by the orientations. This is *not* the case of coupling and bending, where the stacking sequence influences the final behavior along with the orientations. That is why it is much *easier to design laminates with respect to extension* than to bending.

5.2.11 Laminates by the Polar Formalism

The homogenization laws of the laminate tensors are independent from the elasticity algebraic representation, hence *they apply to the polar formalism as well.*

Referring to Fig. 3.1, we need to write tensors $\mathbb{A}$, $\mathbb{B}$, $\mathbb{C}$ and $\mathbb{D}$ in a general frame, represented in the figure by the frame $\{x_1', x_2'\}$, where each ply is rotated of the angle $-\delta_k$. That is why, applying the composition rules (5.18), (5.19), (5.20) and (5.27) to the polar components (4.57), we get (since now on, we denote by T_0^A, T_1^A etc. the polar components of $\mathbb{A}$, and similarly we do for $\mathbb{B}$, $\mathbb{C}$ and $\mathbb{D}$, while T_{0k}, T_{1k} etc. indicate the polar components of the $k-$th layer):

$$
\mathbb{A} \rightarrow
\begin{cases}
T_0^A = \dfrac{1}{h}\displaystyle\sum_{k=1}^{n} T_{0k}(z_k - z_{k-1}),\\
T_1^A = \dfrac{1}{h}\displaystyle\sum_{k=1}^{n} T_{1k}(z_k - z_{k-1}),\\
R_0^A e^{4i\Phi_0^A} = \dfrac{1}{h}\displaystyle\sum_{k=1}^{n} R_{0k}e^{4i(\Phi_{0k}+\delta_k)}(z_k - z_{k-1}),\\
R_1^A e^{2i\Phi_1^A} = \dfrac{1}{h}\displaystyle\sum_{k=1}^{n} R_{1k}e^{2i(\Phi_{1k}+\delta_k)}(z_k - z_{k-1}).
\end{cases}
\tag{5.46}
$$

$$
\mathbb{B} \rightarrow \begin{cases}
T_0^B = \dfrac{1}{h^2}\displaystyle\sum_{k=1}^{n} T_{0k}(z_k^2 - z_{k-1}^2),\\[2ex]
T_1^B = \dfrac{1}{h^2}\displaystyle\sum_{k=1}^{n} T_{1k}(z_k^2 - z_{k-1}^2),\\[2ex]
R_0^B e^{4i\Phi_0^B} = \dfrac{1}{h^2}\displaystyle\sum_{k=1}^{n} R_{0k}e^{4i(\Phi_{0k}+\delta_k)}(z_k^2 - z_{k-1}^2),\\[2ex]
R_1^B e^{2i\Phi_1^B} = \dfrac{1}{h^2}\displaystyle\sum_{k=1}^{n} R_{1k}e^{2i(\Phi_{1k}+\delta_k)}(z_k^2 - z_{k-1}^2).
\end{cases} \tag{5.47}
$$

$$
\mathbb{C} \rightarrow \begin{cases}
T_0^C = \dfrac{1}{h^3}\displaystyle\sum_{k=1}^{n} T_{0k}\left[h^2(z_k - z_{k-1}) - 4(z_k^3 - z_{k-1}^3)\right],\\[2ex]
T_1^C = \dfrac{1}{h^3}\displaystyle\sum_{k=1}^{n} T_{1k}\left[h^2(z_k - z_{k-1}) - 4(z_k^3 - z_{k-1}^3)\right],\\[2ex]
R_0^C e^{4i\Phi_0^C} = \dfrac{1}{h^3}\displaystyle\sum_{k=1}^{n} R_{0k}e^{4i(\Phi_{0k}+\delta_k)}\left[h^2(z_k - z_{k-1}) - 4(z_k^3 - z_{k-1}^3)\right],\\[2ex]
R_1^C e^{2i\Phi_1^C} = \dfrac{1}{h^3}\displaystyle\sum_{k=1}^{n} R_{1k}e^{2i(\Phi_{1k}+\delta_k)}\left[h^2(z_k - z_{k-1}) - 4(z_k^3 - z_{k-1}^3)\right].
\end{cases} \tag{5.48}
$$

$$
\mathbb{D} \rightarrow \begin{cases}
T_0^D = \dfrac{4}{h^3}\displaystyle\sum_{k=1}^{n} T_{0k}(z_k^3 - z_{k-1}^3),\\[2ex]
T_1^D = \dfrac{4}{h^3}\displaystyle\sum_{k=1}^{n} T_{1k}(z_k^3 - z_{k-1}^3),\\[2ex]
R_0^D e^{4i\Phi_0^D} = \dfrac{4}{h^3}\displaystyle\sum_{k=1}^{n} R_{0k}e^{4i(\Phi_{0k}+\delta_k)}(z_k^3 - z_{k-1}^3),\\[2ex]
R_1^D e^{2i\Phi_1^D} = \dfrac{4}{h^3}\displaystyle\sum_{k=1}^{n} R_{1k}e^{2i(\Phi_{1k}+\delta_k)}(z_k^3 - z_{k-1}^3).
\end{cases} \tag{5.49}
$$

Some remarks about the above formulae:

- the isotropic and anisotropic parts of all the tensors remain *separated* in the homogenization of the polar parameters, for all the tensors;
- it is immediately apparent that *special orthotropies are preserved*:

$$
\begin{aligned}
R_{0k} = 0\ \forall k &\Rightarrow R_0^A = R_0^B = R_0^C = R_0^D = 0,\\
R_{1k} = 0\ \forall k &\Rightarrow R_1^A = R_1^B = R_1^C = R_1^D = 0.
\end{aligned} \tag{5.50}
$$

The mechanical meaning of this result is that if a laminate is composed, e.g. by layers that are all square symmetric, though different (it is the case of a laminate obtained using different layers, but all reinforced by balanced fabrics), then all the tensors $\mathbb{A}$, $\mathbb{B}$, $\mathbb{C}$ and $\mathbb{D}$ are square symmetric, regardless of the stacking sequence and of the ply orientations; the same is true for layers that are R_0–orthotropic, but *not* for ordinarily orthotropic layers: generally speaking, ordinary orthotropy is not preserved through the homogenization process.

5.2.12 The Case of Identical Layers: The Lamination Parameters

More results can be obtained for laminates of identical plies. In fact, in such a case, all the layers have the same polar parameters, say T_0, T_1 etc., and an identical thickness, say $h_L = h/n$. Then, Eqs. (5.46)–(5.49) become, after some standard passages (from now on, the superscript L will denote, a polar parameter of the basic layer)

$$\mathbb{A} \rightarrow \begin{cases} T_0^A = T_0^L, \\ T_1^A = T_1^L, \\ R_0^A e^{4i\Phi_0^A} = R_0^L e^{4i\Phi_0^L}(\xi_1 + i\xi_2), \\ R_1^A e^{2i\Phi_1^A} = R_1^L e^{2i\Phi_1^L}(\xi_3 + i\xi_4); \end{cases} \tag{5.51}$$

$$\mathbb{B} \rightarrow \begin{cases} T_0^B = 0 \\ T_1^B = 0 \\ R_0^B e^{4i\Phi_0^B} = R_0^L e^{4i\Phi_0^L}(\xi_5 + i\xi_6), \\ R_1^B e^{2i\Phi_1^B} = R_1^L e^{2i\Phi_1^L}(\xi_7 + i\xi_8); \end{cases} \tag{5.52}$$

$$\mathbb{C} \rightarrow \begin{cases} T_0^C = 0 \\ T_1^C = 0 \\ R_0^C e^{4i\Phi_0^C} = R_0^L e^{4i\Phi_0^L}(\xi_9 + i\xi_{10}), \\ R_1^C e^{2i\Phi_1^C} = R_1^L e^{2i\Phi_1^L}(\xi_{11} + i\xi_{12}); \end{cases} \tag{5.53}$$

$$\mathbb{D} \rightarrow \begin{cases} T_0^D = T_0^L \\ T_1^D = T_1^L \\ R_0^D e^{4i\Phi_0^D} = R_0^L e^{4i\Phi_0^L}(\xi_{13} + i\xi_{14}), \\ R_1^D e^{2i\Phi_1^D} = R_1^L e^{2i\Phi_1^L}(\xi_{15} + i\xi_{16}). \end{cases} \tag{5.54}$$

In the previous equations, the parameters ξ_1 to ξ_{16} are given by the following relations:

$$
\begin{aligned}
\xi_1 + i\xi_2 &= \sum_{j=1}^{n} a_j\, e^{4i\delta_j}, & \xi_3 + i\xi_4 &= \sum_{j=1}^{n} a_j\, e^{2i\delta_j},\\
\xi_5 + i\xi_6 &= \sum_{j=1}^{n} b_j\, e^{4i\delta_j}, & \xi_7 + i\xi_8 &= \sum_{j=1}^{n} b_j\, e^{2i\delta_j},\\
\xi_9 + i\xi_{10} &= \sum_{j=1}^{n} c_j\, e^{4i\delta_j}, & \xi_{11} + i\xi_{12} &= \sum_{j=1}^{n} c_j\, e^{2i\delta_j},\\
\xi_{13} + i\xi_{14} &= \sum_{j=1}^{n} d_j\, e^{4i\delta_j}, & \xi_{15} + i\xi_{16} &= \sum_{j=1}^{n} d_j\, e^{2i\delta_j}.
\end{aligned}
\tag{5.55}
$$

We then can see that for laminates of identical plies, all what has been said in the general case is still valid and in addition[3]:

- the isotropic parts of $\mathbb{A}$ and $\mathbb{D}$ are *identical to that of the basic layer*;
- the isotropic parts of $\mathbb{B}$ and $\mathbb{C}$ *vanishes*: $\mathbb{B}$ and $\mathbb{C}$ are merely anisotropic, with a null mean;
- hence, in a laminate design process *only the anisotropic parts of the laminate can be tailored*, while the choice of the basic layer automatically fixes the anisotropic parts;
- this eliminates from the design problem all the isotropic parts, hence *six design variables can be disregarded.*

This is a particularly important consequence of the polar formalism: the most part of laminate design problems concerns laminates composed by identical plies. In such a case, the property of the polar formalism of splitting any elastic tensor in its isotropic and anisotropic parts reveals to be very helpful in decreasing, automatically, the number of design variables, that are reduced to only 12: R_0^A, R_1^A, R_0^B, R_1^B, R_0^D, R_1^D and the polar angles Φ_0^A, Φ_1^A, Φ_0^B, Φ_1^B, Φ_0^D, Φ_1^D; once $\mathbb{A}$ and $\mathbb{D}$ known, $\mathbb{C}$ is easily calculated.

Looking at Eqs. (5.51)–(5.54), one can observe another decomposition of the elastic tensors of a laminate, that between *material and geometric part*. The first one is fixed once the basic layer chosen and it is represented by the *polar parameters of the basic layer*. The second one is that to be designed, it accounts for the geometry of the laminate, i.e. *orientation angles and stacking sequence*, and is represented by the so-called *lamination parameters*,[4] the terms ξ_1–ξ_{16}.

[3]One should consider that the knowledge of two tensors among $\mathbb{A}$, $\mathbb{D}$ and $\mathbb{C}$ is sufficient to determine also the third one, because $\mathbb{C} = \mathbb{A} - \mathbb{D}$; that is why only 18 variables instead of 24 are needed to characterize elastically a laminate.

[4]The lamination parameters have been introduced for the first time in Tsai and Hahn (1980) and successively widely used in a great number of applications in laminates design, see for instance the pioneer works of Miki (1982, 1983, 1985), or Fukunaga and Vanderplaats (1991), Hammer et al. (1997), Grédiac (1999b). In Eq. (5.55) the lamination parameters are written in complex form and generalize the concept introduced by Tsai and Pagano, introducing the parameters ξ_9 to ξ_{12} for describing tensor $\mathbb{C}$.

To be remarked that this separation is possible *only in the case of identical plies*: the lamination parameters cannot be defined for hybrid laminates, but the polar parameters yes, they are universally valid, so they represent a *more general concept* and tool than that of lamination parameters.

So, for the very common case of identical plies, the use of the polar formalism automatically reduces the number of design variables from 18 to 12, which is not the case if the Cartesian approach is used, where all the variables, representing the isotropic and anisotropic parts, are mixed together.

5.2.13 Geometrical Bounds

A question arises: can the lamination parameters,[5] or, in the end, the mechanical parameters of the laminate, *take any value*? In other words, the bounds on the polar parameters of an elastic tensor, Eq. (4.111), are valid also for the polar parameters of tensors $\mathbb{A}$ and $\mathbb{D}$[6]?

Generally speaking, the answer is no: the polar parameters of $\mathbb{A}$ or $\mathbb{D}$, considered for the while as independent tensors, cannot take any value. Moreover, it has been shown in Vannucci (2013) that tensors $\mathbb{A}$ and $\mathbb{D}$ are a *subset* of elastic tensors describing a behavior which is more bounded than the general elastic behavior. In other words, tensors $\mathbb{A}$ and $\mathbb{D}$ are limited by bounds that are more restrictive than bounds (4.111). Such bounds are called *geometrical bounds*, because they take into account for the geometry of the stack, i.e. of the orientation angles and, for $\mathbb{D}$, also of the stacking sequence.

It is exactly the laminate's geometry that bounds more strictly than the positiveness of the elastic energy the polar parameters of $\mathbb{A}$ and $\mathbb{D}$: the effect of the combination of the orientations, represented by the lamination parameters, is in the sense of shrinking the elastic domain allowed by Eq. (4.111). For instance, if one considers just the value of R_0^A or R_1^A in Eq. (5.51), it is easily seen that $R_0^A \leq R_0$ and $R_1^A \leq R_1$, due to the lamination parameters ξ_1 to ξ_4.

The geometrical bounds can be given in the same way for $\mathbb{A}$ and $\mathbb{D}$ if the following dimensionless quantities are introduced (the superscript L indicates a parameter of the basic layer, while $\mathbb{T}$ one of tensors $\mathbb{A}$ or $\mathbb{D}$):

$$\rho=\frac{R_0^L}{R_1^L},\quad \tau_0=\frac{T_0^L}{R_0^L},\quad \tau_1=\frac{T_1^L}{R_1^L},\quad \rho_0=\frac{R_0^{\mathbb{T}}}{R_0^L},\quad \rho_1=\frac{R_1^{\mathbb{T}}}{R_1^L},\quad c_0=\cos 4\Phi_0^{\mathbb{T}}. \tag{5.56}$$

Then, using the results in Hammer et al. (1997), Vannucci (2013) has shown that the polar independent geometrical bounds are

[5] A general expression of the geometrical bounds can be given only in the case of a lamiaate composed by identical plies. That is why we refer here to lamination parameters.

[6] $\mathbb{B}$ and $\mathbb{C}$ are not defined tensors, so we cannot write bounds on their components using the positiveness of the elastic energy.

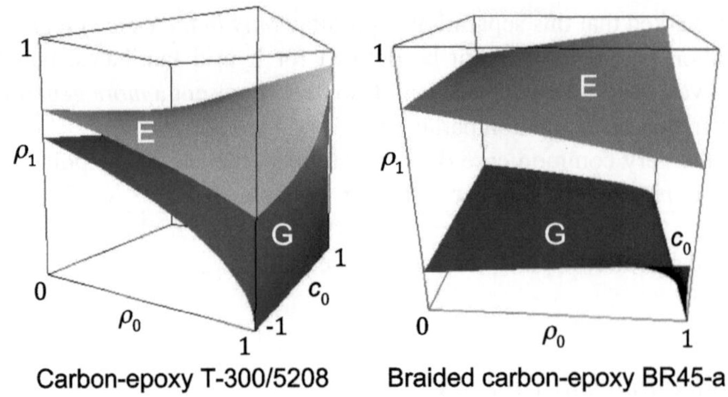

Fig. 5.7 Geometrical bounds for anisotropic laminates made by two different materials in Table 4.9

$$\begin{cases} 0 \leq \rho_0 \leq 1, \\ 0 \leq \rho_1, \\ 2\rho_1^2 \leq \dfrac{1-\rho_0^2}{1-(-1)^{K^L}\rho_0\, c_0}. \end{cases} \tag{5.57}$$

In Fig. 5.7, the bounds for laminates made with identical layers composed by two different materials are shown: the geometrical bounds are denoted by letter G, while the elastic ones by letter E.[7] It is well clear that the geometrical bounds define a smaller admissible set of parameters than the elastic bounds, deriving from the positiveness of the elastic energy.

The above bounds (5.57) and Fig. 5.7 are referred to completely anisotropic $\mathbb{A}$ or $\mathbb{D}$. For an orthotropic tensor, the bounds are simpler. In particular, for an ordinarily orthotropic tensor, $c_0 = K^{\mathbb{T}} \in \{0, 1\}$ and the bounds are just

$$\begin{cases} 0 \leq \rho_0 \leq 1, \\ 0 \leq \rho_1, \\ 2\rho_1^2 - 1 \leq (-1)^{K^{\mathbb{T}}-K^L}\rho_0. \end{cases} \tag{5.58}$$

[7] Using the dimensionless quantities in Eq. (5.56), it is not difficult to show that the elastic bound, corresponding to Eq. (4.111)$_2$, is

$$2\rho_1^2 < \rho\tau_0\tau_1 \frac{1-\left(\frac{\rho_0}{\tau_0}\right)^2}{1-\frac{\rho_0}{\tau_0}c_0},$$

for a completely anisotropic tensor, while that for an orthotropic one becomes

$$\frac{2\rho_1^2}{\rho\,\tau_1} - \tau_0 < (-1)^{K^{\mathbb{T}}}\rho_0.$$

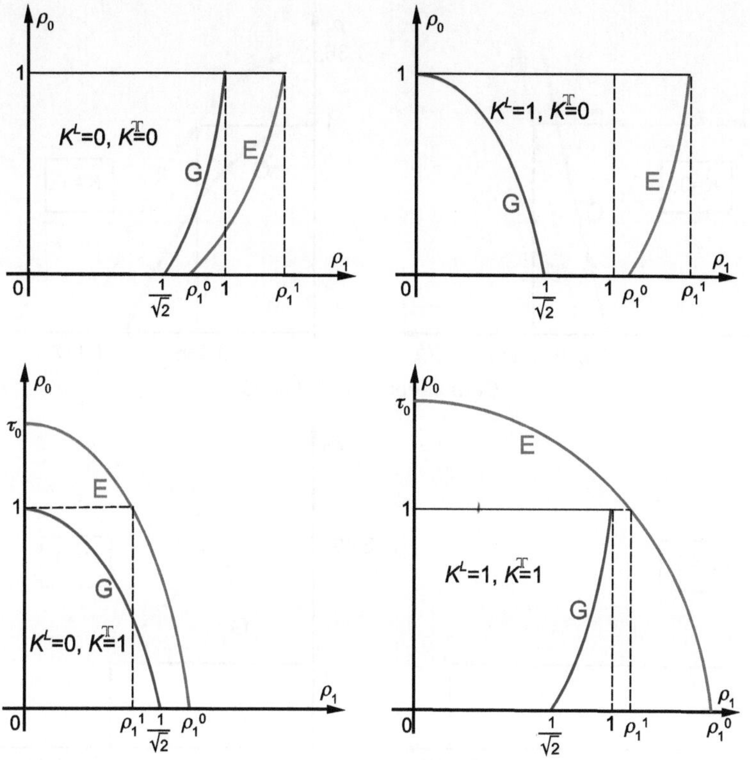

Fig. 5.8 Geometrical bounds for orthotropic laminates: the four possible situations

In Fig. 5.8 the four general possible situations, depending upon the values of K^L and $K^{\mathbb{T}}$, are shown, while Fig. 5.9 is referred to the same two materials of Fig. 5.7, with the first one having $K^L = 0$, while for the second one $K^L = 1$. Again, the curve denoted by E corresponds to the elastic bound, while G indicates the geometrical bound.

Bounds (5.57) and (5.58) concern exclusively the polar parameters of $\mathbb{A}$ or $\mathbb{D}$ *separately*; because they are in a dimensionless form, their expression is identical for the two cases. However, it is important to underline that these bounds for $\mathbb{A}$ and $\mathbb{D}$ *cannot be completely independent*, because they are functions of the same quantities: the polar parameters of the basic layer and the orientation angles, besides the stacking sequence, for $\mathbb{D}$. Hence, it should be preferable to establish the *admissible set of all the laminate's polar parameters*, which is a subset of $\mathbb{R}^6$. Moreover, it is easy to check that bounds (5.57) and (5.58) apply as well to $\mathbb{B}$ and $\mathbb{C}$, but, as already remarked, for these two tensors elastic bounds are not defined.

If one considers the whole elastic behavior of a laminate, it is defined by twelve lamination parameters, ξ_1 to ξ_8 and ξ_{13} to ξ_{16}, because $\mathbb{C}$ depends upon $\mathbb{A}$ and $\mathbb{D}$, and one could consider the general problem of the contemporary design of $\mathbb{A}$, $\mathbb{B}$ and $\mathbb{D}$. In

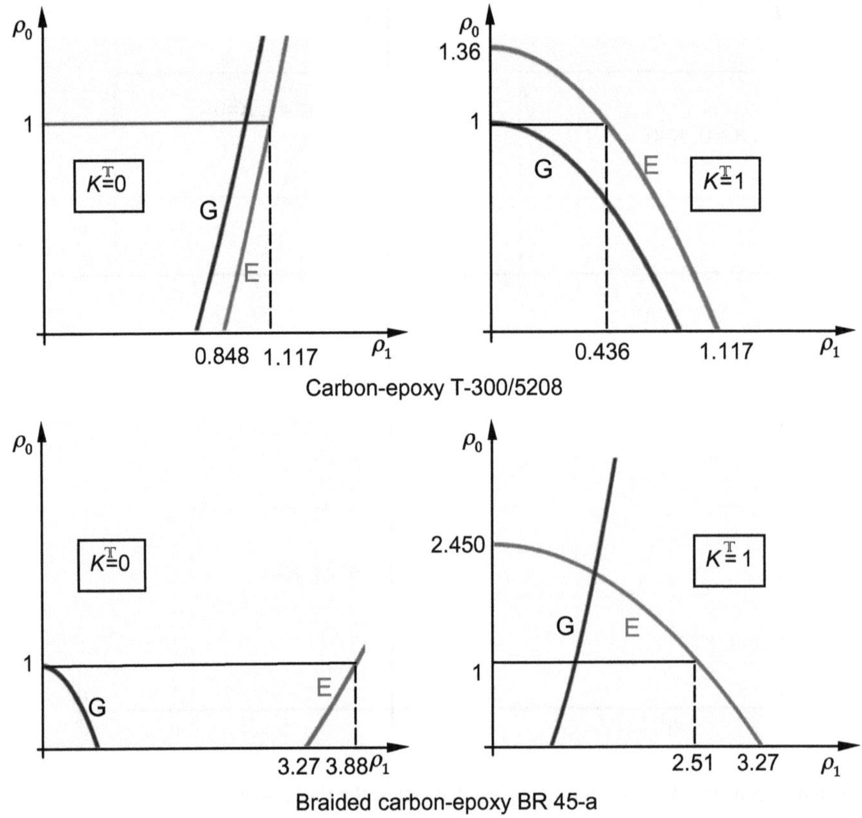

Fig. 5.9 Geometrical bounds for orthotropic laminates made by two different materials in Table 4.9

such a case, the problem depends upon 12 quantities, in the $\mathbb{R}^{12}$ space of the $\xi_i s$, and one should determine the feasible domain as a subset of $\mathbb{R}^{12}$. This is still to be done and probably impossible to be obtained: previous studies on this subject, but done using directly the lamination parameters, never succeeded in solving this still open problem, though some partial solutions have been given, Diaconu et al. (2002a, b).

This point constitutes one of the most serious *mathematical open problems* for a correct formulation of optimization problems of laminates including at the same time extension and bending properties. In some papers, Grenestedt and Gudmundson (1993), Fukunaga and Vanderplaats (1991), some necessary conditions for different combinations of the lamination parameters have been given. Nevertheless, as said above, the complete set of sufficient conditions in $\mathbb{R}^{12}$ for all the set of the twelve lamination parameters is still not known.

For the while there are only three possibilities:

- to consider problems concerning *only* $\mathbb{A}$ *or* $\mathbb{D}$; this is what is commonly done, and the most part of times only $\mathbb{A}$ is designed;
- to *restrict* more or less empirically the set of the design variables; this has been done very rarely, only in some research papers, Foldager et al. (1998), Foldager et al. (1999);
- to bound the research of the solution to the set of *quasi-homogenous laminates*; this approach is *mathematically rigorous*, because the admissible set of design variables is strictly the same for $\mathbb{A}$ and $\mathbb{D}$; this approach has been used in some research studies using a two-step approach, see Sect. 6.6.1, Jibawy et al. (2011), Montemurro et al. (2012), Montemurro et al. (2012b), Catapano et al. (2015).

Some remarks to end this section. First of all, unlike elastic bounds, geometrical bounds (5.57) and (5.58) concern exclusively the anisotropic part of a tensor. Then, the problem formulated with the approach of this section, based upon the use of the polar formalism, is stated using dimensionless intrinsic, i.e. invariant, quantities, unlike the case of a formulation based upon the lamination parameters. This allows, in particular, the reduction of the independent parameters to be considered, from 4 to 3 in the case of a completely anisotropic tensor $\mathbb{A}$ or $\mathbb{D}$, from 3 to only 2 for the orthotropic case, and to have an explicit expression of the bounds. Such a fact is interesting per se and allows, for instance, to completely trace the geometrical bounds, i.e. the admissible domain, like in Figs. 5.7, 5.8 and 5.9, to be compared, e.g., with the Fig. 2 in Hammer et al. (1997).

It is to be noticed, however, an interesting fact: bounds (5.57) define a non-convex domain. This is apparent from Fig. 5.7 and can be easily checked verifying, e.g., that the Hessian of the function in Eq. $(5.57)_3$ is not definite.[8] Unlike the general case, the domain for an orthotropic laminate, defined by bounds (5.58), is always convex. This can be checked simply verifying that for Eq. $(5.58)_3$ $d^2\rho_1/d\rho_0^2 = -1/[4\sqrt{2}(1+(-1)^{K^{\mathrm{T}}-K^L}\rho_0)^{3/2}] < 0$ always,[9] because of Eq. $(5.58)_1$. The convexity of the geometric domain is evident in Figs. 5.8 and 5.9.

This is not the case when the problem is formulated with the lamination parameters. In fact, the convexity of the admissible domain of the lamination parameters in $\mathbb{R}^{12}$ has been proved by Grenestedt and Gudmundson (1993). In the orthotropic case, the same authors have found the admissible domain of $\mathbb{R}^4$ for $\mathbb{A}$ and $\mathbb{D}$, while the relations among the four lamination parameters of $\mathbb{A}$ or $\mathbb{D}$ have been derived by Fukunaga and Sekine (1992); Fukunaga et al. (1994).

Finally, considering the results above, we can see that using the polar formalism in place of the lamination parameters implies a loss of convexity, but the possibility of expliciting better the geometrical bounds and of reducing the number of variables.

---

[8]This verification is actually possible, but the results, obtained making use of the formal code *Mathematica*, are too long to be written here, so they are omitted.

[9]This result, actually, states that ρ_1 is a concave function of ρ_0. Nevertheless, if a reflexion about the line $\rho_1 = 0$ is done, which amounts to change ρ_1 into $-\rho_1$, then Eq. $(5.58)_3$ becomes a convex function and the set defined by bounds (5.58), as intersection of convex sets, is a convex domain. Because this operation does not change the topology of the domain, the set defined by Eq. (5.58) is convex.

5.3 Laminates with Special General Properties

In industrial applications, some kinds of laminates are used more frequently because they have some special properties normally desired by designers or they automatically offer some advantages. Nevertheless, for the design of hybrid laminates, general rules do not exist, apart those seen before and given by the use of the polar formalism, see Sect. 5.2.11. That is why, almost all the rules below can be applied only to laminates with *identical plies* and in the following we will refer only to this case.

Going back to what said in the Introduction, Sect. 5.1, we recall that the design of an anisotropic laminate normally means:

- the design of the number of plies;
- the choice of the material;
- the design with respect to usual mechanical properties, like buckling, strength and so on;
- the design of the material properties of $\mathbb{A}$ and $\mathbb{D}$, which implies that of the elastic symmetries: orthotropy or isotropy or square symmetry etc.;
- normally, the uncoupling requirement: $\mathbb{B} = \mathbb{O}$;
- other properties, like for instance quasi-homogeneity etc.

This shows how much complicate can be the design of laminates. That is why the use of special classes of laminates, able to ensure automatically some general properties, is so common, with all the drawbacks indicated in Sect. 5.1. We list below the principal properties sought for by engineers and the corresponding classes of laminates used in practical applications. We specify since now that all the following properties can be obtained also using a numerical approach, like the one detailed in Sect. 6.4.1.

5.3.1 *Bending-Extension Uncoupling*

Bending-extension uncoupling is the most frequent requirement asked for by designers. As $T_0^B = T_1^B = 0$, Eq. (5.52), a laminate is uncoupled $\iff$

$$R_0^B = R_1^B = 0. \tag{5.59}$$

This happens $\iff$

$$\xi_5 = \xi_6 = \xi_7 = \xi_8 = 0. \tag{5.60}$$

To remark that when stated by the polar formalism, uncoupling corresponds to two conditions to be satisfied, while with the lamination parameters to four and with the Cartesian components to six, the six distinct components of $\mathbb{B}$.

Of course, a particular choice of the orientations δ_k can solve problem (5.59) or (5.60). Nevertheless, there is another possibility, independent from the δ_ks. In fact,

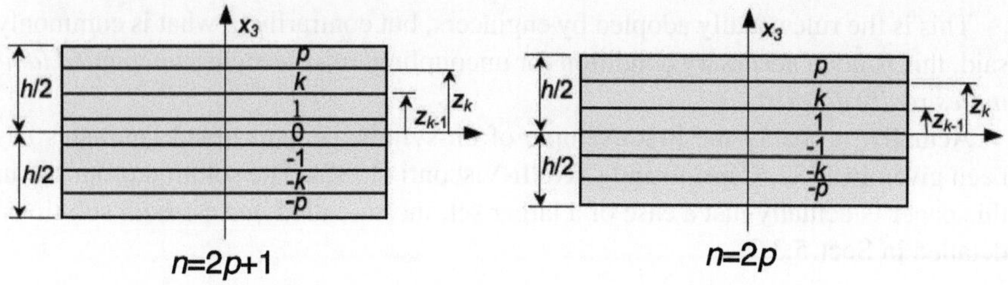

Fig. 5.10 Scheme for the numbering of the layers from the mid-plane

$$b_k = \frac{1}{n^2}(2k - n - 1) \Rightarrow \sum_{k=1}^{n} b_k = 0, \tag{5.61}$$

In addition, it can be checked very easily that the b_k are odd with respect to the mid-plane. In fact, if the plies are numbered starting from the mid-plane like in Fig. 5.10, then it is still $a_k = 1/n$, like in Eq. (5.45), but

$$\text{if } n = 2p + 1 \quad \begin{cases} b_k = \dfrac{2k}{n^2}, \\ c_k = \dfrac{4}{n^3}(p^2 + p - 3k^2), \\ d_k = \dfrac{1}{n^3}(12k^2 + 1), \end{cases} \tag{5.62}$$

while

$$\text{if } n = 2p \quad \begin{cases} b_k = \dfrac{1}{n^2}\left(2k - \dfrac{k}{|k|}\right), \ b_0 = 0, \\ c_k = \dfrac{4}{n^3}(p^2 - 3k^2 + 3|k| - 1), \ c_0 = 0, \\ d_k = \dfrac{1}{n^3}(12k^2 - 12|k| + 4), \ d_0 = 0. \end{cases} \tag{5.63}$$

In both the cases coefficients b_k depend linearly on the position in the stack, k, and they are odd with respect to the mid-plane, i.e.

$$b_{-k} = -b_k, \tag{5.64}$$

so that a sufficient condition for having $\mathbb{B} = \mathbb{O}$ is to use *symmetric stacks*:

$$\delta_{-k} = \delta_k \ \forall k. \tag{5.65}$$

This is the rule usually adopted by engineers, but contrarily to what is commonly said, this is not a necessary condition for uncoupling: *un-symmetric uncoupled laminates really do exist.*

Actually, probably the first example of un-symmetric uncoupled laminates has been given in 1982, Caprino and Crivelli-Visconti (1982). The solution proposed in this paper is actually just a case of a larger set, the so-called *quasi-trivial* solutions, detailed in Sect. 5.3.3.

5.3.2 Quasi-homogeneity

As seen in Sect. 5.2.8, quasi-homogeneity is given by the conditions $\mathbb{B} = \mathbb{C} = \mathbb{O}$. Through Eqs. (5.52) and (5.53) this corresponds to the four polar conditions

$$R_0^B = R_1^B = R_0^C = R_1^C = 0, \tag{5.66}$$

or to the eight conditions on the lamination parameters

$$\xi_5 = \xi_6 = \xi_7 = \xi_8 = \xi_9 = \xi_{10} = \xi_{11} = \xi_{12} = 0, \tag{5.67}$$

or, finally, to the twelve conditions imposing the vanishing of the Cartesian components of $\mathbb{B}$ and $\mathbb{C}$.

Equations (5.62) and (5.63) show that coefficients c_k are an even quadratic function of the position on the stack, k, i.e. they are symmetrically distributed about the mid-plane. Hence, the rule of the symmetric stack cannot give quasi-homogeneity. Nevertheless, also in this case quasi-trivial solutions exist.

5.3.3 Quasi-trivial Solutions

The existence of quasi-trivial solutions has been proved in 1998 for the first time, Vannucci et al. (1998)), Vannucci and Verchery (2001a, b). It is based upon a simple fact: looking at Eqs. (5.45) or (5.62) and (5.63) one can see that[10]

$$\sum_{k=1}^{n} b_k = \sum_{k=1}^{n} c_k = 0. \tag{5.68}$$

---

[10]It is easy to remark also that

$$\sum_{k=1}^{n} a_k = \sum_{k=1}^{n} d_k = 1,$$

which is a basic property of the lamination parameters.

Let us consider the case of quasi-homogeneity. Then, to have a solution to Eq. (5.66) *it is sufficient to choose the same orientation for a group of layers whose sum of the coefficients* b_k *and* c_k *is null.* Such a group is called a *saturated group*; if a sequence is composed by saturated groups, automatically it is quasi-homogeneous, *regardless of the value of the orientation of each saturated group*, i.e. independently from the orientation angles. Such solutions are called *quasi-trivial*, to recall that there is no need to solve equations (5.66) or (5.67). To remark that the search for quasi-trivial solutions needs the satisfaction of only 2 *integer conditions*, instead of the 8 real ones of Eq. (5.67):

$$\sum_{g=1}^{sg}\left(\sum_{k=1}^{n_g} b_k\right)=0,\ \sum_{g=1}^{sg}\left(\sum_{k=1}^{n_g} c_k\right)=0 \tag{5.69}$$

where sg is the number of saturated groups and n_g the number of plies in the $g-$th saturated group.

In fact, conditions (5.66) corresponds, through Eqs. (5.52), (5.53) and (5.55), to

$$\begin{aligned}
\xi_5+i\xi_6&=\sum_{j=1}^{n} b_j\ e^{4i\delta_j}=0,\\
\xi_7+i\xi_8&=\sum_{j=1}^{n} b_j\ e^{2i\delta_j}=0,\\
\xi_9+i\xi_{10}&=\sum_{j=} 1^n c_j\ e^{4i\delta_j}=0,\\
\xi_{11}+i\xi_{12}&=\sum_{j=1}^{n} c_j\ e^{2i\delta_j}=0.
\end{aligned} \tag{5.70}$$

So, because the orientation is the same for all the plies of the same saturated group, it is sufficient to satisfy the two conditions in Eq. (5.69) to obtain a quasi-homogeneous solution. Of course, quasi-trivial solutions exist also for uncoupling, which simply must fulfill only Eq. $(5.69)_1$. It is easy to check that actually the symmetric solution is just a special case of the quasi-trivial set. It can be shown that the number of saturated groups can be at most of $[(n+1)/2]$ for uncoupling and less than $[n/2]$ for quasi-homogeneity (symbol [ ] denotes here the integer part).

Quasi-trivial solutions can be found by an enumeration algorithm; what is interesting is their *high number* already for small values of n, see Tables 5.1 and 5.2 and what is surprising is that the number of un-symmetric solutions is *much greater* than that of the symmetric ones (the symmetric solutions in Table 5.1 are indicated in brackets, while for uncoupling, Table 5.2, there is just one mechanically and

Table 5.1 Quasi-trivial solutions for quasi-homogeneity up to 30 plies; in brackets, the number of symmetric solutions; sg = saturated groups

Plies	2sg	3sg	4sg	5sg	6sg	Total
7	1(1)					1
8	1					1
9						0
10						0
11	3(2)					3
12	1					1
13	2	2				2
14		2(1)				4
15	2	2				4
16	5	3(1)				8
17	15	8				23
18		5				5
19	30	22				52
20	30	9	1			40
21	31	13(2)				44
22	17(2)	98(1)	13	2		130
23	95(1)	499				594
24	140	26	1			167
25	163	2132	57			2352
26	54	1059(2)	354(3)	26(2)	2	1495
27	86(1)	918	256	21	1	1282
28	203	4789(1)	871(2)	33	6	5902
29	61	37747	7546	87		45441
30	53	5552	512(3)	29		6146

mathematically distinct[11] symmetric solution for each n, the one with the highest number of saturated groups).

The solution found by Caprino and Crivelli-Visconti (1982), is actually the 8-ply solution in Table 5.1. If we give the label 0 to the first saturated group, 1 to the second, 2 the third and so on, this solution is an anti-symmetric one: 0 1 1 0 1 0 0 1. Using the same symbolism, some quasi-trivial solutions are given in Tables 5.3, 5.4 and 5.5 for the case of quasi-homogeneity. To end this Section, we remark on the one hand

[11] A solution is *mathematically distinct* if no other quasi-trivial solution with a greater number of saturated groups can be derived from it, while a solution with ng saturated groups is *mechanically distinct* if, for a given set of ng orientations, it has a mechanical behavior distinct from any other solution with the same number of saturated groups (for instance, two un-symmetric solutions that are one the symmetric of the other with respect to the mid-plane, are of course the same solution, mechanically speaking). Only solutions that are at the same time mathematically and mechanically distinct are listed in Tables 5.1, 5.2, 5.3, 5.4 and 5.5.

Table 5.2 Quasi-trivial solutions for uncoupling up to 18 plies; sg = saturated groups

Plies	2sg	3sg	4sg	5sg	6sg	7sg	8sg	9sg	Total
4	1								1
5		1							1
6		1							1
7		1	1						2
8	1		1						2
9		1	2	1					4
10		4		1					5
11			6	4	1				11
12	1	4	9		1				15
13			14	20	6	1			41
14		22	17	17		1			57
15			5	111	48	9	1		174
16		29	168	48	29		1		275
17			1	458	471	90	12	1	1033
18		57	746	686	104	45		1	1639

side that symmetric solutions are not the rule, but more properly an exception, so that using just symmetric stacks for the design is really very limitative, and on the other hand side that the great number of quasi-trivial solutions makes them interesting for applications in design problems.

5.3.4 *Orthotropy*

Normally, completely anisotropic laminates are not used by designers: usually *orthotropy is a minimal requirement* for the elastic symmetries of the laminate. Nevertheless, while it is rather simple to obtain an orthotropic $\mathbb{A}$, it is much more difficult to get orthotropy for $\mathbb{D}$. This is due essentially to the fact that the coefficients d_k *change layer by layer*, while the a_k are constant, see Eq. (5.45), this implying, as already said, that the *stacking sequence influences the properties of* $\mathbb{D}$, *but not of* $\mathbb{A}$. For this reason, the most part of times engineers use some special sequences ensuring the orthotropy of $\mathbb{A}$ but *not*, generally speaking, of $\mathbb{D}$. This problem is generally completely ignored, to such a point that almost the totality of papers on vibrations and buckling of laminates, where the orthotropy of $\mathbb{D}$ matters, are not correct, because orthotropy is imposed just for $\mathbb{A}$, which is not sufficient to ensure that of $\mathbb{D}$. In some texts, it is said that if $n > 6$, than the difference with respect to an exact solution, i.e. where $\mathbb{D}$ is really orthotropic, is negligible; accurate studies show that this is not always the case, Vincenti et al. (2013).

Table 5.3 Examples of quasi-trivial solutions for quasi-homogeneity up to 18 plies; sg size = number of plies in each saturated group

Plies	Stacking sequence	sg size
7	0 1 0 0 0 1 0	5/2
8	0 1 1 0 1 0 0 1	4/4
11	0 1 1 0 1 1 0 0 1 0 1	5/6
	0 1 0 1 0 0 0 1 0 1 0	7/4
	0 1 1 0 0 1 0 0 1 1 0	6/5
12	0 1 0 1 1 1 0 0 0 1 0 1	6/6
13	0 1 1 1 0 1 1 0 0 1 1 0 1	5/8
	0 1 1 0 0 0 0 0 1 1 1 0 0	8/5
	0 1 2 0 0 2 0 1 0 0 1 2 0	7/3/3
	0 1 2 0 2 0 2 1 0 0 1 0 2	6/3/4
14	0 1 2 1 1 2 2 0 0 1 1 0 1 2	4/6/4
	0 1 2 2 0 1 0 0 1 0 2 2 1 0	6/4/4
15	0 1 1 1 0 0 1 1 1 1 0 0 0 1 1	6/9
	0 1 0 2 0 2 0 0 0 1 0 1 0 2 0	9/3/3
	0 1 2 2 2 0 0 0 2 1 0 1 2 2 0	6/3/6
16	0 0 1 1 1 1 0 0 1 1 0 0 0 0 1 1	8/8
	0 1 2 2 0 2 0 1 1 2 0 2 0 0 1 2	6/4/6
	0 1 2 1 2 0 0 1 0 1 1 2 0 2 0 1	6/6/4
	0 1 2 0 2 0 0 1 1 0 0 2 0 2 1 0	8/4/4
17	0 1 0 0 1 1 0 0 0 0 0 1 0 1 1 0 0	11/6
	0 1 1 0 1 1 1 0 0 1 0 1 1 0 1 0 1	7/10
	0 1 1 0 0 1 1 1 0 1 0 0 1 0 1 0 1	8/9
	0 1 2 1 0 0 0 1 2 0 1 0 1 0 2 1 0	8/6/3
	0 1 2 2 0 1 0 1 0 2 0 2 0 1 1 2 0	7/5/5
18	0 1 2 0 1 2 2 2 1 1 0 0 1 0 0 2 2 1	6/6/6
	0 1 2 2 0 1 1 2 0 1 2 0 2 0 1 0 1 2	6/6/6
	0 1 2 2 0 1 1 2 0 2 1 0 1 0 2 0 2 1	6/6/6
	0 1 2 2 1 0 2 0 1 1 0 2 1 2 0 0 2 1	6/6/6
	0 1 1 2 2 2 0 0 2 1 0 0 1 1 1 2 2 0	6/6/6

Orthotropy can be of different types for a laminate. All the interesting cases of ordinary orthotropy are given below[12] and stated with the polar formalism, the most effective way to this purpose[13]:

[12]Only orthotropy of $\mathbb{A}$ and $\mathbb{D}$ is considered and for all the cases it is assumed $\mathbb{B} = \mathbb{O}$. Cases with $\mathbb{B} \neq \mathbb{O}$ are also possible, but disregarded because uninteresting. In such a situation, generally speaking, $\mathscr{A}$ and $\mathscr{D}$ are not orthotropic, despite the fact that $\mathbb{A}$ and $\mathbb{D}$ are. An example of this, is the 10-ply sequence $[-\alpha\ \alpha\ \alpha\ -\alpha\ 0\ 0\ -\alpha\ \alpha\ 0\ 0]$, where $\mathbb{A} = \mathbb{D}$ and they are orthotropic, while $\mathscr{A} = \mathscr{D}$ but they are *not* orthotropic.

[13]Lekhnitskii (1963), p. 47, and successively Pedersen (1990) gave a formula for orthotropy, that written using, for the sake of compactness, the Tsai and Pagano parameters, Eq. (3.29), reads like

Table 5.4 Examples of quasi-trivial solutions for quasi-homogeneity 19–24 plies; sg size = number of plies in each saturated group

Plies	Stacking sequence	sg size
19	0 1 1 1 1 0 0 1 1 0 1 0 1 1 1 0 1 0 1	7/12
	0 1 0 1 1 0 0 0 1 0 0 1 0 0 1 1 1 0 0	11/8
	0 1 2 2 0 0 0 1 2 0 0 2 2 1 0 1 0 0 2	9/4/6
	0 1 2 2 0 0 2 1 2 0 0 0 2 1 2 1 0 0 2	8/4/7
	0 1 2 2 2 2 2 1 0 0 2 0 2 1 2 1 2 0 2	5/4/10
20	0 0 1 1 1 0 0 1 0 0 0 0 0 1 1 0 1 1 0 0	12/8
	0 1 1 1 0 0 0 1 0 0 1 1 0 1 1 1 0 0 0 1	10/10
	0 1 2 2 2 1 1 1 1 2 0 2 2 0 1 0 1 2 2 1	4/8/8
	0 1 2 1 2 2 0 2 0 1 2 0 2 1 0 1 1 2 0 2	6/6/8
	0 1 2 2 3 1 3 1 2 3 0 1 2 0 2 0 1 1 2 3	4/6/6/4
21	0 1 1 0 1 1 0 0 1 1 1 0 1 0 0 1 1 0 1 0 1	9/12
	0 1 0 1 0 1 0 1 0 1 0 0 0 0 1 0 1 1 1 0 0	12/9
	0 1 2 1 2 2 2 0 2 0 1 0 2 0 1 1 2 1 2 2 0	6/6/9
	0 1 2 1 0 2 2 0 0 0 0 1 1 2 0 2 1 1 0 2 0	9/6/6
	0 1 2 1 0 0 2 2 0 2 0 1 0 1 1 0 0 2 1 2 0	9/6/6
22	0 1 1 0 1 1 1 1 0 1 0 1 1 0 0 0 1 1 1 1 1 0	8/14
	0 1 1 2 1 2 1 1 2 2 1 0 1 1 1 0 0 1 2 1 2 1	4/12/6
	0 1 2 3 2 3 3 2 2 1 2 0 3 2 1 0 0 3 1 2 3 2	4/4/8/6
	0 1 2 3 2 4 4 3 2 1 4 0 3 2 1 0 0 2 1 2 3 4	4/4/6/4/4
	0 1 2 3 4 1 3 2 4 4 1 1 0 0 3 2 1 0 2 3 1 4	4/6/4/4/4
23	0 0 1 1 1 1 1 1 0 1 0 1 0 1 0 1 0 1 0 1 0 1 1	9/14
	0 0 1 1 1 1 1 1 0 0 0 1 0 1 0 1 0 1 0 1 1 1 0	10/13
	0 0 1 1 1 0 1 0 1 1 1 0 0 1 0 0 1 0 0 1 1 0 1	11/12
	0 1 2 2 2 2 1 0 2 2 1 2 1 2 0 1 0 2 0 2 2 2 1	5/6/12
	0 1 2 2 1 2 2 0 1 2 2 2 1 1 0 2 0 2 0 1 1 2 2	5/7/11
24	0 1 0 0 0 1 0 0 0 1 0 0 0 0 0 0 0 1 1 1 0 0 0 0	18/6
	0 0 0 1 1 1 1 0 1 1 1 1 0 0 0 0 1 0 0 0 0 1 1 1	12/12
	0 1 2 1 1 2 1 1 0 2 0 1 0 2 1 1 1 0 1 2 1 2 0 1	6/12/6
	0 1 2 1 2 0 0 0 0 2 0 0 0 0 1 1 2 1 2 1 2 0 0 0	12/6/6
	0 1 2 3 2 3 1 3 0 2 0 1 0 1 3 1 2 0 2 3 2 3 0 1	6/6/6/6

(Footnote 13 continued)

$$U_7 U_2^2 - 4U_7 U_6^2 - 4U_6 U_3 U2 = 0.$$

Nevertheless, this condition is of more difficult use than the polar condition (4.12.1) and actually it is not a necessary and sufficient condition for ordinary orthotropy, because it states also, as particular cases, special orthotropies and isotropy.

Table 5.5 Examples of quasi-trivial solutions for quasi-homogeneity, 25–30 plies; sg size = number of plies in each saturated group

Plies	Stacking sequence	**sg size**
25	0 1 0 1 0 1 0 0 0 0 0 1 0 1 0 1 0 1 0 0 0 1 1 0 0	16/9
	0 1 2 1 0 1 1 1 1 0 2 1 1 1 2 0 1 0 1 0 0 1 2 1 1	7/14/4
	0 1 2 0 1 0 0 1 0 1 2 0 0 0 2 0 1 0 1 0 1 1 2 0 0	13/8/4
	0 1 2 3 1 1 1 3 1 0 2 3 1 0 2 3 1 0 1 1 1 0 2 1 3	5/11/4/5
	0 1 2 1 3 3 3 3 3 0 2 3 3 0 2 1 1 0 1 1 3 0 2 3 3	5/6/4/10
26	0 1 0 1 0 1 0 0 0 1 0 1 0 0 0 1 0 1 0 1 0 0 1 1 0 0	16/10
	0 1 2 1 1 2 1 0 1 0 2 0 0 1 0 1 2 1 1 2 1 1 0 2 1 0	8/12/6
	0 1 2 1 2 3 0 0 1 0 0 1 0 2 2 0 1 0 1 2 3 1 1 2 0 0	10/8/6/2
	0 1 2 3 4 2 3 4 0 0 3 2 3 1 1 2 4 0 2 3 0 0 1 4 2 3	6/4/6/6/4
	0 1 2 3 4 5 3 4 0 2 3 0 3 1 1 0 4 0 2 3 5 2 1 4 0 3	6/4/4/6/4/2
27	0 1 0 1 1 1 1 0 1 0 0 0 0 1 1 1 0 1 1 1 0 0 0 1 1 1 0	12/15
	0 1 2 2 1 2 0 2 1 2 2 1 1 0 1 2 2 0 1 2 0 2 0 2 1 1 2	6/9/12
	0 1 2 3 3 1 3 3 1 1 0 1 0 0 3 1 3 0 2 2 3 1 1 3 3 1 0	6/9/3/9
	0 0 1 2 3 0 3 4 4 3 0 0 0 3 2 0 0 0 1 1 3 0 2 0 4 0 3	12/3/3/6/3
	0 1 2 3 4 1 1 5 5 1 0 0 4 1 3 0 0 1 2 2 1 1 3 4 5 1 0	6/9/3/3/3/3
28	0 1 1 1 0 0 1 1 1 0 0 1 0 0 1 1 1 0 1 0 1 1 0 1 1 0 0 1	12/16
	0 1 1 2 2 1 2 2 2 2 2 2 1 2 1 0 1 1 0 2 1 0 1 2 2 2 1 2	4/10/14
	0 1 2 3 2 1 1 3 2 3 2 2 2 1 2 0 3 1 0 3 1 0 1 2 1 2 2 3	4/8/10/6
	0 1 2 3 1 4 4 3 1 3 1 2 1 1 1 0 3 2 0 3 4 0 1 4 2 1 1 3	4/10/4/6/4
	0 1 2 3 4 5 0 4 5 4 3 2 1 4 0 5 3 2 0 1 0 4 1 0 2 3 5 4	6/4/4/4/6/4
29	0 1 1 0 0 0 1 1 1 0 0 0 0 1 1 0 0 1 0 1 1 0 1 0 0 0 1 1 0	16/13
	0 1 2 1 2 2 2 1 2 2 2 1 2 1 1 2 0 2 0 1 2 1 0 1 2 2 2 2 1	4/10/15
	0 1 0 0 2 3 2 3 1 0 3 3 0 0 0 2 0 1 2 0 1 0 0 0 1 3 3 2 0	13/5/5/6
	0 1 2 3 3 3 2 3 1 0 2 2 3 0 3 3 0 1 3 0 1 2 2 3 1 3 2 0 3	6/5/7/11
	0 1 2 3 3 4 4 1 1 2 1 1 3 2 4 1 0 4 0 3 2 3 0 1 1 3 2 1 4	4/9/5/6/5
30	0 1 1 0 1 0 1 1 0 1 0 1 1 1 1 0 1 1 0 1 0 0 1 0 0 1 1 0 1 1	12/18
	0 0 1 2 2 1 1 1 1 0 0 0 1 2 0 1 0 0 0 1 2 2 0 1 1 1 2 0 1 0	12/12/6
	0 1 2 3 3 0 2 0 1 0 0 2 0 1 3 1 0 0 0 3 3 2 2 0 1 2 1 3 0 0	12/6/6/6
	0 1 2 3 4 4 1 3 2 2 4 0 3 0 3 1 0 1 0 2 4 4 1 3 2 2 1 4 3 0	6/6/6/6/6
	0 1 2 3 2 4 4 1 3 3 4 0 1 0 4 2 0 3 0 2 1 1 3 2 2 4 1 4 3 0	6/6/6/6/6

Orth 1 : $\mathbb{A}$ orthotropic:

$$\Phi_0^A - \Phi_1^A = K^A\frac{\pi}{4}, \quad K^A \in \{0, 1\}; \tag{5.71}$$

Orth 2 : $\mathbb{D}$ orthotropic:

$$\Phi_0^D - \Phi_1^D = K^D\frac{\pi}{4}, \quad K^D \in \{0, 1\}; \tag{5.72}$$

Orth 3 : full orthotropy:

$$\begin{cases} \Phi_0^A - \Phi_1^A = K^A\frac{\pi}{4}, & K^A \in \{0, 1\}, \\ \Phi_0^D - \Phi_1^D = K^D\frac{\pi}{4}, & K^D \in \{0, 1\}; \end{cases} \tag{5.73}$$

Orth 4 : full orthotropy with the same type of orthotropy for $\mathbb{A}$ and $\mathbb{D}$:

$$\begin{cases} \Phi_0^A - \Phi_1^A = K^A\frac{\pi}{4}, & K^A \in \{0, 1\}, \\ \Phi_0^D - \Phi_1^D = K^D\frac{\pi}{4}, & K^D \in \{0, 1\}; \\ K^A = K^D; \end{cases} \tag{5.74}$$

Orth 5 : full orthotropy with the same axes for $\mathbb{A}$ and $\mathbb{D}$:

$$\begin{cases} \Phi_0^A - \Phi_1^A = K^A\frac{\pi}{4}, & K^A \in \{0, 1\}, \\ \Phi_0^D - \Phi_1^D = K^D\frac{\pi}{4}, & K^D \in \{0, 1\}, \\ \Phi_1^A - \Phi_1^D = K\frac{\pi}{2}, & K \in \{0, 1\}; \end{cases} \tag{5.75}$$

Orth 6 : full orthotropy with the same axes and type of orthotropy for $\mathbb{A}$ and $\mathbb{D}$:

$$\begin{cases} \Phi_0^A - \Phi_1^A = K^A\frac{\pi}{4}, & K^A \in \{0, 1\}, \\ \Phi_0^D - \Phi_1^D = K^D\frac{\pi}{4}, & K^D \in \{0, 1\}, \\ K^A = K^D, \\ \Phi_1^A - \Phi_1^D = K\frac{\pi}{2}, & K \in \{0, 1\}; \end{cases} \tag{5.76}$$

Orth 7 : full orthotropy with $\mathbb{A} = \mathbb{D}$ (quasi-homogeneous full orthotropy):

$$\begin{cases} \Phi_0^A - \Phi_1^A = K^A \dfrac{\pi}{4}, \quad K^A \in \{0, 1\}, \\ R_0^C = 0, \\ R_1^C = 0, \\ R_0^B = 0, \\ R_1^B = 0. \end{cases} \tag{5.77}$$

We consider now some rules for obtaining orthotropic laminates of one of the types above. It is important to specify that all the rules given below are just *sufficient and not necessary* conditions.

5.3.4.1 Extension Orthotropy

The most used strategy for obtaining extension orthotropy is that of using *balanced laminates*. A laminate is said to be balanced if for each ply at the orientation α there is another ply at the orientation $-\alpha$. Balanced laminates have hence an even number of plies. This simple rule ensures always orthotropy in extension, *regardless of the values of the orientation of the twin layers*. In fact, from Eqs. (5.71) and (5.55), we get that $\mathbb{A}$ is ordinarily orthotropic $\iff$

$$\begin{aligned} \xi_2 &= \sum_{k=1}^{n} \sin 4\delta_k = 0, \\ \xi_4 &= \sum_{k=1}^{n} \sin 2\delta_k = 0. \end{aligned} \tag{5.78}$$

The last condition is immediately get operating a rotation of Φ_1^A in (5.51) and remembering how this is done in the polar method, i.e. simply subtracting Φ_1^A from all the polar angles.

It is evident that a *sufficient* condition for fulfilling the above two equations is that $\forall \delta_k = \alpha, \exists\, \delta_j = -\alpha, \ j \neq k$. This is just a balanced stacking sequence.

A particular case of balanced laminates is that of the *angle-ply* sequences, that have only two possible orientations: α and $-\alpha$.

Another type of sequences ensuring extension orthotropy are *quasi-isotropic* laminates[14]: the only allowed orientations are $0°$, $\pm 45°$, $90°$, with the constraint to have the *same number of plies at* $+45°$ *and* $-45°$. Of course, being the superposition of a cross-ply, see Sect. 5.3.4.3, and an angle-ply, $\mathbb{A}$ *is orthotropic, but not* $\mathbb{D}$. These laminates are very used in aeronautics, because the elastic behavior has not weak directions and the presence of fibers in four directions helps in preventing the propagation of cracks.

[14]They are called in this way because if the number of the plies in each direction is constant, then $\mathbb{A}$ is isotropic, see Sect. 5.3.5.1.

When orthotropy is demanded, balanced, angle-ply and quasi-isotropic laminates are currently used, normally in symmetric stacks to ensure uncoupling.

5.3.4.2 Bending Orthotropy

Bending orthotropy is much more difficult to be obtained, because of the presence of coefficients d_k in the expression of ξ_{13} to ξ_{16}, Eq. (5.55). That is why balanced, angle-ply or quasi-isotropic laminates are not orthotropic in bending, in general.

Actually, because the d_ks are symmetric with respect to the mid-plane, if the twin layers $\alpha/-\alpha$ were disposed symmetrically with respect to the mid-plane, to form hence an *antisymmetric stack*, then the laminate would be orthotropic also in bending, but generally speaking it would be also coupled. Because the primary requirement is uncoupling, this kind of strategy is normally not used, unless in special procedures, see the next Section, and in such a case not exclusively bending, but also extension orthotropy is obtained.

5.3.4.3 Full Orthotropy with Coincident Axes

The most simple way to ensure at the same time extension and bending orthotropy, and with the same couple of orthotropy axes, is to use *cross-ply* laminates, that are laminates with only two orientations: 0° and 90°. It is immediate to check that in such a case it is

$$\xi_2 = \xi_4 = \xi_{14} = \xi_{16} = 0, \tag{5.79}$$

so the laminate is hence *orthotropic at the same time in extension and in bending*; once more, normally the sequences are symmetric. It is not needed that the number of plies at 0° be equal to that of the plies at 90°.

To remark that for an equal number of plies at 0° and 90°,

$$\xi_3 = \sum_{k=1}^{n} a_k \cos 2\delta_k = 0 \Rightarrow R_1^A = 0, \tag{5.80}$$

i.e., the laminate is *square symmetric in extension*, but *not* in bending, due to the presence of the coefficients d_k.

A pioneer and almost overlooked work is that of Bartholomew (1976, 1977), who was the first to publish as early as 1977 a list of fully orthotropic laminates. This approach has been reconsidered more recently by York, who has given, for different problems of laminates anisotropy, a thorough classification of solutions, York (2007, 2008, 2009, 2010, 2011).

Valot (2002), has shown that, if we exclude the case of cross-ply laminates, the only general kind of stacking sequence that gives fully orthotropic laminates, with orthotropy in the same axes for bending and extension, is the antisymmetric one.

Normally, this approach is not used because designers first focus on uncoupling and hence automatically on symmetric stacks. Then, they try to obtain orthotropy, but normally they fail in obtaining bending orthotropy.

Another approach, Valot and Vannucci (2005), consists in first considering balanced antisymmetric stacking sequences, that automatically ensure full orthotropy with the same couple of orthotropy axes in bending and extension, and then in trying to find among these the uncoupled ones: as a matter of fact, uncoupling appears to be a less hard condition to be satisfied than bending orthotropy, so this strategy can be successful.

It has been shown in Valot and Vannucci (2005) that for an antisymmetric sequence, the condition $\mathbb{B} = \mathbb{O}$ can be reduced to the equation ($n = 2p$ or $n = 2p+1$)

$$\left(\sum_{k=2}^{p} b_k \sin 2\delta_k\right)^4 - b_1^2\left(\sum_{k=2}^{p} b_k \sin 2\delta_k\right)^2 + b_1^2\left(\sum_{k=2}^{p} b_k \sin 2\delta_k \cos 2\delta_k\right)^2 = 0, \tag{5.81}$$

with the orientation δ_1 given by

$$\sin 2\delta_1 = -\frac{1}{b_1}\sum_{k=2}^{p} b_k \sin 2\delta_k. \tag{5.82}$$

This equation is highly non linear and it can be solved only numerically. For a small number of layers, the complete set of solutions, that are not unique, can be represented graphically; in Fig. 5.11 the solutions for $n = 6$ and $n = 7$ are presented, while those for $n = 8$ and $n = 9$ are in Fig. 5.12 (for $n = 7$ and $n = 9$ it is assumed that $\delta_0 = 0°$ or $\delta_0 = 90°$).

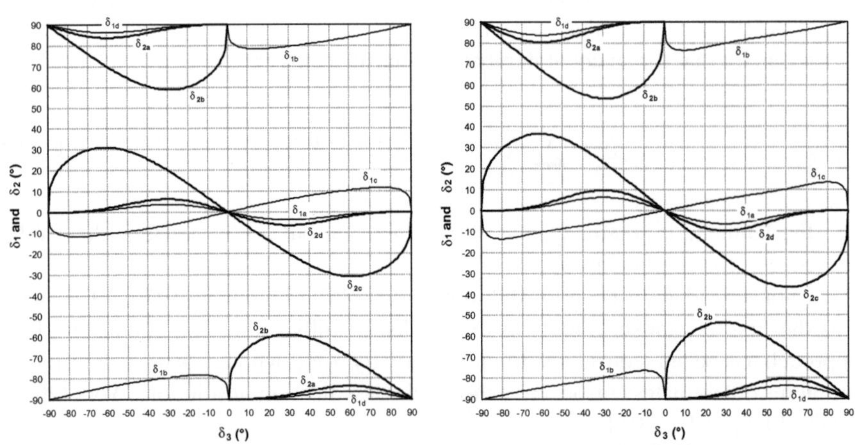

(a) 6 plies (b) 7 plies

Fig. 5.11 Fully orthotropic laminates with coincident axes: antisymmetric uncoupled solutions for $n = 6$ and 7

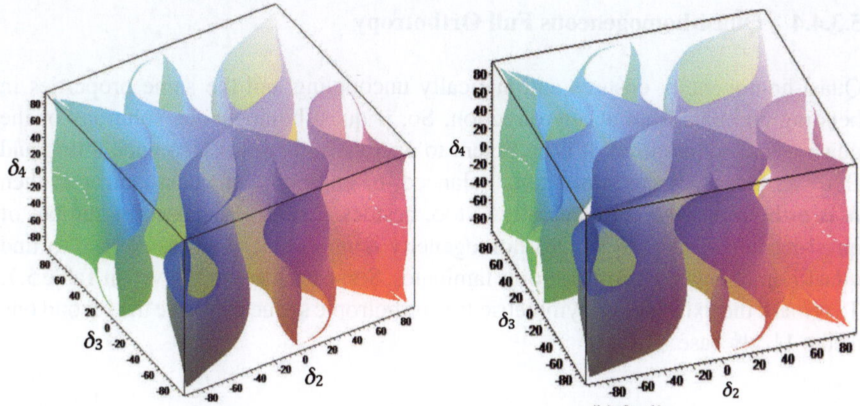

(a) 8 plies (b) 9 plies

Fig. 5.12 Fully orthotropic laminates with coincident axes: antisymmetric uncoupled solutions for $n = 8$ and 9

Table 5.6 Examples of antisymmetric uncoupled laminates, with $\mathbb{A}$ and $\mathbb{D}$ orthotropic and coincident axes

Plies	Stacking sequence
7	$[\alpha/-\alpha_2/0/\alpha_2/-\alpha]$
9	$[0/\alpha/-\alpha_2/0/\alpha_2/-\alpha/0]$
	$[\alpha/-\alpha_2/\alpha/0/-\alpha/\alpha_2/-\alpha]$
	$[\alpha/-\alpha/0/-\alpha/0/\alpha/0/\alpha/-\alpha]$
10	$[\alpha/-\alpha_2/\alpha/0_2/-\alpha/\alpha_2/-\alpha]$
	$[0/\alpha/-\alpha_2/\alpha/-\alpha/\alpha_2/-\alpha\ 0]$
	$[\alpha/-\alpha/0/-\alpha/\alpha/-\alpha/\alpha/0/\alpha/-\alpha]$
11	$[\alpha/-\alpha/0_2-\alpha/0/\alpha/0_2/\alpha/-\alpha]$
	$[0/\alpha/-\alpha/0/-\alpha/0/\alpha/0/\alpha/-\alpha/0]$
	$[\alpha/-\alpha/0/-\alpha/\alpha/0/-\alpha/\alpha/0/\alpha/-\alpha]$
12	$[0/\alpha/0/-\alpha_3/\alpha_2/0/-\alpha/0]$
	$[\alpha/-\alpha/\alpha/-\alpha_3/\alpha_3/-\alpha/\alpha/-\alpha]$
	$[\alpha/-\alpha/0/-\alpha/\alpha/0_2/-\alpha/\alpha/0/\alpha/-\alpha]$

Another strategy, is that of using *uncoupled solutions of the quasi-trivial set*; in such a set, antisymmetric sequences are sought for: they are orthotropic in extension and bending with coincident orthotropy axes and in addition $\mathbb{B} = \mathbb{O}$. Some examples are given in Table 5.6.[15]

[15]Following a common convention, sequences are indicated between square brackets and a subscript indicates the number of adjacent layers with the same orientation.

5.3.4.4 Quasi-homogeneous Full Orthotropy

Quasi-homogeneity ensures automatically uncoupling and the same properties in bending and extension at any direction. So, if quasi-homogenous solutions of the quasi-trivial set are used, it is sufficient to operate on $\mathbb{A}$ to get also uncoupling and $\mathbb{D} = \mathbb{A}$. Hence, if the laminate is balanced, or angle-ply, or quasi-isotropic, then $\mathbb{A}$ is orthotropic and automatically $\mathbb{D}$ too, besides $\mathbb{B} = \mathbb{O}$. Because the number of quasi-trivial solutions for quasi-homogeneity is important, it is rather easy to find quasi-homogeneous full orthotropic laminates. Some examples are given in Table 5.7. To remark the existence of symmetric full-orthotropic sequences, like the second one of the 11-ply case.

5.3.4.5 Special Orthotropies

In the above Sections, we have seen some sufficient conditions to have ordinary orthotropy. Besides this, there are also two types of special orthotropies: R_0-orthotropy, Sect. 4.12.2, and square symmetry, Sect. 4.12.4.

A sufficient condition to obtain extension R_0-orthotropy can be obtained easily: from Eqs. (5.45), (5.51) and (5.55) we can see that $\mathbb{A}$ is R_0-orthotropic if

$$\sum_{j=1}^{n} \cos 4\delta_j + i \sin 4\delta_j = 0. \tag{5.83}$$

To satisfy this condition, it is sufficient to put half of the stack at α and the other half at $\alpha + \pi/4$. Of course, this rule works strictly only for n even and if applied in conjunction to one of the above strategies used for ordinary orthotropy, namely using quasi-trivial sequences, it can give also full or quasi-homogeneous R_0-orthotropy.

A case of extension square-symmetry, i.e. of $R_1^A = 0$, has been already given in Sect. 5.3.4.3: Eq. (5.80) is a general sufficient condition to obtain extension square-symmetry. Roughly speaking, it is sufficient to put half of the stack at α and the other half at $\alpha + \pi/2$ to get $R_1^A = 0$. If this simple rule is used in conjunction with one of the strategies specified for ordinary orthotropy and making use of the quasi-trivial set, also bending or full square-symmetry can be obtained.

5.3.5 *Isotropy*

In some cases, isotropy is sought for; namely, this solution can be interesting when the forces do not have a precise direction but a composite laminate is preferred for its lightness. Also in this case, we must distinguish between extension, bending and full isotropy.

Table 5.7 Examples of quasi-homogeneous fully orthotropic laminates obtained from sequences in Tables 5.3, 5.4 and 5.5

Plies	Stacking sequence
7	$[0/90/0_3/90/0]$
8	$[0/90_2/0/90/0_2/90]$
	$[\alpha/-\alpha_2/\alpha/-\alpha/\alpha_2/-\alpha]$
11	$[0/90_2/0/90_2/0_2/90/0/90]$
	$[0/90/0/90/0_3/90/0/90/0]$
12	$[\alpha/-\alpha/\alpha/-\alpha_3/\alpha_3/-\alpha/\alpha_2]$
13	$[0/90_2/0_5/90_3/0_2]$
	$[0/\alpha/-\alpha/0_2/-\alpha/0/\alpha/0_2/\alpha/-\alpha/0]$
14	$[0/\alpha/-\alpha_2/0/\alpha/0_2/\alpha/0/-\alpha_2/\alpha/0]$
15	$[90/0_3/90_2/0_4/90_3/0_2]$
	$[\alpha/0/-\alpha_3/\alpha_3/-\alpha/0/\alpha/0/-\alpha_2/\alpha]$
	$[0/\alpha/0/-\alpha/0/-\alpha/0_3/\alpha/0/\alpha/0/-\alpha/0]$
16	$[\alpha_2/-\alpha_4/\alpha_2/-\alpha_2/\alpha_4-\alpha_2]$
	$[0/\alpha/-\alpha/0/-\alpha/0_2/\alpha_2/0_2/-\alpha/0/-\alpha/\alpha/0]$
17	$[0/90/0_2/90_2/0_5/90/0/90_2/0_2]$
	$[0/\alpha/-\alpha_2/0/\alpha/0/\alpha/0/-\alpha/0/-\alpha/0/\alpha_2/-\alpha/0]$
18	$[0/\alpha/-\alpha/0/\alpha/-\alpha_3/\alpha_2/0_2/-\alpha/0_2/-\alpha_2/0]$
	$[0/-\alpha_3/0/\alpha_2/-\alpha/0/\alpha/-\alpha/0/-\alpha/0/\alpha/0/\alpha/-\alpha]$
19	$[0/90/0/90_2/0_3/90/0_2/90/0_2/90_3/0_2]$
20	$[\alpha/-\alpha_3/\alpha_3/-\alpha/\alpha_2/-\alpha_2/\alpha/-\alpha_3/\alpha_3/-\alpha]$
	$[\alpha/-\alpha/0/-\alpha/0_2/\alpha/0/\alpha/-\alpha/0/\alpha/0/-\alpha/\alpha/-\alpha_2/0/\alpha/0]$
	$[45/0/90_2/-45/0/-45/0/90/-45/45/0/90/45/90/45/0_2/90/-45]$
	$[\alpha/\beta/-\beta_2/-\alpha/\beta/-\alpha/\beta/-\beta/-\alpha/\alpha/\beta/-\beta/\alpha/-\beta/\alpha/\beta_2/-\beta/-\alpha]$
21	$[0/90/0/90/0/90/0/90/0/90/0_4/90/0/90/3/0_2]$
	$[0/\alpha/-\alpha/\alpha/0_2/-\alpha_2/0/-\alpha/0/\alpha/0/\alpha_2/0_2/-\alpha/\alpha/-\alpha/0]$
22	$[\alpha/-\alpha/0/90/0/90_2/0_2/-\alpha/0/\alpha/90/0/-\alpha/\alpha_2/90/-\alpha/0/90/0]$
	$[\alpha/0/-\alpha/\beta/-\beta/0/\beta/-\alpha/-\beta_2/0_2/\alpha_2/\beta/-\alpha/0/\alpha/-\alpha/\beta/0/-\beta]$
	$[\alpha/-\alpha/0/\beta/0/-\beta_2/\beta/0/-\alpha/-\beta/\alpha/\beta/0/-\alpha/\alpha_2/0/-\alpha/0/\beta/-\beta]$
23	$[0_2/90_3/0/90/0/90_3/0_2/90/0_2/90/0_2/90_2/0/90]$
	$[0_2/90_6/0/90/0/90/0/90/0/90/0/90/0/90/0/90_2]$
24	$[\alpha_3/-\alpha_4/\alpha/-\alpha_4/\alpha_4/-\alpha/\alpha_4/-\alpha_3]$
	$[0/\alpha/-\alpha/\alpha/-\alpha/0_4/-\alpha/0_4/\alpha_2/-\alpha/\alpha/-\alpha/\alpha/-\alpha/0_3]$
	$[\alpha/-\alpha/\beta/-\beta/\beta/-\beta/-\alpha/-\beta/\alpha/\beta/\alpha/-\alpha/\alpha/-\alpha/-\beta/-\alpha/\beta/$ $\alpha/\beta/-\beta/\beta/-\beta/\alpha/-\alpha]$
25	$[0/90/0/90/0/90/0_5/90/0/90/0/90/0/90/0_3/90_2/0_2]$
	$[\alpha/0/90/-\alpha/0_3/-\alpha/0/\alpha/90/-\alpha/0/\alpha/90/-\alpha/0/\alpha/0_3/\alpha/90/0/-\alpha]$

(continued)

Table 5.7 (continued)

Plies	Stacking sequence
26	$[0/90/0/90/0/90/0_3/90/0/90/0_3/90/0/90/0/90/0_2/90_2/0_2]$
	$[0/\alpha/\beta/-\beta/-\alpha/\beta/-\beta/-\alpha/0_2/-\beta/\beta/-\beta/\alpha_2/\beta/-\alpha/0/\beta/-\beta/0_2/\alpha/-\alpha/\beta/-\beta]$
	$[\alpha/\beta/-\beta/-\alpha/0/90/-\alpha/0/\alpha/-\beta/-\alpha/\alpha/-\alpha/\beta_2/\alpha/0/\alpha/-\beta/-\alpha/90/-\beta/\beta/0/\alpha/-\alpha]$
27	$[0/90/0/90_4/0/90/0_4/90_3/0/90_3/0_3/90_3/0]$
	$[0/\alpha/90/-\alpha_2/\alpha/-\alpha_2/\alpha_2/0/\alpha/0_2/-\alpha/\alpha/-\alpha/0/90_2/-\alpha/\alpha_2/-\alpha_2/\alpha/0]$
	$[90/0/\alpha/\beta/-\alpha/0_2/-\beta_2/0/90_2/-\alpha/0/\beta/90_2/0/\alpha_2/0_2/\beta/-\alpha/-\beta/0/90]$
28	$[0/90_3/0_2/90_3/0_2/90/0_2/90_3/0/90/0/90_2/0/90_2/0/90_2/0_2/90]$
	$[\alpha/\beta/\gamma/-\beta/-\alpha/-\gamma/\alpha/-\alpha/-\gamma/-\alpha/-\beta/\gamma/\beta/-\alpha/\alpha/-\gamma/-\beta/\gamma/\alpha/\beta/\alpha/-\alpha/\beta/\alpha/\gamma/-\beta/-\gamma/-\alpha]$
29	$[0/90_2/0_3/90_3/0_4/90_2/0_2/90/0/90_2/0/90/0_3/90_2/0]$
	$[0/\alpha/0_2/-\alpha/90/-\alpha/90/\alpha/0/90_2/0_3/-\alpha/0/\alpha/-\alpha/0/\alpha/0_3/\alpha/90_2/-\alpha]$
30	$[0/90_2/0/90/0/90_2/0/90/0/90_4/0/90_2/0/90/0_2/90/0_2/90_2/0/90_2]$
	$[\alpha_2/-\alpha/0_2/-\alpha_4/\alpha_3/-\alpha/0/\alpha/-\alpha/\alpha_3/-\alpha/0_2/\alpha/-\alpha_3/0/\alpha/-\alpha/\alpha]$
	$[0/90/\alpha/-\alpha_2/0/\alpha/0/90/0_2/\alpha/0/90/-\alpha/90/0_3/-\alpha_2/\alpha_2/0/90/\alpha/90/-\alpha/0_2]$
	$[0/\alpha/\beta/-\alpha/\beta/-\beta_2/\alpha/-\alpha_2/-\beta/0/\alpha/0/-\beta/\beta/0/-\alpha/0/\beta/\alpha_2/-\alpha/\beta_2/-\beta/\alpha/-\beta/-\alpha/0]$

5.3.5.1 Extension Isotropy

The search for isotropic laminates dates back to 1953, when Werren and Norris (1841), published a first study, giving a sufficient condition to obtain an isotropic tensor $\mathbb{A}$:

> if the n plies of a laminate are subdivided into $g \geq 3$ groups, each one having the same number n/g of layers, and if these groups have orientations that differ by an angle equal to $2\pi/g$, then the laminate is isotropic in extension.

Possible combinations are:

- $0°,\ 60°,\ -60°$,
- $0°,\ 45°,\ -45°,\ 90°$,
- $0°,\ 72°,\ 144°,\ 216°,\ 288°$,

and so on. In particular, the second of these solutions belongs to the set of quasi-isotropic laminates, introduced in Sect. 5.3.4.1; the reason of the name *quasi-isotropic* is now clear.

Normally, engineers use the above combinations with symmetric stacks, to ensure uncoupling. Nevertheless, we have already seen, discussing about orthotropy, that this is not at all necessary. It is possible, for instance, to use the quasi-trivial solutions

Table 5.8 Examples of un-symmetric uncoupled laminates with $\mathbb{A}$ isotropic, obtained using quasi-trivial uncoupled solutions

Plies	Stacking sequence
12	$[0/-60/60/0/60_2/-60_2/-60/0_2/60]$
	$[0/-60/60/0/-60/60/-60/60_2/0_2/-60]$
	$[0/-60/60_2/0/-60/60/-60/0/-60/0/60]$
	$[0/-60/60_2/-60/0/60/0/-60/-60/0/60]$
16	$[0/45/-45/0/45/90_3/-45_2/45/-45/90/0_2/45]$
	$[0/45_2/0/-45/90_2/-45/90/-45_2/90/45/0_2/45]$
	$[0/45/-45/0/90/-45/45/90/45/90_2/-45_2/0_2/45]$
	$[0/45/-45/0/90_2/45/-45/45/90/-45_2/90/0_2/45]$
	$[0/45/-45/0/90/45/90/-45/90/45/-45_2/90/0_2/45]$
	$[0/45/-45/0/90/-45/45/90/-45/90_2/45_2/0_2/-45]$
	$[0/45/-45/0/45/90_2/-45/90/-45/45/90/-45/0_2/45]$
	$[0/45/-45/0/-45/90_2/45/90/-45/45/90/45/0_2/-45]$
18	$[0/60/0/60/-60_3/60/0/-60_2/0/60_2/0/60/0/-60]$
	$[0/60/0/-60_2/60/0/-60_2/60_3/-60/0_2/60/0/-60]$
	$[0/60/0/60/-60_3/0/60/-60_2/60/0/60/0/60/0/-60]$
	$[0/60/0/-60_2/60_2/0/-60/60/-60_2/0/60/0/-60/0/60]$
	$[0_2/60/-60_2/60/-60/60_2/-60/0_2/-60/60/0_2/60/-60]$
	$[0/60/0/-60/60/-60/60/0/60/-60_3/0/-60/0/60/0/60]$
	$[0_2/60/-60_2/60_2/-60/60/-60/0/-60/0_2/60/0/60/-60]$
	$[0/60/0/-60/60/-60/0/-60_2/60_2/-60/60/0_2/60/0/-60]$
	$[0/60/0/-60/60/-60_2/0/60/-60/60_2/0/-60/0/-60/0/60]$
	$[0/60/0/60/-60_2/0/-60/60/-60/60/-60/60/0_2/-60/0/60]$
	$[0_2/60/-60/60/-60/60/-60/60/-60/0/-60/0_2/60/0/-60/60]$
	$[0_2/60/-60/60/-60/60/-60/-60/60/0/-60/0_2/60/0/60/-60]$

of uncoupling; some of these solutions are listed in Table 5.8. To remark that all of them are un-symmetric; for the sake of shortness, only some solutions are listed (actually, there are 57 solutions for the 16-ply case and 55 for the 18-ply one).

In Fig. 5.13 we show the directional diagram of the Young's moduli in extension and in bending for any of the 18-ply sequences in Table 5.8, when the basic layer is the carbon-epoxy T300/5208 in Table 4.9. The isotropic behavior in extension is self-evident, while in bending the behavior is completely anisotropic. All the sequences share the same isotropic behavior, but not the bending one, because the stacking sequence affects the bending but not the extension response of the laminate.

Other approaches are possible; for instance, Verchery et al. (2000), showed a simple and exact method to find isotropic laminates with respect to extension, with only four orientations, 0°, 90° and $\pm\varphi$ but not of the Werren and Norris type. They found 10 different solutions, with at least 30 layers, but all of them with bending-extension coupling.

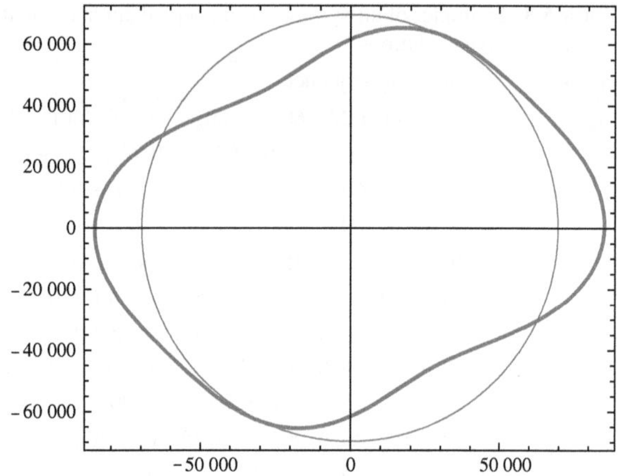

Fig. 5.13 Directional diagram of the Young's moduli E of the 18-ply first sequence in Table 5.8; *thin line*: E for extension, *thick line*: E for bending

5.3.5.2 Bending Isotropy

Just like orthotropy, and for the same reasons, bending isotropy is much more difficult to be obtained. The first study concerning such problem has been done by Paradies. In 1996, he proposed an approach, partially derived from a work of Fukunaga (1990), to find isotropic laminates with respect to bending or to extension, and showed solutions with at least 16 identical or 12 different layers, Paradies (1996).

5.3.5.3 Full Isotropy

Full isotropy means that $\mathbb{A}$ and $\mathbb{D}$ are isotropic and that $\mathbb{B} = \mathbb{O}$. Such a situation corresponds to have

$$R_0^A = R_1^A = R_0^B = R_1^B = R_0^D = R_1^D = 0. \tag{5.84}$$

From the polar equations (5.51) and (5.54), we get immediately that *full isotropy necessarily implies quasi-homogeneity*: $\mathbb{A} = \mathbb{D}$. It is not possible to have a laminate that is contemporarily isotropic in extension and bending and with different elastic moduli for the two cases.

Fully isotropic laminates made of composite materials are interesting for applications: mechanically they behave like a homogeneous isotropic plate with the advantage of being considerably lighter than metallic plates, so their use is preferred whenever lightness is a fundamental requirement, like in aircraft constructions or in aerospace applications. That is why their fabrication has attired the attention of researchers since the eighties.

Verchery and Vong, have been the first, as early as 1986, to find an exactly fully isotropic laminate, composed of 48 layers, Verchery and Vong (1986). Successively, Kandil and Verchery proposed a method to find laminates which are extensionally isotropic, uncoupled and which are nearly isotropic in bending, and they proposed also a non-dimensional measure of nearly-isotropy, Kandil and Verchery (1988).

In 1990, Fukunaga proposed a method to find fully isotropic laminates, and he showed the existence of six different solutions for a 40-layer laminate, Fukunaga (1990). Each solution had four different orientations, but not equally spaced as in the Werren and Norris condition. The solutions of Fukunaga are of the Caprino and Crivelli-Visconti type, Caprino and Crivelli-Visconti (1982); it is in this way that Fukunaga obtained uncoupling.

In 1992, Wu and Avery discovered some fully-isotropic laminates with 36 plies, Wu and Avery (1992). They used an approach previously proposed by Wu (1979), where bending-extension coupling is eliminated by the use of symmetric stacking sequences: this is exactly why they did not succeed in finding solutions with a lower number of plies.

Grédiac found in 1999, by a numerical procedure, a nearly fully isotropic 12-layer laminate, with 12 different and unusual orientations, Grédiac (1999a, b), while Vannucci and Verchery demonstrated in 2000 that isotropy in two dimensional elasticity corresponds to the minimum of a positive definite square symmetric form of the polar elastic constants of the laminate, so giving a minimum characterization of isotropy having a physical meaning, Vannucci and Verchery (2000).

A simple strategy for obtaining fully isotropic laminates is similar to the one already used for quasi-homogeneous full orthotropy: to apply the Werren and Norris rule to a quasi-homogeneous sequence of the quasi-trivial set. Some examples are given in Table 5.9; all of them are un-symmetric (for the sake of shortness, only some cases are shown: there are, in fact, 219 different solutions for the 27-ply case, while for the 30-ply the solutions are 29, Vannucci and Verchery 2002).

The six 18-ply sequences in Table 5.9, that actually are a particular case of the sequences in Table 5.3, are the exact solutions for full isotropy with the lowest number of layers found so far, Vannucci and Verchery (2002). The reason of this result basically is in the idea of not limiting the search for isotropic laminates to the set of symmetric stacks; this has allowed to divide by 2 the number of plies with respect to the solutions found by Wu and Avery. Such an example clearly shows the importance, for the design of light structures, of using un-symmetric stacks.

5.3.6 Sensitivity to Orientation Errors

All of the solutions to the problems considered in the previous Sections are *deterministic*, i.e. they are obtained considering that all the quantities can be realized, in practice, perfectly in agreement with the theoretical design values. Of course, reality is slightly different: errors always affect inexorably each quantity or dimension when one passes to practical realization. It is hence interesting to ponder what are the effects of such errors on the final result.

When dealing with these problems, one needs to fix the property to be analyzed and the type of error to be considered. In the case of laminates, the most interesting

Table 5.9 Examples of un-symmetric fully isotropic laminates, obtained using quasi-homogeneous solutions of the quasi-trivial set

Plies	Stacking sequence
18	$[0/60_2/-60_3/0_2/-60/60/0_2/60_3/-60_2/0]$
	$[0/60/-60/0/60/-60_2/-60/60_2/0_2/60/0_2/-60_2/60]$
	$[0/60/-60_2/60/0/-60/0/60_2/0/-60/60/-60/0_2/-60/60]$
	$[0/60/-60_2/0/60_2/-60/0/-60/60/0/60/0/-60/0/-60/60]$
	$[0/60/-60_2/0/60_2/-60/0/60/-60/0/-60/0/60/0/60/-60]$
24	$[0/-45/90/45/90/45/-45/45/0/90/0/-45/0/-45/45/-45/90/0/90/45/90/45/0/-45]$
27	$[0/60/-60/0/-60_2/60/0/60_2/-60/60/0/60/-60/0_2/-60/0/-60_2/60/0/60/0/60/-60]$
	$[0/60/-60/0/-60/60_2/0/-60/60/-60_2/0/60/-60/0_2/60/0/-60/60_2/0/-60/0/60/-60]$
	$[0/60_2/0/-60_4/0/60/0/60/-60/60/-60/60/0_3/60/0/-60_2/60_2/0/-60]$
	$[0/60/0/-60_2/60/-60/60/-60/60/0/60/-60/0_4/-60/60_2/-60/0/-60/60_2/-60/0]$
	$[0/60/0/-60_2/60/-60/60/-60/60/0/-60/60/0_4/60_2/-60_2/0/-60/60/-60/60/0]$
	$[0/60/-60/0/-60/60/-60/60/0/60/0/-60/60/-60_2/0_2/60/-60/60/0/60/0_2/-60_2/60]$
30	$[0/72/144/216_2/0/288_2/72/144/288_2/144_2/216/72_2/216/72/0_4/144/216_2/288_2/144/72]$
	$[0/72/144/216/288/216/0/144/72/288_2/144/288/72/216/72/216/0/144/0_2/216/144/288/72/0/72/144/216/288]$
	$[0/72/144/216/288_2/0/144/72/216_2/288/144/72/216/72/144/0/288/0_2/288/216/144/72/0/72/144/288/216]$
	$[0/72/144/216/288/216/0/144/72/288_2/144/288/216/72_2/216/0/144/0_2/72/144/288/216/0/216/144/72/288]$

error is that affecting ply orientations. About properties, different choices can be done. In (Vannucci 2001), Vannucci has considered a simple problem: which is the consequence of an orientation error of a single layer on the coupling of a laminate designed to be uncoupled? The problem is stated introducing a suitable measure of the coupling and then analyzing the effects on it of an error on a single layer. The effects of material parameters, condensed in the anisotropy ratio ρ, is then studied. A short account of this approach is given in Sect. 5.3.6.1.

Successively, the same approach has been extended in Vincenti et al. (2003) to the sensitivity of quasi-homogeneity to an orientation error, cf. Sect. 5.3.6.2. Finally, the effects of a random distribution of orientation errors on all the layers of a stack is numerically studied in Belaïd et al. (2001) and still in Vincenti et al. (2003), cf. Sect. 5.3.6.3. All these results have been obtained for laminates with identical plies, because for hybrid laminates no general rule can be given.

5.3.6.1 Effects on Coupling of a Single Orientation Error

To assess the effect of the error of orientation of one layer on tensor $\mathbb{B}$ of a laminate designed to be uncoupled, we introduce a measure of the coupling in the form of a scalar called the *degree of coupling* β defined as

$$\beta=\frac{B}{B_{max}}, \tag{5.85}$$

where B is a suitable norm of $\mathbb{B}$ and B_{max} is highest possible value. Of course, $\beta\in[0,1]$ and $\beta=0$ corresponds to uncoupling, while $\beta=1$ to the highest possible coupling for the laminate. We take for the expression of B that proposed in Kandil and Verchery (1988):

$$B=\sqrt{T_0^{B2}+2T_1^{B2}+R_0^{B2}+4R_1^{B2}}, \tag{5.86}$$

that for a laminate composed by identical layers becomes, see Eq. (5.52),

$$B=\sqrt{R_0^{B2}+4R_1^{B2}}. \tag{5.87}$$

Using Eqs. (5.52), (5.62) and (5.63), with some standard passages we can give to the last equation the form

$$B=\frac{h^2}{2}\sqrt{\begin{array}{l}\left(R_0^{L2}+4R_1^{L2}\right)\sum\limits_{j=-p}^{p}b_j^2+\\ 2\sum\limits_{l=-p}^{p}\sum\limits_{m=l+1}^{p}b_lb_m\left[R_0^{L2}\cos 4(\delta_l-\delta_m)+4R_1^{L2}\cos 2(\delta_l-\delta_m)\right]\end{array}}. \tag{5.88}$$

The above expression allows also to find B_{max}: the term

$$2\sum_{l=-p}^{p}\sum_{m=l+1}^{p}b_lb_m\left[R_0^{L2}\cos 4(\delta_l-\delta_m)+4R_1^{L2}\cos 2(\delta_l-\delta_m)\right] \tag{5.89}$$

must be maximized, because the rest depends upon only the basic layer, not upon the orientations. This expression can be rewritten as

$$R_1^{L2}\sum_{l=-p}^{p}\sum_{m=l+1}^{p}b_lb_m\mu(\rho,\delta_{lm}),\quad \delta_{lm}=\delta_l-\delta_m, \tag{5.90}$$

where

$$\mu(\rho,\delta_{lm})=\rho^2\cos 4\delta_{lm}+4\cos 2\delta_{lm}. \tag{5.91}$$

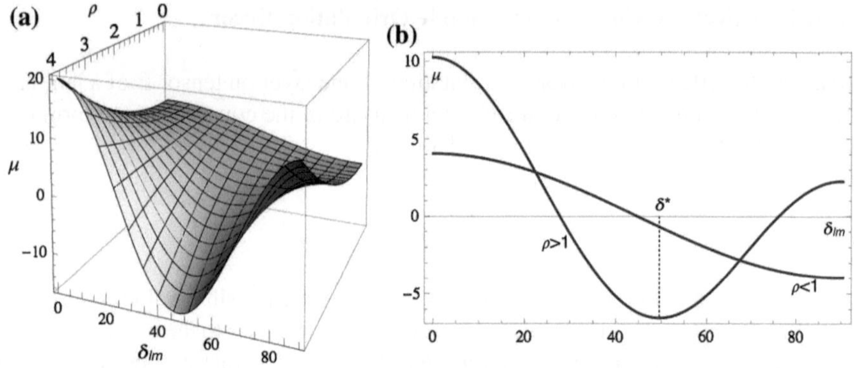

Fig. 5.14 Function $\mu(\rho, \delta_{lm})$; **a** 3D view, **b** 2D view for two values of ρ

Because coefficients b_j are odd with respect to the mid-plane, $b_l b_m > 0 \iff$ the plies l and m are on the same half of the stack with respect to the mid-plane. Hence in this case, to maximize B, the function $\mu(\rho, \delta_{lm})$ must take the maximum value for each possible couple of layers, while in the other case of layers on the opposite sides of the mid-plane, it must take the minimum.

Analyzing function $\mu(\rho, \delta_{lm})$, see Fig. 5.14, one can see that:

- if $\rho \leq 1$, $\mu(\rho, \delta_{lm})$ is maximum for $\delta_{lm} = 0$ and minimum for $\delta_{lm} = \pi/2$;
- if $\rho > 1$, $\mu(\rho, \delta_{lm})$ has a global maximum in $\delta_{lm} = 0$, a local maximum in $\delta_{lm} = \pi/2$ and a minimum for $\delta_{lm} = \delta^*$, where

$$\delta^* = \frac{1}{2} \arccos\left(-\frac{1}{\rho^2}\right). \tag{5.92}$$

So, finally, we can conclude that $B = B_{max}$ when the mid-plane divides the stacks into two parts, in each one of them the orientation is unique for all the plies and the two orientations differ of the angle $\pi/2$ if $\rho \leq 1$, of δ^* if $\rho > 1$. The value of B_{max} can then be easily calculated:

$$B_{max} = \begin{cases} 2h^2 R_1^L \sum_{j=1}^{p} b_j = \frac{1}{2} h^2 R_1^L (n^2 - n \bmod 2) & \text{if } \rho \leq 1, \\ 2h^2 R_1^L \frac{1+\rho^2}{2\rho} \sum_{j=1}^{p} b_j = \frac{1}{2} h^2 R_1^L \frac{1+\rho^2}{2\rho} (n^2 - n \bmod 2) & \text{if } \rho > 1. \end{cases} \tag{5.93}$$

A similar procedure allows for calculating B_ε, i.e. the norm of $\mathbb{B}$ when only one the orientation of the m-th layer is affected by an orientation error, say ε_m:

$$B_\varepsilon = \frac{1}{\sqrt{2}} h^2 R_1^L |b_m| \sqrt{\rho^2(1 - \cos 4\varepsilon_m) + 4(1 - \cos 2\varepsilon_m)}. \tag{5.94}$$

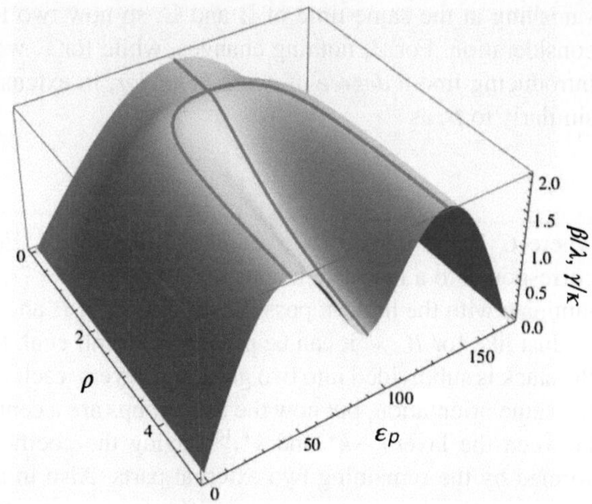

Fig. 5.15 Function β/λ or γ/κ; the curve is the locus of the stationary values of the function

Because the coefficients b_j take the highest absolute value for the two external layers, i.e. for $m = p$, we get that the highest value of β is obtained for an error of orientation affecting one of the two outer layers. It is then easy to calculate β:

$$\beta = \begin{cases} \dfrac{\lambda}{\sqrt{2}}\sqrt{\rho^2(1-\cos 4\varepsilon_p)+4(1-\cos 2\varepsilon_p)} \text{ if } \rho \leq 1, \\ \dfrac{\lambda}{\sqrt{2}}\dfrac{2\rho}{1+\rho^2}\sqrt{\rho^2(1-\cos 4\varepsilon_p)+4(1-\cos 2\varepsilon_p)} \text{ if } \rho > 1, \end{cases} \tag{5.95}$$

with

$$\lambda = \frac{1}{2}\frac{b_p}{\sum_{j=1}^{p} b_j} = 2\frac{n-1}{n^2 - \text{mod}2}. \tag{5.96}$$

The last two equations show that the sensitivity β to a single defect decreases with n and that the orientation angles do not matter. It is then easy to show that the angle that maximizes β is $\pi/2$ if $\rho \leq 1$ and δ^* if $\rho > 1$; in both the cases,

$$\beta_{max} = 2\lambda. \tag{5.97}$$

We show in Fig. 5.15 the function β/λ. We observe that the materials with $\rho \to \infty \Rightarrow R_1^L = 0$, i.e. square symmetric layers, are the most sensitive to errors, while the less sensitive are those with $\rho = 0 \Rightarrow R_0^L = 0$, i.e. layers that are R_0-orthotropic.

5.3.6.2 Extension to Quasi-Homogeneity

The extension to quasi-homogeneity can be done rather easily, adapting to tensor $\mathbb{C}$ the approach outlined in the previous Section. In fact, quasi-homogeneity needs the vanishing at the same time of $\mathbb{B}$ and $\mathbb{C}$, so now two tensors need to be taken into consideration. For $\mathbb{B}$, nothing changes, while for $\mathbb{C}$ we follow the same procedure, introducing first a *degree of equal behavior*, in extension and bending, γ, defined, similarly to β, as

$$\gamma = \frac{C}{C_{max}}, \tag{5.98}$$

where C is the norm of $\mathbb{C}$, calculated like for B, Eq. (5.87). Of course, $\gamma = 0$ corresponds to a laminate having $\mathbb{C} = \mathbb{O}$, i.e. $\mathbb{A} = \mathbb{D}$, while $\gamma = 1$ characterizes a laminate with the highest possible difference of $\mathbb{A}$ and $\mathbb{D}$.

Just like for B_{max}, it can be proved, Vincenti et al. (2003), that C_{max} is get when the stack is subdivided into two groups of layers, each one composed by plies having the same orientation, but now the two groups are a central one, formed by the layers between the layers $-k^*$ and k^*,[16] having the coefficient $c_k = 0$, and the group formed by the remaining two external parts. Also in this case, C_{max} is obtained if the orientations of the two groups differ for an angle equal to $\pi/2$ if $\rho \leq 1$, of δ^* if $\rho > 1$. Of course, this rule implies that it is impossible to get at the same time B_{max} and C_{max}. The value of C_{max} can the be calculated:

$$C_{max} = \begin{cases} 4R_1^L \sum\limits_{k=-k^*}^{k^*} c_k =, 4R_1^L \text{ if } \rho \leq 1, \\[2ex] 4R_1^L \dfrac{1+\rho^2}{2\rho} \sum\limits_{k=-k^*}^{k^*} c_k \text{ if } \rho > 1. \end{cases} \tag{5.99}$$

Actually, the calculation of $\pm k^*$ is unnecessary, because, thanks to the properties of coefficients c_k, Eqs. (5.62) and (5.63), it can be proved that

$$\sum_{k=-k^*}^{k^*} c_k = \begin{cases} n^2 - 1 + \sum\limits_{k=1}^{p} \left|n^2 - 12k^2 - 1\right| \text{ for } n \text{ odd}, \\[2ex] \sum\limits_{k=1}^{p} \left|n^2 - 12k^2 + 12k - 4\right| \text{ for } n \text{ even}. \end{cases} \tag{5.100}$$

Once more, the maximum value of γ is obtained for an orientation error ε_p affecting one of the two outer layers. The expression of γ is rather surprisingly formally identical to that of β, Eq. (5.95), but with the coefficient λ replaced by the coefficient

[16] Actually, it can happen that the layers $\pm k^*$ are not real, but this is not a problem, because, as said in the text, it is not needed to know $\pm k^*$ for calculating γ.

$$\kappa = \frac{1}{2}\frac{|c_p|}{\sum_{k=-k^*}^{k^*} c_k} = \begin{cases} \dfrac{(n-2)(n-1)}{n^2 - 1 + \sum_{k=1}^{p} |n^2 - 12k^2 - 1|}, & \text{for } n \text{ odd}, \\ \dfrac{(n-2)(n-1)}{\sum_{k=1}^{p} |n^2 - 12k^2 + 12k - 4|} & \text{for } n \text{ even}. \end{cases} \tag{5.101}$$

Also for γ it is

$$\gamma_{max} = 2\kappa, \tag{5.102}$$

and Fig. 5.15 represents the function γ/κ as well.

Finally, a synthetic assessment of the consequences of an orientation defect on quasi-homogeneity can be obtained considering the vector (β, γ). Its norm

$$\xi = \sqrt{\beta^2 + \gamma^2} \tag{5.103}$$

is a measure of the deviation from quasi-homogeneity, while the angle θ that the vector forms with the axis of β,

$$\tan\theta = \frac{\gamma}{\beta}, \tag{5.104}$$

indicates the importance of γ relatively to β.

5.3.6.3 Case of Randomly Distributed Orientation Errors

In the real world, errors can affect the orientation of all the layers. In such a case, it is impossible to obtain analytical results like in the preceding Sections: a numerical approach must be used.

Proceeding like in Vincenti et al. (2003), we introduce a vector $Err = (\varepsilon_1, \ldots, \varepsilon_n)$ of angular defects, with the ε_j statistically independent random variables. Representing errors, the probability density function of the ε_js is a Gaussian distribution with null mean and standard deviation

$$\sigma = 1.96\Psi, \tag{5.105}$$

with Ψ a characteristic angle; it is chosen in such a way that the probability that $\varepsilon_j \in [-\Psi, \Psi] = 95\%$.

The orientation of each ply is then computed as

$$\delta_j^{real} = \delta_j^{theor} + \varepsilon_j, \ \forall j = 1, \ldots, n. \tag{5.106}$$

Then a campaign of numerical tests can be conducted. In (Vincenti et al. 2003), a set of quasi-homogeneous sequences with only two saturated groups has been used, with a group orientated at 0 and the other at α. In this way, tests have been conducted in order to evaluate the influence of α, Ψ, ρ and n. For each set of these parameters,

10,000 numerical tests have been performed using vectors Err randomly selected by a Monte Carlo method. For each set of parameters, the ratios ξ and β have hence been calculated as averages on the population of 10,000 tests.

The results clearly show that ξ increases linearly with Ψ, while θ remains constant, which means that β and γ increase linearly and in the same proportion with Ψ. On the other hand, ξ and θ are insensitive to α. This confirms what seen also in the previous Sections, that the theoretical orientations do not matter. This result can be extended to generic sequences as a conjecture, that has been confirmed also by tests performed on other types of sequences, Belaïd et al. (2001), Vincenti et al. (2003).

The number of plies n makes decrease ξ according to a power law, while θ decreases with n only if this is odd and it is constant if n is even. Finally, ρ makes vary both ξ and θ, with a threshold variation in a log-log scale.

The numerical results described here are shown in Fig. 5.16, for what concerns the variation of ξ with ρ and n. Because we have seen that ξ depends linearly upon Ψ, we can put

$$\xi = \Psi f(n, \rho), \tag{5.107}$$

with

$$\ln f(n, \rho) = a_1 + \frac{a_2}{\sqrt{(\rho + a_3)^2 + a_4 \rho}}. \tag{5.108}$$

The coefficients a_1 to a_5 have been determined numerically for the two cases of n even and odd and they are reported in Table 5.10.

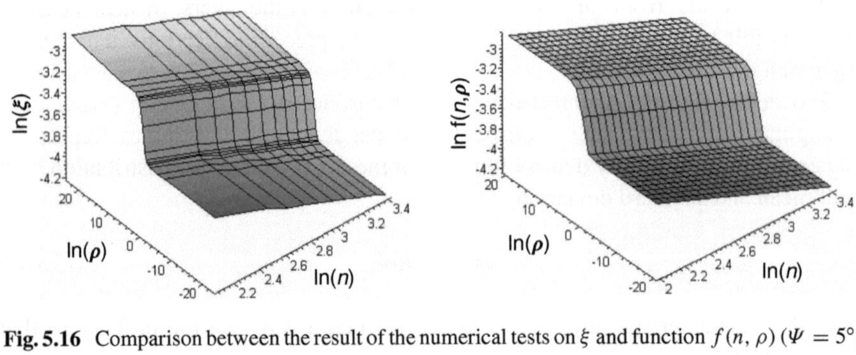

Fig. 5.16 Comparison between the result of the numerical tests on ξ and function $f(n, \rho)$ ($\Psi = 5°$, n even)

Table 5.10 Coefficients a_i of function $\ln f(n, \rho)$

	a_1	a_2	a_3	a_4	a_5
n even	−0.38	−0.06	−0.49	0.85	−0.10
n odd	−0.34	−0.07	−0.50	0.78	−0.11

5.4 Thermal and Hygral Properties

We extend here the CLPT to the case of a temperature field through the thickness of the laminate. To this end, we assume that the total strain tensor $\boldsymbol{\varepsilon}$ is given by the sum of the mechanical and thermal parts

$$\boldsymbol{\varepsilon} = \boldsymbol{\varepsilon}_m + \boldsymbol{\varepsilon}_t, \tag{5.109}$$

with $\boldsymbol{\varepsilon}_m$ given by the converse of the Hooke's law,

$$\boldsymbol{\varepsilon}_m = \mathbb{Z}\boldsymbol{\sigma}, \tag{5.110}$$

while for $\boldsymbol{\varepsilon}_t$ we assume

$$\boldsymbol{\varepsilon}_t = t\,\boldsymbol{\alpha}, \tag{5.111}$$

with t the temperature, evaluated with respect to that of a conventionally unstrained state, while $\boldsymbol{\alpha}$ is the *tensor of thermal expansion coefficients*; from the previous equations, we can see that $\boldsymbol{\alpha}$ is a second-rank symmetric tensor.

Inverting Eq. (5.110) taking into account of Eqs. (5.109) and (5.111), we get the *Hooke–Duhamel* constitutive law

$$\boldsymbol{\sigma} = \mathbb{E}(\boldsymbol{\varepsilon} - t\boldsymbol{\alpha}). \tag{5.112}$$

For the case of a laminate, any linear thermal field characterized by a temperature t_{up} and t_{bot}, respectively on the upper and lower surfaces of the plate, can be decomposed in the sum of two different fields:

- a constant field of value

$$t = \frac{t_{up} + t_{bot}}{2}; \tag{5.113}$$

- an antisymmetric linear field characterized by the constant gradient

$$\nabla t = \frac{t_{up} - t_{bot}}{h}, \tag{5.114}$$

where the temperatures are evaluated with respect to the *manufacturing temperature*, i.e. the temperature at which the laminate is unstrained. The existence of a linear field of temperature is correct, rigorously speaking, only in the case of a steady heat flow through the thickness of the plate and for laminates whose thermal conductivity is constant for all the layers, which is necessarily the case if the layers are identical.

5.4.1 *The Fundamental Law of Laminates in Thermo-Elasticity*

With the above assumptions, the stress state in the $k-$th ply of the laminate is

$$\boldsymbol{\sigma}_k = \mathbb{Q}_k(\delta_k)(\boldsymbol{\varepsilon}^0 + x_3\,\boldsymbol{\kappa}) - (t + \nabla t\, x_3)\,\mathbb{Q}(\delta_k)\boldsymbol{\alpha}_k(\delta_k). \tag{5.115}$$

If now this equation is used in the expression of tensors **N** and **M**, Eqs. (5.15), we get, proceeding like in Sect. 5.2.5,

$$\left\{\frac{\mathbf{N}}{\mathbf{M}}\right\} = \left[\begin{array}{c|c} h\mathbb{A} & \frac{h^2}{2}\mathbb{B} \\ \hline \frac{h^2}{2}\mathbb{B} & \frac{h^3}{12}\mathbb{D}\end{array}\right]\left\{\frac{\boldsymbol{\varepsilon}^0}{\boldsymbol{\kappa}}\right\} - t\left\{\frac{h\mathbf{U}}{\frac{h^2}{2}\mathbf{V}}\right\} - \nabla t\left\{\frac{\frac{h^2}{2}\mathbf{V}}{\frac{h^3}{12}\mathbf{W}}\right\}. \tag{5.116}$$

This equation extend to thermo-elasticity the fundamental law of the CLPT, Eq. (5.21). The second-rank symmetric tensors **U**, **V** and **W** are the thermal corresponding of tensors $\mathbb{A}$, $\mathbb{B}$ and $\mathbb{D}$: **U** is the tensor of in-plane actions per unit of temperature t uniform through the thickness and **W** the tensor of bending moments per unit of thermal gradient ∇t. **V** is a coupling tensor that describes at the same time the in-plane actions produced by $\nabla t = 1$ and the bending moments produced by a field $t = 1$.

Unlike the case of $\mathbb{B}$, we will see in the following that defining a thermo-elastically uncoupled laminate is not trivial: the condition $\mathbf{V} = \mathbf{O}$ is not, generally speaking, sufficient to ensure a complete thermo-elastic uncoupling.

The expressions of **U**, **V** and **W** are quite similar to those of $\mathbb{A}$, $\mathbb{B}$ and $\mathbb{D}$: if we introduce the second-rank symmetric tensor

$$\boldsymbol{\gamma}(\delta_k) = \mathbb{Q}(\delta_k)\boldsymbol{\alpha}(\delta_k),\;\; k = 1, \ldots, n, \tag{5.117}$$

then it is

$$\begin{aligned}
\mathbf{U} &= \frac{1}{h}\sum_{k=1}^{n}\int_{z_{k-1}}^{z_k}\boldsymbol{\gamma}_k(\delta_k)dx_3 = \frac{1}{h}\sum_{k=1}^{n}(z_k - z_{k-1})\boldsymbol{\gamma}_k(\delta_k),\\
\mathbf{V} &= \frac{2}{h^2}\sum_{k=1}^{n}\int_{z_{k-1}}^{z_k}x_3\boldsymbol{\gamma}_k(\delta_k)dx_3 = \frac{1}{h^2}\sum_{k=1}^{n}(z_k^2 - z_{k-1}^2)\boldsymbol{\gamma}_k(\delta_k),\\
\mathbf{W} &= \frac{12}{h^3}\sum_{k=1}^{n}\int_{z_{k-1}}^{z_k}x_3^2\boldsymbol{\gamma}_k(\delta_k)dx_3 = \frac{4}{h^3}\sum_{k=1}^{n}(z_k^3 - z_{k-1}^3)\boldsymbol{\gamma}_k(\delta_k).
\end{aligned} \tag{5.118}$$

If the layers are all identical, the above results are simpler; proceeding like in the case of $\mathbb{A}$, $\mathbb{B}$ and $\mathbb{D}$, we get

$$\mathbf{U} = \sum_{k=1}^{n}a_k\boldsymbol{\gamma}(\delta_k),\;\; \mathbf{V} = \sum_{k=1}^{n}b_k\boldsymbol{\gamma}(\delta_k),\;\; \mathbf{W} = \sum_{k=1}^{n}d_k\boldsymbol{\gamma}(\delta_k) \tag{5.119}$$

with the coefficients a_k, b_k and d_k that are still given by Eq. (5.45) or by Eqs. (5.62) and (5.63).

Composite materials with organic matrices are subjected also to another occurrence, the moisture absorption. This phenomenon happens, usually, with a different time scale with respect to thermal changes, so these two occurrences are normally unrelated. Nevertheless, moisture absorption is ruled by equations that are completely similar to those concerning thermal effects. In particular, in the presence of a linear variation of the moisture concentration c across the thickness, we obtain, proceeding in the same way, analogous results: formally, the phenomenon of moisture absorption in a laminated plate is hence ruled by equations that are algebraically completely identical to those governing the phenomenon of thermal expansion.

That is why in the following we continue with only the thermal part: for the hygral one, everything can be rephrased *verbatim* once the temperature t replaced by the moisture absorption c.

5.4.2 *The Inverse Fundamental Law of Laminates in Thermo-Elasticity*

To obtain the converse of Eq. (5.116), we proceed in the following way: first, we put

$$\left\{\frac{\mathbf{N}^*}{\mathbf{M}^*}\right\} = t\left\{\frac{h\mathbf{U}}{\frac{h^2}{2}\mathbf{V}}\right\} + \nabla t\left\{\frac{\frac{h^2}{2}\mathbf{V}}{\frac{h^3}{12}\mathbf{W}}\right\}, \tag{5.120}$$

to obtain

$$\left\{\frac{\mathbf{N}+\mathbf{N}^*}{\mathbf{M}+\mathbf{M}^*}\right\} = \left[\begin{array}{c|c} h\mathbb{A} & \frac{h^2}{2}\mathbb{B} \\ \hline \frac{h^2}{2}\mathbb{B} & \frac{h^3}{12}\mathbb{D} \end{array}\right]\left\{\frac{\boldsymbol{\varepsilon}^0}{\boldsymbol{\kappa}}\right\}. \tag{5.121}$$

This relation is formally identical to Eq. (5.21) and can be inverted in the same way outlined in Sect. 5.2.9, getting

$$\left\{\frac{\boldsymbol{\varepsilon}^0}{\boldsymbol{\kappa}}\right\} = \left[\begin{array}{c|c} \frac{1}{h}\mathscr{A} & \frac{2}{h^2}\mathscr{B} \\ \hline \frac{2}{h^2}\mathscr{B}^\top & \frac{12}{h^3}\mathscr{D} \end{array}\right]\left\{\frac{\mathbf{N}+\mathbf{N}^*}{\mathbf{M}+\mathbf{M}^*}\right\}, \tag{5.122}$$

Injecting now Eq. (5.120) in the above equation, we get finally

$$\left\{\frac{\boldsymbol{\varepsilon}^0}{\boldsymbol{\kappa}}\right\} = \left[\begin{array}{c|c} \frac{1}{h}\mathscr{A} & \frac{2}{h^2}\mathscr{B} \\ \hline \frac{2}{h^2}\mathscr{B}^\top & \frac{12}{h^3}\mathscr{D} \end{array}\right]\left\{\frac{\mathbf{N}}{\mathbf{M}}\right\} + t\left\{\frac{\mathbf{u}}{\mathbf{v}_1}\right\} + \nabla t\left\{\frac{\mathbf{v}_2}{\mathbf{w}}\right\}, \tag{5.123}$$

where, using Eq. (5.33),

$$\begin{aligned}
\mathbf{u} &= \mathscr{A}\mathbf{U} + \mathscr{B}\mathbf{V} = (\mathbb{A} - 3\mathbb{B}\mathbb{D}^{-1}\mathbb{B})^{-1}(\mathbf{U} - 3\mathbb{B}\mathbb{D}^{-1}\mathbf{V}),\\
\mathbf{v}_1 &= \frac{2}{h}(\mathscr{B}^\top\mathbf{U} + 3\mathscr{D}\mathbf{V}) = \frac{6}{h}(\mathbb{D} - 3\mathbb{B}\mathbb{A}^{-1}\mathbb{B})^{-1}(\mathbf{V} - \mathbb{B}\mathbb{A}^{-1}\mathbf{U}),\\
\mathbf{v}_2 &= \frac{h}{6}(3\mathscr{A}\mathbf{V} + \mathscr{B}\mathbf{W}) = \frac{h}{2}(\mathbb{A} - 3\mathbb{B}\mathbb{D}^{-1}\mathbb{B})^{-1}(\mathbf{V} - \mathbb{B}\mathbb{D}^{-1}\mathbf{W}),\\
\mathbf{w} &= \mathscr{B}^\top\mathbf{V} + \mathscr{D}\mathbf{W} = (\mathbb{D} - 3\mathbb{B}\mathbb{A}^{-1}\mathbb{B})^{-1}(\mathbf{W} - 3\mathbb{B}\mathbb{A}^{-1}\mathbf{V}).
\end{aligned} \tag{5.124}$$

Only apparently the inverse law depends upon four tensors, because actually, as it can be easily checked,

$$\mathbf{v}_2 = \frac{h^2}{12}\mathscr{A}\left[\mathscr{D}^{-1}\mathbf{v}_1 + \frac{6}{h}\mathbb{B}(\mathbb{A}^{-1}\mathbf{U} - \mathbb{D}^{-1}\mathbf{W})\right]. \tag{5.125}$$

To remark the physical meaning of the above second-rank tensors:

- $\mathbf{u}$ is the tensor of the coefficients of thermal expansion of the laminate for a uniform change of temperature t; its SI units are °C^{-1};
- $\mathbf{v}_1$ is the tensor of the coefficients of thermal expansion of the laminate for a gradient of temperature ∇t; its SI units are $(m\ ^\circ\mathrm{C})^{-1}$;
- $\mathbf{v}_2$ is the tensor of the coefficients of thermal curvature of the laminate for a uniform change of temperature t; its SI units are m °C^{-1};
- $\mathbf{w}$ is the tensor of the coefficients of thermal curvature of the laminate for a gradient of temperature ∇t; its SI units are °C^{-1}.

5.4.3 *The Polar Formalism for the Thermo-Elastic Tensors*

The thermo-elastic second-rank symmetric tensors $\mathbf{U}$, $\mathbf{V}$ and $\mathbf{W}$ can be expressed, of course, using the polar formalism; if T^γ, R^γ and Φ^γ denote the polar components of tensor $\boldsymbol{\gamma}$, then

$$\mathbf{U} \rightarrow \begin{cases} T^U = \dfrac{1}{h}\displaystyle\sum_{k=1}^{n} T_k^\gamma (z_k - z_{k-1}),\\[2ex] R^U e^{2i\Phi^U} = \dfrac{1}{h}\displaystyle\sum_{k=1}^{n} R_k^\gamma e^{2i(\Phi_k^\gamma + \delta_k)}(z_k - z_{k-1}); \end{cases} \tag{5.126}$$

$$\mathbf{V} \rightarrow \begin{cases} T^V = \dfrac{1}{h^2}\displaystyle\sum_{k=1}^{n} T_k^\gamma (z_k^2 - z_{k-1}^2),\\[2ex] R^V e^{2i\Phi^V} = \dfrac{1}{h^2}\displaystyle\sum_{k=1}^{n} R_k^\gamma e^{2i(\Phi_k^\gamma + \delta_k)}(z_k^2 - z_{k-1}^2); \end{cases} \tag{5.127}$$

$$\mathbf{W} \rightarrow \begin{cases} T^W = \dfrac{4}{h^3}\displaystyle\sum_{k=1}^{n} T_k^\gamma (z_k^3 - z_{k-1}^3), \\ R^W e^{2i\Phi^W} = \dfrac{4}{h^3}\displaystyle\sum_{k=1}^{n} R_k^\gamma e^{2i(\Phi_k^\gamma+\delta_k)}(z_k^3 - z_{k-1}^3). \end{cases} \tag{5.128}$$

For the case of identical layers, we get

$$\mathbf{U} \rightarrow \begin{cases} T^U = T^\gamma, \\ R^U e^{2i\Phi^U} = R^\gamma e^{2i\Phi^\gamma}(\xi_3 + i\xi_4); \end{cases} \tag{5.129}$$

$$\mathbf{V} \rightarrow \begin{cases} T^V = 0, \\ R^V e^{2i\Phi^V} = R^\gamma e^{2i\Phi^\gamma}(\xi_7 + i\xi_8); \end{cases} \tag{5.130}$$

$$\mathbf{W} \rightarrow \begin{cases} T^W = T^\gamma, \\ R^W e^{2i\Phi^W} = R^\gamma e^{2i\Phi^\gamma}(\xi_{15} + i\xi_{16}). \end{cases} \tag{5.131}$$

We can notice that, like in the case of the elastic tensors $\mathbb{A}$ and $\mathbb{D}$, the isotropic part of $\mathbf{U}$ and $\mathbf{W}$ is equal to that of $\boldsymbol{\gamma}$, while, like for $\mathbb{B}$, that of $\mathbf{V}$ is null.

5.4.4 Thermally Uncoupled Laminates

As previously said, the condition of elastic uncoupling, $\mathbb{B} = \mathbb{O}$, is *not* sufficient to ensure also thermal-elastic uncoupling. This can be easily seen comparing Eq. (5.118)$_2$ with Eq. (5.19). Anyway, if $\mathbb{B} = \mathbb{O}$ but $\mathbf{V} \neq \mathbf{O}$, we get from Eq. (5.124)

$$\mathbf{u} = \mathbb{A}^{-1}\mathbf{U}, \quad \mathbf{v}_1 = \frac{6}{h}\mathbb{D}^{-1}\mathbf{V}, \quad \mathbf{v}_2 = \frac{h}{2}\mathbb{A}^{-1}\mathbf{V}, \quad \mathbf{w} = \mathbb{D}^{-1}\mathbf{W}. \tag{5.132}$$

Of course, if also $\mathbf{V} = \mathbf{O}$, then we get immediately that $\mathbf{u} = \mathbf{v}_1 = \mathbf{v}_2 = \mathbf{w} = \mathbf{O}$, i.e. a laminate is thermally uncoupled if and only if $\mathbb{B} = \mathbb{O}$ *and* $\mathbf{V} = \mathbf{O}$.

What said above is valid in the most general case of non identical layers. When all the layers are equal, then the results are slightly different. This fact can be easily seen using the polar expression of $\mathbf{U}$, $\mathbf{V}$ and $\mathbf{W}$, Eqs. (5.129)–(5.131). In fact, if the layers are identical,

$$\mathbb{B} = \mathbb{O} \Rightarrow \xi_7 + i\xi_8 = 0 \Rightarrow \mathbf{V} = \mathbf{O}. \tag{5.133}$$

Hence, unlike the general case of non-identical layers, now elastic uncoupling implies thermal uncoupling: $\mathbb{B} = \mathbb{O} \Rightarrow \mathbf{V} = \mathbf{O}$ and, by Eq. (5.132), also $\mathbf{v}_1 = \mathbf{v}_2 = \mathbf{O}$.

The converse, however, is not true: $\mathbf{V}=\mathbf{O} \nRightarrow \mathbb{B}=\mathbb{O}$, because of Eq. (5.52)$_3$. Hence, it is possible to have laminates composed by identical layers that are coupled but thermally uncoupled. This property can have interesting practical applications, e.g. in realizing wind turbine blades with a passive control of the pitch, like sketched in Fig. 5.5, or other morphing devices; about the importance of thermal uncoupling for elastically coupled laminates, see Sect. 6.9.2.

In such a case of $\mathbb{B} \neq \mathbb{O}$ and $\mathbf{V}=\mathbf{O}$,

$$\begin{aligned}
\mathbf{u} &= \mathscr{A}\mathbf{U} = (\mathbb{A}-3\mathbb{B}\mathbb{D}^{-1}\mathbb{B})^{-1}\mathbf{U},\\
\mathbf{v}_1 &= \frac{2}{h}\mathscr{B}^\top\mathbf{U} = -\frac{6}{h}(\mathbb{D}-3\mathbb{B}\mathbb{A}^{-1}\mathbb{B})^{-1}\mathbb{B}\mathbb{A}^{-1}\mathbf{U},\\
\mathbf{v}_2 &= \frac{h}{6}\mathscr{B}\mathbf{W} = -\frac{h}{2}(\mathbb{A}-3\mathbb{B}\mathbb{D}^{-1}\mathbb{B})^{-1}\mathbb{B}\mathbb{D}^{-1}\mathbf{W},\\
\mathbf{w} &= \mathscr{D}\mathbf{W} = (\mathbb{D}-3\mathbb{B}\mathbb{A}^{-1}\mathbb{B})^{-1}\mathbf{W},
\end{aligned} \tag{5.134}$$

i.e. $\mathbf{v}_1$ and $\mathbf{v}_2$ do not vanish necessarily: a change of temperature t makes the plate warp but does not produce bending moments and a gradient ∇t stretches the plate but does not give rise to membrane forces.

5.4.5 *Thermally Quasi-homogeneous Laminates*

The concept of quasi-homogeneity can be extended to the thermal properties too: we say that a laminate is *thermally quasi-homogeneous* if

$$\mathbf{V}=\mathbf{Y}=\mathbf{O}, \tag{5.135}$$

where

$$\mathbf{Y}:=\mathbf{U}-\mathbf{W} \tag{5.136}$$

is the *thermal homogeneity tensor*. From Eq. (5.118) we get

$$\mathbf{Y}=\frac{1}{h^3}\sum_{k=1}^{n}[h^2(z_k-z_{k-1})-4(z_k^3-z_{k-1}^3)]\boldsymbol{\gamma}_k(\delta_k), \tag{5.137}$$

and for the case of identical layers

$$\mathbf{Y}=\sum_{k=1}^{n}c_k\boldsymbol{\gamma}_k(\delta_k), \tag{5.138}$$

with the coefficients c_k given by Eq. (5.45)$_3$ or by Eqs. (5.62)$_2$ and (5.63)$_2$. In this case, an interesting result can be readily found using the polar formalism. In fact, the polar components of $\mathbf{Y}$ are

$$\mathbf{Y} \rightarrow \begin{cases} T^Y = 0, \\ R^Y e^{2i\Phi^Y} = R^Y e^{2i\Phi^Y}(\xi_{11} + i\xi_{12}). \end{cases} \tag{5.139}$$

For a laminate composed of identical plies elastically quasi-homogenous we have already seen that $\mathbb{B} = \mathbb{O} \Rightarrow \mathbf{V} = \mathbf{O}$, Eq. (5.133); in addition,

$$\mathbb{C} = \mathbb{O} \rightarrow \xi_{11} + i\xi_{12} = 0 \Rightarrow \mathbf{Y} = \mathbf{O}. \tag{5.140}$$

Hence, elastic quasi-homogeneity implies thermal quasi-homogeneity, but the converse is not true, because of Eqs. (5.52)$_3$ and (5.53)$_3$: $\mathbf{Y} = \mathbf{O} \nRightarrow \mathbb{C} = \mathbb{O}$. In other words, it is possible to have laminates with an equal thermo-elastic anisotropic behavior in extension and in bending (i.e. with the coefficients of thermal expansion and curvature equal in all the directions) but a different in- and out of-plane elastic behaviors.

As a last remark, because for laminates with identical plies $\mathbb{B} = \mathbb{C} = \mathbb{O} \Rightarrow \mathbf{V} = \mathbf{Y} = \mathbf{O}$, the sequences in Tables 5.3, 5.4 and 5.5 are also thermally quasi-homogeneous, while those in Table 5.9 have equal and isotropic coefficients of thermal expansion and curvature.

5.5 Higher-Order Laminate Theories and the Polar Formalism

We have seen in Sect. 5.2.3 the most important drawback of the Kirchhoff kinematical assumptions for laminates: the impossibility to obtain the transversal shear components σ_5 and σ_5. This problem can be solved using higher order plates theories.

The most well-known higher-order theory is the so-called *Reissner–Mindlin theory*, after the name of the two scientists that independently proposed it, Reissner (1945), Mindlin (1951).[17] This theory is based upon a kinematical model that is the two-dimensional counterpart of the Timoshenko's beam theory; when applied to laminated plates, it is often called the *First-order Shear Deformation Theory*, FSDT in short; this model is briefly outlined in Sect. 5.5.1.

Though improving the results of the Kirchhoff model, the FSDT model is not yet satisfactory, mechanically speaking: if the predictions in terms of displacements are really improved with respect to the CLPT, which gives good results only for thin plates, some mechanical inconsistencies still subsist also in the FSDT. That is why finer laminate models and approaches have been proposed in the recent past. This topic is still an active research field and the literature on the matter is extremely huge, though too often proposing just small improvements of some numerical model.

[17] Strictly speaking, in the theories of Reissner and Mindlin the assumptions are not exactly the same; the subject has been widely treated in the literature, see e.g. Reddy (1999), Wang et al. (2001). However, the discrepancies are of a secondary importance, so that usually the theory goes under the name of Reissner–Mindlin plate theory.

It is not the objective of a text on anisotropy to enter deeply this subject, and the interested reader is addressed to the vast literature on the subject. A rigorous, recent textbook on heterogeneous anisotropic plates is Sab and Lebée (2016), where a modern introduction to CLPT and FSDT is also given. We just give in Sect. 5.5.2 a brief introduction to the Reddy's *Third-order Shear Deformation Theory*, TSDT in short, a well known model that improves the FSDT by a finer kinematical assumption.

In three interesting works appeared in 2015, Montemurro has extended the polar formalism to FSDT and TSDT, Montemurro (2015a, b, c), showing interesting properties of this method especially for what concerns the design procedures with higher order laminate theories. We briefly outline this approach in the next two Sections.

5.5.1 The First-Order Shear Deformation Theory of Laminated Plates

Basically, the difference between the CLPT and the FSDT is that any segment initially orthogonal to the plate's mid-plane still remains undeformed but no more perpendicular to the bent surface of the mid-plane. In this case, the displacement field (5.4) becomes, still in the assumption of small strains,

$$\begin{aligned} u(x_1,x_2,x_3)&=u_0(x_1,x_2)-x_3\beta_1(x_1,x_2),\\ v(x_1,x_2,x_3)&=v_0(x_1,x_2)-x_3\beta_2(x_1,x_2),\\ w(x_1,x_2)&=w_0(x_1,x_2), \end{aligned}\tag{5.141}$$

with β_1 and β_2 the rotation angles, respectively in the planes (x_1, x_3) and (x_2, x_3), of the rotated segment with respect to the axis x_3. Now, $\beta_1(x_1, x_2)$ and $\beta_2(x_1, x_2)$ are unknowns, together with $u_0(x_1, x_2)$, $v_0(x_1, x_2)$ and $w_0(x_1, x_2)$.

The consequence of Eq. (5.141) is that the strain field (5.5) becomes

$$\begin{aligned} \varepsilon_{11}&=\frac{\partial u_0(x_1,x_2)}{\partial x_1}-x_3\frac{\partial \beta_1(x_1,x_2)}{\partial x_1},\\ \varepsilon_{12}&=\frac{1}{2}\left[\frac{\partial u_0(x_1,x_2)}{\partial x_2}+\frac{\partial v_0(x_1,x_2)}{\partial x_1}-x_3\left(\frac{\partial \beta_1(x_1,x_2)}{\partial x_1}+\frac{\partial \beta_2(x_1,x_2)}{\partial x_2}\right)\right],\\ \varepsilon_{22}&=\frac{\partial v_0(x_1,x_2)}{\partial x_2}-x_3\frac{\partial \beta_2(x_1,x_2)}{\partial x_2},\\ \varepsilon_{13}&=\frac{1}{2}\left[\frac{\partial w_0(x_1,x_2)}{\partial x_1}-\beta_1(x_1,x_2)\right],\\ \varepsilon_{23}&=\frac{1}{2}\left[\frac{\partial w_0(x_1,x_2)}{\partial x_2}-\beta_2(x_1,x_2)\right],\\ \varepsilon_{33}&=0. \end{aligned}\tag{5.142}$$

So, we remark that now the shear strains ε_{13} and ε_{23} are *not* null and they are constant through the plate's thickness. In addition, now Eq. (5.8) is replaced by

$$\boldsymbol{\kappa}=\left\{\begin{array}{c}\kappa_1\\ \kappa_2\\ \kappa_6\end{array}\right\}=-\left\{\begin{array}{c}\frac{\partial\beta_1(x_1,x_2)}{\partial x_1}\\ \frac{\partial\beta_2(x_1,x_2)}{\partial x_2}\\ \sqrt{2}\left(\frac{\partial\beta_1(x_1,x_2)}{\partial x_2}+\frac{\partial\beta_2(x_1,x_2)}{\partial x_1}\right)\end{array}\right\}. \tag{5.143}$$

The definition of the curvature tensor $\boldsymbol{\kappa}$ is different, but proceeding like in Sects. 5.2.4 and 5.2.5 one easily realizes that nothing has changed for tensors $\mathbb{A}$, $\mathbb{B}$ and $\mathbb{D}$ defining the mechanical behavior of the laminate: the anisotropy of the laminate is hence independent from the kinematical plate model, of course, as announced in the Introduction of this chapter, Sect. 5.1.

However, because now $\varepsilon_{13}\neq 0$ and $\varepsilon_{23}\neq 0$, transversal shear forces are present. In fact, through the constitutive law we get, for the $k-$th ply,

$$\left\{\begin{array}{c}\sigma_4\\ \sigma_5\end{array}\right\}_k=\left[\begin{array}{cc}C'_{44} & C'_{45}\\ C'_{45} & C'_{55}\end{array}\right]_k\left\{\begin{array}{c}\varepsilon_4\\ \varepsilon_5\end{array}\right\}. \tag{5.144}$$

with C'_{44}, C'_{45} and C'_{55} given by Eq. $(5.10)_{3,4,5}$ with $\theta=-\delta_k$.

This relation shows that the shear stresses are discontinuous at the ply interfaces, because generally speaking

$$\left[\begin{array}{cc}C'_{44} & C'_{45}\\ C'_{45} & C'_{55}\end{array}\right]_k\neq\left[\begin{array}{cc}C'_{44} & C'_{45}\\ C'_{45} & C'_{55}\end{array}\right]_{k+1}, \tag{5.145}$$

as the ply materials can be different or simply because $\delta_k\neq\delta_{k+1}$. In addition, because $\varepsilon_4=\sqrt{2}\varepsilon_{23}$ and $\varepsilon_5=\sqrt{2}\varepsilon_{13}$ are constant for each layer, σ_4 and σ_5 are constant too. These two results are mechanical inconsistencies of the FSDT: constant transversal shear stresses do not satisfy the local equilibrium equations, and in addition they should be continuous at the interfaces and null on the upper and lower surfaces, in absence of tangential surface loads.

A possible, classically used, partial remedy to the above inconsistencies, Reddy (2003), consists in introducing a *corrective function* $\zeta(x_3)$ quadratic in x_3; normally,

$$\zeta(x_3)=\frac{5}{4}\left[1-\left(\frac{2\,x_3}{h}\right)^2\right]. \tag{5.146}$$

In this way, local equilibrium and vanishing of the shear stresses at the lower and upper plate's surfaces are satisfied, but the continuity at the ply interfaces *not* yet.

With the introduction of the corrective function $\zeta(x_3)$, after integration of σ_4 and σ_5 on the plate's thickness we get the following constitutive law for the transversal shear forces:

$$\left\{\begin{array}{c}F_4\\ F_5\end{array}\right\}=h\left[\begin{array}{cc}H_{44} & H_{45}\\ H_{45} & H_{55}\end{array}\right]\left\{\begin{array}{c}\varepsilon_4\\ \varepsilon_5\end{array}\right\}, \tag{5.147}$$

where

$$H_{ij} = \frac{5}{4h}\sum_{k=1}^{n}\left[z_k - z_{k-1} - \frac{4}{3}\frac{z_k^3 - z_{k-1}^3}{h^2}\right](C'_{ij})_k, \quad i,j = 4,5. \tag{5.148}$$

For laminates composed of identical plies, the last equation becomes

$$H_{ij} = \sum_{k=1}^{n} h_k C'_{ij}, \quad i,j = 4,5, \tag{5.149}$$

with

$$h_k = \frac{5}{12}\left(\frac{3}{n} - d_k\right), \tag{5.150}$$

with the d_ks given by Eq. (5.45)$_4$. Equation (5.147) completes the constitutive law of laminates (5.21) in the FSDT; it can be rewritten in compact form as

$$\{F\} = h[H]\{\gamma_0\}, \tag{5.151}$$

with a clear correspondence of the symbols; $\{\gamma_0\}$ is called the *transverse shear strain vector of the laminate*. Equations (5.10)$_{3,4,5}$ show that the $C'_{ij},\ i,j = 4,5$ behave like the components of a second-rank tensor when rotated of $-\theta$ about the axis x_3. As a consequence, this is true also for $[H]$ so that, passing to the polar formalism, we can write

$$\begin{aligned} H_{44} &= T^H + R^H \cos 2\Phi^H, \\ H_{45} &= R^H \sin 2\Phi^H, \\ H_{55} &= T^H - R^H \cos 2\Phi^H, \end{aligned} \tag{5.152}$$

with, for the case of identical plies,

$$\begin{aligned} T^H &= \frac{5}{6}T^C, \\ R^H e^{2i\Phi^H} &= R^C e^{2i\Phi^C}(\xi_{17} + i\xi_{18}). \end{aligned} \tag{5.153}$$

In the last equation, the superscript C indicates a polar component of the matrix $[C'_{ij}],\ i,j = 4,5$, in eq.while

$$\xi_{17} + i\xi_{18} = \sum_{k=1}^{n} h_k e^{-2i\delta_k} \tag{5.154}$$

defines the two additional lamination parameters for $[H]$. Montemurro (2015a, b), has shown that (the superscript L indicates, as usual, a quantity of the basic layer)

$$R^H e^{2i\Phi^H} = \frac{R^C}{R_1^L} e^{2i(\Phi^C - \Phi_1^L)} (3R_1^A e^{2i\Phi_1^A} - R_1^D e^{2i\Phi_1^D}). \tag{5.155}$$

This relation links the anisotropic part of $[H]$ with those of $\mathbb{A}$ and $\mathbb{D}$, i.e. the transverse anisotropy of the laminate with the in-plane one, in the polar formalism. It is rather interesting because in design problems this allows to continue to use planar polar variables, i.e. those of $\mathbb{A}$, $\mathbb{B}$ and $\mathbb{D}$, also when the FSDT is used. In fact, the terms T^C, R^C, R_1^L, Φ^C and Φ_1^L are fixed quantities, once the basic layer chosen, so finally the anisotropic part of $[H]$ is simply a linear combination of those of $\mathbb{A}$ and $\mathbb{D}$ with index 1 and the number of design variables is not increased passing from the CLPT to the FSDT. Different numerical cases are treated with this approach in Montemurro (2015a).

5.5.2 The Third-Order Shear Deformation Theory of Laminated Plates

The TSDT has been introduced and successively developed by Reddy (1984, 1987, 1990, 2003). The objective of the TSDT, and also of other higher order laminate theories, is to remedy to the inconsistencies of the FSDT seen in the previous Section without introducing an artificial corrective function like $\zeta(x_3)$. The drawback is that this can be done to the price of the introduction of higher order stress resultants and stiffness matrices whose physical meaning is not evident nor immediate.

Basically, in the TSDT it is assumed that the orthogonal segment to the mid-plane is not only rotated, but also deformed according to a cubic law. Because strains, and hence stresses, depend upon the derivatives of the displacement field, they are quadratic in x_3, which corresponds to the exact solution for isotropic homogeneous plates. Without entering in the details, that are rather technical, we bound ourselves to a rapid survey of the TSDT and to the use of the polar formalism in this case.

The displacement field is assumed to be of the type

$$\begin{aligned} u(x_1, x_2, x_3) &= u_0(x_1, x_2) + \beta_1(x_1, x_2)\left(x_3 - \frac{4x_3^3}{3h^2}\right) - \frac{4x_3^3}{3h^2}\frac{\partial w_0}{\partial x_1},\\ v(x_1, x_2, x_3) &= v_0(x_1, x_2) + \beta_2(x_1, x_2)\left(x_3 - \frac{4x_3^3}{3h^2}\right) - \frac{4x_3^3}{3h^2}\frac{\partial w_0}{\partial x_2},\\ w(x_1, x_2) &= w_0(x_1, x_2). \end{aligned} \tag{5.156}$$

To remark that with this displacement field, the requirement of traction-free boundaries for the lower and upper surfaces of the plate is met automatically.

Moreover, the constitutive law can be written as[18]

[18] With a slight abuse of notation, and for the sake of simplicity, we indicate matrices by the symbols usually reserved to tensors; here, however, $\mathbb{E}$ is *not* the elastic tensor of the material.

$$\left\{\begin{array}{c}\mathbf{N}\\\hline\mathbf{M}\\\hline\mathbf{P}\end{array}\right\}=\left[\begin{array}{c|c|c}h\mathbb{A}&\frac{h^2}{2}\mathbb{B}&\frac{h^4}{8}\mathbb{E}\\\hline\frac{h^2}{2}\mathbb{B}&\frac{h^3}{12}\mathbb{D}&\frac{h^5}{80}\mathbb{F}\\\hline\frac{h^4}{8}\mathbb{E}&\frac{h^5}{80}\mathbb{F}&\frac{h^7}{448}\mathbb{G}\end{array}\right]\left\{\begin{array}{c}\boldsymbol{\varepsilon}^0\\\hline\boldsymbol{\varepsilon}^1\\\hline\boldsymbol{\varepsilon}^3\end{array}\right\},$$
$$\left\{\begin{array}{c}\mathbf{F}\\\hline\mathbf{G}\end{array}\right\}=\left[\begin{array}{c|c}h\mathbb{R}&\frac{h^3}{12}\mathbb{S}\\\hline\frac{h^3}{12}\mathbb{S}&\frac{h^5}{80}\mathbb{T}\end{array}\right]\left\{\begin{array}{c}\boldsymbol{\gamma}^0\\\hline\boldsymbol{\gamma}^2\end{array}\right\}. \tag{5.157}$$

$\mathbb{A}$, $\mathbb{B}$ and $\mathbb{D}$ are still defined like in Sect. 5.2.5, while, for the case of identical layers,

$$\mathbb{E}=\sum_{k=1}^{n}e_k\mathbb{Q}(\delta_k),\ \mathbb{F}=\sum_{k=1}^{n}f_k\mathbb{Q}(\delta_k),\ \mathbb{G}=\sum_{k=1}^{n}g_k\mathbb{Q}(\delta_k),$$
$$\mathbb{R}=\sum_{k=1}^{n}a_k\mathbb{C}'(\delta_k),\ \mathbb{S}=\sum_{k=1}^{n}d_k\mathbb{C}'(\delta_k),\ \mathbb{T}=\sum_{k=1}^{n}f_k\mathbb{C}'(\delta_k), \tag{5.158}$$

where $\mathbb{C}'$ denotes the matrix $[C'_{ij}]$, $i, j = 4, 5$, in Eq. (5.144). Coefficients a_k and d_k are still given by Eq. $(5.45)_{1,4}$ while

$$\begin{aligned}
e_k &= \frac{1}{n^4}[8k^3-12k^2(n+1)+2k(3n^2+6n+4)-(n^3+3n^2+4n+2)],\\
f_k &= \frac{1}{n^5}[80k^4-160k^3(n+1)+40k^2(3n^2+6n+4)-\\
&\quad 40k(n^3+3n^2+4n+2)+5n^4+20n^3+40n^2+40n+16],\\
g_k &= \frac{1}{n^7}[448k^6-1344k^5(n+1)+560k^4(3n^2+6n+4)-\\
&\quad 1120k^3(n^3+3n^2+4n+2)+84k^2(5n^4+20n^3+40n^2+40n+16)-\\
&\quad 28(3n^5+15n^4+40n^3+60n^2+48n+16)+7n^6+42n^5+\\
&\quad 140n^4+280n^3+336n^2+224n+64].
\end{aligned} \tag{5.159}$$

Montemurro (2015c), has expressed the above tensors in the polar formalism; the result is

$$\mathbb{E}\rightarrow\begin{cases}T_0^E=0,\\ T_1^E=0,\\ R_0^Ee^{4i\Phi_0^E}=R_0^Le^{4i\Phi_0^L}\displaystyle\sum_{k=1}^{n}e_ke^{4i\delta_k},\\ R_1^Ee^{2i\Phi_1^E}=R_1^Le^{2i\Phi_1^L}\displaystyle\sum_{k=1}^{n}e_ke^{2i\delta_k};\end{cases} \tag{5.160}$$

$$\mathbb{F} \rightarrow \begin{cases} T_0^F = T_0^L, \\ T_1^F = T_1^L, \\ R_0^F e^{4i\Phi_0^F} = R_0^L e^{4i\Phi_0^L} \sum_{k=1}^{n} f_k e^{4i\delta_k}, \\ R_1^F e^{2i\Phi_1^F} = R_1^L e^{2i\Phi_1^L} \sum_{k=1}^{n} f_k e^{2i\delta_k}; \end{cases} \tag{5.161}$$

$$\mathbb{G} \rightarrow \begin{cases} T_0^G = T_0^L, \\ T_1^G = T_1^L, \\ R_0^G e^{4i\Phi_0^G} = R_0^L e^{4i\Phi_0^L} \sum_{k=1}^{n} g_k e^{4i\delta_k}, \\ R_1^G e^{2i\Phi_1^G} = R_1^L e^{2i\Phi_1^L} \sum_{k=1}^{n} g_k e^{2i\delta_k}; \end{cases} \tag{5.162}$$

$$\mathbb{R} \rightarrow \begin{cases} T^R = T^C, \\ R^R e^{2i\Phi_1^R} = R^C e^{2i\Phi^C} \sum_{k=1}^{n} a_k e^{-2i\delta_k}; \end{cases} \tag{5.163}$$

$$\mathbb{S} \rightarrow \begin{cases} T^S = T^C, \\ R^S e^{2i\Phi_1^S} = R^C e^{2i\Phi^C} \sum_{k=1}^{n} d_k e^{-2i\delta_k}; \end{cases} \tag{5.164}$$

$$\mathbb{T} \rightarrow \begin{cases} T^T = T^C, \\ R^T e^{2i\Phi_1^T} = R^C e^{2i\Phi^C} \sum_{k=1}^{n} f_k e^{-2i\delta_k}; \end{cases} \tag{5.165}$$

We can remark that the isotropic part of $\mathbb{E}$ is null, like for $\mathbb{B}$, while that of $\mathbb{F}$ and $\mathbb{G}$ is equal to that of the basic layer, like for $\mathbb{A}$ and $\mathbb{D}$.

In addition, similarly to what happens in the FSDT, Montemurro has proved that

$$\begin{aligned} R^R e^{2i\Phi_1^R} &= \frac{R^C}{R_1^L} R_1^A e^{2i(\Phi^C+\Phi_1^L-\Phi_1^A)}, \\ R^S e^{2i\Phi_1^S} &= \frac{R^C}{R_1^L} R_1^D e^{2i(\Phi^C+\Phi_1^L-\Phi_1^D)}, \\ R^T e^{2i\Phi_1^T} &= \frac{R^C}{R_1^L} R_1^F e^{2i(\Phi^C+\Phi_1^L-\Phi_1^F)}. \end{aligned} \tag{5.166}$$

Hence, once more, in a design procedure only the design variables defining the anisotropy in the plane of the plate are needed, not also those describing the transversal properties, that can be expressed by the (5.166), see Montemurro (2015c) for more details and numerical examples.

Before ending with this short presentation of higher order laminate theories, a last remark: the reader has certainly remarked how much theoretical developments, and also numerical calculations, become harder and harder increasing the order of the theory. If this is true for isotropic homogeneous plates, this is much more evident for anisotropic laminates.

References

P. Bartholomew, Ply stacking sequences for laminated plates having in-plane and bending orthotropy. Technical report 76003, Royal Aircraft Establishment, Farnborough, UK (1976)

P. Bartholomew, Ply stacking sequences for laminated plates having in-plane and bending orthotropy. Fibre Sci. Technol. **10**, 239–253 (1977)

F. Belaïd, P. Vannucci, G. Verchery, Numerical investigation of the influence of orientation defects on bending-tension coupling of laminates, in *Proceedings of ICCM13 - International Conference on Composite Materials* (Beijing, PRC, 2001)

C. Caprino, I. Crivelli-Visconti, A note on specially orthotropic laminates. J. Compos. Mater. **16**, 395–399 (1982)

A. Catapano, B. Desmorat, P. Vannucci, Stiffness and strength optimization of the anisotropy distribution for laminated structures. J. Optim. Theory Appl. **167**, 118–146 (2015)

C.G. Diaconu, M. Sato, H. Sekine, Feasible region in general design space of lamination parameters for laminated composites. AIAA J. **40**, 559–565 (2002a)

C.G. Diaconu, M. Sato, H. Sekine, Layup optimization of symmetrically laminated thick plates for fundamental frequencies using lamination parameters. Struct. Multidiscip. Optim. **24**, 302–311 (2002b)

J.P. Foldager, J.S. Hansen, N. Olhoff, A general approach forcing convexity of ply angle optimization in composite laminates. Struct. Optim. **16**, 201–211 (1998)

J.P. Foldager, J.S. Hansen, N. Olhoff, A convex formulation in ply-angle ptimization, in *Proceedings of WCSMO3 - World Congress on Structural and Multidisciplinary Optimization* (Buffalo - NY, 1999)

H. Fukunaga, On isotropic laminate configurations. J. Compos. Mater. **24**, 519–535 (1990)

H. Fukunaga, H. Sekine, Stiffness design method of symmetric laminates using lamination parameters. AIAA J. **30**, 2791–2793 (1992)

H. Fukunaga, G.N. Vanderplaats, Stiffness optimization of orthotropic laminated composites using lamination parameters. AIAA J. **29**, 641–646 (1991)

H. Fukunaga, H. Sekine, M. Sato, Optimum design of symmetric laminated plates for fundamental frequency. J. Sound Vib. **171**, 219–229 (1994)

M. Grédiac, Un programme de conception de stratifiés presentant des rigidités données, in *Proc. of JNC11 - Journées Nationales sur les Composites 11*, vol. 1 (Arcachon, France, 1998), pp. 197–204

M. Grédiac, Design of laminated plates with required in-plane, coupling and bending stiffnesses using non-standard ply angles, in *Proceedings of ICCM12 - International Conference on Composite Materials* (Paris, France, 1999a)

M. Grédiac, A procedure for designing laminated plates with required stiffness properties. Application to thin quasi-isotropic quasi-homogeneous uncoupled laminates. J. Compos. Mater. **33**, 1939–1956 (1999b)

M. Grédiac, On the stiffness design of thin woven composites. Compos. Struct. **51**, 245–255 (2001)
J.L. Grenestedt, P. Gudmundson, Layup optimization of composite material structures, in *IUTAM Symposium on Optimal Design with Advanced Materials*, ed. by P. Pedersen (Elsevier, Amsterdam, The Netherlands, 1993), pp. 311–336
V.B. Hammer, M.P. Bendsoe, R. Lipton, P. Pedersen, Parametrization in laminate design for optimal compliance. Int. J. Solids Struct. **34**, 415–434 (1997)
A. Jibawy, C. Julien, B. Desmorat, A. Vincenti, F. Léné, Hierarchical structural optimization of laminated plates using polar representation. Int. J. Solids Struct. **48**, 2576–2584 (2011)
N. Kandil, G. Verchery, New methods of design for stacking sequences of laminates, in *Proceedings of CADCOMP88 - Computer Aided Design in Composite Materials 88* (Southampton, UK, 1988), pp. 243–257
S.G. Lekhnitskii, Theory of elasticity of an anisotropic elastic body. English translation (1963) by P. Fern (Holden-Day, San Francisco, CA, 1950)
M. Miki, Material design of composite laminates with required in-plane elastic properties, in *Proceedings of ICCM 4 - Fourth International Conference on Composite Materials* (Tokio, Japan, 1982), pp. 1725–1731
M. Miki, A graphical method for designing fibrous laminated composites with required in-plane stiffness. Trans. Japanese Soc. Compos. Mater. **9**, 51–55 (1983)
M. Miki, Design of laminated fibrous composite plates with required flexural stiffness, in *Recent advances in composites in the USA and Japan - ASTM STP 864*, ed. by J.R. Vinson, M. Taya (American Society for Testing and Materials, Philadelphia, PA, 1985), pp. 387–400
R.D. Mindlin, Influence of rotatory inertia ans shear on flexural motions of isotropic, elastic plates. J. Appl. Mech. **18**, 31–38 (1951)
M. Montemurro, An extension of the polar method to the First-order shear deformation theory of laminates. Compos. Struct. **127**, 328–339 (2015a)
M. Montemurro, Corrigendum to "An extension of the polar method to the First-order shear deformation theory of laminates". Compos. Struct. **131**, 1143–1144 (2015b)
M. Montemurro, The polar analysis of the third-order shear deformation theory of laminates. Compos. Struct. **131**, 775–789 (2015c)
M. Montemurro, A. Vincenti, P. Vannucci, A two-level procedure for the global optimum design of composite modular structures - application to the design of an aircraft wing. Part 1: theoretical formulation. J. Optim. Theory Appl. **155**, 1–23 (2012a)
M. Montemurro, A. Vincenti, P. Vannucci. A two-level procedure for the global optimum design of composite modular structures - application to the design of an aircraft wing. Part 2. J. Optim. Theory Appl. **155**, 24–53 (2012b)
R. Paradies, Designing quasi-isotropic laminates with respect to bending. Compos. Sci. Technol. **56**, 461–472 (1996)
P. Pedersen, Combining material and element rotation in one formula. Commun. Appl. Numer. Methods **6**, 549–555 (1990)
J.N. Reddy, A simple higher order theory for laminated composite plates. J. Appl. Mech. **51**, 745–752 (1984)
J.N. Reddy, A small strain and moderate rotation theory of laminated anisotropic plates. J. Appl. Mech. **54**, 623–626 (1987)
J.N. Reddy, A general non-linear third-order theory of plates with moderate thickness. Int. J. Nonlinear Mech. **25**, 677–686 (1990)
J.N. Reddy, *Theory and Analysis of Leastic Plates* (Taylor & Francis, Philadelphia, PA, 1999)
J.N. Reddy, *Mechanics of Composite Laminated Plates and Shells: Theory and Analysis* (CRC Press, Boca Raton, FL, 2003)
E. Reissner, The effect of transverse shear deformation on the bending of elastic plates. J. Appl. Mech. **12**, 68–77 (1945)
K. Sab, A. Lebée, *Homogenization of Heterogeneous Thin and Thick Plates* (Wiley, New York, NY, 2016)
S.W. Tsai, T. Hahn, *Introduction to Composite Materials* (Technomic, Stamford, CT, 1980)

E. Valot, *Anisotropie du comportement à rupture de composites à structure isotrope*. PhD thesis, ISAT - University of Burgundy, Nevers, France (2002)
E. Valot, P. Vannucci, Some exact solutions for fully orthotropic laminates. Compos. Struct. **69**, 157–166 (2005)
P. Vannucci, On bending-tension coupling of laminates. J. Elast. **64**, 13–28 (2001)
P. Vannucci, A note on the elastic and geometric bounds for composite laminates. J. Elast. **112**, 199–215 (2013)
P. Vannucci, G. Verchery, Caractérisation de l'isotropie et autres propriétés des stratifiés comme problème de minimum, in *Proceedings of JNC 12* (Cachan, France, 2000), pp. 333–342
P. Vannucci, G. Verchery, A special class of uncoupled and quasi-homogeneous laminates. Compos. Sci. Technol. **61**, 1465–1473 (2001a)
P. Vannucci, G. Verchery, Stiffness design of laminates using the polar method. Int. J. Solids Struct. **38**, 9281–9294 (2001b)
P. Vannucci, G. Verchery, A new method for generating fully isotropic laminates. Compos. Struct. **58**, 75–82 (2002)
P. Vannucci, X.J. Gong, G. Verchery, Determination des stratifies quasi-homogènes par l'approche polaire, in *Proc. of JNC11 - Journées Nationales sur les Composites* (Arcachon, France, 1998)
G. Verchery, Search of simple criteria for designing composite structures, in *Proceedings of CDW-12* (Saint-Maximin-la-Saint-Baume, France, 2007)
G. Verchery, Design rules for the laminate stiffness. Mech. Compos. Mater. **47**, 47–58 (2011)
G. Verchery, T.S. Vong, Une méthode d'aide graphique à la conception des séquences d'empilement dans les stratifiés, in *Proc. of JNC5 - Cinquièmes Journées Nationales sur les Composites* (Paris, France, 1986), pp. 267–280
G. Verchery, P. Vannucci, V. Person, The polar method as a tool for solving inverse problems of the classical laminated plate theory, in *Inverse Problems in Engineering Mechanics II*, ed. by M. Tanaka, G.S. Dulikravich (Elsevier Science, UK, 2000), pp. 317–326
P. Villaggio, *Mathematical Models for Elastic Structures* (Cambridge University Press, Cambridge, UK, 1997)
A. Vincenti, P. Vannucci, G. Verchery, Influence of orientation errors on quasi-homogeneity of composite laminates. Compos. Sci. Technol. **63**, 739–749 (2003)
A. Vincenti, P. Vannucci, M.R. Ahmadian, Optimization of laminated composites by using genetic algorithm and the polar description of plane anisotropy. Mech. Adv. Mater. Struct. **20**, 242–255 (2013)
C.M. Wang, G.T. Lim, J.N. Reddy, K.H. Lee, Relationships between bending solutions of Reissner and Mindlin plate theories. Eng. Struct. **23**, 838–849 (2001)
F. Werren, C.B. Norris, Mechanical properties of a laminate designed to be isotropic. Technical Report, US Forest Products Laboratory, Madison. W **I**, 1953 (1841)
K.M. Wu, Isotropic composite plates. Technical Report EM-429, General Motors Research Laboratories, Warren, Michigan, USA (1979)
K.M. Wu, B.L. Avery, Fully isotropic laminates and quasi-homogeneous laminates. J. Compos. Mater. **26**, 210–2117 (1992)
C.B. York, Characterization and ply mixing rules for non-symmetric forms of fully orthotropic laminates, in *Proceedings of 48th AIAA/ASME/ASCE/AHS/ASC Structures, Structural Dynamics, and Materials Conference* (Honolulu, HA, 2007)
C.B. York, Stacking sequences for extensionally isotropic, fully isotropic and quasi-homogeneous orthotropic laminates, in *Proceedings of 49th AIAAA/ASME/ASCE/AHS/ASC Structures, Structural Dynamics and Materials* (Schaumburg, IL, 2008)
C.B. York, Characterization of nonsymmetric forms of fully orthotropic laminates. J. Aircraft **46**(4), 1114–1125 (2009)
C.B. York, Unified approach to the characterization of coupled composite laminates: Benchmark configurations and special cases. J. Aerosp. Eng. **23**(4), 219–242 (2010)
C.B. York, Coupled quasi-homogeneous orthotropic laminates. Mech. Compos. Mater. **47**, 405–426 (2011)

Chapter 6
Design Problems and Methods of Anisotropic Structures

Abstract An introduction to some problems and methods of optimal design of anisotropic structures is presented in this chapter. The point of view is rather unconventional, some recent developments are presented, several examples are included in the chapter and the polar formalism is widely used. An original presentation of the optimal material orientation is first given. Then, a classification of anisotropic structure design problems in three different types is proposed, and the influence of the material anisotropy on the optimal solutions analyzed. A unified polar formulation of the problems of the first type is given and the importance of the non uniqueness of the stacking sequence argued. For the second-type problems, some considerations on the constraint handling are proposed while the problems of the third type are formulated as two subsequent and related problems: the structural and the constitutive law problem. The optimization of the anisotropy fields and that of modular systems are then treated. The numerical tools are also introduced, with a particular attention to metaheuristics. The chapter ends with some multiphysics problems: the tailoring of the thermo-elastic and piezoelectric properties and the thermally stable laminates.

6.1 Introduction

The design of anisotropic laminates is quite different from that of classical isotropic plates. It is actually more complicate, because of the number of parameters to be determined: in the design of a steel plate, the unique parameter to be fixed is its thickness, while when the same plate must be realized using an anisotropic laminate, the engineer, once decided the material to be employed, must determine not only the thickness, and hence the number of layers, but also the orientation of all the layers, what is called the *stacking sequence*.

It is important to understand that designing the mechanical properties of a laminate does not only mean to dimension or optimize a given parameter, e.g. a natural vibration frequency or a buckling load, but also to fix the global mechanical response of the structure: the type of anisotropy, the kind of behavior for extension and bending, that are normally different, the bending-extension coupling, usually unwanted, etc.

P. Vannucci, *Anisotropic Elasticity*, Lecture Notes in Applied and Computational Mechanics 85, DOI 10.1007/978-981-10-5439-6_6

In addition, the mechanical performances required to such structures are higher and higher in modern applications.

Faced with so complicated problems, the common strategy of engineers has been *simplification*: laminates to be used have been restricted, also for technological reasons, to few possible classes, each one of them possessing automatically some general properties. The most well known example is the class of symmetric stacks, automatically ensuring bending-extension uncoupling. In this way, the design problems are normally drastically simplified and in some cases solutions can be found rather easily.

A huge number, almost the totality, of research works on laminates makes use of this sort of *shortcuts* to design; such simplifying rules have largely influenced the work of engineers and researchers, to such an extent that these *shortcuts*, always just sufficient and never necessary conditions to obtain a given property, have influenced also technical norms, especially for aircraft constructions.

Unfortunately, the most part of times the mathematical effect of such *shortcuts* is that of shrinking very drastically the set of feasible solutions, so much that in several cases the true optimal solution remains excluded. Almost never solutions considered as optimal ones, and often presented in research works, are really optimal if they have been calculated adopting simplifying rules. A classical example is that of the minimum weight design: when this is done assuming only symmetric stacks, with the purpose of eliminating bending-extension coupling, the solution is hardly the lightest one, because the sequence is doubled to ensure symmetry.

This kind of approach has largely influenced the research on the mechanical design of laminates, because it has dismissed the researchers' attention from a fundamental problem: is it possible to formulate, in the most general way, i.e. without facilitating a priori *shortcuts*, the design problem of a laminate? And, once this done, how the solution can be found?

Because of the difficulty of the matter, engineers and researchers need to have recourse to a robust, general mathematical approach. The only possible one, able to preserve the generality of the problem and to give a precise theoretical frame, with experimented and reliable methods, is that of *structural optimization*. Each laminate design problem must be stated as an optimum one, typically a minimum problem, which is, in the end, the real purpose of any design process. Once the problem put in this form, the designer has the possibility of finding its real optimal solution.

As said above, historically this has not been the way normally followed by engineers and researchers, and industry usually was satisfied by sufficiently, suboptimal, appropriate solutions. Few are hence the works in the direction sketched before: formulate a design problem without facilitating *shortcuts* and resolving it as a minimum problem. The following of this chapter is exactly devoted to give an account of the research done in this direction. Some methods and examples of problems concerning laminates formulated and solved in the most general way, using current or dedicated structural optimization methods, are presented and discussed.

The literature on laminates design is, as already said, very wide and composed for the most part of times of papers that are merely technical contributions. A rather recent state of the art on stiffness optimization for laminates, though far from being complete, can be found in the two papers from Ghiasi et al. (2009, 2010). A still interesting panorama on the matter is also given in Abrate (1994), while a classical and rather complete textbook on laminates design is the one from Gürdal et al. (1999).

In this chapter, we begin considering a basic and general problem of anisotropy design, that of the optimal directions of anisotropy. Then, we propose a classification of design problems of laminates, along with a unified formulation for a wide class of them. We pass afterwards to introduce some considerations about the numerical methods and we tackle two more recent problems: the optimization of anisotropy fields, i.e. the case of structures whose anisotropy is a function of the position, and that of modular structures. We then end the chapter with some problems concerning multiphysics properties.

All the examples in this chapter are solved using the polar formalism. It is just as a mathematical tool for the design of the mechanical properties that this method gives its most important contribution: using as design parameters exclusively invariants and angles reveals to be a true advantage for the optimal design of plane anisotropic structures. For the case of laminates design, the polar method reveals to be even more interesting, for three different reasons: (*i*) the polar invariants help in representing effectively symmetries and in stating them, (*ii*) some of the polar invariants are preserved by the laminate, as we have seen in the previous chapter, (*iii*) the separation between the isotropic and anisotropic phases allows for better understanding the design process and to eliminate redundant design variables.

6.2 A Basic Problem: The Optimal Orientation of Anisotropy

The very peculiar property of anisotropic bodies is that their mechanical response is direction dependent. Hence, the following question is quite natural: in which way to orientate anisotropy?

More specifically, for a given problem, i.e. for a strain or stress state, which is the best way to direct the anisotropy axes, e.g. the orthotropy directions for an orthotropic body, in order to obtain the best mechanical response? The answer to this question depends, of course, upon the criterion, i.e. upon the property to be optimized. The most common criteria are the stiffness and the strength, the two cases considered also in this chapter, for which some general results can be found, but other criteria can be considered too, e.g. vibration frequencies, buckling loads, fatigue strength.

6.2.1 A Short Account of the State of the Art

The problem of orientating planar anisotropy has been tackled first in the pioneer works of Banichuk (1981, 1983), concerning the optimal shape and anisotropy of a twisted beam and the orientation of an anisotropic body in planar elasticity. In the same period, a fundamental property of optimal anisotropic structures with respect to stiffness has been proved in Seregin and Troitskii (1981), for bodies having some elastic symmetry, using the technique of Lagrangian multipliers: stress and strain must be coaxial (using the polar formalism, this is easily proved to be true in 2D for any kind of elastic tensor, also completely anisotropic, cf. Sect. 6.2.2.3). A fundamental contribution has been given by Kartvelishvili and Kobelev (1984); their paper concerns the maximization of the stiffness and of the natural frequencies for bent plates, and in this article a classical condition of stationarity of the strain energy, already found by Banichuk (1983, p. 196), appears for the first time in a form currently used afterwards.

In the field of 3D anisotropy, Cowin (1994), looked for the stationarity of the strain energy in a set of orientations and showed that the cubic symmetry is the only syngony for which a stationary point can exist for each possible stress state, while stationarity depends upon the stress state for the other cases of anisotropy. Successively, Vianello (1996a) and Sgarra and Vianello (1997a, b) showed the existence of at least two rotations of the material axes, with respect to the principal directions of strain, that render stationary the strain energy for any syngony. This only apparently contradicts the result of Cowin, because in these two papers the strain state is fixed, while in the work of Cowin, a general strain state is considered (Cowin 1997). Later, Cowin and Yang (2000), considered the problem of the stationary points of the strain energy with respect to the material orientation and syngony for an arbitrary stress state. Special classes of elastic symmetries have been considered by Rovati and Taliercio (2003). In 1996, Banichuk (1996), used spectral methods of tensor analysis for discussing general qualitative properties of the solutions not only of stationary strain energy, but also of dynamic stiffness and distortion energy.

The extension to finite elasticity has been considered by Blume (1994), Vianello (1996b), and Majak and Pohlak (2010b). In particular, Vianello showed that the properties of stationary points do not change with respect to the linear case.

Due to the practical importance, many researches concentrated mainly on the plane problem of the optimal orientation of orthotropic layers. In this field a seminal paper has been published in 1989 by Pedersen (1989), further developed in other contributions, e.g. Sacchi-Landriani and Rovati (1991), Sacchi-Landriani et al. (1993), Rovati and Taliercio (1991), Majak (2001), where non-orthotropic planar materials are studied (Majak and Pohlak 2010a; Vincenti and Desmorat 2011), this last being the only paper using the polar method.

Less treated is the problem of finding the material orientations that maximizes the strength. In Majak and Hannus (2003), Majak and Hannus formulated the Tsai–Hill and Tsai–Wu failure criteria in terms of strains for 3D and 2D orthotropic materials. In both the cases, the failure index was assumed as the objective function and an analytical method, in the case of 2D orthotropic materials, was proposed to calculate its minimum. In Majak and Pohlak (2010a) the authors show that if the strength criterion is based upon an elasticity-like tensor, e.g. the Hill criterion or the Tsai–Hill one, then the maximization of the strength is formally analogous to that of the stiffness. A more complete work in the field is that of Catapano et al. (2012), concerning orthotropic layers. This approach, based upon the polar formalism applied to strength criteria, cf. Sect. 4.17.3, is briefly recalled in Sect. 6.2.3.

In the next sections, we bound ourselves to consider the general problem of the local optimal orientation of the anisotropy in the planar case. The approach is based upon the polar formalism and concerns completely anisotropic, specially orthotropic and ordinarily orthotropic layers. We prefer first to introduce the general case of completely anisotropic 2D bodies, which can be the case, if not of single layers, of laminated plates and then to pass, as a special case, to the most interesting case of orthotropic bodies. This gives a more general insight in the matter.

6.2.2 A Polar Approach to the Maximization of the Stiffness

The approach followed in this section rephrases, in an original way, the papers of Pedersen (1989) and, in part, of Vincenti and Desmorat (2011). The use of the polar method allows to find easily and directly the fundamental results, while an original point of view on the matter is given.

6.2.2.1 Formalizing the Optimum Problem

We consider the problem of orientating the material in such a way to maximize the local stiffness.[1] This can be formalized classically as a problem of minimum, see e.g. Banichuk (1983):

$$\min_{\Psi} V = \min_{\Psi} \frac{1}{2}\boldsymbol{\varepsilon} \cdot \mathbb{E}\boldsymbol{\varepsilon} : \tag{6.1}$$

the least the strain energy density V, the highest the stiffness. It is worth noting that this approach consists in optimizing the stiffness for a given strain state, like in Pedersen (1989), while in Vincenti and Desmorat (2011) a dual but equivalent approach, minimizing the stress energy, is used. In the above equation, Ψ is the angle between the direction of principal strain ε_I and the direction of the x_1-material axis, and it is the design variable of the optimum problem. In the polar formalism, the above problem reads like, cf. Eq. (4.89),

$$\min_{\Psi} V = \min_{\Psi} \{4T_1 t^2 + 8R_1 \cos 2(\Phi_1 - \varphi) r\, t + 2\,[T_0 + R_0 \cos 4(\Phi_0 - \varphi)]\, r^2\}, \tag{6.2}$$

where t, r and φ are the polar parameters of the strain tensor $\boldsymbol{\varepsilon}$, see Eq. (4.52) that gives, for $\varphi = 0$, the value of the principal strains:

$$\varepsilon_I = t + r, \quad \varepsilon_{II} = t - r, \tag{6.3}$$

[1] Actually, also the opposite problem, i.e. that of minimizing the stiffness, can be interesting, for example in impact engineering, for improving the crashworthiness of a structure. The two problems being opposite, the best solution of one is the worst for the other one, so, it is sufficient to continue to deal with the problem of maximum stiffness.

and conversely

$$t=\frac{\varepsilon_I+\varepsilon_{II}}{2},\quad r=\frac{\varepsilon_I-\varepsilon_{II}}{2}. \tag{6.4}$$

Let us transform slightly Eq. (6.2): after considering that the design variable Ψ is, by definition, equal to

$$\Psi=\Phi_1-\varphi, \tag{6.5}$$

we use the anisotropy ratio ρ defined in Eq. (4.131) and introduce the dimensionless parameter ϑ

$$\vartheta=\frac{t}{r}, \tag{6.6}$$

so we get for V the expression

$$V=2r^2\{T_0+2T_1\vartheta^2+R1[4\vartheta\cos 2\Psi+\rho\cos 4(\Phi+\Psi)]\}, \tag{6.7}$$

where

$$\Phi=\Phi_0-\Phi_1 \tag{6.8}$$

is the difference of the polar angles and as such a tensor invariant of $\mathbb{E}$, cf. Sect. 4.5.2. It is worth noting that, because independent from Ψ, the isotropic part of V does not take part to the optimum problem, i.e. it does not influence the optimal orientation; this was to be expected, but it is interesting to remark that this part of V depends at the same time from t and r, i.e. on the spherical and deviatoric parts of $\boldsymbol{\varepsilon}$, cf. Eq. (4.53).

The dimensionless parameter ϑ has an important mechanical meaning: it is the *deviatoric ratio*, i.e. it represents the ratio between the norms of the spherical (with the sign) and deviatoric parts of ε. We introduce, for further developments, two other ratios. First, we introduce

$$\gamma=\frac{\vartheta}{\rho}, \tag{6.9}$$

the ratio between the deviatoric and the anisotropy ratios, so measuring, in some sense, the importance of the *balance* between the spherical and deviatoric parts of $\boldsymbol{\varepsilon}$ with respect to the *balance* of the anisotropic phases of $\mathbb{E}$; hence, it relates, through invariants, the strain and the elastic fields.

Then, we introduce also the ratio between the two principal strains, like done in Pedersen (1989):

$$\eta=\frac{\varepsilon_{II}}{\varepsilon_I}. \tag{6.10}$$

The two dimensionless parameters ϑ and η are different quantities that can be used, with related but different meanings, for studying problem (6.2). The use of ϑ allows a complete polar approach, i.e. where the polar formalism is used to represent both $\boldsymbol{\varepsilon}$ and $\mathbb{E}$, while the use of η corresponds to a semi-complete polar approach, as $\boldsymbol{\varepsilon}$ is represented through its Cartesian components in the principal frame. We will develop in the following both the approaches for the orthotropic case, while for the completely anisotropic one only γ will be used, as more effective. In fact, parameter γ allows for a very concise description of the results of problem (6.2), and it will be used in the two cases.

6.2.2.2 The Completely Anisotropic Case

The stationary points of V can be found differentiating Eq. (6.7) with respect to Ψ:

$$\frac{dV}{d\Psi} = -8R_1r^2[2\vartheta \sin 2\Psi + \rho \sin 4(\Phi + \Psi)]. \tag{6.11}$$

Putting equal to zero the above quantity, after having introduced the ratio γ, gives the equation

$$2\gamma \sin 2\Psi + \sin 4(\Phi + \Psi) = 0, \tag{6.12}$$

to be solved with respect to Ψ. This equation generalizes that found for the first time in Kartvelishvili and Kobelev (1984). Unfortunately, this equation does not have any general solution, and it should be solved numerically. We introduce also the second derivative, useful in the following:

$$\frac{d^2V}{d\Psi^2} = -32R_1\, r^2[\vartheta \cos 2\Psi + \rho \cos 4(\Phi + \Psi)]. \tag{6.13}$$

A quick glance at Eq. (6.12) is sufficient to see that the principal direction of strain ε_I can never be aligned with the x_1-material axis, because of the presence of the term Φ, which, for a completely anisotropic body, is always different from 0 and from any multiple of $\pi/4$. In fact, if $\gamma \neq 0$ and $\Psi = 0$ or $\pi/2$, Eq. (6.12) reduces to the condition

$$\sin 4\Phi = 0 \Rightarrow \Phi = K\frac{\pi}{4}, \tag{6.14}$$

which is never true for a completely anisotropic body, because this is the invariant condition for a material to be ordinarily orthotropic, cf. Sect. 4.12.1.

It is however possible to give a graphical representation of the locus of the roots of Eq. (6.12) in the space (γ, Φ, Ψ), see Fig. 6.1.

In addition, we can consider some special cases for which trivial solutions can be easily found:

- $\vartheta = 0 \Rightarrow t = 0$, i.e. an isochoric strain state (completely deviatoric). Then Eq. (6.12) is satisfied $\iff$

$$\Psi = \alpha\frac{\pi}{4} - \Phi, \quad \alpha \in \{0, 1\}, \tag{6.15}$$

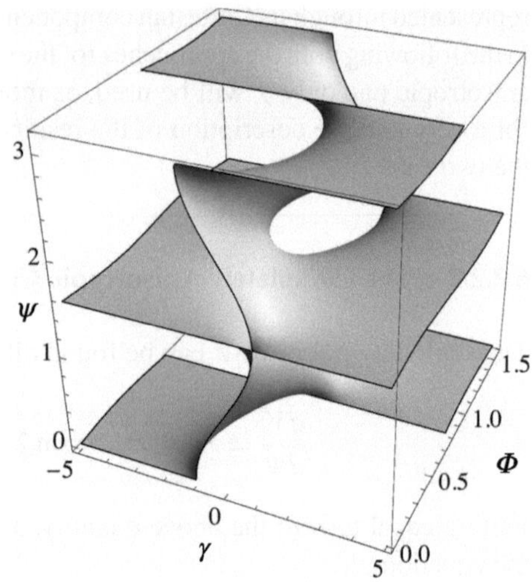

Fig. 6.1 Locus of the solutions of Eq. (6.12)

that is, the directions of stationary V are inclined on the x_1-material axis of a multiple of $\pi/4$ minus a constant term, corresponding to the invariant Φ. In this case, Eq. (6.13) becomes

$$\frac{d^2V}{d\Psi^2} = -(-1)^{\alpha} 32 R_0\, r^2 > 0 \iff \alpha = 1. \tag{6.16}$$

Hence, V has a minimum if the x_1-material direction is rotated counterclockwise of $\pi/4 - \Phi$ with respect to the direction of ε_I, a maximum if the rotation angle is $-\Phi$.

- $\vartheta \to \infty \Rightarrow r = 0$, i.e. a perfectly spherical strain state. Then Eq. (6.12) cannot be used anymore and one must go back to Eq. (6.11), to be rewritten as

$$\frac{dV}{d\Psi} = -8R_1 r[2t \sin 2\Psi + \rho r \sin 4(\Phi + \Psi)] = 0, \tag{6.17}$$

which is of course satisfied $\forall\Psi$ when $r = 0$. This is quite obvious: because the strain state is spherical, the strain energy is not affected by the material orientation and its value is

$$V = 4T_1 t^2 \ \ \forall\Psi. \tag{6.18}$$

- another special, though certainly very particular, case is that of $\rho = 1$ and $\vartheta = \pm 1/2$. This happens when the two anisotropic invariants R_0 and R_1 have the same value and the norm of the deviatoric part of $\boldsymbol{\varepsilon}$ is twice that of the spherical one. In such a case, Eq. (6.12) becomes

$$\sin 2\Psi \pm \sin 4(\Phi + \Psi) = 0 \Rightarrow \begin{cases} \Psi = -2\Phi - \dfrac{\pi}{2} \text{ if } \vartheta = \dfrac{1}{2}, \\ \Psi = -2\Phi \text{ if } \vartheta = -\dfrac{1}{2}. \end{cases} \tag{6.19}$$

In both the cases, we get

$$\frac{d^2V}{d\Psi^2} = -16R_1r^2\cos 4\Phi > 0 \iff \frac{\pi}{8} < \Phi < \frac{3}{8}\pi, \tag{6.20}$$

i.e. V has a minimum if the material has an invariant $\Phi \in [\pi/8, 3\pi/8]$, a maximum for Φ elsewhere in $[0, \pi/2]$.

6.2.2.3 A Polar Proof of the Necessity and Sufficiency of Coaxiality of $\boldsymbol{\sigma}$ and $\boldsymbol{\varepsilon}$

We can now give a simple proof, based on the polar formalism, that the coaxiality of $\boldsymbol{\sigma}$ and $\boldsymbol{\varepsilon}$ is a necessary and sufficient condition for V to be stationary, and this for any type of planar anisotropy.[2]

To this purpose, we write the component σ_{12}:

$$\sigma_{12} = E_{1211}\varepsilon_1 + E_{1222}\varepsilon_{22} + 2E_{1212}\varepsilon_{12}, \tag{6.21}$$

that in the polar formalism reads like

$$\begin{aligned} \sigma_{12} = &(R_0 r \sin 4\Phi_0 + 2R_1 t \sin 2\Phi_1)(t + r\cos 2\varphi) + \\ &(-R_0 r \sin 4\Phi_0 + 2R_1 t \sin 2\Phi_1)(t - r\cos 2\varphi) + \\ &2(T_0 - R_0\cos 4\Phi_0) r \sin 2\varphi. \end{aligned} \tag{6.22}$$

The condition for $\boldsymbol{\sigma}$ and $\boldsymbol{\varepsilon}$ to be coaxial, i.e. for having the same principal directions, is of course that $\sigma_{12} = 0$ when written in the principal frame for $\boldsymbol{\varepsilon}$, i.e. for $\varphi = 0$, which can be made easily subtracting φ by all the polar angles in the last expression. Remembering Eqs. (6.5), (6.6), (6.8) and (6.9), we obtain

$$2\gamma \sin 2\Psi + \sin 4(\Phi + \Psi) = 0, \tag{6.23}$$

that is exactly the condition for V being stationary, Eq. (6.12). Hence, the same values of Ψ that superpose the principal directions of $\boldsymbol{\sigma}$ and $\boldsymbol{\varepsilon}$ also render V stationary. This is a strong mechanical characterization of the extrema of V for an anisotropic body and could be used as optimality condition.

---

[2] As said in Sect. 6.2.1, this general property has been proved in Seregin and Troitskii (1981) for 3D orthotropic bodies.

6.2.2.4 Special Orthotropies

The results of Sect. 6.2.2.2 allows for solving easily the two cases of special orthotropy (in both of them, we can put $\Phi = 0$). Let us first consider the case of R_0-orthotropy, i.e. $R_0 = 0 \Rightarrow \rho = 0$. Then, Eq. (6.11) reduces to

$$\vartheta \sin 2\Psi = 0 \Rightarrow \Psi = \beta\frac{\pi}{2}, \quad \beta \in \{0, 1\}. \tag{6.24}$$

Hence, there are only two stationary solutions, $\Psi = 0$ and $\Psi = \pi/2$, i.e. in both the cases, one of the two principal directions is aligned with the x_1-material direction. If now we compute the second derivative, Eq. (6.13), we get

$$\frac{d^2V}{d\Psi^2} = -32R_1rt\cos 2\Psi = \begin{cases} -32R_1rt > 0 & \iff \quad t < 0, \text{ for } \Psi = 0, \\ 32R_1rt > 0 & \iff \quad t > 0, \text{ for } \Psi = \dfrac{\pi}{2}. \end{cases} \tag{6.25}$$

Only apparently the solution $\Psi = 0$ is a minimum for V if the strain state globally corresponds to a contraction (negative spherical part of $\boldsymbol{\varepsilon}$, $t < 0$), and a maximum if it corresponds to a dilatation (positive spherical part of $\boldsymbol{\varepsilon}$, $t > 0$). In fact, one should keep in mind that in this approach, the angle Ψ is measured from the direction of ε_I, and when $t < 0$, $\varepsilon_I < \varepsilon_{II}$. So, finally, the orientation of the material does not change with the sign of t with respect to the highest principal strain. This situation is not unique to R_0-orthotropy, as we will see also in the following, and actually it can happen also for completely anisotropic bodies.

A particular situation is that of a R_0-orthotropic body with $\vartheta = 0 \Rightarrow t = 0$, which is the case of a purely deviatoric strain state. In such a case, the value of the strain energy becomes

$$V = 2T_0r^2, \tag{6.26}$$

which is independent from Ψ: like in the case of a purely spherical strain state, V is insensitive to the body orientation.

It is worth noting (Vannucci 2010), that the duality with the r_0-orthotropy gives as a result that everything what said here for V and R_0-orthotropy is valid also for the stress energy and r_0-orthotropy. This means, e.g. that the stress energy is stationary for each orientation of a sheet of paper when submitted to a pure shear, cf. Sect. 4.12.3.

If now we consider square symmetric layers, i.e. the case $R_1 = 0 \Rightarrow \rho \to \infty$, we write Eq. (6.12) as follows

$$2\gamma \sin 2\Psi + \sin 4(\Phi + \Psi) = 0 \rightarrow 2R_1\vartheta \sin 2\Psi + R_0 \sin 4(\Phi + \Psi) = 0, \tag{6.27}$$

so that, for $R_1 = 0$, we get for V the stationarity condition

$$\sin 4(\Phi + \Psi) = 0 \Rightarrow \Psi = \xi\frac{\pi}{4} - \Phi, \quad \xi \in \{0, 1\}. \tag{6.28}$$

Now $\Phi = \Phi_0$, i.e. it is not an invariant of $\mathbb{E}$, so, putting for the sake of simplicity $\Phi_0 = 0$, the above result implies that the stationarity of V is get when the direction of ε_I is rotated of an angle 0 or $\pi/4$ with respect to the x_1-orthotropy direction. The second derivative becomes

$$\frac{d^2V}{d\Psi^2} = -32R_0r^2\cos 4\Psi = \begin{cases} -32R_0r^2 < 0 \text{ for } \Psi = 0, \\ 32R_0r^2 > 0 \text{ for } \Psi = \dfrac{\pi}{4}. \end{cases} \tag{6.29}$$

So, V is minimum when the angle between the directions of ε_I and of x_1 is $\pi/4$, maximum when it is 0.

6.2.2.5 Ordinary Orthotropy: Solution as Function of γ

When a material is ordinarily orthotropic, cf. Sect. 4.12.1 and Eqs. (4.128), (6.7), (6.11) and (6.13) become respectively

$$\begin{aligned} V &= 2r^2\{T_0 + 2T_1\vartheta^2 + R_0[4\gamma\cos 2\Psi + (-1)^K\cos 4\Psi]\}, \\ \frac{dV}{d\Psi} &= -8R_0r^2[2\gamma\sin 2\Psi + (-1)^K\sin 4\Psi], \\ \frac{d^2V}{d\Psi^2} &= -32R_0\, r^2[\gamma\cos 2\Psi + (-1)^K\cos 4\Psi]. \end{aligned} \tag{6.30}$$

From Eq. $(6.30)_2$, the condition of stationarity of V corresponding to the more general Eq. (6.12) can be easily rewritten as

$$\sin 2\Psi[\gamma + (-1)^K\cos 2\Psi] = 0, \tag{6.31}$$

whose solutions are

$$\Psi = 0, \quad \Psi = \frac{\pi}{2}, \quad \Psi^* = \frac{1}{2}\arccos[-(-1)^K\gamma] \text{ for } |\gamma| < 1. \tag{6.32}$$

Putting these values in the expression of the second derivative of V, Eq. $(6.30)_3$, we get

- for $\Psi = 0$,

$$\frac{d^2V}{d\Psi^2} = -32R_0\, r^2[\gamma + (-1)^K] > 0 \iff \gamma < -(-1)^K; \tag{6.33}$$

- for $\Psi = \dfrac{\pi}{2}$,

$$\frac{d^2V}{d\Psi^2} = -32R_0\, r^2[-\gamma + (-1)^K] > 0 \iff \gamma > (-1)^K; \tag{6.34}$$

- for $\Psi = \Psi^*$,

$$\frac{d^2V}{d\Psi^2} = -32(-1)^K R_0\, r^2(\gamma^2 - 1) > 0 \iff K = 0 \quad (|\gamma| < 1). \tag{6.35}$$

In some cases, $\Psi = 0$ and $\Psi = \pi/2$ can be both minima or maxima for V, so we need to distinguish between global or local extremum. To this purpose, we calculate

$$\Delta V = V(\Psi = 0) - V(\Psi = \frac{\pi}{2}) = 16R_0 r^2\gamma > 0 \iff \gamma > 0. \tag{6.36}$$

We can now establish the complete list of possible situations, summarized in Table 6.1 and sketched in Fig. 6.2.

As a last remark, we notice that the solutions $\Psi = 0$ and $\Psi = \pi/2$ make coincide the orthotropy directions with the principal directions of $\boldsymbol{\sigma}$ and $\boldsymbol{\varepsilon}$, but this is no more true for $\Psi = \Psi^*$; in such a case, $\boldsymbol{\sigma}$ and $\boldsymbol{\varepsilon}$ are still coaxial, but the orthotropy axes do not coincide with their principal directions.

These results should be compared with those given in Pedersen (1989, 1990), with the care that the parameter γ used by Pedersen actually corresponds to $(-1)^K\gamma$. In

Table 6.1 Optimal solutions for V for the case of ordinary orthotropy, ordered by γ

		$\gamma < -1$	$-1 < \gamma < 0$	$0 < \gamma < 1$	$\gamma > 1$
$K = 0$	$\Psi = 0$	glob. min.	loc. max.	glob. max.	glob. max.
	$\Psi = \Psi^*$	–	glob. min.	glob. min.	–
	$\Psi = \frac{\pi}{2}$	glob. max.	glob. max.	loc. max.	glob. min.
$K = 1$	$\Psi = 0$	glob. min.	glob. min.	loc. min.	glob. max.
	$\Psi = \Psi^*$	–	glob. max.	glob. max.	–
	$\Psi = \frac{\pi}{2}$	glob. max.	loc. min.	glob. min.	glob. min.

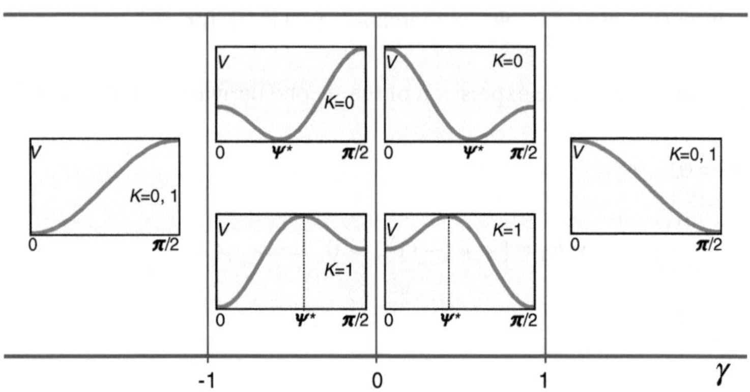

Fig. 6.2 Scheme of the optimal solutions for V for the case of ordinary orthotropy, ordered by γ

particular, also in this case we can remark that a change in the sign of the spherical invariant t, i.e. passing from a state where the average strain is a contraction to one where it is a dilatation, only apparently changes the quality of the result, because once more ε_I passes from the lowest to the highest principal strain, so the optimal orientation of the material does not change with respect to the highest principal strain.

6.2.2.6 Ordinary Orthotropy: Solution in the Plane ($\vartheta - \rho$)

The above results can be outlined also using the parameters ϑ and ρ, i.e. splitting γ in its two constitutive parts. This can be useful for in such a way the orthotropy characteristics, ρ, are separated from the strain field ones, ϑ.

Of course, the solutions of Eq. (6.12) do not change: $\Psi = 0$, $\Psi = \pi/2$ and $\Psi = \Psi^*$, that can be rewritten as

$$\Psi^* = \frac{1}{2}\arccos\left[-(-1)^K\frac{\vartheta}{\rho}\right], \quad \text{for } \rho > |\vartheta|. \tag{6.37}$$

The study of the sign of the second derivative of V for the three above solutions gives:

- for $\Psi = 0$,

$$\frac{d^2V}{d\Psi^2} = -32R_1\, r^2[\vartheta + (-1)^K\rho] > 0 \iff (-1)^K\rho < -\vartheta; \tag{6.38}$$

- for $\Psi = \dfrac{\pi}{2}$,

$$\frac{d^2V}{d\Psi^2} = -32R_1\, r^2[-\vartheta + (-1)^K\rho] > 0 \iff (-1)^K\rho < \vartheta; \tag{6.39}$$

- for $\Psi = \Psi^*$,

$$\frac{d^2V}{d\Psi^2} = -32(-1)^K\, r^2\frac{R_1}{\rho}(\vartheta^2 - \rho^2) > 0 \iff K = 0 \quad (\rho > |\vartheta|). \tag{6.40}$$

A complete scheme of the solutions in the plane ($\vartheta - \rho$) is given in Fig. 6.3. Also in this case, one should keep in mind that $\vartheta < 0 \Rightarrow t < 0$, i.e. $\varepsilon_I < \varepsilon_{II}$.

6.2.2.7 Ordinary Orthotropy: Solution in the Plane ($\eta - \rho$)

A last description of the solutions to Eq. (6.12) can be given using, instead of the ratio ϑ between the spherical and deviatoric parts of $\boldsymbol{\varepsilon}$, the ratio η, between ε_{II} and ε_I, Eq. (6.10). From Eqs. (6.3), (6.6), (6.9) and (6.10), we obtain

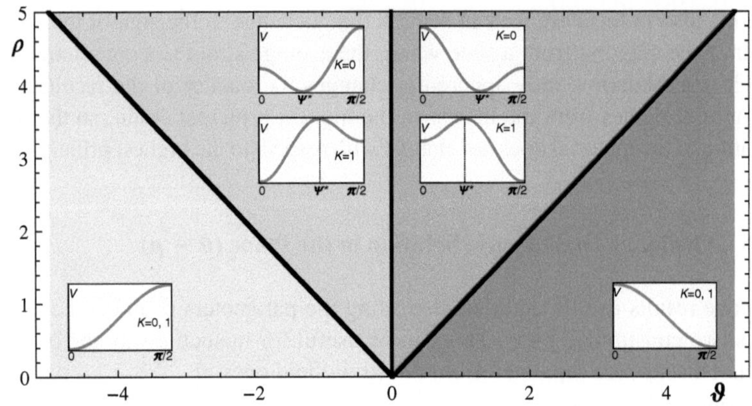

Fig. 6.3 Scheme of the optimal solutions for V in the plane $(\vartheta - \rho)$ for the case of ordinary orthotropy

$$\vartheta = \frac{1+\eta}{1-\eta} \Rightarrow \gamma = \frac{1}{\rho}\frac{1+\eta}{1-\eta}, \tag{6.41}$$

so that

$$\Psi^* = \frac{1}{2}\arccos\left[-(-1)^K \frac{1}{\rho}\frac{1+\eta}{1-\eta}\right], \quad \text{for } \rho > \left|\frac{1+\eta}{1-\eta}\right| \tag{6.42}$$

and

$$\frac{d^2V}{d\Psi^2} = -8R_1\varepsilon_I^2(1-\eta)[(1+\eta)\cos 2\Psi + (-1)^K(1-\eta)\rho\cos 4\Psi]. \tag{6.43}$$

The study of the sign of $d^2V/d\Psi^2$ for the three solutions gives hence:

- for $\Psi = 0$,

$$\frac{d^2V}{d\Psi^2} = -8R_1\,\varepsilon_I^2\left[\frac{1+\eta}{1-\eta} + (-1)^K\rho\right] > 0 \iff (-1)^K\rho < -\frac{1+\eta}{1-\eta}; \tag{6.44}$$

- for $\Psi = \dfrac{\pi}{2}$,

$$\frac{d^2V}{d\Psi^2} = -8R_1\,\varepsilon_I^2\left[-\frac{1+\eta}{1-\eta} + (-1)^K\rho\right] > 0 \iff (-1)^K\rho < \frac{1+\eta}{1-\eta}; \tag{6.45}$$

- for $\Psi = \Psi^*$,

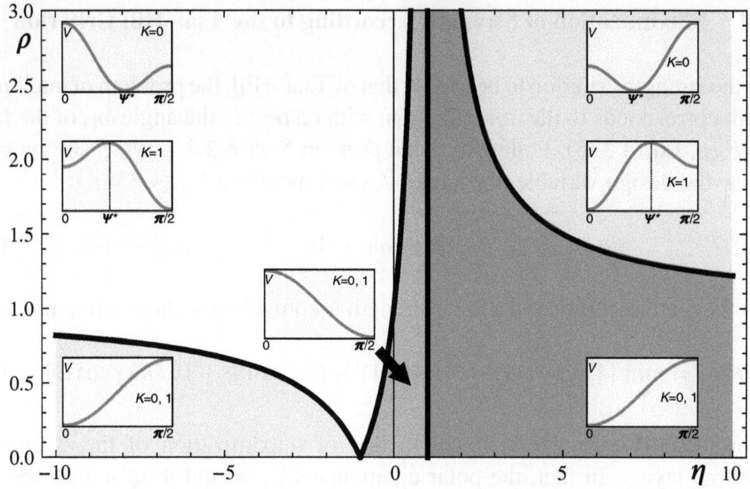

Fig. 6.4 Scheme of the optimal solutions for V in the plane $(\eta - \rho)$ for the case of ordinary orthotropy

$$\frac{d^2V}{d\Psi^2} = -8(-1)^K \frac{R_1}{\rho}\,\varepsilon_I^2\left[\left(\frac{1+\eta}{1-\eta}\right)^2 - \rho^2\right] > 0 \iff$$
$$K = 0 \quad \left(\rho > \left|\frac{1+\eta}{1-\eta}\right|\right). \tag{6.46}$$

A complete scheme of the solutions in the plane $(\eta - \rho)$ is shown in Fig. 6.4.

6.2.3 A Polar Approach to the Maximization of the Strength

A simple way to maximize the local strength is to consider that the least the failure index, the highest the strength: maximizing local strength corresponds to minimize the failure index. This last depends upon the strength criterion. We have introduced the polar, invariant, expression of the failure indexes for the tensorial criteria of Tsai–Hill, in Sect. 4.17.3.1, Hoffman, in Sect. 4.17.3.2, and Tsai–Wu, in Sect. 4.17.3.3. We now use them to solve the problem of the best orientation of the material for maximizing local strength in the three cases. The procedure is quite similar to that used above for the case of stiffness. However, because strength criteria in anisotropy are known only for orthotropic materials, only this case will be considered in the following. The developments shown in the next sections are a succinct presentation of the results found in Catapano et al. (2012), presented here in the same way used for the stiffness in the previous sections.

6.2.3.1 Maximization of Strength According to the Tsai–Hill Criterion

When the strength criterion to be used is that of Tsai–Hill, the problem of maximizing strength corresponds to the minimization, with respect to the angle ω_1, of the failure index F_{Hill}, Eq. (4.336). Following what done in Sect. 6.2.2.1, we easily recognize that now the design variable, the angle Ψ, corresponds in Eq. (4.336), to[3]

$$\Psi = \omega_1 - \Phi, \tag{6.47}$$

so that the maximization of the local strength is equivalent to the minimum problem:

$$\min_{\Psi} F_{Hill} = \min_{\Psi} \{4\gamma_0 R^2 + 8\gamma_1 T^2 + 4(-1)^l \lambda_0 R^2 \cos 4\Psi + 16TR\lambda_1 \cos 2\Psi\}. \tag{6.48}$$

This problem is formally identical to that of maximization of the stiffness for orthotropic layers. In fact, the polar components T, R and Φ of $\boldsymbol{\sigma}$ take the place of those of $\boldsymbol{\varepsilon}$, t, r and φ, while the polar components of $\mathbb{E}$, T_0, T_1, $(-1)^K R_0$, R_1 and Φ_1 are replaced by those of $\mathbb{F}$, γ_0, γ_1, $(-1)^l\lambda_0$, λ_1 and ω_1.

In fact, if we introduce, in complete analogy with what done in Sect. 6.2.2.1 for stiffness, the ratios

$$\Lambda = \frac{\lambda_0}{\lambda_1}, \quad \Theta = \frac{T}{R}, \quad \Gamma = \frac{\Theta}{\Lambda}, \tag{6.49}$$

the corresponding of parameters ρ, ϑ and γ respectively, then we get for F_{Hill} and its derivatives with respect to Ψ the expressions

$$\begin{aligned}
F_{Hill} &= 4R^2\{\gamma_0 + 2\gamma_1\Theta^2 + \lambda_0[4\Gamma \cos 2\Psi + (-1)^l \cos 4\Psi]\},\\
\frac{dF_{Hill}}{d\Psi} &= -16\lambda_0 R^2[2\Gamma \sin 2\Psi + (-1)^l \sin 4\Psi],\\
\frac{d^2F_{Hill}}{d\Psi^2} &= -64\lambda_0 R^2[\Gamma \cos 2\Psi + (-1)^l \cos 4\Psi].
\end{aligned} \tag{6.50}$$

Apart a factor 2, which has no consequences on the optimal solutions, the above equations are completely analogous to those in Eq. (6.30); the discussion done for V can hence be repeated *verbatim* for F_{Hill}; in particular, the solutions are the angles

$$\Psi = 0, \quad \Psi = \frac{\pi}{2}, \quad \Psi^* = \frac{1}{2}\arccos[-(-1)^l\Gamma] \text{ if } |\Gamma| < 1. \tag{6.51}$$

---

[3] The reader should consider that now the angle Φ is not the angle defined in Eq. (6.8), but it denotes the polar angle of tensor $\boldsymbol{\sigma}$, cf. Sect. 4.5.1, and that parameter γ is not the one defined by Eq. (6.9) but the polar component of tensor $\mathbf{f}$, cf. Sect. 4.17.3.2.

6.2.3.2 Maximization of Strength According to the Hoffman and Tsai–Wu Criteria

The failure indexes for the Hoffman and Tsai–Wu criteria are formally identical, cf. Sects. 4.17.3.2 and 4.17.3.3. We can hence consider a unique problem of local strength maximization. If we assume that in Eqs. (4.338) and (4.339) it is

$$\omega_1 = \omega \Rightarrow \omega_1 - \Phi = \omega - \Phi = \Psi, \tag{6.52}$$

i.e. that the weakness tensors $\mathbb{F}$ and $\mathbf{f}$ have the same orientation, because they are related to the same layer, then the maximization of strength corresponds to the minimum problem

$$\begin{aligned}\min_{\Psi} F_{Hoff/TW} = \min_{\Psi} \{&4\gamma_0 R^2 + 8\gamma_1 T^2 + 4(-1)^l \lambda_0 R^2 \cos 4\Psi + \\ &16TR\lambda_1 \cos 2\Psi + 2\gamma T + 2\lambda R \cos 2\Psi\}.\end{aligned} \tag{6.53}$$

Differentiating we get

$$\begin{aligned}\frac{dF_{Hoff/TW}}{d\Psi} &= -16\lambda_0 R^2 [2\Gamma_1 \sin 2\Psi + (-1)^l \sin 4\Psi], \\ \frac{d^2 F_{Hoff/TW}}{d\Psi^2} &= -64\lambda_0 R^2 [\Gamma_1 \cos 2\Psi + (-1)^l \cos 4\Psi],\end{aligned} \tag{6.54}$$

where

$$\Gamma_1 = \frac{\lambda + 8\lambda_1 T}{8\lambda_0 R}. \tag{6.55}$$

There is hence a complete analogy between the previous case and this one, provided that the ratio Γ is replaced by Γ_1. This time, hence, the solution Ψ^*, Eq. (6.51) does not depend uniquely on Θ and Γ, but also on λ, the deviatoric part of tensor $\mathbf{f}$. Taken this into account, the discussion of the solution is completely analogous to that of V or F_{Hill}.

As a last remark, it is worth noting that, for both the cases of F_{Hill} and $F_{Hoff/TW}$, the best solution for strength coincides with the best one for stiffness whenever the solution is $\Psi = 0$ or $\Psi = \pi/2$. However, if the solution is Ψ^* at least for one of the two problems, then, generally speaking, the two optima do not coincide: the stiffest solution is not the strongest one and viceversa. Nevertheless, the condition of coaxiality of $\boldsymbol{\sigma}$ and $\boldsymbol{\varepsilon}$ is always valid also for the strength problem, as it can be easily recognized comparing Eqs. $(6.50)_2$ and $(6.54)_1$ with (6.23).

The theoretical problem of the best material orientation concerns not only single layers, but laminates as well. We switch hence now to the mechanics of laminates, for considering more general design problems of anisotropic structures, and we will find again, in the following, the problem of the optimal local orientation of the material in a given structure. Before, a short recall of the mechanics of laminates is necessary.

6.3 Design Problems of Anisotropic Structures

The design of anisotropic structures, as already said in the Introduction, Sect. 6.1, is considerably much more complicate than that of classical, i.e. metallic, structures. In fact, while for classical materials structures, once chosen the material, the object of the design is constituted uniquely by the structural dimension, typically thicknesses for plates and shells, the cross section dimensions for beams, for anisotropic, laminated structures the designer must not only consider these problems, but at the same time he has to manage the general properties of the laminate: bending-extension coupling, type of anisotropy of $\mathbb{A}$ and $\mathbb{D}$, material orientation and so on.

Any problem of anisotropic laminated structures design appears hence as an articulated design problem, with in some cases interacting effects between different requirements. In such complicate cases of design, it is useful to formulate the problem as an optimization one, so exploiting the formal mathematical framework of optimization techniques and methods.

Unfortunately, as said in the Introduction, designers have tried to simplify bounding the search for a solution into a set of special laminates, having automatically some special properties, in order to eliminate from the design a part of the whole structural problem. Typical is the case of the use of balanced symmetric stacks, ensuring automatically, i.e. regardless of the layer orientation angles, bending-extension coupling and in-plane orthotropy, two requirements almost always sought by designers.

Because in this way the feasible domain is considerably narrowed, the final solution will be, the most part of times, a sub-optimal and not a global optimal one. It is hence interesting an approach where the design problem is stated in the most general way, in order to have the possibility to obtain a true global optimum.

In the following, we propose an approach based upon the use of the polar formalism and global optimization techniques. Unless otherwise indicated, all the problems are related to the design of laminates composed by identical plies of a chosen material, the most common case of design problem for laminated structures.

Because there are several different design problems for such structures, we start giving a classification of them, essentially based upon the type of objective function. In all the cases, the problem is formulated as a minimum one, which is, of course, always possible to be done (any maximum problem can be transformed into a minimum one simply changing the sign of the objective function).

6.3.1 Different Types of Design Problems

We distinguish between three classes of design problems for anisotropic laminated structures (Vannucci et al. 2012; Vannucci 2013b):

Problems of the first type: unconstrained constitutive law problem design: in these problems, just the constitutive law of the laminate is the object of the design. Such problems, hence, concern only the design of basic elastic properties, e.g.

bending-extension uncoupling, orthotropy, isotropy and so on, but other requirements can be added, to take into account, for instance, of hygral and thermal properties.

Problems of the second type: constrained constitutive law problem design: in such a kind of problems, the constitutive law is still the object of the design, but now some constraints are imposed to the problem. Such constraints can be of different nature: mechanical, technological etc. What mainly changes in this situation is the mathematical approach and the solution method, as now the constraints enter the mathematical optimization problem.

Problems of the third type: structural optimization: in this kind of problems, a given objective function for a structure to be designed must be minimized along with a prescribed constitutive law sought for the final design. In this case, hence, the objective function is no more of the type of problems of the first or of the second class, but a specific objective: compliance, strength, vibration frequencies, buckling loads and so on. A boundary value problem is, usually, to be solved and the requirement for a prescribed constitutive law becomes now not an objective function, but a constraint to the structural problem. We will see in Sect. 6.6.1 that such a type of problems can be split into two sequential and linked optimization problems, with the second one of the first or of the second type.

These three different types of anisotropic laminated structures design problems are detailed in Sects. 6.4–6.6, but before we ponder on a general question: which is the influence of anisotropy on optimal solutions?

6.3.2 Influence of Anisotropy on Optimal Solutions

A subject that has been largely unconsidered by literature is the influence that the anisotropy of the material, and not of the final laminate, has on the solution of a given optimum problem. The matter is not simple; a first question is: what parameter describing anisotropy is of some importance in examining the dependence of a solution on anisotropy of the material? The advantage of the polar formalism is that it is based upon material invariants, so these ones can be used as intrinsic quantities characterizing the material in order to consider their influence on an optimal solution. This point constitutes by itself an intrinsic advantage with respect to the use of frame-dependent quantities. More specifically, we will see in the following that interesting dimensionless invariant parameters can be introduced using the polar invariants.

A first problem of this type has been already introduced in Sect. 4.12.1, when the problem of maximizing the shear modulus G_{12} for a laminate has been tackled. In particular, we have seen that the parameter K^L of the basic layer strongly affects the result: changing K^L from 0 to 1 changes the best solution into the worst one, and viceversa, and it determines also the uniqueness of the solution.

We consider here a rather unusual problem in laminate design: the flexural optimal design. Normally, in fact, literature pays a lot of attention to extension design of laminates, while only a very small part of it regards the flexural properties. This because the flexural design is much more complicate than the extensional one. The results presented hereafter are more widely exposed in Vannucci (2009b).

More specifically, we consider here a problem already introduced in Sect. 4.16.2: the flexural behavior of a bending-extension uncoupled laminate composed of identical layers. The plate is rectangular, with sides' length a and b, respectively along the axes x and y. The plate is simply supported, its mass per unit area is μ and the following condition is imposed to the anisotropy of the laminate: tensor $\mathbb{D}$ has to be orthotropic and the axes of orthotropy aligned with the axes of the plate.

In the above assumptions, cf. Vannucci (2009b), Jones (1999) (we just recall here, for the sake of convenience, Eqs. (4.277)–(4.279) already introduced in Sect. 4.16.2):

- the compliance J of the plate loaded by a sinusoidally distributed force orthogonal to its surface:

$$J = \frac{\gamma_{pq}}{p^4 h^3 \left(1+\chi^2\right)^2 \sqrt{R_0^2+R_1^2}} \frac{1}{\varphi\left(\xi_{13},\xi_{15}\right)}; \tag{6.56}$$

- the buckling load multiplier λ_{pq} for the mode pq when the plate is loaded by in-plane forces N_x and N_y:

$$\lambda_{pq} = \frac{\pi^2 p^2 h^3}{12a^2} \frac{\left(1+\chi^2\right)^2 \sqrt{R_0^2+R_1^2}}{N_x+N_y\chi^2} \varphi\left(\xi_{13},\xi_{15}\right); \tag{6.57}$$

- the frequency of transversal vibrations ω_{pq} for the mode pq:

$$\omega_{pq}^2 = \frac{\pi^4 p^4 h^3}{12\mu a^4} \left(1+\chi^2\right)^2 \sqrt{R_0^2+R_1^2} \varphi\left(\xi_{13},\xi_{15}\right). \tag{6.58}$$

We recall also that in the above equations γ_{pq} is a coefficient depending on the geometry of the plate and on the loading, p and q are the number of half-waves in the directions x and y respectively, while the dimensionless parameter χ is the ratio of the wavelengths in the two directions, Eq. (4.280). Finally, function $\varphi(\xi_{13}, \xi_{15})$ is given by Eq. (4.281):

$$\varphi\left(\xi_{13},\xi_{15}\right) = \tau + \frac{1}{\sqrt{1+\rho^2}} \left[(-1)^{K^L} \rho c_0(\chi)\xi_{13} + 4c_1(\chi)\xi_{15}\right], \tag{6.59}$$

with

$$c_0(\chi) = \frac{\chi^4 - 6\chi^2 + 1}{(1+\chi^2)^2}, \quad c_1(\chi) = \frac{1-\chi^2}{1+\chi^2}, \tag{6.60}$$

and it is a dimensionless function of dimensionless parameters and variables: τ is the *isotropy-to-anisotropy ratio* introduced in Eq. (4.282), while ρ is the anisotropy ratio already used several times previously and ξ_{13} and ξ_{15} are the bending lamination parameters for orthotropic laminates, Eq. (5.55); they define the feasible set in the form of a parabolic sector shown in Fig. 4.7.

From the above equations, it is apparent that in the three cases, regarding all the possible situations concerning the flexural response of the plate, this response is always a function of $\varphi(\xi_{13}, \xi_{15})$; in this dimensionless function, the isotropic part is well separated from the anisotropic one.

The question here is the dependence of the optimal solutions upon the values of τ and ρ. It can be shown that all the optimal solutions can be of only two types: cross-ply or angle-ply laminates. Let us first consider cross-ply solutions, for which $\xi_{13} = 1$ and $-1 \leq \xi_{15} \leq 1$. The conditions to be satisfied for (6.59) to be a maximum are:

$$(-1)^{K^L} c_0(\chi) > 0 \iff \begin{cases} K^L = 0 \text{ and } \chi \in \left[0, \sqrt{2}-1\right] \text{ or } \chi > \sqrt{2}+1, \\ K^L = 1 \text{ and } \sqrt{2}-1 < \chi < \sqrt{2}+1, \end{cases}$$
$$\frac{c_1(\chi)}{\sqrt{1+\rho^2}} = 0 \iff \chi = 1 \text{ or } \rho = \infty. \tag{6.61}$$

The corresponding highest values of the function, hereafter denoted by φ_{cp}, are

$$\varphi_{cp} = \tau - \frac{(-1)^{K^L}\rho}{\sqrt{1+\rho^2}} \quad \text{for } \chi = 1,$$
$$\varphi_{cp} = \tau + (-1)^k c_0(\chi) \quad \text{for } \rho = \infty. \tag{6.62}$$

The maximum value of φ_{cp} is

$$\varphi_{cp}^{\max} = \left[\tau - \frac{(-1)^{K^L}\rho}{\sqrt{1+\rho^2}}\right]_{K^L=1,\ \rho=\infty} = \left[\tau + (-1)^{K^L} c_0(\chi)\right]_{K^L=0,\ \chi\in\{0,\ \infty\}} =$$
$$\left[\tau + (-1)^{K^L} c_0(\chi)\right]_{K^L=1,\ \chi=1} = \tau + 1. \tag{6.63}$$

Some remarks can be done; first of all, cross-ply solutions exist only in presence of a generalized square symmetry, either of the material, i.e. $\rho = 0$ (i.e. basic layers reinforced by a balanced fabric), or of the geometry and mode, i.e. $\chi = 1$ (e.g. a square plate with $p = q$). Then, we see easily that changing the type of ordinary orthotropy, i.e. changing K^L from 0 to 1, corresponds once again to transform an optimal into an anti-optimal (i.e. from the best into the worst) solution. Finally, as the lamination parameter ξ_{15} disappears from the above equations, this means that any laminate composed by a combination of layers at 0° and at 90° is an optimal solution: uniqueness of the solution is lost for the optimal (or anti-optimal) case when cross-ply laminates are sought for.

Let us now consider angle-ply solutions; in such a case, it is an easy task to see that the two lamination parameters are linked by the relation, cf. Eq. (5.55),

$$\xi_{13} = 2\xi_{15}^2 - 1, \tag{6.64}$$

and the objective function becomes

$$\varphi(\xi_{15}) = \tau + \frac{1}{\sqrt{1+\rho^2}}\left[(-1)^{K^L}\rho\, c_0(\chi)(2\xi_1^2-1) + 4\, c_1(\chi)\xi_1\right], \tag{6.65}$$

whose maxima and minima can be only

$$\begin{aligned}
\varphi_{11} &= \varphi(\xi_{15}=1) = \tau + \frac{1}{\sqrt{1+\rho^2}}\left[(-1)^{K^L}\rho c_0(\chi) + 4c_1(\chi)\right],\\
\varphi_{22} &= \varphi(\xi_{15}=-1) = \tau + \frac{1}{\sqrt{1+\rho^2}}\left[(-1)^{K^L}\rho c_0(\chi) - 4c_1(\chi)\right],\\
\varphi_{\delta\delta} &= \varphi(\xi_{15}=\xi_{15}^{st}) = \tau - \frac{(-1)^{K^L}}{\sqrt{1+\rho^2}}\frac{\rho^2 c_0^2(\chi)+2c_1^2(\chi)}{\rho\, c_0(\chi)},
\end{aligned} \tag{6.66}$$

with

$$\left.\frac{\partial\varphi}{\partial\xi_{15}}\right|_{\xi_{15}^{st}} = 0 \Rightarrow \xi_{15}^{st} = -\frac{(-1)^{K^L}}{\rho}\frac{c_1(\chi)}{c_0(\chi)}. \tag{6.67}$$

Actually, only $\varphi_{\delta\delta}$ is an angle-ply solution, because the orientation angle corresponding to $\xi_{15} = 1$ is $\delta = 0°$ while to $\xi_{15} = -1$ corresponds $\delta = 90°$, so these two cases determine laminates with all the layers aligned along the x or y direction, respectively. The orientation angle giving solution $\varphi_{\delta\delta}$ is

$$\delta = \frac{1}{2}\arccos\xi_{15}^{st}. \tag{6.68}$$

Angle-ply solutions do not exist always, because of the bounds $\xi \in [-1, 1]$, imposed to each lamination parameter, hence to ξ_{15}^{st} too. These bounds are satisfied for

$$\begin{aligned}
&\rho>1 \ \text{ and } \ 0<\chi\leq\chi_1(\rho),\\
&\rho\geq 0 \ \text{ and } \ \chi_2(\rho)<\chi\leq\chi_3(\rho),\\
&\rho>1 \ \text{ and } \ \chi\geq\chi_4(\rho),
\end{aligned} \tag{6.69}$$

where the functions $\chi_i(\rho),\ i = 1, \ldots, 4$, are the solutions of the equations $\xi_{15}^{st} = \pm 1$, with $K^L \in \{0, 1\}$:

$$\begin{aligned}
\chi_1(\rho) &= \sqrt{\frac{3\rho-\sqrt{8\rho^2+1}}{\rho+1}}, \quad \chi_2(\rho) = \sqrt{\frac{3\rho-\sqrt{8\rho^2+1}}{\rho-1}},\\
\chi_3(\rho) &= \sqrt{\frac{3\rho+\sqrt{8\rho^2+1}}{\rho+1}}, \quad \chi_4(\rho) = \sqrt{\frac{3\rho+\sqrt{8\rho^2+1}}{\rho-1}}.
\end{aligned} \tag{6.70}$$

Conditions (6.69) determine, in the plane (ρ, χ) the region where angle-ply solutions exist. In Fig. 6.5, this region is shaded; inside it, the optimal solution is an angle-ply laminate, outside the optimal solution is a unidirectional laminate, either at 0° or at 90°.

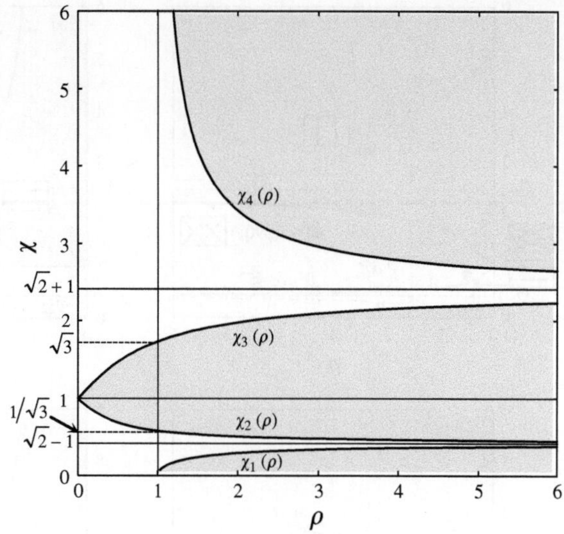

Fig. 6.5 Domain of the plane (ρ, χ) where solution $\varphi_{\delta\delta}$ exists (*shaded*)

Computing the second derivative of function in Eq. (6.65) gives

$$\frac{\partial^2\varphi}{\partial\xi_1^2} = 4(-1)^k\frac{\rho}{\sqrt{1+\rho^2}}c_0(\chi). \tag{6.71}$$

Once more, optimal solutions become anti-optimal whenever K^L changes from 0 to 1, and viceversa. Namely, by Eq. (6.60), it can be easily seen that:

- if $(\rho > 1, 0 < \chi \le \chi_1(\rho))$ or $(\rho > 1, \chi \ge \chi_4(\rho))$, then $\varphi_{\delta\delta} = \varphi_{max}$ for $K^L = 1$, $\varphi_{\delta\delta} = \varphi_{min}$ for $K^L = 0$;
- if $(\rho \ge 0, \chi_2(\rho) < \chi \le \chi_3(\rho))$, then $\varphi_{\delta\delta} = \varphi_{max}$ for $K^L = 0$, $\varphi_{\delta\delta} = \varphi_{min}$ for $K^L = 1$.

So, anisotropy affects optimal solutions also in this case: the parameter K^L, i.e. the type of ordinary orthotropy, changes each optimal solution into an anti-optimal one and viceversa, whilst parameter ρ, along with the geometric parameter χ, determines the region where the optimal solution is of the angle-ply type, i.e. the region where optimisation is meaningful. Actually, unidirectional laminates are limit solutions, and in such cases an optimization procedure, i.e. the computation of angle δ by Eq. (6.68), becomes unnecessary. Nevertheless, the value of these three solutions φ_{11}, φ_{22} and $\varphi_{\delta\delta}$ can be compared. It is not difficult to find the following result:

$$\begin{aligned}
&\varphi_{11}>\varphi_{22} \iff \chi>1,\\
&\varphi_{\delta\delta}>\begin{cases}\varphi_{11}\\ \varphi_{22}\end{cases} \iff \begin{cases}\sqrt{2}-1<\chi<\sqrt{2}+1 \text{ if } K^L=0,\\ 0<\chi\sqrt{2}-1 \text{ or } \chi>\sqrt{2}+1 \text{ if } K^L=1.\end{cases}
\end{aligned} \tag{6.72}$$

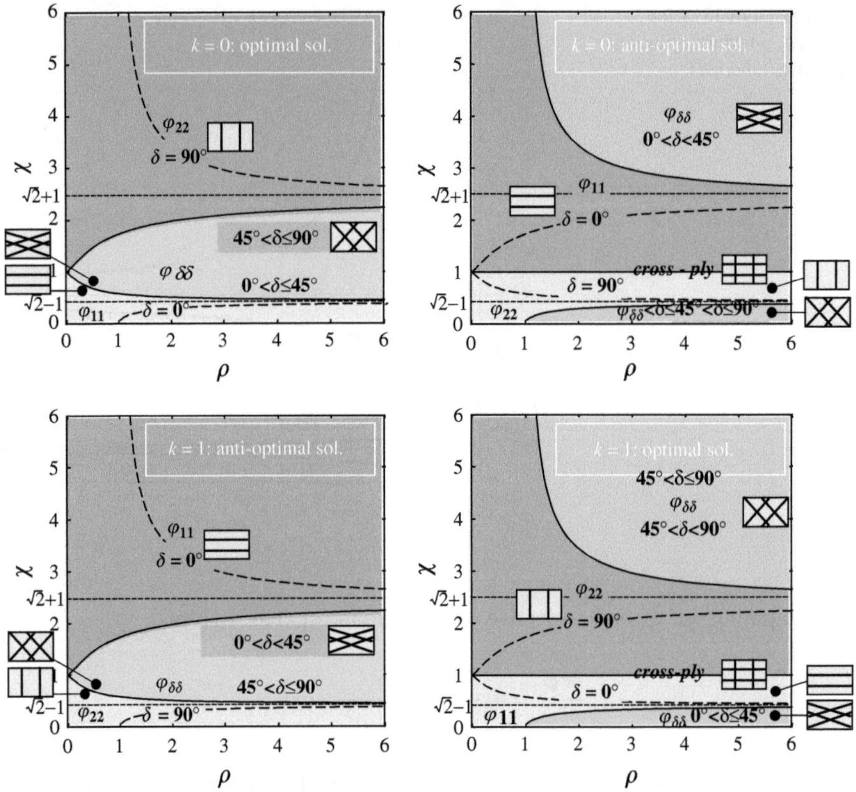

Fig. 6.6 Map of the optimal and antioptimal solutions

These results are summarized in Fig. 6.6, which completely shows, along with Eqs. (6.66)–(6.68), the influence of the two anisotropic dimensionless parameters K^L and ρ on the optimal, and anti-optimal, solutions; in Fig. 6.6, the layer orientations of the different solutions are indicated by simple schemes.

More details can be found in Vannucci (2009b); finally, these cases show that the material's anisotropy has really a great influence on the solution of an optimization problem: it can affects the uniqueness of the solution, it can switch an optimum into an anti-optimum, it determines the size and shape of the feasible domain, it affects, of course, the value of the minimum or of the maximum.

6.4 Design Problems of the First Type

The objective of the problems of this type is the design of the general properties of tensors $\mathbb{A}$, $\mathbb{B}$ and $\mathbb{D}$. In other words, the question is to design the elasticity of the laminate, i.e. to obtain an isotropic, orthotropic, uncoupled etc. behavior. However, the optimization of specific properties, i.e. for instance the maximization of the stiffness in one direction, is not comprehended in this type of problems, but in those of the third type.

We begin with giving a general unified approach, useful for numerical implementation, of the problems of the first type, and then we go back to some classical considerations, mainly for putting in evidence an important topic in laminates design: the non uniqueness of the solution in the space of layer orientations.

6.4.1 Unified Polar Formulation of the Optimum Problem

It is possible to give a unified formulation for all these problems in the framework of the polar formalism and using a quadratic approach (Vannucci 2006). In fact, any problem concerning the design of the constitutive law can be formulated as the minimization of a quadratic function of the polar parameters of $\mathbb{A}$, $\mathbb{B}$ and $\mathbb{D}$:

$$\min_{\delta_k} \mathscr{I}(\mathscr{P}_j(\delta_k)) = \{\mathscr{P}\}^\top[H]\{\mathscr{P}\} = H_{rs}\mathscr{P}_r\mathscr{P}_s,\ \ j,\ r,\ s = 1, \dots, 18, \tag{6.73}$$

where H_{rs} are the components of a 18×18 symmetric matrix $[H]$, whose entries determine the objective of the problem to be solved, cf. Vannucci (2006). The terms $\mathscr{P}_j$ are the eighteen polar parameters of the laminate, representing tensors $\mathbb{A}$, $\mathbb{B}$ and $\mathbb{D}$, condensed in the vector (the δ_ks are the ply orientations)

$$\begin{aligned}\mathscr{P} =&\{\mathscr{P}_j(\delta_k),\ j = 1, \dots, 18\} = \\ &\left\{\frac{1}{\mathscr{M}}(T_0, T_1, R_0, R_1), \Phi - (\Phi_0 - \Phi_1), \Phi_1^* - \Phi_1\right\}_{A,B,D}.\end{aligned} \tag{6.74}$$

In the above equation, $\mathscr{M}$ is a modulus used for having a dimensionless function $\mathscr{I}(\mathscr{P}_j)$; it can be chosen in different ways, e.g. Kandil and Verchery (1988):

$$\mathscr{M} = \frac{1}{n}\sum_{j=1}^{n}\sqrt{T_{0_j}^2 + 2T_{1_j}^2 + R_{0_j}^2 + 4R_{1_j}^2}. \tag{6.75}$$

Moreover, Φ is the polar angular invariant sought for a tensor, $\mathbb{A}, \mathbb{B}$ or $\mathbb{D}$, and in particular $\Phi = K\pi/4,\ K \in \{0, 1\}$ for orthotropy, while Φ_1^* is the angle that serves to fix an orientation.

The form (6.73) is valid for any situation, also for hybrid laminates, and in this case it is a positive definite form. If identical layers laminates are concerned, the isotropic part of $\mathbb{B}$ is null, cf. Eq. (5.5.2), while those of $\mathbb{A}$ and $\mathbb{D}$ are equal to that of the basic layer, Eqs. (5.51) and (5.54), and are not affected by the optimization process. So, they can be eliminated from the design procedure putting equal to zero the corresponding components of the matrix $[H]$, and the design problem is reduced to a maximum of 12 design variables, the anisotropic polar parameters of $\mathbb{A}$, $\mathbb{B}$ and $\mathbb{D}$. In this case of identical layers, $\mathscr{I}(\mathscr{P}_j)$ is a *positive semi-definite quadratic function, whose minima are zero-valued.*

Hence, the design of the elastic characteristics of a laminate is reduced to the search of the minima, zero-valued if the layers are identical, of a quadratic function. The form of this quadratic function depends, of course, upon the problem at hand, which determines the value of the components H_{rs}. Different typical cases are detailed in Vannucci (2006). For instance, if we order the $\mathscr{P}_j$s as

$$\begin{aligned}
&\mathscr{P}_1 = \frac{T_0^A}{\mathscr{M}},\ \mathscr{P}_2 = \frac{T_1^A}{\mathscr{M}},\ \mathscr{P}_3 = \frac{R_0^A}{\mathscr{M}},\ \mathscr{P}_4 = \frac{R_1^A}{\mathscr{M}},\\
&\mathscr{P}_5 = \Phi^A - (\Phi_0^A - \Phi_1^A),\ \mathscr{P}_6 = \Phi_1^{A*} - \Phi_1^A,\\
&\mathscr{P}_7 = \frac{T_0^B}{\mathscr{M}},\ \mathscr{P}_8 = \frac{T_1^B}{\mathscr{M}},\ \mathscr{P}_9 = \frac{R_0^B}{\mathscr{M}},\ \mathscr{P}_{10} = \frac{R_1^B}{\mathscr{M}},\\
&\mathscr{P}_{11} = \Phi^B - (\Phi_0^B - \Phi_1^B),\ \mathscr{P}_{12} = \Phi_1^{B*} - \Phi_1^B,\\
&\mathscr{P}_{13} = \frac{T_0^D}{\mathscr{M}},\ \mathscr{P}_{14} = \frac{T_1^D}{\mathscr{M}},\ \mathscr{P}_{15} = \frac{R_0^D}{\mathscr{M}},\ \mathscr{P}_{16} = \frac{R_1^D}{\mathscr{M}},\\
&\mathscr{P}_{17} = \Phi^D - (\Phi_0^D - \Phi_1^D),\ \mathscr{P}_{18} = \Phi_1^{D*} - \Phi_1^D,
\end{aligned} \tag{6.76}$$

then it is easy to see that, cf. Sect. 5.3:

- for bending-extension uncoupling, $\mathbb{B} = \mathbb{O}$, one must put $H_{7\,7} = H_{8\,8} = H_{9\,9} = H_{10\,10} = 1$ and all the other components $H_{rs} = 0$;
- for extension isotropy one must put $H_{3\,3} = H_{4\,4} = 1$ and all the other components $H_{rs} = 0$;
- for bending isotropy one must put $H_{15\,15} = H_{16\,16} = 1$ and all the other components $H_{rs} = 0$;
- for extension ordinary orthotropy, one must put $H_{5\,5} = 1$, and all the other components $H_{rs} = 0$; in addition, the value of $K^A \in \{0, 1\}$ must be specified;
- for bending ordinary orthotropy, one must put $H_{17\,17} = 1$ and all the other components $H_{rs} = 0$; in addition, the value of $K^D \in \{0, 1\}$ must be specified;
- for R_0-orthotropy in extension, one must put $H_{3\,3} = 1$ and all the other components $H_{rs} = 0$;
- for R_0-orthotropy in bending, one must put $H_{15\,15} = 1$ and all the other components $H_{rs} = 0$;
- for square symmetry in extension, one must put $H_{4\,4} = 1$ and all the other components $H_{rs} = 0$;
- for square symmetry in bending, one must put $H_{16\,16} = 1$ and all the other components $H_{rs} = 0$.

Of course, other situations can be considered too and different requirements can be added at the same time; many examples are given in Vannucci (2006), where however the definition of the $\mathscr{P}_j$s is slightly different, though substantially equivalent. That given in Eq. (6.74) allows to obtain always a positive definite (semi-definite, for identical layers) quadratic function $\mathscr{I}(\mathscr{P}_j)$.

By its same nature, cfr. Eqs. (5.51)–(5.54), problem (6.73) is *nonconvex* in the n-dimensional space of layer orientations δ_i, $i = 1, \ldots, n$, and several minima can exist. So, a suitable numerical strategy is needed to effectively search for the minima. This point is discussed in Sect. 6.4.3.

The problems of the first class formulated as in (6.73) concern exclusively elastic properties; nevertheless, by a similar approach, also other properties, like the thermal, hygral or piezoelectric ones, can be handled in a similar manner, cf. Sect. 6.9.

6.4.2 Identical Layers Laminates: The Lamination Set, Non Uniqueness of the Stacking Sequence

Looking at Eqs. (6.76) and (5.51)–(5.54), it is easy to recognize that, for laminates composed by identical layers, in the polar formalism the *material part*, represented by the polar parameters of the basic layer, is separated from the *geometric part*, that representing the geometry of the stack (layer positions and orientations), condensed in the lamination parameters, Eq. (5.55), cf. Sect. 5.2.12. In addition, the isotropic part is well separated by the anisotropic one and it is not represented by the lamination parameters, but uniquely by *material polar* parameters, those of the basic layer. These properties do not exist with a Cartesian representation.[4]

The separation of the material and geometric parts, i.e. of the polar parameters of the basic layer and of the lamination parameters representing the stack, clearly indicates that, once the material of the basic layer chosen, the only parameters to be considered for the design are the lamination parameters. This approach has been largely used in the literature since more than thirty years, though almost exclusively in conjunction with the Tsai and Pagano parameters, cf. Miki (1982, 1983, 1985), Fukunaga and Vanderplaats (1991), Fukunaga and Sekine (1992), Hammer et al. (1997), Autio (2000, 2001), Diaconu et al. (2002), Liu et al. (2004), Liu and Haftka (2004), but the literature on this subject is really plethoric, a general introduction to this kind of approach can be found in Gürdal et al. (1999).

The design, in the case of laminates composed by identical layers, can then be reduced to the space of the lamination parameters, and these are of course 12, because they represent the anisotropic part of the laminate.[5]

---

[4]In particular, it is worth noting that the separation of the isotropic from the anisotropic part is particularly interesting. As already specified in the previous section, the isotropic part of a laminate composed by identical layers cannot be modified by the optimization process. With the polar formalism, it is easy to eliminate this part from the design procedure, it is sufficient to disregard it, simply. This is, however, impossible using a Cartesian representation, where isotropic and anisotropic parts cannot be separated. In such a case, any optimization problem is formulated using 6 quantities for each one of the tensors $\mathbb{A}$, $\mathbb{B}$ and $\mathbb{D}$, while only 12 quantities are really needed, the four anisotropic polar parameters of the three above tensors. Any problem of optimization of design of laminates composed by identical layers is hence formally ill-formulated in the Cartesian framework.

[5]The four lamination parameters of $\mathbb{C}$ are superfluous if those of $\mathbb{A}$ and $\mathbb{D}$ are used, cf. Note 3 of Chap. 5.

The discussion about the space of the lamination parameters has already been done in Sect. 5.2.13. Here, we just recall that they are not independent, as apparent from Eq. (5.55), which is natural, considering that they all depend upon the same stacking sequence. Let us here consider just a simple case, that of the design of the properties of an uncoupled orthotropic $\mathbb{A}$ or $\mathbb{D}$. In such a case, the *lamination set* Ω, i.e. the domain of the admissible values for the lamination parameters, just depend upon only two of them: ξ_1 and ξ_3 for $\mathbb{A}$, ξ_{13} and ξ_{15} for $\mathbb{D}$. Because the lamination parameters are normalized quantities, the problem is formally identical in the two cases, and the lamination set corresponds to the sector of parabola shown, for the case of $\mathbb{D}$, in Fig. 4.7 (for $\mathbb{A}$, it is sufficient to replace ξ_{13} with ξ_1 and ξ_{15} with ξ_3).

A *lamination point* is a point of the lamination set; of course, a tensor, $\mathbb{A}$ or $\mathbb{D}$, is *completely determined by a lamination point*, i.e. there is a one-to-one correspondence between a tensor and a lamination point: the knowledge of the lamination parameters uniquely determine the elastic response of a laminate composed by identical layers. This is clear looking at Eqs. (5.51)–(5.54).

So, it is natural that a lot of approaches to the design of laminates is based upon the use of the lamination parameters, whose number do not depends on the number of layers n and, in addition, cf. Sect. 5.2.13 and Foldager et al. (1998), the optimization is convex. Nevertheless, one should be aware of the fact that determining the lamination point do not ends the design process. In fact, eventually the ply orientations δ_j are to be fixed. This is a key point, that is automatically solved with the approach detailed in the previous section, because the δ_js are the design variables, but not with the approach using the lamination parameters as primary design variables.

The way to determine the orientations once the lamination point known[6] is not unique. We give here a brief account of a simple method, proposed in Miki (1982). To this purpose, we use the results listed in Sect. 4.16.2 about the correspondence between some special points of the lamination set and the type of stacks (angle-ply, cross-ply, isotropic etc.). However, it is worth noting that the method outlined below is valid only for $\mathbb{A}$, because in the case of $\mathbb{D}$ the solution is affected by the position of the layers on the stacks through coefficients d_j, cf. Eq. (5.54).[7]

It is possible to put into relation lamination points on the contour of the lamination set with those inside the domain. In fact, let us consider balanced laminates; in this case, we can write

$$\xi_1 = \sum_{k=1}^{n_g} \nu_k \cos 4\delta_k, \quad \xi_3 = \sum_{k=1}^{n_g} \nu_k \cos 2\delta_k, \tag{6.77}$$

[6]The determination of the lamination point, i.e. of the optimal elastic behavior of the laminate, is, generally speaking, the first part of a problem of the third type, cf. Sect. 6.6.1.

[7]Miki has extended his method also to some cases of flexural design in Miki (1985).

where

$$\nu_k = \frac{n_k}{n} \tag{6.78}$$

is the *volume fraction* of layers of the group k among the n_g groups of orientations.

The lamination parameters are hence linear functions of the ν_k. So, the lamination parameters of the points of a segment linking two points of Ω represent laminates with volume fractions *proportional to its distances* from these two points. In other words, the vector of the volume fractions of a lamination point Q located between P and R, distant ℓ, is

$$\mathbf{v} = (1 - r)\mathbf{v}_P + r\mathbf{v}_R, \tag{6.79}$$

where $r\ell$ is the distance QP and $\mathbf{v}_P$ and $\mathbf{v}_R$ are the volume fraction vectors of P and R. That is why points on the line AB represent cross-ply laminates.

For instance, point P = (0.2, 1) represents a cross-ply with 60% of plies at 0° and 40% at 90°; it is the case, e.g. of laminates $[0_3/90_2]_s$.[8]

Nevertheless, any point of Ω belongs to infinite lines. This means that there is *not a bijective correspondence between lamination points and stacking sequences*. The same elastic properties can be obtained by different sequences: the solution is *never unique* in terms of the stacking sequence. This is a key point in laminate design: a given elastic behavior can be obtained with different laminates. This property, which actually enriches considerably the search for optimal laminates, will be thoroughly exploited in the problems of the third type, cf. Sect. 6.6.

As an example, we know that the origin O corresponds to isotropic laminates ($\xi_1 = \xi_3 = 0$). This lamination point can be obtained in different ways:

- segment AG: OA $= 2/3\ell \rightarrow 2/3$ of the plies at $\pm 60°$ and $1/3$ at $0°$; so $\mathbb{A}$ is isotropic if an equal number of plies is distributed on the orientations 0°, 60° and −60°;
- segment BH: same solution, rotated of 30°;
- segment CD: O is at equal distance form C and D, hence an isotropic laminate is obtained putting an equal number of plies at 0°, 45°, −45°, 90°;
- segment EF: same solution, rotated of 22.5°.

All these solutions are of the Werren and Norris type, cf. Sect. 5.3.5.1, but of course other solutions, more general, can be found. Evidently, these considerations apply not only to isotropic sequences, but to any other problem.

It is worth noting that, because any interior point of the lamination set can be determined as the intersection of two segments, it is sufficient to use balanced stacks with only two different unknown orientations $\pm\delta_1$ and $\pm\delta_2$ to realize any possible elastic state. Analytically, for a laminate of the type $[(\pm\delta_1)_{n_1}, (\pm\delta_2)_{n_2}]$, it is

$$\xi_1 = \nu_1 \cos 4\delta_1 + \nu_2 \cos 4\delta_2, \quad \xi_3 = \nu_1 \cos 2\delta_1 + \nu_2 \cos 2\delta_2, \tag{6.80}$$

---

[8] The subscript s after a stack list indicates that the laminate is symmetric and that the stack must be repeated symmetrically from s.

with, of course,

$$\nu_1 = \frac{n_1}{n}, \quad \nu_2 = \frac{n_2}{n}, \quad \nu_2 = 1 - \nu_1. \tag{6.81}$$

Hence, one must fix three parameters to have laminate corresponding to a given lamination point. From the above equations, one can obtain two parameters if the third one is chosen. For instance, if ν_1 is chosen, we get

$$\delta_1 = \frac{1}{2} \arccos Q_1, \quad \delta_2 = \frac{1}{2} \arccos Q_2, \tag{6.82}$$

with

$$\mathbb{Q}_1 = \frac{2\nu_1\xi_3 \pm \sqrt{2\nu_1\nu_2(1+\xi_1-2\xi_3^2)}}{2\nu_1}, \quad Q_2 = \frac{\xi_3 - \nu_1 Q_1}{\nu_2}. \tag{6.83}$$

Miki (1982), has also considered the case of laminate with 3 orientation groups: $0°$, $\pm\delta_1$, $\pm\delta_2$ (membrane orthotropy is of course preserved). In this case,

$$\xi_1 = \nu_1 \cos 4\delta_1 + \nu_2 \cos 4\delta_2 + \nu_3, \quad \xi_3 = \nu_1 \cos 2\delta_1 + \nu_2 \cos 2\delta_2 + \nu_3, \tag{6.84}$$

with

$$\nu_1 = \frac{n_1}{n}, \quad \nu_2 = \frac{n_2}{n}, \quad \nu_3 = 1 - \nu_1 - \nu_2. \tag{6.85}$$

For a choice of two parameters, one gets hence:

$$\delta_1 = \frac{1}{2} \arccos \frac{\xi_3 - \nu_2\Theta - \nu_3}{\nu_1}, \quad \delta_2 = \frac{1}{2} \arccos \Theta, \tag{6.86}$$

where

$$\begin{aligned} \Theta &= \frac{-K_b \pm \sqrt{K_b^2 - K_a K_c}}{K_a}, \\ K_a &= 2\nu_2(\nu_1 + \nu_2), \\ K_b &= 2\nu_2(\nu_3 - \xi_3), \\ K_c &= 2\nu_3^2 + 2\nu_3(\nu_1 - 2\xi_3) - \nu_1(1+\xi_1) + 2\xi_3^2. \end{aligned} \tag{6.87}$$

Anyway, this procedure often conflicts with the need of having volume fractions ν_k that are rational numbers that can be obtained dividing the n plies into n_g groups composed of an integer number of plies, which is often difficult to be done. That is why balanced laminates with only two or three set of orientations are not always used, though a method to find a compromise using sub-optimal solutions has been given in Miki and Sugiyama (1991). Examples of the application of the method of Miki are given in Sects. 6.4.3.2 and 6.5.2.

We end this section with a remark: computing the norms of Eq. $(5.51)_{3,4}$ one gets

$$\begin{aligned} R_0^A &= R_0^L\sqrt{\xi_1^2+\xi_2^2},\\ R_1^A &= R_1^L\sqrt{\xi_3^2+\xi_4^2}, \end{aligned} \tag{6.88}$$

and hence, for the case of $\mathbb{A}$ orthotropic,

$$\xi_1 = \frac{R_0^A}{R_0^L},\quad \xi_3 = \frac{R_1^A}{R_1^L}. \tag{6.89}$$

The lamination parameters represent hence the ratio between corresponding polar invariants of the laminate and of the basic layer; the same, of course, is true also for $\mathbb{B}$ and $\mathbb{D}$, as it can be easily recognized.

6.4.3 Numerical Approaches

As said in Sect. 6.4.1, the objective function (6.73) is nonconvex. This means that the global minimum is not unique and, generally speaking, for this type of problems there are many local and global minima. The advantage of using an approach like that outlined in Sect. 6.4.1 is that, for the case of laminates of identical layers, the value of the global minima is known: 0. This helps in understanding whether or not the result of a numerical optimization procedure is a true solution for the elastic problem (if a local minimum is found, this does not correspond to a laminate having the elastic properties to be obtained, because the objective function must be zero valued in correspondence of a solution).

Classical optimization methods, as descent strategies like gradient or conjugate gradient methods, are not suited for solving this kind of problems, because they are sensitive to the initial point, cf. Rao (2009).

More suited methods for nonconvex optimization are *metaheuristics*. The common feature of these methods is the fact that they do not improve a single initial evaluation of the objective function, but they work simultaneously on a *population of individuals*, i.e. of candidates to the solution. In this way, the exploration of the feasible domain is much more effective and the chances to find a true global minimum are greatly improved with respect to classical descent methods. Another advantage, very interesting for the design of laminates, is that metaheuristics are zero-order methods, i.e. they need just the evaluation of the objective function, not of its derivatives too. Hence, they can be used also when variables are discrete and not continuous, which is often the case with ply angles, for technological reasons.

Each metaheuristic is inspired by a given context: social, biological, physical paradigms are used to define a metaheuristic. The most well known of them, but not the only ones, are[9]:

- *simulated annealing*, proposed in Kirkpatrick et al. (1983) and inspired by the metallurgic procedure of annealing for metallic alloys, successfully employed in laminates design, see for instance (Erdal and Sonmez 2005; Javidrad and Nouri 2011);
- *genetic algorithms*, perhaps the most widely used metaheuristic, whose first idea can be traced back to the pioneer works of Rechenberg and Holland, in the middle of the sixties (Rechenberg 1965; Holland 1975); classical textbooks on the matter are Goldberg (1991), Michalewicz (1992) and for an introduction to their use in laminates design (Gürdal et al. 1999). Genetic algorithms are based upon Darwinian selection of the best fitted individuals and on genetic recombination; they are very robust algorithms and have proven their effectiveness in several fields. They have been widely used in laminates design, see e.g. Le Riche and Haftka (1993, 1995), Kogiso et al. (1994), Nagendra et al. (1996), Sadagopan and Pitchumani (1998), Todoroki and Haftka (1998), Kim et al. (1999), Liu et al. (2000), Muc and Gurba (2001), Soremekun et al. (2001a, 2002, 2001b), Tabakov (2001), Antonio (2002), Walker and Smith (2003), Antonio (2004), but the list is far from being exhaustive, as very numerous are the contributions, though often only technical papers. The author and co-workers have also worked extensively in this domain (Vincenti 2002; Vannucci 2002; Vannucci and Vincenti 2007; Vincenti et al. 2010, 2013; Ahmadian et al. 2011; Montemurro 2012; Montemurro et al. 2011, 2012a, c, d, e, 2015; Catapano 2013; Catapano et al. 2015), and they have produced a special genetic algorithm, the code BIANCA (Biologically Inspired ANalysis of Composite Assemblages) (Montemurro et al. 2012b), which is detailed further in Sect. 6.8.1;
- *particle-swarms*: introduced by Eberhart and Kennedy in 1995 (Eberhart and Kennedy 1995), this metaheuristic is based upon the paradigm of the social behavior of groups of individuals acting without a leadership, like the flocks of birds or fish shoals. The particle swarm optimization is based upon a very simple algorithm, whose best quality is rapidity. There are in the literature different works on particle swarm in laminates design, e.g. Omkar et al. (2008, 2009), Vannucci (2009a), Chang et al. (2010), Bargh and Sadr (2012), Ertas (2012). A brief account of this metaheuristic is given in Sect. 6.4.3.1, where a code, ALE-PSO, is briefly presented.
- *ant colonies*: this metaheuristic is inspired by the search strategy of ants; it has been proposed by Dorigo in 1992 (Dorigo 1992), and successively applied to different problems, also to laminates design, e.g. Aymerich and Serra (2008), Wang et al. (2010), Sebaey et al. (2011, 2013).

[9]In Colorni et al. (1996) different metaheuristics are presented and compared. The list in this section is far to be complete, other metaheuristics exist: neural networks, artificial immune algorithms, tabu search, the algorithm of flies and so on.

6.4.3.1 The Code ALE-PSO

The code ALE-PSO (Adaptive Local Evolution-PSO) is a particle swarm optimization algorithm (PSO, in short) which has the possibility to tune all the parameters of a standard PSO algorithm and to let evolve the main numerical coefficients along the computation steps through a power law specified by the user. This is done to improve the convergence capability of the code towards a true global minimum. Each coefficient can be updated independently by the others, so as to optimize the trade-off between the exploration and exploitation capabilities of the numerical procedure with respect to a given objective function. ALE-PSO can also handle constrained problems, by a technique which is essentially a barrier method.

Basically, the standard PSO algorithm can be condensed in the following updating rule:

$$\begin{aligned} u_{t+1}^k &= r_0\, c_0\, u_t^k + r_1\, c_1 (p_t^k - x_t^k) + r_2\, c_2 (p_t^g - x_t^k), \\ x_{t+1}^k &= x_t^k + u_{t+1}^k, \quad k = 1, \dots, m, \ t = 1, \dots, s, \end{aligned} \tag{6.90}$$

where:

- x_{t+1}^k: vector representing the position, in the n-dimensional problem space, of the k-th particle in a swarm of m particles, at the time-step $t+1$ of s total steps;
- u_{t+1}^k: displacement (often called *velocity*) of the particle x^k from its position x_t^k at the time step t to its updated position x_{t+1}^k at the time-step $t+1$;
- p_t^k: vector recording the best position occupied so far by the k-th particle (*personal best position*); in a minimization problem for the objective function $f(x)$, p_t^k is updated as follows:

$$p_{t+1}^k = \begin{cases} p_t^k & \text{if } f(x_{t+1}^k) \geq f(p_t^k), \\ x_{t+1}^k & \text{if } f(x_{t+1}^k) < f(p_t^k); \end{cases} \tag{6.91}$$

- p_t^g: vector recording the best position occupied so far by any particle in the swarm (*global best position*); in a minimization problem for the objective function $f(x)$, p_t^g is updated as follows:

$$p_{t+1}^g \in \{p_{t+1}^k,\ k = 1, \dots, m\} \text{ such that } f(p_{t+1}g) = \min_{k=1,\dots,m} f(p_{t+1}^k); \tag{6.92}$$

- $r_0,\ r_1,\ r_2$: independent random coefficients, uniformly distributed in the range $[0, 1]$;
- $c_0,\ c_1,\ c_2$: real coefficients called respectively *inertial*, *cognition* and *social parameter*.

Hence, it is clear that the algorithm explore the domain search using a *swarm*, i.e. m distinct *particles*, each one representing a point of the feasible domain. The updating of the position of each particle is formed through a linear combination of the previous velocity, the distance of the particle from its historical best position and its distance from the global best position found so far.

The use of the random coefficients $r_0,\ r_1,\ r_2$, along with an original swarm randomly generated, gives a stochastic nature to the algorithm, weighted, in some way, by the presence of the coefficients $c_0,\ c_1,\ c_2$. Their value has a paramount importance in the convergence, stability and search domain exploration of the algorithm. Only a good choice of these parameters can determine the success of the algorithm, but unfortunately their best choice is function-dependent, so sometimes it is difficult to found a suitable set of parameters. That is why in ALE-PSO a controlled variation of the coefficients is possible, based upon a power-law defined by the user. Other numerical strategies are used to improve the stability of the algorithm, namely controlling the maximum velocity, the exploration of the domain, using a suitable trajectory of the coefficients in the Trelea's deterministic domain (Trelea 2003), etc.

The interested reader is addressed to the paper (Vannucci 2009a) for a deeper insight in the code ALE-PSO. What is interesting to notice is the true simplicity of a PSO algorithm: the core of the method is based upon just two updating rules, which makes this type of algorithm particularly rapid. To be noticed that no derivatives are to be calculated, so discrete variables and discontinuous objective functions can be handled as well.

6.4.3.2 Some Examples of Design Problems of the First Type

Example 1

This example concerns the research of a 12-ply laminate, composed by carbon-epoxy layers of the type T300/5208 in Table 4.9, that must be uncoupled and fully isotropic. It is a sort of benchmark in laminate design, because it seems impossible to obtain a fully isotropic laminate composed by less than 12 identical plies (Grédiac 1999; Vincenti et al. 2003). In this case, the objective function to be minimized is obtained combining the cases of uncoupling, extension and bending isotropy, cf. Sect. 6.4.1, so it is determined by the set of H_{rs}

$$H_{3\,3} = H_{4\,4} = H_{9\,9} = H_{10\,10} = H_{15\,15} = H_{16\,16} = 1, \tag{6.93}$$

and hence the problem corresponds to the minimization of the objective function[10]

$$\mathscr{I}(\mathscr{P}_j) = \mathscr{P}_3^2 + \mathscr{P}_4^2 + \mathscr{P}_9^2 + \mathscr{P}_{10}^2 + \mathscr{P}_{15}^2 + \mathscr{P}_{16}^2, \tag{6.94}$$

with $\mathscr{P}_j(\delta_k),\ k = 1, \ldots, n$: the design variables are the layer orientations δ_k.

This problem has been solved using the code ALE-PSO: a series of numerical tests have been performed, each one composed by 100 runs of the code with the same initial swarm, composed of 100 particles. For each test, two quantities obtained in a run of 300 iterations are considered: the best value of the average on the swarm and

[10] In all the examples of this chapter, the modulus $\mathscr{M}$ has been fixed using Eq. (6.75) applied to the basic layer composing the laminate. For carbon-epoxy T300/5208, often used in the following, it is $\mathscr{M} = 64.592$ GPa.

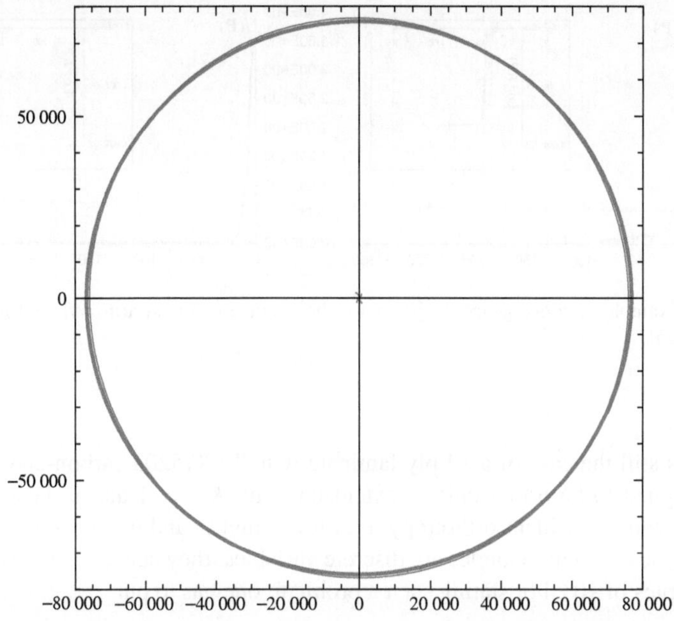

Fig. 6.7 Example 1: directional diagrams of A_{1111}, *thin line*, B_{1111}, in the *center*, and D_{1111}, *thick line* (units: MPa)

the value of the best particle in the swarm (i.e. the particle having the lowest value of the objective function).

The best solution so found corresponds to the stacking sequence

$$[-16.4°, -70.3°, 66.7°, 31.9°, 31.7°, -59.0°,\\ 78.6°, -12.1°, -12.3°, -47.0°, -90.0°, 36.0°],$$

for which the final value of the objective function is $\mathscr{I}(\mathscr{P}_j) = 0.023$. Of course, like in any other numerical approach, this value is to be considered as a residual (theoretically, the result corresponds to $\mathscr{I}(\mathscr{P}_j) = 0$); such a residual, from a practical point of view, is acceptable. This can be checked in Fig. 6.7, where the directional diagrams of A_{1111}, B_{1111} and D_{1111} are plotted[11]: A_{1111} and D_{1111} are practically circular, i.e. isotropic, while B_{1111} is reduced to a point in the center, i.e. $B_{11} \simeq 0$. In Fig. 6.8 we show the convergence diagrams for the average of the swarm and for the best particle. It is apparent the rapidity of the convergence, especially for the best particle.

[11]For the sake of shortness, in this and in the following examples, we just plot the directional diagrams of A_{1111}, B_{1111} and D_{1111} to show the quality of the results; anyway, other tensor components could be traced, showing the same quality of the results.

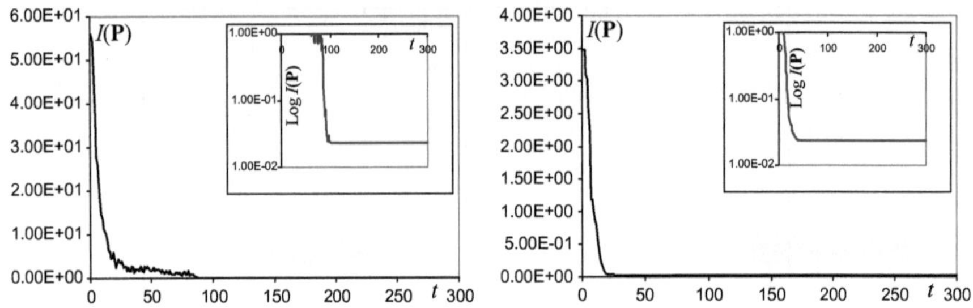

Fig. 6.8 Example 1: convergence diagrams for the average of the swarm, *left*, and for the best particle, *right*

Example 2

This is still the case of a 12-ply laminate with T300/5208 carbon-epoxy layers, now designed to be orthotropic in extension, with $K^A = 1$ and in bending, with $K^D = 0$, with coincident orthotropy axes for $\mathbb{A}$ and $\mathbb{D}$ and uncoupled, $\mathbb{B} = \mathbb{O}$. In addition, the orientation angles are discrete variables: they can take only values that are multiples of 10°. For stating such a problem, one has to put

$$
\begin{aligned}
&H_{5\,5} = H_{6\,6} = H_{9\,9} = H_{10\,10} = H_{17\,17} = H_{18\,18} = 1,\\
&\Phi^A = 45°,\ \Phi^D = 0°,\ \Phi_1^{A^*} = \Phi_1^{D^*},
\end{aligned}
\tag{6.95}
$$

i.e. the objective function to be minimized is

$$
\mathcal{I}(\mathcal{P}_j) = \mathcal{P}_5^2 + \mathcal{P}_6^2 + \mathcal{P}_9^2 + \mathcal{P}_{10}^2 + \mathcal{P}_{17}^2 + \mathcal{P}_{18}^2. \tag{6.96}
$$

This problem has been solved using the genetic code BIANCA. The best solution found after 200 iterations on a population of 300 individuals,[12] is

$$
[0°, 10°, 40°, -40°, -20°, 50°, -20°, 30°, -40°, 30°, 10°, -10°],
$$

to which corresponds a residual $\mathcal{I}(\mathcal{P}_j) = 5.08 \times 10^{-4}$. Also for this example, we show in Fig. 6.9 the directional diagrams of A_{1111}, B_{1111} and D_{1111}.

Looking at Fig. 6.9, one can see that the laminate is practically uncoupled (the diagram of B_{1111} is extremely small compared to that of A_{1111} and D_{1111}) and that $\mathbb{A}$ and $\mathbb{D}$ are really orthotropic with practically coincident orthotropy axes: the main orthotropy axis of $\mathbb{A}$ is turned of 3.33° with respect to the abscissae axis, while that of $\mathbb{D}$ of 3.90°. These last are not aligned with the coordinate axes because this requirement has not be asked for (the condition $\Phi_1^{A^*} = \Phi_1^{D^*} = 0$ should be added for that).

[12]This corresponds to a duration of the run of ~3 s on a standard PC. Several other examples of this kind can be found in Vannucci (2002), p. 152.

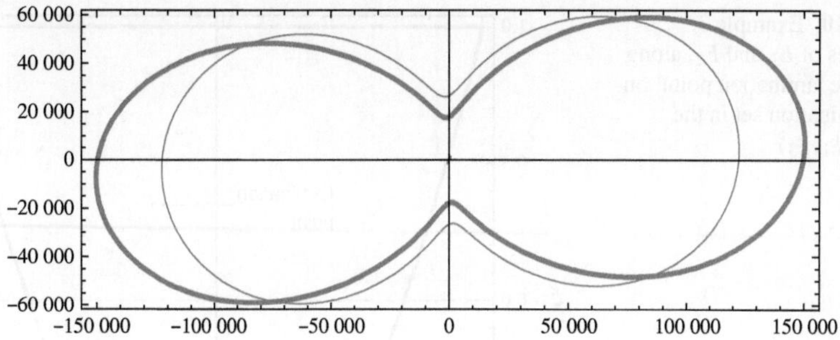

Fig. 6.9 Example 2: directional diagrams of A_{1111}, *thin line*, B_{1111}, in the *center*, and D_{1111}, *thick line* (units: MPa)

Example 3

Let us consider in this example the design of a laminate that must be uncoupled and orthotropic in extension, with given values of the Young's modulus in the x_1- and x_2-direction. This kind of problems can be solved using the approach detailed in Sect. 6.4.2.

Because the laminate is uncoupled and orthotropic in extension, only two lamination parameters are concerned, ξ_1 and ξ_3, and it is

$$E_1 = \frac{1}{\mathscr{A}_{1111}} = A_{1111} - \frac{A_{1122}^2}{A_{2222}}, \quad E_2 = \frac{1}{\mathscr{A}_{2222}} = A_{2222} - \frac{A_{1122}^2}{A_{1111}}, \tag{6.97}$$

so that with the polar formalism, Eq. (4.60), we get

$$\begin{aligned} E_1 &= \frac{8[T_0^L T_1^L + (-1)^{K^L} T_1^L R_0^L \xi_1 - 2R_1^{L^2} \xi_3^2]}{T_0^L + 2T_1^L + (-1)^{K^L} R_0^L \xi_1 - 4R_1^L \xi_3}, \\ E_2 &= \frac{8[T_0^L T_1^L + (-1)^{K^L} T_1^L R_0^L \xi_1 - 2R_1^{L^2} \xi_3^2]}{T_0^L + 2T_1^L + (-1)^{K^L} R_0^L \xi_1 + 4R_1^L \xi_3}. \end{aligned} \tag{6.98}$$

The intersection of the two iso-lines of E_1 and E_2 in the lamination set corresponds to the lamination point. Using then the rules specified in Sect. 6.4.2 to obtain a laminate with two or three different orientation groups, a stacking sequence can be obtained. To be specific, let us consider again the case of a laminate made by T300/5208 carbon-epoxy layers, designed to have $E_1 = 100$ GPa, $E_2 = 60$ GPa, with two orientation groups and the volume fraction $v_1 = 0.60$. Then we get, solving numerically Eq. (6.98) and using Eqs. (6.82) and (6.83) we obtain

$$\xi_1 = 0.346, \; \xi_3 = 0.242 \; \rightarrow \; \delta_1 = 14.02°, \; \delta_2 = 67.92°.$$

The corresponding iso-lines of E_1 and E_2, determining the lamination point, are shown in Fig. 6.10.

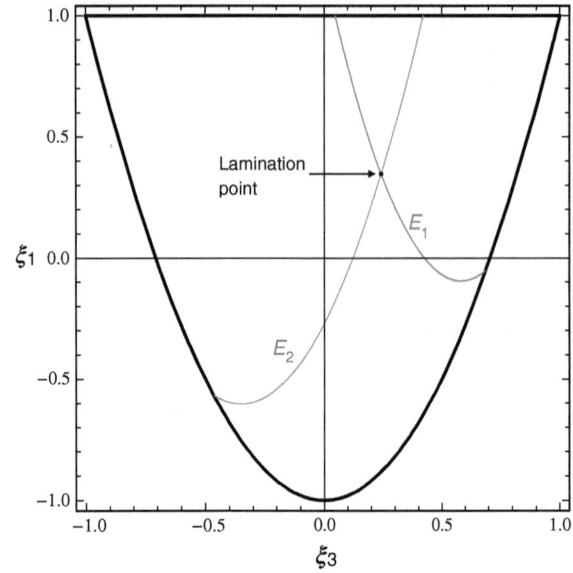

Fig. 6.10 Example 3: iso-lines of E_1 and E_2, along with the lamination point, on the lamination set in the plane (ξ_1, ξ_3)

Because the orientations must be $\pm\delta_1$ and $\pm\delta_2$, considering the value of ν_1 and the fact that the laminate has to be uncoupled, a possible solution is a 20-ply laminate of the type $[\delta_1, -\delta_1, \delta_2, -\delta_2, \delta_1, -\delta_1, \delta_2, -\delta_2, \delta_1, -\delta_1]_s$.

The directional diagrams of E for extension and bending are plotted in Fig. 6.11. It is apparent that the diagram for extension corresponds to an orthotropic behavior, while that of bending is completely anisotropic. It is possible, however, to obtain an orthotropic behavior in bending exactly equal to that in extension if, instead of the above sequence, a quasi-trivial solution of the quasi-homogeneous type, with $\nu_1 = 0.6$, is used. This is the case, e.g. of the first solution for the 20-ply given in Table 5.4.

Still using the quasi-trivial solutions for uncoupling, a great reduction of the number of plies can be obtained, because it is no more needed to double the stack in order to ensure uncoupling. There exist different distinct solutions of this type, e.g.:

$$[\delta_1, -\delta_1, \delta_2, -\delta_2, \delta_1, \delta_2, -\delta_1, \delta_1, -\delta_2, -\delta_1],$$
$$[\delta_1, -\delta_1, \delta_2, -\delta_2, \delta_2, \delta_1, -\delta_1, \delta_1, -\delta_1, -\delta_2],$$
$$[\delta_1, \delta_2, -\delta_2, \delta_2, -\delta_1, -\delta_2, \delta_1, \delta_2, -\delta_1, -\delta_2],$$
$$[\delta_1, \delta_2, -\delta_2, \delta_1, -\delta_1, -\delta_1, \delta_2, \delta_1, -\delta_1, -\delta_2].$$

All of them are un-symmetric stacks of 10 plies, i.e. half the number of plies used above, and all of them share the same diagram of E for extension, exactly the one shown in Fig. 6.11, but they differ for that in bending. This simple example shows how much can be interesting, whenever the reduction of the weight is sought, to use un-symmetric stacks. If one changes the requirements on E_1 and E_2, without changing ν_1, the above stacks will still give an uncoupled orthotropic laminate, matching the specified requirement provided that the corresponding values of δ_1 and δ_2 are used.

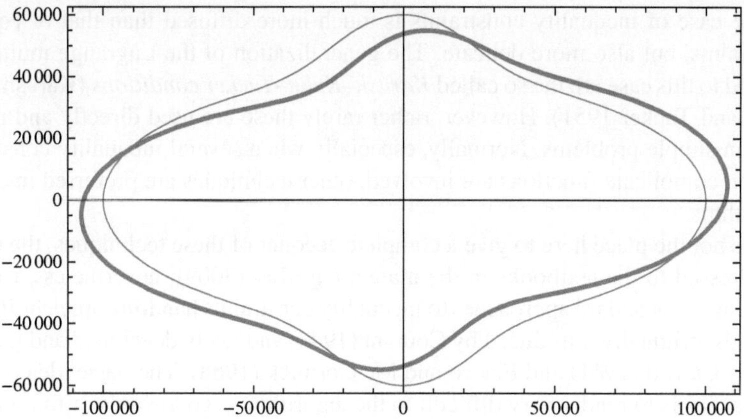

Fig. 6.11 Example 3: directional diagrams of E for extension, *thin line*, and bending, *thick line* (units: MPa)

6.5 Design Problems of the Second Type

What basically changes in the problems of the second type with respect to those of the first one, is the fact that now some constraints, that can be of different nature, are added to the optimum problem, that generally speaking can be stated as

$$\begin{aligned} &\min_{\mathscr{P}_j} \mathscr{I}(\mathscr{P}_j) = \{\mathscr{P}\}^\top [H]\{\mathscr{P}\} = H_{rs}\mathscr{P}_r\mathscr{P}_s, \ \ j,\ r,\ s = 1, \ldots, 18, \\ &\text{with} \\ &g_i(\mathscr{P}_j) \leq 0, \ \ i = 1, \ldots, n_g \\ &\text{and} \\ &h_l(\mathscr{P}_j) = 0, \ \ l = 1, \ldots, n_h. \end{aligned} \tag{6.99}$$

Here, there are n_g inequality constraints in the standard form $g_i(\mathscr{P}_j) \leq 0$ and n_h equality constraints $h_l(\mathscr{P}_j) = 0$ to be satisfied.

6.5.1 *Methods for Handling Constraints*

There are different ways to handle constraints in an optimization problem and the most part of them are well suited for a specific numerical approach. For what concerns equality constraints, the use of the classical Lagrange multipliers approach can be widely used and sometimes direct substitution techniques are also possible, this depends upon the mathematical structure of the objective and constraint functions.

The case of inequality constraints is much more diffused than that of equality constraints, but also more delicate. The generalization of the Lagrange multipliers method to this case are the so called *Karush–Kuhn–Tucker conditions* (Karush 1939; Kuhn and Tucker 1951). However, rather rarely these are used directly and principally in simple problems. Normally, especially when several inequality constraints and, or, complicate functions are involved, other techniques are preferred in coding algorithms.

It is not the place here to give a complete account of these techniques, the reader is addressed to the textbooks in the matter, e.g. Rao (2009); nevertheless, roughly speaking, the standard approaches to inequality constraints handling are *penalization* methods, originally introduced by Courant (1943) and lately developed and generalized by Carroll (1961) and Fiacco and Mc Cormick (1968). The basic idea of these approaches is to render very difficult to the algorithm to go towards or to remain in unfeasible regions of the search domain. This can be done penalizing the objective function, i.e. adding some artificial terms in order to increase greatly its value, so rendering such points uninteresting for the search procedure: the minimum cannot be there.

The adaptation of such ideas to an optimization method depends strongly upon the type of algorithm. Because we have seen that the most suited algorithms for solving laminates design minimum problems are metaheuristics, able to counter non-convexity, we briefly introduce and discuss here penalization methods for metaheuristics like genetic algorithms. A detailed survey of this matter can be found in different works, e.g. Michalewicz (1992), Michalewicz and Schoenauer (1996), Dasgupta and Michalewicz (1997), Coello-Coello (2002).

The basic idea of penalization methods is to transform the constrained problem into an unconstrained one through the addition of some terms to the objective function, based on the amount of constraint violation, which depends upon the point in the search space. Penalty methods are well suited for a wide class of problems; however the question is how to find a proper penalty factor when balancing between feasible and infeasible solutions. We briefly introduce here a technique, called ADP-*Automatic Dynamic Penalization*, first introduced in Vincenti et al. (2010) and then further developed in Montemurro et al. (2013),[13] where the choice of the penalty factor is automatically done by the algorithm, that lets also evolve penalization during iterations, in order to adapt it to the different phases of exploration of the search space.

Generally speaking, the original constrained minimum problem (6.99) is transformed into

---

[13] A discussion of penalization methods as well as a detailed analysis of the functioning of the ADP method and different examples can be found in this reference; for the sake of shortness, we address the reader to it for a deeper insight in the matter.

$$\min_{\mathcal{P}_j} \mathcal{I}_p(\mathcal{P}_j)$$

where

$$\mathcal{I}_p(\mathcal{P}_j) = \begin{cases} \mathcal{I}(\mathcal{P}_j), & \text{if } g_i(\mathcal{P}_j) \leq 0 \ \forall j \text{ and } h_l(\mathcal{P}_j) = 0 \ \forall l, \\ \mathcal{I}(\mathcal{P}_j) + \sum_{i=1}^{n_g} c_i G_i(\mathcal{P}_j) & \text{if } g_i(\mathcal{P}_j) > 0 \text{ and } h_l(\mathcal{P}_j) = 0 \ \forall l, \\ \mathcal{I}(\mathcal{P}_j) + \sum_{k=1}^{n_h} q_k H_k(\mathcal{P}_j) & \text{if } g_i(\mathcal{P}_j) \leq 0 \ \forall j \text{ and } h_l(\mathcal{P}_j) \neq 0, \\ \mathcal{I}(\mathcal{P}_j) + \sum_{i=1}^{n_g} c_i G_i(\mathcal{P}_j) + \sum_{k=1}^{n_h} q_k H_k(\mathcal{P}_j) & \\ \qquad \text{if } g_i(\mathcal{P}_j) > 0 \text{ and } h_l(\mathcal{P}_j) \neq 0. \end{cases} \tag{6.100}$$

In the above equation, $\mathcal{I}_p(\mathcal{P}_j)$ is the penalized objective function, while c_i and q_k are the *penalty coefficients*, for inequality and equality constraints respectively. Moreover,

$$\begin{aligned} G_i(\mathcal{P}_j) &= \max\{0,\ g_i(\mathcal{P}_j)\}, \quad i = 1, \ldots, n_g, \\ H_k(\mathcal{P}_j) &= \max\{0,\ |h_k(\mathcal{P}_j)| - \varepsilon\}, \quad k = 1, \ldots, n_h. \end{aligned} \tag{6.101}$$

It is worth noting that in this way quality constraints are transformed, as usually done, into inequality ones of the form

$$|h_k(\mathcal{P}_j)| - \varepsilon \leq 0, \tag{6.102}$$

which is numerically acceptable if the *tolerance* ε is sufficiently small so as to not affect the quality of the solution. In this way, all the constraints are of the same nature, i.e. inequalities, and they can be handled by the same technique.

It is well known, see e.g. Coello-Coello (2002), that the penalty factors should have a low value, just above the limit below which infeasible solutions are optimal; this is the so-called *minimum penalty rule*. This is why if the penalty is too high or too low, then the genetic algorithm (GA) could experience some difficulties in finding an appropriate feasible optimal solution.

One should consider that in GAs' exploration of the search domain, some infeasible individuals can be anyway important to drive the exploration towards interesting zones of the feasible domain, namely when the optimum point lies on its boundary, i.e. on an active constraint (which is often the case in structural optimization). That is why in the ADP strategy infeasible points are not automatically excluded from the population and are used to dynamically update the penalty coefficients in an automatic way, i.e. without the intervention of the user. This is especially important at the early stages of the search in order to widely explore the whole search space.

About the penalty coefficients c_i and q_k, their choice is of the paramount importance for the correct functioning of the search procedure. However, to choose appropriately these coefficients is very difficult and it could be useful to have an automatic estimation of them and to adjust penalty pressure along the generations by tuning these coefficients. This is done in the ADP method putting, at each generation t,

$$\begin{aligned} c_i(t) &= \frac{|\mathscr{I}(\mathscr{P}_j)^F_{best} - \mathscr{I}(\mathscr{P}_j)^{NF}_{best}|}{G_i(\mathscr{P}_j)^{NF}_{best}}, \quad i = 1, \dots, n_g, \\ q_k(t) &= \frac{|\mathscr{I}(\mathscr{P}_j)^F_{best} - \mathscr{I}(\mathscr{P}_j)^{NF}_{best}|}{H_k(\mathscr{P}_j)^{NF}_{best}}, \quad k = 1, \dots, n_h, \end{aligned} \tag{6.103}$$

where the superscripts F and NF stand for feasible and non-feasible, respectively, while $\mathscr{I}(\mathscr{P}_j)^F_{best}$ and $\mathscr{I}(\mathscr{P}_j)^{NF}_{best}$ are the values of the objective function for the best individuals in the feasible and the infeasible parts of the search domain. Finally, $G_i(\mathscr{P}_j)^{NF}_{best}$ and $H_k(\mathscr{P}_j)^{NF}_{best}$ represent the violated inequality and equality constraints for the best infeasible solution.

This strategy allows to well explore the search domain also in the infeasible part and especially around the boundaries of the feasible domain and has given excellent results in handling problems of the second and the third types for laminates. More details can be found in Montemurro et al. (2013), we just give in the next section an example of application of the ADP method, as implemented in the code BIANCA, to laminates design.

Other constraint handling strategies can be appropriate to metaheuristics other than GAs. In ALE-PSO a *barrier* method has been used: differently from the ADP method, now the swarm must be feasible at each iteration, i.e. it must be composed by particles respecting all the constraints imposed to the computation. Hence, a numerical barrier is erected along the boundary of the feasible domain in order to prevent particles from going out of it. To do this, the initial swarm is composed by feasible particles and its updates at iteration $t+1$ are guaranteed to be feasible by the following procedure

$$x^k_{t+1} = x^k_t + \frac{u^k_{t+1}}{2^m}, \quad m = 0, 1, 2, \dots; \tag{6.104}$$

the iterations are stopped when the constraints are all satisfied. As x^k_t is feasible, this bisection method ensures that, for a sufficiently large m, x^k_{t+1} will be feasible. This is a simple and robust procedure and numerical tests have shown that also minima on the boundary of the feasible domain can be reached (Vannucci 2009a). The use of a barrier method is rather important for a PSO algorithm, because the evolution of the swarm depends upon two points, p_k and p_g, which must be feasible. An example of the use of this barrier method in the code ALE-PSO is given below.

For what concerns the approach of Miki, outlined in Sect. 6.4.2, in problems of the second type, a procedure particular to this method is illustrated in *Example 3* of the next section.

6.5.2 Some Examples of Design Problems of the Second Type

Example 1

We consider the case of a rectangular simply supported 16-ply laminate, which must be designed to be uncoupled and orthotropic in extension and in bending, with the axes of orthotropy aligned with the plate's sides. In addition, we impose that the orientation angles must be integer numbers and that $E_1^A \geq E_1^0,\ E_2^A \geq E_2^0,\ \lambda_{pq}^{crit} \geq \lambda_0$, where E_1^A and E_2^A are the in-plane Young's moduli in the two directions x_1 and x_2 of the plate's axes, and λ_{pq}^{crit} is the buckling multiplier of the pq mode, i.e. the factor by which the applied loads must be multiplied to buckle the plate with p half-waves along x_1 and q along x_2. The objective function is determined by a set of H_{rs} similar to that of *Example 2* in Sect. 6.4.3.2:

$$\begin{gathered} H_{5\,5} = H_{6\,6} = H_{9\,9} = H_{10\,10} = H_{17\,17} = H_{18\,18} = 1, \\ \Phi^A = K^A \frac{\pi}{4},\ \Phi^D = K^D \frac{\pi}{4},\ \Phi_1^{A^*} = \Phi_1^{D^*} = 0, \end{gathered} \tag{6.105}$$

and the objective function to be minimized is hence

$$\mathcal{I}(\mathcal{P}_j) = \mathcal{P}_5^2 + \mathcal{P}_6^2 + \mathcal{P}_9^2 + \mathcal{P}_{10}^2 + \mathcal{P}_{17}^2 + \mathcal{P}_{18}^2. \tag{6.106}$$

To this objective function, one must add the constraints, given here in the standard form,

$$E_1^0 - E_1^A \leq 0,\ \ E_2^0 - E_2^A \leq 0,\ \ \lambda_0 - \lambda_{pq}^{crit} \leq 0. \tag{6.107}$$

The Young's moduli are computed according to Eq. (6.98) and the buckling multiplier by Eq. (6.57). We just remark that, cf. Eqs. (5.54), (5.55), (6.76) and (6.89),

$$\begin{aligned} \xi_1 &= \frac{R_0^A}{R_0^L} = \frac{\mathcal{M}\,\mathcal{P}_3}{R_0^L}, \quad \xi_3 = \frac{R_1^A}{R_1^L} = \frac{\mathcal{M}\,\mathcal{P}_4}{R_1^L}, \\ \xi_{13} &= \frac{R_0^D}{R_0^L} = \frac{\mathcal{M}\,\mathcal{P}_{15}}{R_0^L}, \quad \xi_{15} = \frac{R_1^D}{R_1^L} = \frac{\mathcal{M}\,\mathcal{P}_{16}}{R_1^L}, \end{aligned} \tag{6.108}$$

so that $E_1^A = E_1^A(\mathcal{P}_j),\ E_2^A = E_2^A(\mathcal{P}_j),\ \lambda_{pq}^{crit} = \lambda_{pq}^{crit}(\mathcal{P}_j)$ too.

To be specific, we consider a square plate whose side is $a = b = 100$ mm, loaded by in-plane uniformly distributed normal forces along the boundary $N_1 = N_2 = 1$ N/mm, composed by carbon-epoxy T300/5208 layers. Moreover, we put $p = q = 1,\ \lambda_0 = 80,\ E_1^0 = 60$ GPa and $E_2^0 = 30$ GPa. The problem has been solved using the GA BIANCA, with a population of 500 individuals and for 120 generations, while the ADP method has been used for handling the constraints.

The best solution so found is the following one:

$$\begin{gathered}[-2^\circ, -9^\circ, 28^\circ, 5^\circ, -73^\circ, -12^\circ, -13^\circ, 30^\circ, \\ 82^\circ, -21^\circ, -88^\circ, 9^\circ, -10^\circ, 21^\circ, 7^\circ, -14^\circ],\end{gathered}$$

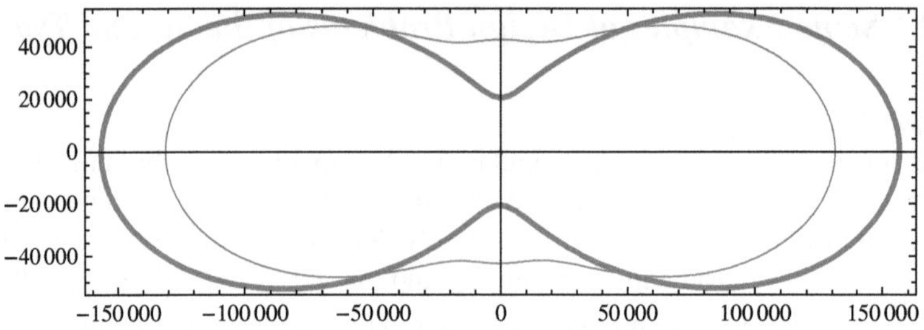

Fig. 6.12 Example 1: directional diagrams of A_{1111}, *thin line*, B_{1111}, in the *center*, and D_{1111}, *thick line* (units: MPa)

to which correspond $\lambda_{11}^{crit} = 84.4$, $E_1^A = 127.7$ GPa and $E_2^A = 41.6$ GPa, for a residual of $\sim 8 \times 10^{-5}$, hence practically null.

Also for this example, we show in Fig. 6.12 the directional diagrams of A_{1111}, B_{1111} and D_{1111}. Once more Fig. 6.12 clearly shows that the laminate is practically uncoupled (the diagram of B_{1111} is extremely small compared to that of A_{1111} and D_{1111}) and that $\mathbb{A}$ and $\mathbb{D}$ are really orthotropic with coincident orthotropy axes and aligned with the plate's sides. We only remark that $\mathbb{A} \neq \mathbb{D}$, as obvious, because quasi-homogeneity has not been required.

Example 2

We consider a 12-plies rectangular laminate, made of carbon-epoxy T300/5208, cf. Table 4.9, designed to be uncoupled and orthotropic in bending, with the axes of orthotropy aligned with the plate's sides. The plate is simply supported, with side lengths a and b. The additional requirement is that the frequency of the first mode of the bending vibrations must be greater than a minimum specified value ω_0.

The objective function is determined by the choice

$$H_{9\,9} = H_{10\,10} = H_{17\,17} = H_{18\,18} = 1, \ \ \Phi^D = K^D \frac{\pi}{4}, \ \ \Phi_1^{D^*} = 0, \tag{6.109}$$

and the constrained minimum problem becomes hence

$$\begin{aligned} &\min_{\mathscr{P}_j} \ (\mathscr{P}_9^2 + \mathscr{P}_{10}^2 + \mathscr{P}_{17}^2 + \mathscr{P}_{18}^2), \\ &\text{with} \\ &\omega_0 - \omega_{pq}(\mathscr{P}_j) \leq 0, \end{aligned} \tag{6.110}$$

where the vibration frequency of the mode pq, having p alf-waves along the x_1-axis and q along the x_2 one, is given by Eq. (6.58). We notice that, like in the case of the previous example for λ_{pq}^{crit}, also in this case and for the same reasons, $\omega_{pq} = \omega_{pq}(\mathscr{P}_j)$ too.

This problem has been solved using the code ALE-PSO and the barrier method described in the previous section. To be specific, we have considered a plate with sides $a = 1$ m, $b = 0.5$ m, designed to have $\omega_{11} \geq \omega_0 = 150$ s^{-1}; the mass per unit area is $\mu = 2.4$ kg/m^2. The best result, obtained using the same procedure indicated in Sect. 6.4.3.2, is the laminate whose stacking sequence is

$$[-66.1°, 66.5°, 44.8°, -36.4°, 48.5°, -5.7°,\\ 77.6°, -90.0°, -84.5°, 38.4°, 57.5°, -46.0°],$$

to which corresponds a residual very close to zero, $\mathscr{I}(\mathscr{P}_j) = 7.2 \times 10^{-35}$ and the frequency $\omega_{11} = 162.72$ s^{-1}.

We show in Fig. 6.13 the directional diagrams of A_{1111} and D_{1111}; the diagram of B_{1111} this time is invisible, because the solution is really uncoupled, due to the good quality of the solution. We can remark that, while $\mathbb{D}$ satisfies the requirements, i.e. it is orthotropic, in particular with $K^D = 1$, and with the axes of orthotropy aligned with the plate's sides, the diagram of A_{1111} clearly shows that $\mathbb{A}$ is completely anisotropic. We have hence here the case of an un-symmetric uncoupled laminate which is orthotropic in bending, but not in extension.

Finally, in Fig. 6.14 we show the convergence diagrams for the average of the swarm and for the best particle.

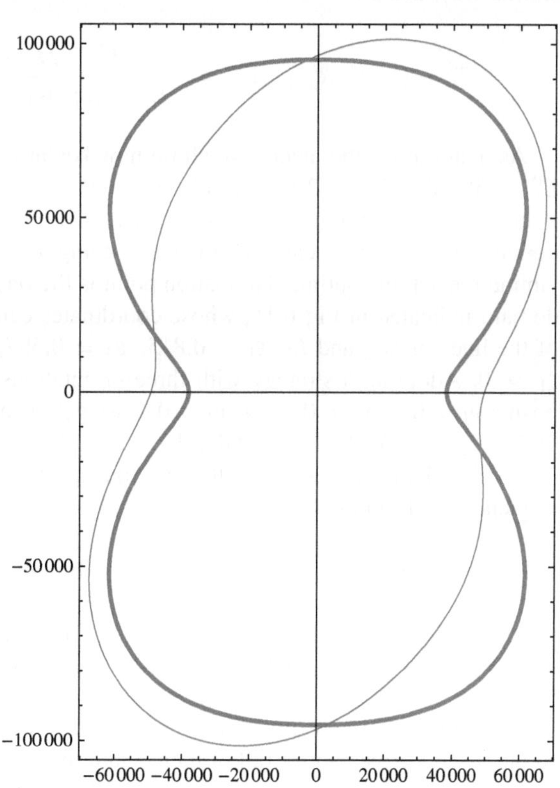

Fig. 6.13 Example 2: directional diagrams of A_{1111}, *thin line*, and D_{1111}, *thick line* (units: MPa)

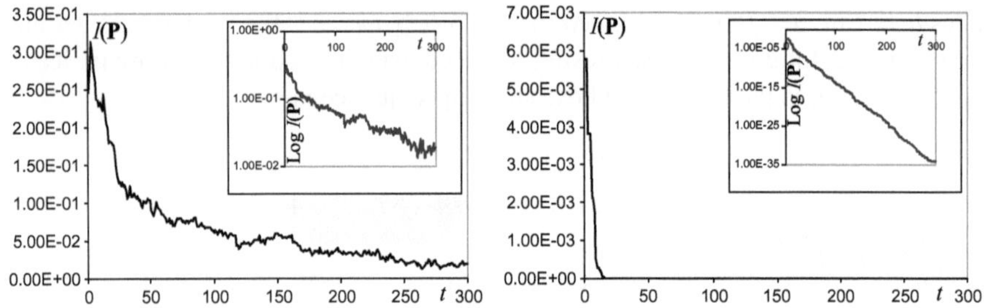

Fig. 6.14 Example 2: convergence diagrams for the average of the swarm, *left*, and for the best particle, *right*

Example 3

Let us consider now the case of a laminate designed to be uncoupled, orthotropic in extension, for which the in-plane Young's modulus E_1 must be maximized under the constraints $E_1 \geq E_1^0$, $E_2 \geq E_2^0$, $G_{12} \geq G_{12}^0$ and $\nu_{12} \leq \nu_{12}^0$.

This situation is quite similar to that of *Example 3*, Sect. 6.4.3.2, and E_1 and E_2 are still given by Eq. (6.98). For what concerns G_{12} and ν_{12}, these can be found in a similar way, and are

$$G_{12} = T_0^L - (-1)^{K^L} R_0^L \xi_1, \quad \nu_{12} = -\frac{T_0^L - 2T_1^L + (-1)^{K^L} R_0^L \xi_1}{T_0^L + 2T_1^L + (-1)^{K^L} R_0^L \xi_1 - 4R_1^L \xi_3}. \tag{6.111}$$

As material for the layers, we chose now kevlar-epoxy 149 in Table 4.9 and put $E_1^0 = 50$ GPa, $E_2^0 = 20$ GPa, $G_{12}^0 = 4$ GPa and $\nu_{12}^0 = 0.3$. With these values, tracing the corresponding iso-lines for E_1, E_2, G_{12} and ν_{12}, we get the feasible domain in Fig. 6.15. Because E_1 increases going towards the upper right part of the lamination set, the optimal lamination point is the one on the border of the feasible domain indicated in Fig. 6.15, whose coordinates can be found as the intersection of the lines of G_{12} and E_2: $\xi_1 = 0.815$, $\xi_3 = 0.597$, and hence $K^A = 0$, because $\xi_1 > 0$. Adopting a strategy with three orientations groups, with $\delta_3 = 0°$, after posing $\nu_1 = 0.2$, $\nu_2 = 0.4 \rightarrow \nu_3 = 0.4$, we get, using Eqs. (6.86) and (6.87), $\delta_1 = 76.3°$, $\delta_2 = 10.3°$, $E_1 = 68.4$ GPa, $E_2 = 20$ GPa, $G_{12} = 4$ GPa and $\nu_{12} = 0.18$.

Possible laminates are the following four quasi-trivial uncoupled un-symmetric sequences of 10 plies:

$$[\delta_1, 0, \delta_2, -\delta_2, 0, \delta_2, 0_2, -\delta_2, -\delta_1],$$
$$[0, \delta_1, \delta_2, -\delta_2, \delta_2, 0_3, -\delta_1, -\delta_2],$$
$$[0, \delta_2, \delta_1, -\delta_2, 0, -\delta_2, 0, -\delta_1, 0, \delta_2],$$
$$[0, \delta_2, -\delta_2, 0, \delta_1, -\delta_1, \delta_2, 0_2, -\delta_2].$$

We show in Fig. 6.16 the directional diagram of E for extension and bending, referred to the first sequence above (all of them share the same diagram of E for extension,

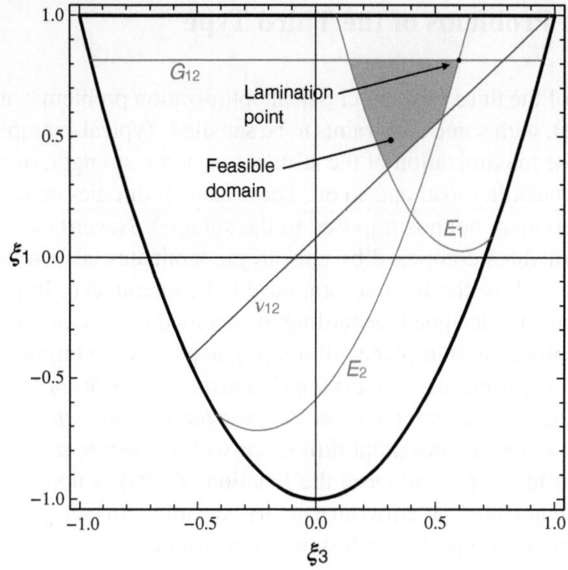

Fig. 6.15 Example 3: feasible domain and lamination point on the lamination set in the plane (ξ_1, ξ_3)

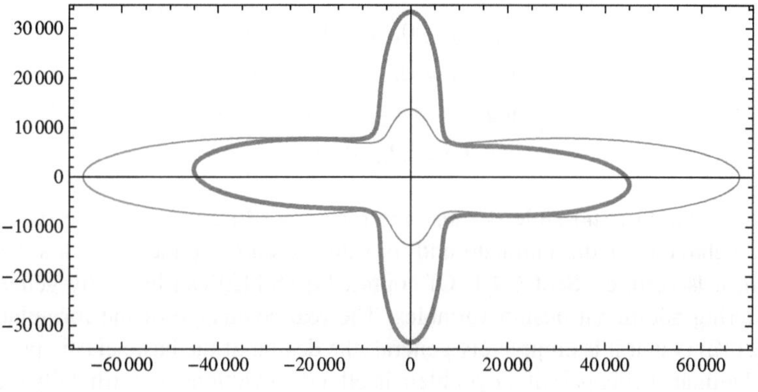

Fig. 6.16 Example 3: directional diagrams of E for extension, *thin line*, and bending, *thick line* (units: MPa)

but not for bending). It is apparent that the requirements on orthotropy, E_x^0 and E_y^0 are satisfied, while the flexural behavior is completely anisotropic.

Other laminates can be used; if, as usually done, a symmetric stack is used, then one needs at least 20 plies, and a possible solution is, e.g.

$$[\delta_1, 0, \delta_2, 0, -\delta_2, 0, \delta_2, 0, -\delta_2, -\delta_1]_s,$$

but many others, fulfilling the condition on the ply volume fractions, can be found as well, all of them sharing the same in-plane elastic behavior.

6.6 Design Problems of the Third Type

The problems of the third type are classical optimization problems: an objective has to be minimized, with some constraints to be satisfied. Typical examples of objective functions are the maximization of the stiffness or of the strength, of a vibration frequency or of a buckling load and so on. These same properties concern also typical minimal, or maximal, bounds imposed to the solution. Nevertheless, in the particular case of structures composed by anisotropic laminates, also some basic elastic properties, selected for the final design, have to be guaranteed. In practice, tensors $\mathbb{A}$, $\mathbb{B}$ and $\mathbb{D}$ must be designed, according to specified requirements: type of elastic symmetry (usually at least in-plane orthotropy), material orientation, uncoupling etc. In other words, *requirements concerning the elastic properties of the laminate enter the design process along with the objective function and the imposed constraints.*

There is, however, a substantial difference with respect to problems of the first or second type: in the present case, the function $\mathscr{I}(\mathscr{P}_j)$ is no more the objective function, but it enter the problem as an equality constraint. In fact, generally speaking, a problem of the third type can be formulated as follows:

$$\begin{aligned}
&\min_{\mathscr{P}_j} f(\mathscr{P}_j),\\
&\text{with}\\
&g_i(\mathscr{P}_j) \leq 0, \quad i = 1, \ldots, n_g,\\
&h_l(\mathscr{P}_j) = 0, \quad l = 1, \ldots, n_h,\\
&\text{and}\\
&\mathscr{I}(\mathscr{P}_j) = H_{rs}\mathscr{P}_r\mathscr{P}_s = 0.
\end{aligned} \tag{6.112}$$

This formulation is possible because the form $\mathscr{I}(\mathscr{P}_j)$ determines the type of the elastic behavior for the laminate and its value in correspondence of a solution is known: it is zero, cf. Sect. 6.4.1. Of course, Eq. (6.112) can be easily generalized considering additional design variables. The real advantage of the formulation in Eq. (6.112) is that it is completely general: the design of the basic elastic properties of the laminate being part of the problem itself, no restrictions are artificially imposed on the solutions, due to some simplifying assumptions. Any problem of laminates optimal design is in this way formulated without restrictions, so allowing, in principle, to obtain a true global optimum solution.

6.6.1 A Problem Naturally Sequential: The Two-Step Approach

The question now is: how to solve a problem of this type? Usually, numerical tools are needed for solving both the boundary value problem normally linked to the design one[14] and this last. Nevertheless, the particular mathematical structure of problem (6.112) allows, *thanks to the non uniqueness of the stacking sequence*, cf. Sect. 6.4.2, the decomposition of problem (6.112) into *two sequential and linked optimization problems*, with the second one of the first or of the second type (Vannucci 2013b).

Basically, this two-step approach goes back to an old idea: because the elastic behavior of a laminate is uniquely determined once the 18 variables $\mathscr{P}_j$ are fixed,[15] i.e. once the lamination points of each one of tensors $\mathbb{A}$, $\mathbb{B}$ and $\mathbb{D}$ are known, the structural problem, i.e. the determination of the optimal structure, comprehending that of the optimal anisotropic behavior, can be solved using as design variables the $\mathscr{P}_j$s.

In a subsequent phase, a suitable stacking sequence, able to realize the optimal material properties determined as solution of the first problem, is searched. Let us see more in detail the two steps of this approach.

6.6.2 Step 1: Structural Anisotropy Optimization Problem

The first step concerns exclusively the minimization of the objective function. Because this minimization can be obtained acting, also though not exclusively, on the elastic behavior, the design of anisotropy characteristics is part of this problem. That is why we call this first step *structural anisotropy optimization problem*. Generally speaking, this problem can be stated as:

$$\begin{aligned}
&\min_{\mathscr{P}_j} f(\mathscr{P}_j)\\
&\text{with}\\
&g_i(\mathscr{P}_j) \leq 0, \quad i = 1, \ldots, n_g,\\
&h_l(\mathscr{P}_j) = 0, \quad l = 1, \ldots, n_h.
\end{aligned} \tag{6.113}$$

Some remarks about this first step:

- problem (6.113) is defined on the set of parameters $\mathscr{P}_j$, that are the true *design variables for this step* besides, in some cases, other quantities, like for instance ply thicknesses or other mechanical and technical parameters; however, for the sake of simplicity, we continue to indicate just the $\mathscr{P}_j$s;
- the number of variables is independent from the layers' number n, but it is fixed: 18 or 12 if the layers are identical; hence, the dimension of the step 1 problem is independent from n; in the most part of times, the number of $\mathscr{P}_j$s really involved

[14] Of course, this is problem depending; we give in Sect. 6.6.5 an example where the structural problem is solved by hand, one using an analytical expression of the objective function and the GA BIANCA and another one where a numerical evaluation of the objective function through a finite element code is needed.

[15] The number of parameter reduces to 12 in the case of identical layers, cf. Sect. 6.4.1. One of the advantages of using the polar formalism, i.e. the $\mathscr{P}_j$s, instead of lamination parameters, is that the former can be always defined, while the latter exist only for laminates composed by identical layers.

in a problem is much less than 18 (e.g. if only $\mathbb{A}$ is to be designed, parameters $\mathscr{P}_j$, $j = 13, \ldots, 18$, defining $\mathbb{D}$, can be ignored);

- among the inequality constraints $g_i(\mathscr{P}_j) \leq 0$, one has also to put those concerning the *geometrical bounds on the polar parameters of the laminate*, i.e. Eq. (5.57). In fact, the optimal elastic behavior that is determined as solution of this first step problem has to be, in practice, realized through a laminate. We have seen in Sect. 5.2.13 that laminates belong to a restricted elastic class, defined by the so-called geometrical bounds. If such constraints are not taken into consideration in the first step, one could obtain an optimal lamination point impossible to be realized by a laminate, because outside the geometrical bounds. So, such bounds must be accounted for since this first step, in order to ensure the practical realization of the structure through laminates. The first step is hence, always, a constrained optimization problem[16];
- a delicate problem is that of determining the feasible domain of variables $\mathscr{P}_j$; the reader is addressed to Sect. 5.2.13, where this question has been treated;
- the resolution of step 1 is problem dependent and general rules cannot be given; in some cases, optimality criteria ensure and suggest a possible procedure, the most part of times a numerical one, because analytical solutions are possible only for very simple problems. We will see in Sect. 6.7, for instance, an approach particular to a given problem. From now on, we indicate by $\mathscr{P}_j^{opt}$ the set of parameters $\mathscr{P}_j$ that are the solution of the first step problem. They completely and uniquely determine the elastic behavior of the structure.

6.6.3 Step 2: Constitutive Law Problem

Once the set of optimal parameters $\mathscr{P}_j^{opt}$ determined, the problem becomes to find a suitable laminate, i.e. a stacking sequence with parameters $\mathscr{P}_j = \mathscr{P}_j^{opt}$, $i = 1, \ldots, 18$. Also this problem can be formulated as the search for the minimum of a quadratic function:

$$\min_{\delta_k} \sum_{j=1}^{18} (\mathscr{P}_j(\delta_k) - \mathscr{P}_j^{opt})^2, \quad k = 1, \ldots, n. \tag{6.114}$$

Some remarks about the second step:

- this problem is formally similar to that of a laminate design problem of the first type, Eq. (6.73): the search of the zeros, i.e. of the global minima, of a non-convex positive semi-definite quadratic function;

[16]If the structure were realized by a hypothetical homogenized single layer anisotropic material, the first step should still be a constrained problem, because in this case the bounds (4.111), ensuring the positive definiteness of the strain energy, should replace the geometrical bounds (5.57).

- problem (6.114) typically defines a *distance problem* in the n-dimensional space of ply orientations δ_k: a *target behavior*, defined by parameters $\mathscr{P}_j^{opt}$, is to be realized by a laminate;
- because the set of parameters $\mathscr{P}_j^{opt}$ determines the elastic behavior of the laminate, this second step is a *constitutive law problem*;
- Equation (6.114) states an optimization problem that is always non-convex; suitable numerical methods are hence metaheuristics, as discussed in Sect. 6.4.3;
- the non uniqueness of the stacking sequence corresponding to a given $\mathscr{P}_j^{opt}$ can now be interpreted as the consequence of the non convexity of the objective function of problem (6.114) with respect to the orientations δ_k;
- normally, problem (6.114) is unconstrained; however, in some cases, it is possible that some constraints are to be added, taking into account for special requirements of the problem under consideration.

6.6.4 Final Commentaries on the Two-Step Approach

The two-step approach, separating the *structural* from the *constitutive law* problem, transforms an original constrained optimum problem into two linked sub-problems: the first one, normally simpler than the original because ruled by a lower number of design variables, the second one, a (normally) unconstrained non-convex problem with a classical mathematical structure, that of a distance problem in a n-dimensional space.

It is worth now to mention another aspect of this approach: some general properties concerning the elastic behavior, like e.g. uncoupling, quasi-homogeneity, orthotropy, can be considered as *automatically acquired* in the first step and then imposed in the second one. This greatly simplifies the problem; if, for instance, uncoupling is sought, as usually done, during the first step one has to consider simply that $\mathbb{B} = \mathbb{O}$, so that the $\mathscr{P}_j$, $j = 7, \ldots, 12$, defining $\mathbb{B}$ simply do not take part to problem (6.113). In the following second step, it is necessary to take care of the fact that $\mathscr{P}_j^{opt} = 0\ \forall j = 7, \ldots, 10$, (actually, in the case of identical layers it is sufficient to put $\mathscr{P}_9^{opt} = \mathscr{P}_{10}^{opt} = 0$).

A last question remains: how can we be sure that the second step will give a laminate whose lamination point is determined by the set $\mathscr{P}_j^{opt}$ and that, in addition, respect also all the requirements tacitly imposed in the first step to the elastic behavior? This question is tackled in Sect. 6.8.2; we can only say, now, that it is precisely at this point that the non uniqueness of the stacking sequence plays a crucial role: because the number of laminates sharing the same number of layers n and the same lamination point, i.e. the $\mathscr{P}_j^{opt}$s, is usually large, and rapidly increasing with n, it is very probable to find, among them, at least a sequence satisfying also the requirements tacitly considered in the first step and usually concerning the type of elastic behavior.

Finally, so far we have considered, for the sake of simplicity, only elastic properties; nevertheless, in a similar way other requirements, e.g. on hygral or thermal properties, can be considered too. Some problems of this type are considered in Sect. 6.9.

6.6.5 Some Examples of Design Problems of the Third Type

Example 1

This first example is somewhat similar to *Example 1* of Sect. 6.5.2: the buckling multiplier λ_{pq}^{crit} of a simply supported rectangular plate is to be maximized for the case of two compressive uniformly distributed in-plane loads N_1 and N_2 acting along the plate's sides a and b. The plate must be uncoupled and orthotropic in bending, with $K^A = K^D = 1$ and the axes of orthotropy aligned with the plate's sides.

Some additional mechanical and technological constraints are part of the problem: $E_1^A \geq E_1^0$, $E_2^A \geq E_2^0$ for the in-plane Young's moduli along the x_1 and x_2 directions and the orientation angles δ_k must be integer numbers.

The true difference with *Example 1* of Sect. 6.5.2 is that now the objective function to be optimized is the buckling multiplier λ_{pq}^{crit}, while before it was a constraint, and the requirements on the elastic properties are no more the objective function, but a constraint to the step 1 problem, that can be automatically and easily considered, and a part of the objective function of the step 2 problem. In fact, we can formulate the problem as follows (constraints are in standard form):

Step 1:

$$\begin{aligned}
&\min(-\lambda_{pq}^{crit})\\
&\text{with}\\
&E_1^0 - E_1^A \leq 0,\\
&E_2^0 - E_2^A \leq 0,\\
&-\mathscr{P}_3 \leq 0,\\
&\frac{\mathscr{M}\,\mathscr{P}_3}{R_0^L} - 1 \leq 0,\\
&-\mathscr{P}_4 \leq 0,\\
&2\left(\frac{\mathscr{M}\,\mathscr{P}_4}{R_1^L}\right)^2 - (-1)^{(K^A-K^L)}\frac{\mathscr{M}\,\mathscr{P}_3}{R_0^L} - 1 \leq 0,\\
&-\mathscr{P}_{15} \leq 0,\\
&\frac{\mathscr{M}\,\mathscr{P}_{15}}{R_0^L} - 1 \leq 0,\\
&-\mathscr{P}_{16} \leq 0,\\
&2\left(\frac{\mathscr{M}\,\mathscr{P}_{16}}{R_1^L}\right)^2 - (-1)^{(K^D-K^L)}\frac{\mathscr{M}\,\mathscr{P}_{15}}{R_0^L} - 1 \leq 0.
\end{aligned} \tag{6.115}$$

Function λ_{pq}^{crit} is still given by Eq. (6.57), E_1^A and E_2^A by Eq. (6.98), with $E_1^A = E_1^A(\mathscr{P}_j)$, $E_2^A = E_2^A(\mathscr{P}_j)$, $\lambda_{pq}^{crit} = \lambda_{pq}^{crit}(\mathscr{P}_j)$, cf. Eq. (6.108).
Because of the requirements imposed to the elastic behavior, we do not need to take care, in this phase, of uncoupling, as well as of the orthotropy of $\mathbb{A}$

and $\mathbb{D}$: they are considered as automatically acquired. Constraints 3 to 10 corresponds, in terms of $\mathscr{P}_j$s, to the geometrical bounds for $\mathbb{A}$ and $\mathbb{D}$ in the case of orthotropy, Eq. (5.58). For this problem, step 1 depends hence upon only 4 design variables: $\mathscr{P}_3,\ \mathscr{P}_4,\ \mathscr{P}_{15},\ \mathscr{P}_{16}$.

Step 2:

$$\min_{\delta_k}\left[(\mathscr{P}_3-\mathscr{P}_3^{opt})^2+(\mathscr{P}_4-\mathscr{P}_4^{opt})^2+\mathscr{P}_5^2+\mathscr{P}_6^2+\mathscr{P}_9^2+\mathscr{P}_{10}^2+\right.$$
$$\left.(\mathscr{P}_{15}-\mathscr{P}_{15}^{opt})^2+(\mathscr{P}_{16}-\mathscr{P}_{16}^{opt})^2+\mathscr{P}_{17}^2+\mathscr{P}_{18}^2\right], \tag{6.116}$$

with $\Phi_A=\Phi_D=\pi/4,\ \Phi_1^{A^*}=\Phi_1^{D^*}=0$ in Eq. (6.76) and, of course, $\mathscr{P}_j^{opt},\ j=3,4,15,16$, the solution to the first step problem. This is un unconstrained minimization problem of a positive semi-definite quadratic form depending upon the n variables δ_k, because $\mathscr{P}_j=\mathscr{P}_j(\delta_k)$.

To give a specific case, be $a/b=1.5,\ p=q=1, N_1=N_2=1$ N/mm, $E_1^0=60$ GPa, $E_2^0=30$ GPa and the laminate is composed by 16 T300/5208 carbon-epoxy layers, cf. Table 4.9. Both step 1 and step 2 problems have been solved using the GA BIANCA with the ADP method for handling the constraints. The best solution so found is ($\mathscr{M}=64.592$ GPa, cf. Footnote 10)

$$\mathscr{P}_3^{opt}=0.172,\ \mathscr{P}_4^{opt}=0.081,\ \mathscr{P}_{15}^{opt}=0.197,\ \mathscr{P}_{16}^{opt}=0.103$$

or, in terms of polar parameters,

$$R_0^A=11.084 \text{ GPa}, R_1^A=5.234 \text{ GPa}, R_0^D=12.70 \text{ GPa}, R_1^D=6.666 \text{ GPa},$$

for step 1, which gives $\lambda_{11}^{crit}=6.86\times10^6$, $E_1=60.61$ GPa and $E_2=31.16$ GPa, and the sequence

$$[-24°, 39°, -47°, 37°, 32°, -47°, -6°, -47°,$$
$$55°, 59°, 18°, -38°, -38°, 19°, -40°, 42°]$$

for step 2. We plot in Fig. 6.17 the directional diagrams of A_{1111}, B_{1111} and D_{1111}; it is apparent that the laminate is in- and out of- plane orthotropic and that it is practically uncoupled. In Fig. 6.18 we plot the convergence diagrams for the objective function λ_{11}^{crit}; the calculation has been done on a population of 400 individuals and the best solution has been found after 150 generations.

As a last remark about this example, we notice that in this case, thanks to the peculiarities of the problem, we dispose of an analytical expression for the objective function, Eq. (6.57). This is possible thanks to the assumption of uncoupling and bending orthotropy with axes aligned with the plate's side and to the absence of shear loads on the edges of the plate.

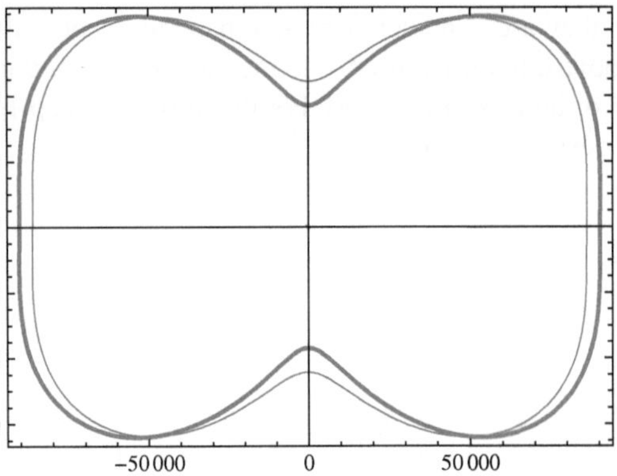

Fig. 6.17 Example 1: directional diagrams of A_{1111}, *thin line*, B_{1111}, in the *center*, and D_{1111}, *thick line* (units: MPa)

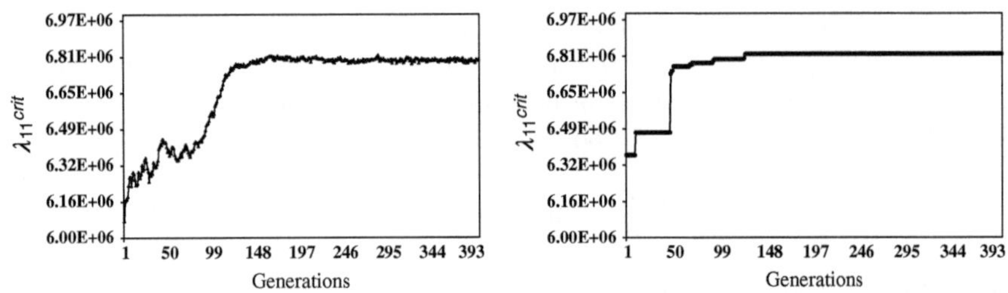

Fig. 6.18 Example 1: evolution of the objective function λ_{11}^{crit} along generations: average, *left*, and best individual, *right*

Example 2

We consider now the problem of a *Standard Piezoelectric Laminate* (in short, SPL): this is an uncoupled laminate composed by identical layers and by two identical isotropic piezoelectric actuators bonded to the upper and lower surfaces, see Fig. 6.19.

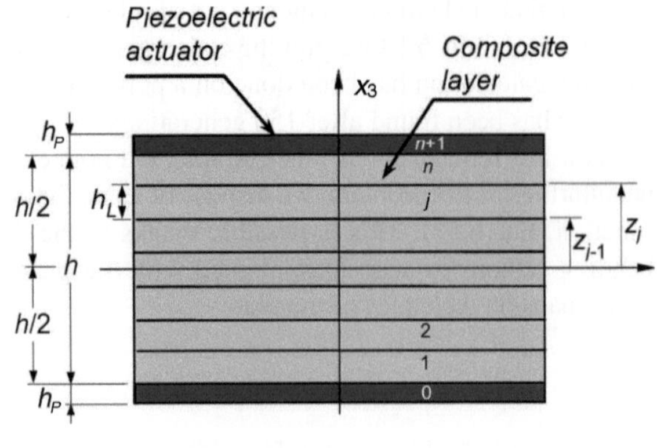

Fig. 6.19 Example 2: sketch of the SPL

When a difference of electric potential of opposite signs is applied to the two piezoelectric actuators, the plate bends; we want to design the stacking sequence in order to maximize the curvature in one direction and have a null curvature in the orthogonal one. In addition, the laminate must be orthotropic in bending, with the directions of orthotropy coincident with the directions of maximum and null bending. In other words, we are looking for the most performing uncoupled orthotropic laminate for cylindrical bending along the axes of orthotropy.

Applying classical piezoelectric theory to the present case (Vannucci 2011), we get the following components of the curvature tensor $\boldsymbol{\kappa}$:

$$\kappa_{11} = \beta(\mathscr{D}_{1111} + \mathscr{D}_{1122}),\ \kappa_{22} = \beta(\mathscr{D}_{1122} + \mathscr{D}_{2222}),\ \kappa_{12} = 0, \tag{6.117}$$

where β is a problem depending constant:

$$\beta = e_3 V(h + h_P), \tag{6.118}$$

with e_3 the piezoelectric coefficient of the isotropic actuators, V the applied electric potential difference, h the total thickness of the composite laminate and h_A that of an actuator, see Fig. 6.19. Because β cannot be modified by an optimization procedure, it will be neglected in the following.

In Eq. (6.117), the $\mathscr{D}_{mnpq}$s are the Cartesian components of the compliance bending tensor $\mathscr{D}$. Because the laminate is uncoupled, $\mathscr{D} = \mathbb{D}^{-1}$, Eq. (5.36), and $\mathscr{D}_{mnpq} = \mathscr{D}_{mnpq}(\mathscr{P}_j)$, $j = 13, \ldots, 18$. In particular, because the composite layers are identical and the plate is orthotropic, $\mathscr{D}_{mnpq} = \mathscr{D}_{mnpq}(\mathscr{P}_{15}, \mathscr{P}_{16})$ uniquely: all the problem depends upon only two variables, $\mathscr{P}_{15}$ and $\mathscr{P}_{16}$. We can hence formalize the two-step problem:

Step 1:

$$\begin{aligned}
&\min_{\mathscr{P}_j} f(\mathscr{P}_{15}, \mathscr{P}_{16})\\
&\text{with}\\
&g(\mathscr{P}_{15}, \mathscr{P}_{16}) = 0,\\
&-\mathscr{P}_{15} \leq 0,\\
&\frac{\mathscr{M}\,\mathscr{P}_{15}}{R_0^L} - 1 \leq 0,\\
&-\mathscr{P}_{16} \leq 0,\\
&2\left(\frac{\mathscr{M}\,\mathscr{P}_{16}}{R_1^L}\right)^2 - (-1)^{(K^D - K^L)}\frac{\mathscr{M}\,\mathscr{P}_{15}}{R_0^L} - 1 \leq 0;
\end{aligned} \tag{6.119}$$

inequality constraints correspond to the geometrical bounds on $\mathbb{D}$; once more, we do not need to take care, at this stage, of uncoupling, as well as of the orthotropy of $\mathbb{D}$: they are considered as automatically acquired. The objective function $f(\mathscr{P}_{15}, \mathscr{P}_{16})$ and the equality constraint function $g(\mathscr{P}_{15}, \mathscr{P}_{16})$ are given by

$$
\begin{aligned}
f(\mathscr{P}_{15}, \mathscr{P}_{16}) &= -(\mathscr{D}_{1122} + \mathscr{D}_{2222}) = \frac{D_{1111} - D_{1122}}{D_{1122}^2 - D_{1111}D_{2222}},\\
g(\mathscr{P}_{15}, \mathscr{P}_{16}) &= \mathscr{D}_{1111} + \mathscr{D}_{1122} = \frac{D_{1122} - D_{2222}}{D_{1122}^2 - D_{1111}D_{2222}}.
\end{aligned}
\tag{6.120}
$$

Step 2:

$$
\min_{\delta_k} \left[\mathscr{P}_9^2 + \mathscr{P}_{10}^2 + (\mathscr{P}_{15} - \mathscr{P}_{15}^{opt})^2 + (\mathscr{P}_{16} - \mathscr{P}_{16}^{opt})^2 + \mathscr{P}_{17}^2 + \mathscr{P}_{18}^2\right],
\tag{6.121}
$$

with $\Phi_D = K^D\pi/4$, $\Phi_1^{D^*} = 0$ in Eq. (6.76) and $\mathscr{P}_j^{opt}$, $j = 15, 16$, the solution to the first step problem.

The main difference with respect to the previous example is the presence, in the first step, of an equality constraint; anyway, using the approach outlined in Sect. 6.5.1, this can be transformed into an inequality constraint once a suitable tolerance ε fixed. Also this case can hence be treated like the previous one.

Nevertheless, we prefer here another approach, less general but much simpler, leading to a resolution by hand of the first step problem. In fact, using Eqs. (5.54) and (5.55), and considering that

$$
R_0^D = \mathscr{M}\ \mathscr{P}_{15} = R_0^L\xi_{13}, R_1^D = \mathscr{M}\ \mathscr{P}_{16} = R_1^L\xi_{15},
\tag{6.122}
$$

we obtain the objective f and the constraint g as functions of the lamination parameters ξ_{13} and ξ_{15}:

$$
\begin{aligned}
f(\xi_{13}, \xi_{15}) &= -\frac{T_0^L + (-1)^{K^L}R_0^L\xi_{13} + 2R_1^L\xi_{15}}{4\left\{T_1^L[T_0^L + (-1)^{K^L}R_0^L\xi_{13}] - 2R_1^{L^2}\xi_{15}^2\right\}},\\
g(\xi_{13}, \xi_{15}) &= \frac{T_0^L + (-1)^{K^L}R_0^L\xi_{13} - 2R_1^L\xi_{15}}{4\left\{T_1^L[T_0^L + (-1)^{K^L}R_0^L\xi_{13}] - 2R_1^{L^2}\xi_{15}^2\right\}} = 0.
\end{aligned}
\tag{6.123}
$$

We then use Eq. $(6.123)_2$ to eliminate ξ_{15} from f, so that finally we can state the first step problem in the new form

$$
\begin{aligned}
&\min_{\xi_{13}} f^*(\xi_{13}) = -\frac{1}{2T_1^L - T_0^L - (-1)^{K^L}R_0^L\xi_{13}},\\
&\text{with}\\
&g(\xi_{13}, \xi_{15}) = T_0^L + (-1)^{K^L}R_0^L\xi_{13} - 2R_1^L\xi_{15} = 0,\\
&-1 - \xi_{15} \le 0,\\
&\xi_{15} - 1 \le 0,\\
&\xi_{13} \le 0,\\
&2\xi_{15}^2 - \xi_{13} - 1 \le 0.
\end{aligned}
\tag{6.124}
$$

The four inequalities in standard form corresponds to the geometrical bounds in terms of lamination parameters and define the lamination set, the usual parabolic sector already encountered before, this time in the plane (ξ_{13}, ξ_{15}), that of the bending properties.

The equality constraint corresponds to a straight line in the plane (ξ_{13}, ξ_{15}) and the objective function f^* decreases with ξ_{13} if $K^L = 0$ and increases if $K^L = 1$. Hence, the optimal solution is necessarily on the boundary of the lamination set and the lamination point can be easily found as the intersection between the line $g(\xi_{13}, \xi_{15}) = 0$ and the contour of the lamination set.

We remark that, on one hand, it is the constraint $g(\xi_{13}, \xi_{15}) = 0$ that determines the elastic solution of the problem and, on the other hand, the solution to this first step problem depends uniquely upon the characteristics of the basic layer. As a consequence, the same solution is shared by all the laminates that must comply with the same condition $g(\xi_{13}, \xi_{15}) = 0$ and that differ, for instance, in other respects, namely for the number of the layers, objective function, additional constraints and so on.

As a numerical application, let us consider the case of a laminate composed by T300/5208 carbon-epoxy layers; in this case, we obtain ($K^L = 0$) that the lamination point has coordinates

$$\begin{aligned} &\xi_{13} = 2\xi_{15}^2 - 1 = 0.55, \\ &\xi_{15} = \frac{R_1^L + \sqrt{R_1^{L2} - 2(-1)^{K^L}[T_0^L - (-1)^{K^L} R_0^L]R_0^L}}{2(-1)^{K^L} R_0^L} = 0.88, \end{aligned} \tag{6.125}$$

which gives the minimum of the objective function $f^*_{min} = -0.085$. We show in Fig. 6.20 the lamination point determined on the boundary of the lamination set by the constraint $g(\xi_{13}, \xi_{15})$ in Eq. (6.124).

To the lamination point correspond, through Eq. (6.122) the parameters $\mathscr{P}_{15} = 0.168$, $\mathscr{P}_{16} = 0.292$; we can now go to the second step problem (6.121), that has been solved using the code ALE-PSO, run 10 times with a swarm of 100 particles for 500 iterations. In Vannucci (2011) this has been done for different number of layers n, we just give the result for the case $n = 7$; in that case, the best solution so found is the sequence

$$[-14.0°, 14.5°, 13.6°, -4.0°, -12.3°, -14.9°, 14.1°],$$

with a residual of $\sim 1.2 \times 10^{-7}$. As usual, we plot in Fig. 6.21 the directional diagrams of A_{1111}, B_{1111}, and D_{1111}. Once more, while we can see that $\mathbb{D}$ is orthotropic, $\mathbb{A}$ is completely anisotropic, because no requirements have been given on it; the component B_{1111} is practically invisible, confirming that $\mathbb{B} = \mathbb{O}$. We give in Fig. 6.22 the evolution of the objective function along iterations, in the average for the whole swarm and that of the best particle, for the best run of ALE-PSO among the ten done.

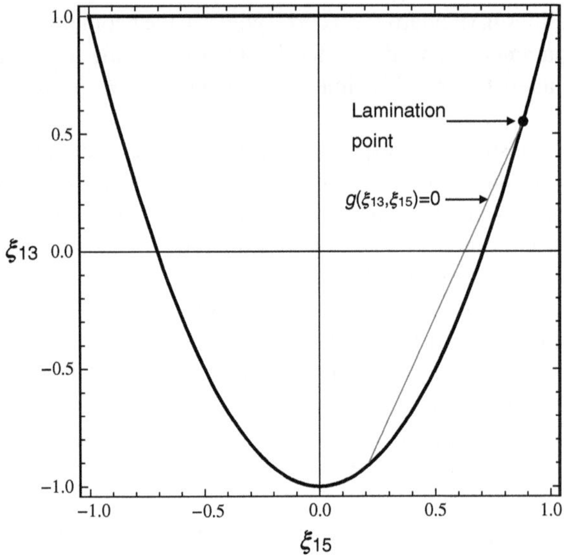

Fig. 6.20 Example 2: lamination point

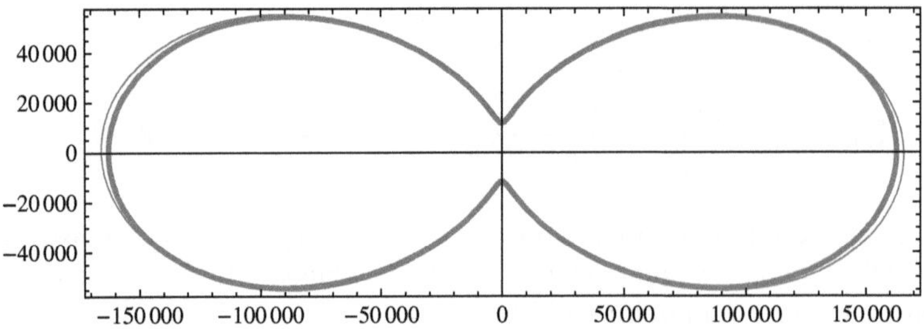

Fig. 6.21 Example 2: directional diagrams of A_{1111}, *thin line*, B_{1111}, in the *center*, and D_{1111}, *thick line* (units: MPa)

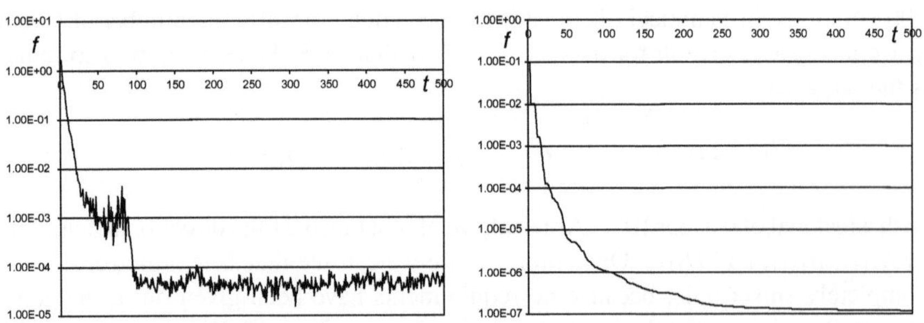

Fig. 6.22 Example 2: evolution of the objective function f along iterations: average, *left*, and best individual, *right* (log scale)

Example 3

This last example concerns a more involved problem: the design of a laminated plate whose damping given by a set of elastomeric patches bonded to its upper surface must be maximized. The problem has been tackled in Montemurro et al. (2015) and it is by far more complicate than the two previous ones: not only the patches are to be designed in number, position and dimensions, but also the laminate has to be completely determined, in terms of layer orientations and also number of plies. This kind of problems can be dealt with by a numerical strategy suited for modular systems presented in Sect. 6.8; this method is based upon the GA BIANCA, specially conceived for dealing also with these problems.

The objective function is the sum of the first N_f modal loss factors η of the structure. The *modal loss factor* is defined as

$$\eta = \frac{E_I(\omega)}{E_R(\omega)}, \tag{6.126}$$

with the Young's modulus E depending on the frequency of vibrations ω according to

$$E(\omega) = E^R(\omega) + i\,E^I(\omega), \tag{6.127}$$

where

$$\begin{aligned} E^R(\omega) &= E_s^R + E_d^R \log\frac{\omega}{\omega^0},\\ E^I(\omega) &= E_s^I + E_d^I \log\frac{\omega}{\omega^0}, \end{aligned} \tag{6.128}$$

with $\omega_0 = 1$ Hz a reference value for frequencies. We remark that the mechanical problem is nonlinear, as the material properties, the Young's modulus, depend upon the structural response through the frequency. Hence, a nonlinear numerical approach has to be used to solve the mechanical problem and obtain the modal loss factors; in this case, hence, the objective function has to be determined by a numerical approach, namely using a nonlinear finite elements method.

The problem treated in Montemurro et al. (2015) is sketched in Fig. 6.23; circular patches are used as damping devices.

We remark also that in this case the design process is complete: the damping devices *and* the material properties are to be designed simultaneously. In particular, for the laminate the requirements are that it has to be quasi-homogeneous and orthotropic, with the orthotropy axes aligned with the plate's sides. Different constraints, of geometrical, technological and mechanical origin, complete the design. The complete description of this problem is very long and outside the scope of this text, so we address the reader to Montemurro et al. (2015) for more details. Here, we bound the description to the results.

The best result is found with a system composed of 28 layers and 7 patches; the first five vibration modes of the structure so designed are shown in Fig. 6.24. The second step problem is

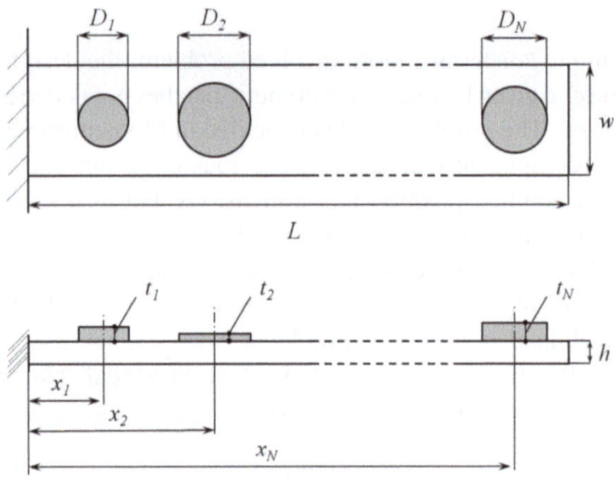

Fig. 6.23 Example 3: scheme of the structure to be optimized

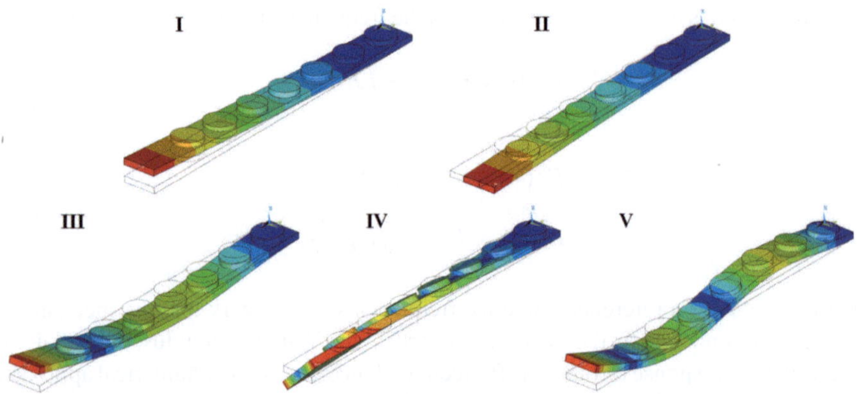

Fig. 6.24 Example 3: first five vibration modes of the optimized structure

$$\begin{aligned}\min_{\delta_k} \Big[&\mathscr{P}_5^2 + \mathscr{P}_6^2 + \mathscr{P}_9^2 + \mathscr{P}_{10}^2 + (\mathscr{P}_3 - \mathscr{P}_3^{opt})^2 + (\mathscr{P}_4 - \mathscr{P}_4^{opt})^2 + \\ &(\mathscr{P}_3 - \mathscr{P}_{15})^2 + (\mathscr{P}_3 - \mathscr{P}_{16})^2 + (\mathscr{P}_5 - \mathscr{P}_{17})^2 + (\mathscr{P}_6 - \mathscr{P}_{18})^2)\Big],\end{aligned} \tag{6.129}$$

with $\mathscr{P}_3^{opt}$ and $\mathscr{P}_4^{opt}$ the optimal values of the only two parameters involved in step 1 problem and found as solution of this last. Running BIANCA with the second step problem gives as best solution the following stacking sequence (integer values for the orientations were demanded):

$$[0°, -1°, 0°, -2°, 0°_2, 17°, -1°, 2°, -4°, 0°, -2°, -1°,$$
$$-6°, -16°, -1°_2, 17°, 2°, 17°, -4°, -3°, -1°, 0°_2, -2°, 0°_2]$$

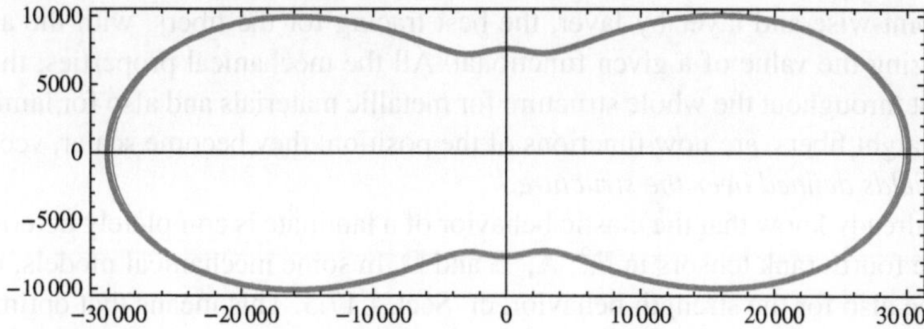

Fig. 6.25 Example 3: directional diagrams of A_{1111}, *thin line*, B_{1111}, in the *center*, and D_{1111}, *thick line* (units: MPa)

Fig. 6.26 Example 3: convergence diagram of the best individual for the second step problem (log scale)

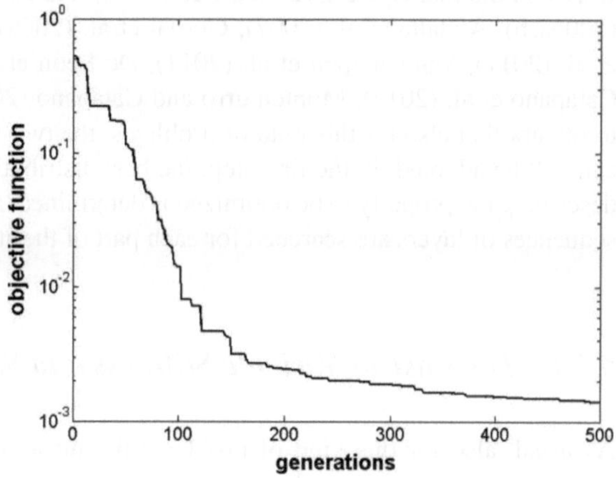

For this example too we plot the directional diagrams of A_{1111}, B_{1111} and D_{1111} in Fig. 6.25; also in this case B_{1111} is practically invisible, confirming uncoupling, and $A_{1111} \sim D_{1111}$, i.e. the laminate is really practically quasi-homogeneous and clearly orthotropic.

Finally, we show in Fig. 6.26 the evolution along generations of the objective function of step 2 problem; the calculation has been done on a population of 500 individuals and for 500 generations.

6.7 Optimization of Anisotropy Fields

Modern technologies of *automatized fiber placement* allow the manufacturing of a new kind of laminates, where the fibers reinforcing each layer are placed along *given arbitrary trajectories*. In this way, the designer can achieve a further step towards performing structures: to taylor point wise the mechanical response of the structure to best meet the requirements. Of course, this opens a new optimization problem: to find, point-wise and layer by layer, the best tracing for the fibers, with the aim of minimizing the value of a given functional. All the mechanical properties, that are constant throughout the whole structure for metallic materials and also for laminates with straight fibers, are now functions of the position: they become scalar, vector or tensor *fields defined over the structure*.

We already know that the elastic behavior of a laminate is completely determined by three fourth rank tensors in $\mathbb{R}^2$: $\mathbb{A}$, $\mathbb{B}$ and $\mathbb{D}$. In some mechanical models, this is the same also for the strength behavior, cf. Sect. 4.17.3. This means that optimizing locally the stiffness or strength of laminates means to optimize the distribution of some *anisotropic tensor fields*.

This subject of research in anisotropic structures design is rather recent; some works on the matter are Hyer and Lee (1991), Parnas et al. (2003), Setoodeh et al. (2006a, b), Abdalla et al. (2007), Gürdal et al. (2008), Blom et al. (2010), Jibawy et al. (2011), Van Campen et al. (2011), De Leon et al. (2012), Catapano (2013), Catapano et al. (2015), Montemurro and Catapano (2016). It is however important to remark that also for this kind of problems, the two-step approach outlined above can still be adopted: in the first step, the best distribution of the anisotropic tensors describing the property to be optimized is determined, and in the second one suitable sequences of layers are searched for each part of the structure.

6.7.1 The Case of Variable Stiffness and Strength

As usual, also for this kind of problems the most considered objective has been the maximization of strength, but not exclusively. We give here a brief account of a more involved problem: the simultaneous optimization of stiffness *and* strength. This problem has been tackled in Catapano (2013), Catapano et al. (2015), and generalizes a previous approach concerning only stiffness (Jibawy et al. 2011).

6.7.1.1 Statement of the Problem

We need to specify some points in order to have a well posed specific optimization problem:

1. we consider a laminated anisotropic structure, whose *geometry, boundary conditions and applied loads are known and fixed*;
2. the stacking sequence is composed by *identical plies*, i.e. by plies all composed of the same material and having the same thickness;
3. the material composing the basic plies is *chosen a priori once and for all*; hence, it does not take part in the optimization process;
4. each ply is constituted by an *orthotropic material*, whose behavior is sensibly *linearly elastic up to its ultimate load*;

5. the total thickness h of the laminate is fixed and is the same everywhere; as a consequence, by assumption 2, the number of layers is *fixed and known a priori* as well: it cannot be modified during the optimization process; this assumption is the so-called *iso-perimetric constraint*, necessary to obtain a meaningful problem (otherwise, the stiffest and strongest structure would be just the one having an infinite thickness);
6. the structure is subject to only *in-plane loads*[17];
7. the laminate has to be *orthotropic everywhere*;
8. the laminate has to be *extension-bending uncoupled everywhere*;
9. the design variables describing stiffness are *independent* from those describing strength;
10. the only common parameter of stiffness and strength is the *direction*: we assume that the laminate is orthotropic not only for the stiffness, but also for the strength, and that the *direction of the orthotropy axes is exactly the same (or turned of $\pi/2$) for both the cases*;
11. the strength properties of the laminate can be condensed in the *components of an elasticity-like fourth order plane tensor*, cf. Sect. 4.17.3;
12. the relevant mechanical parameters for the optimization of the laminated structure stiffness and strength are the *polar invariants* of the tensors describing the different mechanical properties;
13. we take the CLPT as kinematic model for the laminate, cf. Sect. 5.2;
14. the limit state is modeled by the Tsai–Hill failure criterion, cf. Sect. 4.17.3.1.

While assumptions 1–8 are rather common and do not deserve special commentary, assumption 9 is rather important; mathematically, it allows to *completely split stiffness from strength*, except for the direction of orthotropy, see assumption 10, and hence to precisely define the dimensionality of the optimum problem.

About assumption 10, this is based upon a physical reasonable rationale: for a simple ply, the geometrical disposition of the matter or of the phases, i.e. of the reinforcing fibers, naturally induces orthotropy for both the stiffness and strength properties, and with the same axes. So inspired, we enforce such a circumstance also for the whole laminate.

Assumption 11 is a rationale suggested by what has been discussed in Sect. 4.17.3: because the strength criteria for orthotropic plies are actually based upon such an assumption, we enforce it also for the strength of the laminate considered as a whole, a homogenized plate. In fact (Catapano 2013), we introduce a *homogenized failure index* as objective function for strength. This is a local functional giving a measure of the mechanical state of stress in a point. Such homogenized failure index, based upon the Tsai–Hill criterion, for an uncoupled laminate subjected to only in-plane loads, is defined as

$$F_{TH}^{lam} = \frac{1}{h}\int_0^h F_{TH}^{ply} dz = \boldsymbol{\varepsilon}^0 \cdot \mathbb{G}^A \boldsymbol{\varepsilon}^0 \leq 1, \tag{6.130}$$

[17]The case of bending loads can be treated in the same way.

with

$$\mathbb{G}^A = \frac{1}{h}\sum_{j=1}^{n}(z_j - z_{j-1})\mathbb{G}(\delta_j), \ \ \mathbb{G} = \mathbb{QFQ}. \tag{6.131}$$

In the above equation, $\mathbb{F}$ is the fourth-order elasticity-like tensor defined in Sect. 4.17.3, while, with a slight abuse of notation, we have indicated by $\mathbb{Q}$ the reduced stiffness matrix $[Q]$, Eq. (3.61) (the plane stress state is assumed, for the thinness of the plate).

Because $\mathbb{G}$ is the composition of elasticity-type tensors, it is itself a fourth-rank tensor of the type of elasticity, and $\mathbb{G}^A$ as well. As such, they can be expressed in the polar formalism, be hence $\Gamma_0^L,\ \Gamma_1^L,\ \Lambda_0^L,\ \Lambda_1^L,\ \Omega_0^L,\ \Omega_1^L$ the polar parameters of $\mathbb{G}$, i.e. relative to the basic layer, and $\Gamma_0^A,\ \Gamma_1^A,\ \Lambda_0^A,\ \Lambda_1^A,\ \Omega_0^A,\ \Omega_1^A$ those of $\mathbb{G}^A$, relative to the in-plane strength of the laminate. For strength, only orthotropy can be modeled, because strength criteria are formulated only for this case of elastic symmetry; we put then, like for elasticity, $\Omega_0 - \Omega_1 = H\pi/4,\ H \in \{0, 1\}$. For what concerns the stiffness, the objective function is the compliance.

6.7.1.2 The First Step Problem: Optimization of the Anisotropy Fields

The problem is now clear: to determine, point wise, the optimal orthotropy directions and anisotropy polar parameters of stiffness and strength for maximizing simultaneously the stiffness and strength of the structure. In the approach followed in Catapano et al. (2015), in fact, stiffness and strength are *not* competitive objectives: one of the two is taken as the *leading objective*, but the other one will be part of the design process too, and not as a constraint nor as a competitor to the leading objective. Two cases are considered, according to whether the stiffness or the strength is the leading objective. Because the parameters describing stiffness are independent from those of strength, all this is meaningful, mechanically speaking. There is, however, a link between these two objectives: the tensor fields to be optimized are $\mathbb{G}^A$ and $\mathbb{A}$ and they are linked by the assumption that the direction of orthotropy is the same, which means that it must be everywhere $\Phi_1^A = \Omega_1^A$.

The great redundancy of laminates corresponding to the same $\mathbb{A}$, cf. Sect. 6.4.2, allows then, in the second step, to find a laminate that realizes both the stiffness and the strength optimal values. In this case too, hence, the non uniqueness of the stacking sequence plays a crucial role.

From a mechanical point of view the optimization of the strength along with stiffness can interpreted as follows: let us consider the case of stiffness as leading objective. Then, a local optimization of its strength means to look for the structure that, still conserving its best stiffness, maximizes, pointwise, its strength. Hence, it means to look for the strongest structure among all the possible ones realizing the same optimal stiffness. This is an engineering relevant problem: how to better strengthen a structure designed with respect to stiffness? Dually, when a stiffness optimization is performed on an anisotropy field giving the strongest structure, one obtains the stiffest solution for a fixed optimum strength. For the engineer, this means to improve at its best the stiffness of a structure designed to withstand some applied loads.

The first step is hence, for this kind of problems, decomposed into two linked sub-problems, one for the leading property to be optimized, the other one for the secondary one. The only link between the two sub-problems composing the first step is the *conservation of the orthotropy direction*.

The numerical procedure for the solution of the first step problem is really technical, the reader is addressed to Catapano (2013) for a complete description of the matter. We just cite the fact that the minimization is done by a modified version of an alternate direction algorithm introduced by Allaire and Kohn (1993; 1997), based on the complementary energy minimum. Originally conceived for the minimization of the compliance, it has been adapted in Catapano (2013) to include also strength and the convergence proved. The second step problem is, on its side, classical and solved using the GA BIANCA.

In Catapano (2013), the local optimization is actually done after a discretization of the structure in finite elements, and the optimization is done for each element. In a more recent and promising approach (Montemurro and Catapano 2016), the discretization of the anisotropy fields is replaced by a field function describing the design variables, i.e. the anisotropy parameters. This field function is described by a finite and limited set of parameters, that are the new design variables; such an approach allows a better continuity of the fiber placement.

6.7.1.3 Some Examples

Example 1

The first example (Catapano et al. 2015), is that of a holed square plate submitted to in-plane forces uniformly distributed along the edges, see Fig. 6.27. For the symmetry of the problem, just a quarter of the plate can be modeled.

The total thickness h of the plate, made of carbon-epoxy T300/5208, is fixed to 3.75 mm, i.e. to 30 plies. The loading values are $F_x = 500\,\text{N/mm}$ and $F_y = 250\,\text{N/mm}$. As starting point of the first step problem, the stiffness polar parameters are those of a cross-ply laminate: $(-1)^{K^A} R_0^A = (-1)^{K^L} R_0^L$ and $R_1^A = 0$, while the initial strength polar parameters are chosen equal to those of the basic ply: $(-1)^{H^A} \Lambda_0^A = (-1)^{H^A} \Lambda_0^L$ and $\Lambda_1^A = \Lambda_1^L$; finally, $\Phi_1^A = \Omega_1^A = 0$.

In Fig. 6.28 we show the local distribution of the compliance for the initial and final states. The solution is computed in the two cases of stiffness and strength leading objective. It is apparent that the final state corresponds to a stiffer structure. In Fig. 6.29 we show the local distribution of the Tsai–Hill failure index F_{TH}^{lam}, computed again for the two cases of leading objective. Also in this case, the improving of the strength is really clear.

Actually, the optimization if confirmed by the numerical results: the reduction of the compliance is the $\sim$32% of the initial value, while that of F_{TH}^{lam} is of $\sim$94%, and the differences between the two cases of leading objective are really small (at most 0.3%).

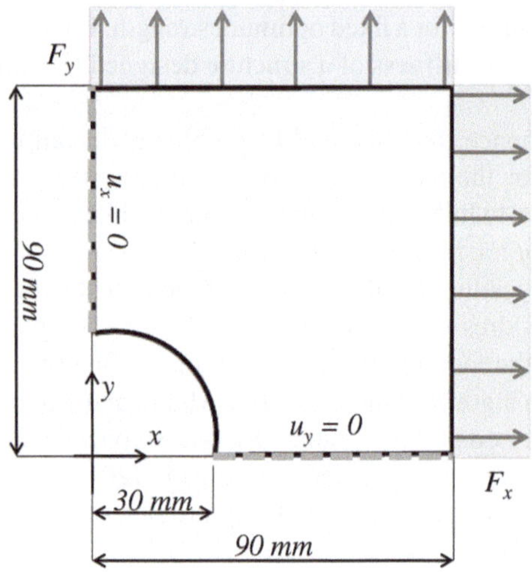

Fig. 6.27 Example 1: scheme of the problem

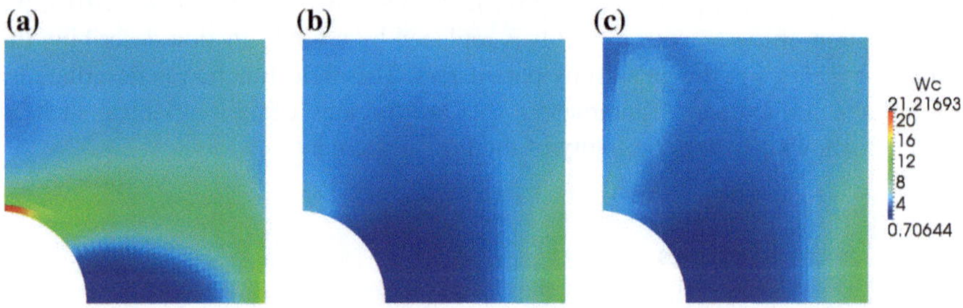

Fig. 6.28 Example 1: local distribution of the compliance for: **a** the initial state, **b** the final state with stiffness as leading objective and **c** the final state with strength as leading objective

The optimal distribution of the stiffness parameters $(-1)^{K^A} R_0^A$, R_1^A and Φ_1^A are plotted in Fig. 6.30; it can be observed that $(-1)^{K^A} R_0^A = (-1)^{K^L} R_0^L$ everywhere, while R_1^A changes. In the second step, a suitable sequence of plies is found locally, i.e. for each finite element by an approach similar to that described in Sect. 6.6.3, now including also the polar parameters of $\mathbb{G}^A$, i.e. $(-1)^{H^A} \Lambda_0^A$ and Λ_1^A. In this way, a distribution of fibers orientations is found layer by layer; the result for the first four plies is plotted in Fig. 6.31. It is apparent that discontinuities in adjacent elements is an issue for practical applications. As already said, a promising way to counter this problem is that proposed in Montemurro and Catapano (2016) but a complete integration in the design process of anisotropic structures of technological requirements is still an open problem.

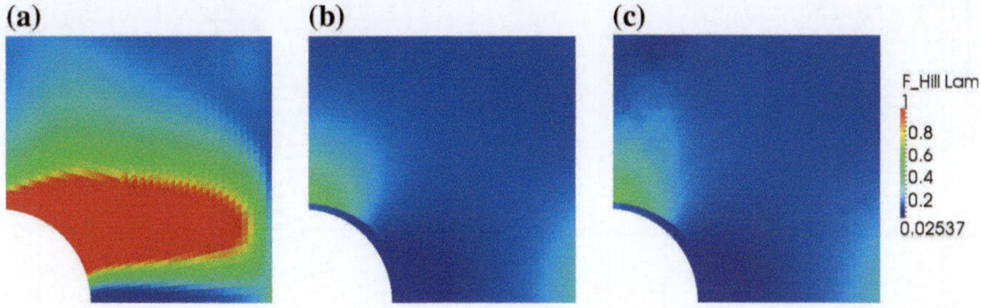

Fig. 6.29 Example 1: local distribution of the failure index F_{TH}^{lam} for: **a** the initial state, **b** the final state with stiffness as leading objective and **c** the final state with strength as leading objective

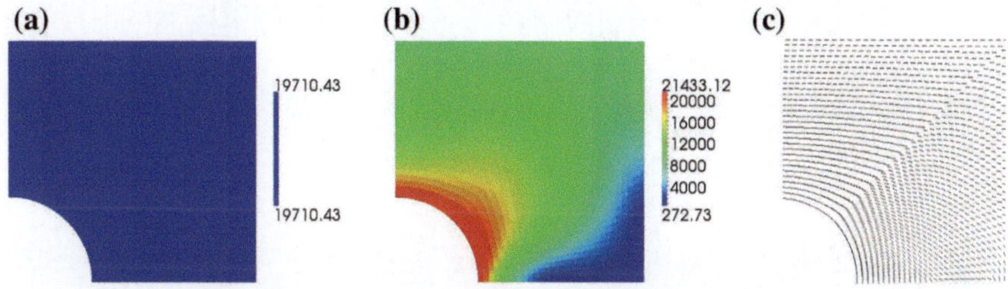

Fig. 6.30 Example 1: local distribution of optimal $(-1)^{K^A}R_0^A$, R_1^A and Φ_1^A

Example 2

The second example concerns the design of a wedge like in Fig. 6.32 (Catapano 2013). We consider a circular sector, subject to in-plane shear loading conditions. The geometry of the plate and the boundary conditions are shown in Fig. 6.32. The loading value is $F_y = 150$ N/mm. The initial values of the stiffness polar parameters correspond to a unidirectional lamina oriented along the x axis having the properties of the basic ply: $(-1)^{K^A}R_0^A = (-1)^{K^L}R_0^L, R_1^A = R_1^L$ and $\Phi_1^A = 0$ everywhere. The tip of the sector has been truncated to avoid a singular point; F_y is uniformly distributed on the vertical side of the wedge. Let us just consider the results of the first step problem when the leading property is stiffness. Figures 6.33 and 6.34 show respectively the change of the distribution of the compliance and of the failure index F_{TH}^{lam} from the initial to the final, optimal structure. The reduction of the both the quantities is apparent and actually the reduction of the compliance is of 55.6% and that of F_{TH}^{lam} of 96.5%.

The optimal distribution of the polar parameters is shown in Fig. 6.35 for $(-1)^{K^A}R_0^A$ and R_1^A, in Fig. 6.36 for $(-1)^{H^A}\Lambda_0^A$ and Λ_1^A and in Fig. 6.37 for the material orientation $\Phi_1^A = \Omega_1^A$.

Also in this case, we can see that the optimum is reached when $(-1)^{K^A}R_0^A = (-1)^{K^L}R_0^L$ everywhere in the structure. Looking at Fig. 6.34, we can appreciate how much the strength of the wedge is improved: in the optimized structure, almost

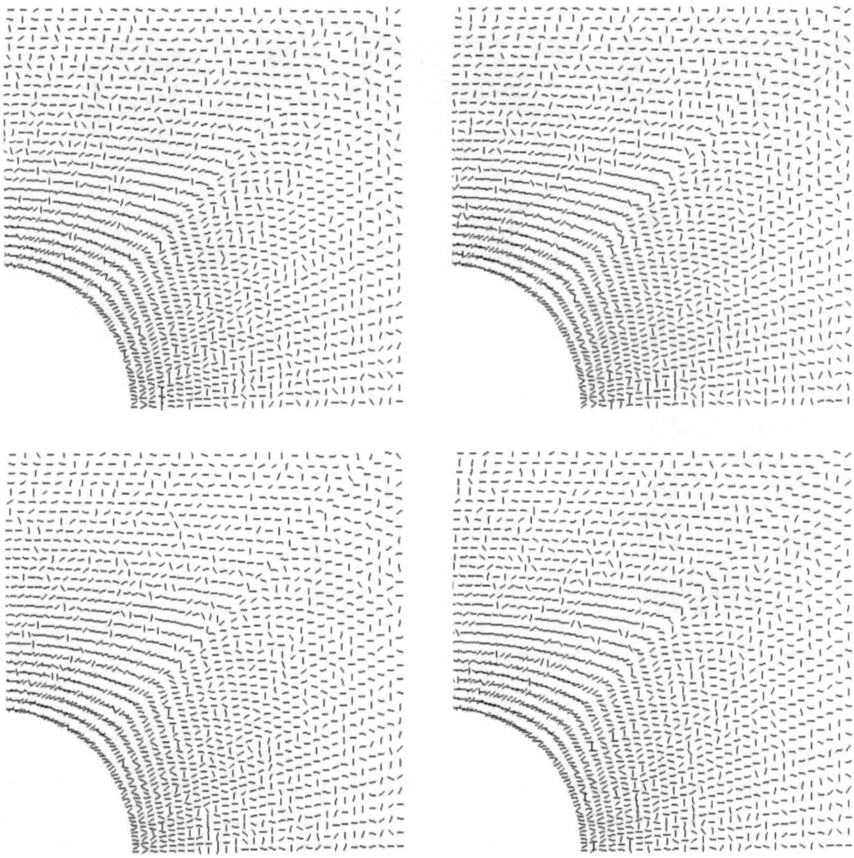

Fig. 6.31 Example 1: distribution of the orientation for the first four plies

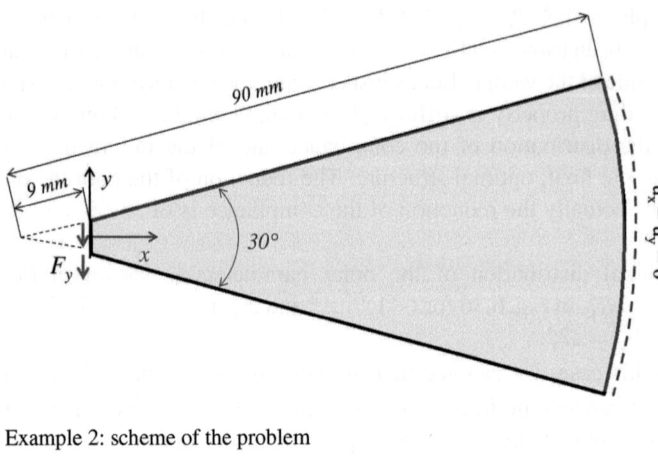

Fig. 6.32 Example 2: scheme of the problem

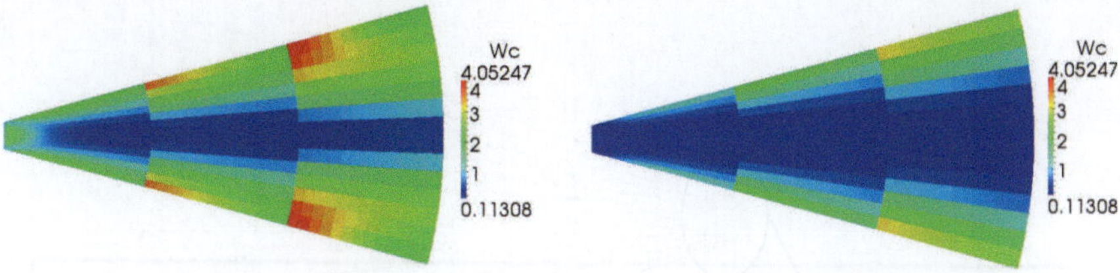

Fig. 6.33 Example 2: distribution of the compliance: *left*, initial, *right*, final

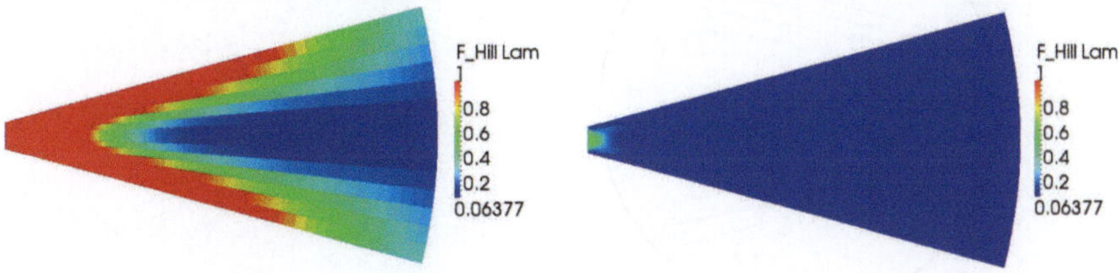

Fig. 6.34 Example 2: distribution of the failure index F_{TH}^{lam}: *left*, initial, *right*, final

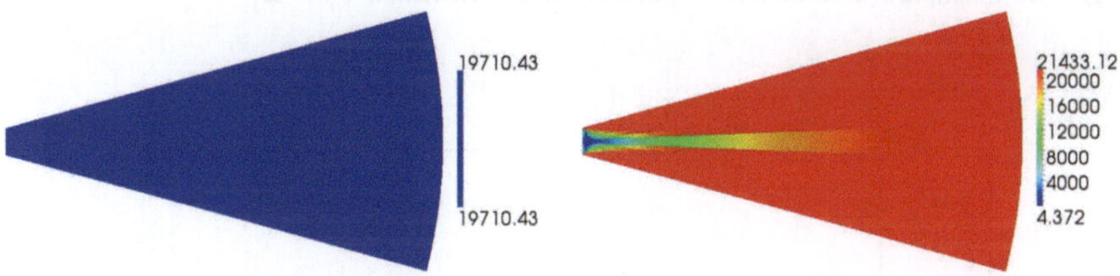

Fig. 6.35 Example 2: distribution of optimal $(-1)^{K^A} R_0^A$, *left*, and R_1^A, *right*

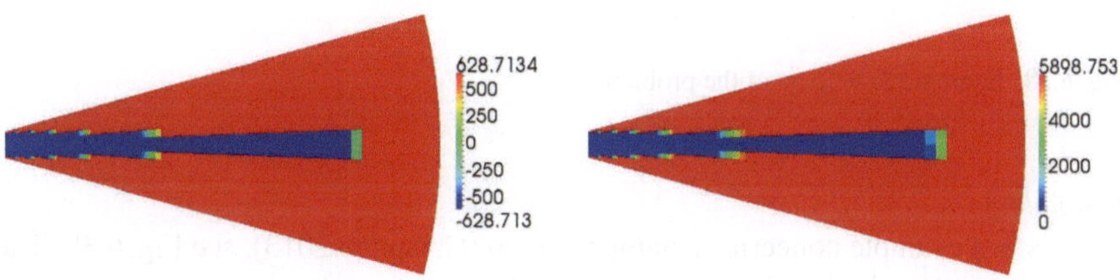

Fig. 6.36 Example 2: distribution of optimal $(-1)^{H^A} \Lambda_0^A$, *left*, and Λ_1^A, *right*

everywhere in the plate F_{TH}^{lam} is far below 1, while in the original state many parts of the wedge were in a critical state.

Looking at Fig. 6.37, we can see that the solution so found is very similar to the analytical one found in Royer-Carfagni (2000), Pedersen (2003) for the optimal fiber layout of a composite anisotropic wedge.

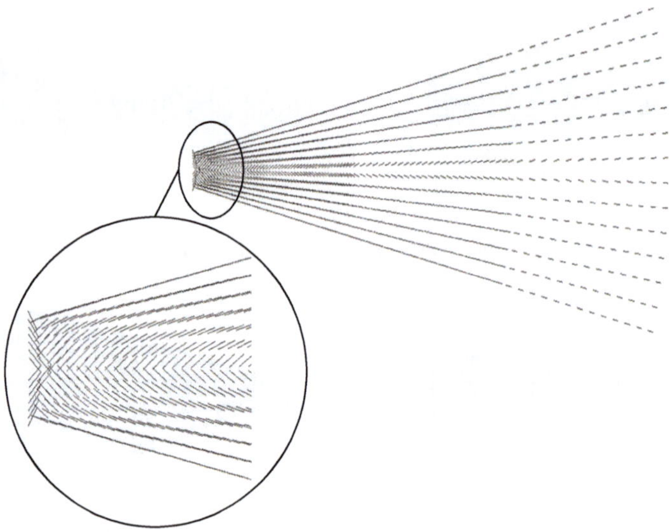

Fig. 6.37 Example 2: distribution of optimal material orientation $\Phi_1^A = \Omega_1^A$

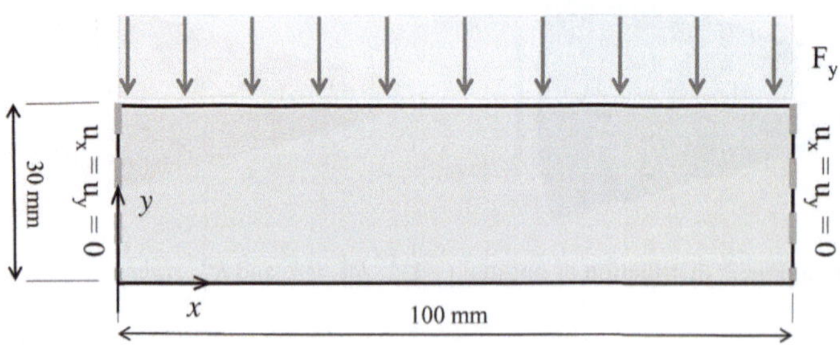

Fig. 6.38 Example 3: scheme of the problem

Example 3

This last example concerns a clamped beam (Catapano 2013), see Fig. 6.38. The loading is uniformly distributed, with $F_y = 200$ N/mm, and the initial configuration for the stiffness polar parameters is that of a balanced cross-ply laminate oriented at $\pm 45°$ with respect to the x axis: $(-1)^{K^A} R_0^A = (-1)^{K^L} R_0^L$ and $R_1^A = 0$ everywhere. In this case, we just present the solution when the strength is the leading objective.

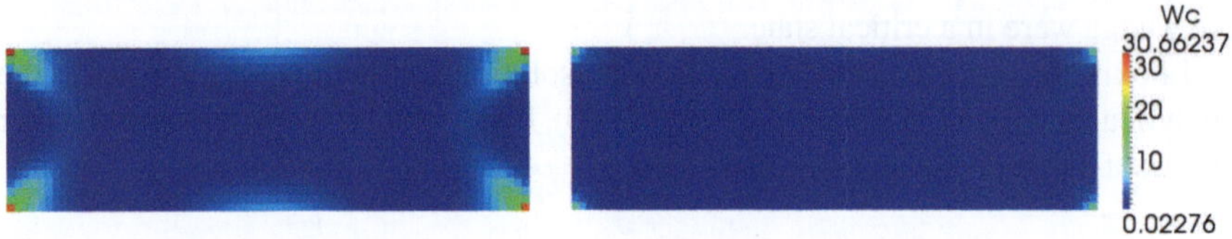

Fig. 6.39 Example 3: distribution of the compliance: *left*, initial, *right*, final

The reduction of the compliance, with respect to the initial condition, is of 61.1%, while that of the failure index F_{TH}^{lam} of 95.1% and the maximum displacement is reduced of ~65%. Figure 6.39 shows the change of the distribution of the compliance from the initial to the final, optimal configuration of the material parameters, while the same change for the failure index F_{TH}^{lam} is shown in Fig. 6.40. Also in this case, it is really clear the effect of the anisotropy optimization: the critical zones disappear in the final configuration. Finally, we plot in Fig. 6.41 the optimal distribution of the stiffness parameters $(-1)^{K^A} R_0^A$ and R_1^A, in Fig. 6.42 that of the strength parameters $(-1)^{H^A} \Lambda_0^A$ and Λ_1^A and the optimal distribution of the material orientation $\Phi_1^A = \Omega_1^A$ in Fig. 6.43.

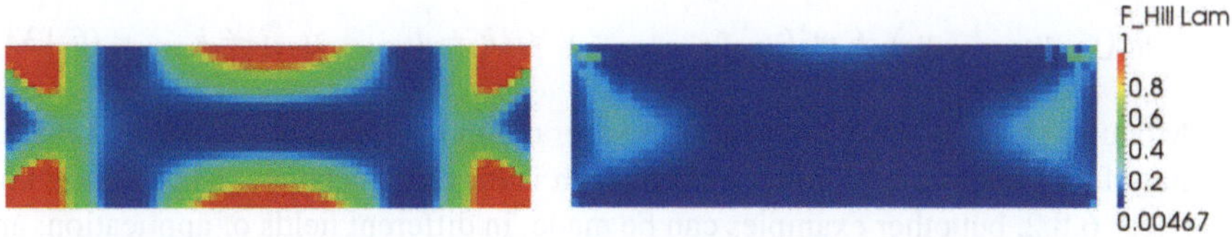

Fig. 6.40 Example 3: distribution of the failure index F_{TH}^{lam}: *left*, initial, *right*, final

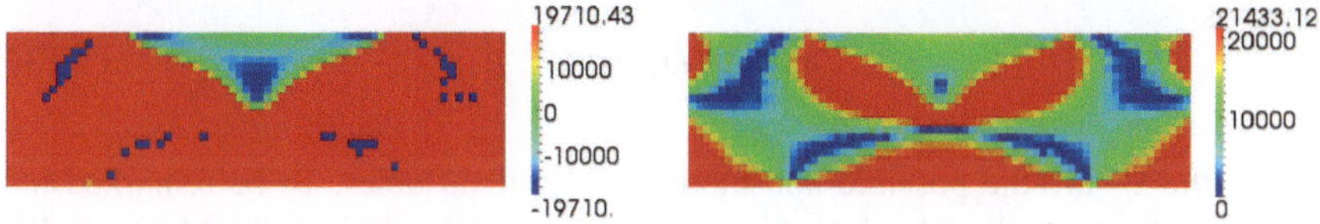

Fig. 6.41 Example 3: distribution of optimal $(-1)^{K^A} R_0^A$, *left*, and R_1^A, *right*

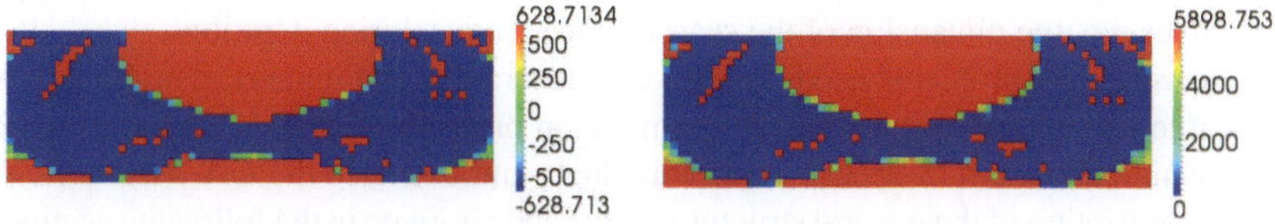

Fig. 6.42 Example 3: distribution of optimal $(-1)^{H^A} \Lambda_0^A$, *left*, and Λ_1^A, *right*

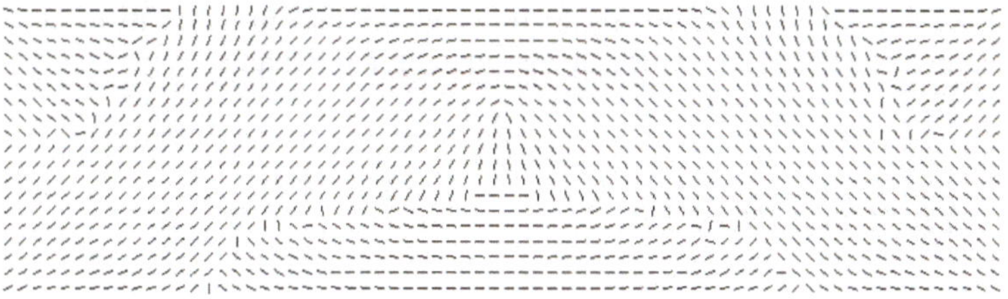

Fig. 6.43 Example 3: distribution of optimal material orientation $\Phi_1^A = \Omega_1^A$

6.8 Optimization of Modular Systems

A system composed by some some repetitive parts is said to be *modular* and the parts that are repeated in the system are the *modules*. Mathematically, each module m_i is a vector of $\mathbb{R}^\nu$:

$$m_i = m_i(y_1, y_2, \ldots, y_\nu), \quad i = 1, \ldots, \mu, \tag{6.132}$$

with ν the number of variables describing a module and μ the number of modules composing the system. Generally speaking, all the modules are determined by the same set of variables y_j but the values of these variables can be different for each module, i.e.

$$m_i(y_1, y_2, \ldots, y_\nu) \neq m_k(y_1, y_2, \ldots, y_\nu) \ \forall i, k = 1, \ldots, \mu, \ i \neq k. \tag{6.133}$$

Modular systems are encountered in several applications and a typical example is that of an anisotropic laminate, where each layer can be considered as a module, see Sect. 6.8.2, but other examples can be made, in different fields of application: an example is given in Sect. 6.8.3.

Generally speaking, the optimization problem of a modular system can be formulated as

$$\min_{x_l, m_i(y_j), \mu} \ f(x_l, m_i(y_j), \mu), \tag{6.134}$$

where the x_ls are the possible non-modular design variables. To be remarked that such a problem implies the determination not only of the design variables x_l and y_j, for all the modules, but *also* the optimal number of modules to be used in the system. As a consequence, these are problem with *variable dimension*: the number of design variables, i.e. the dimension of the search space, varies during iterations.

Solving such problems needs hence suitable strategy and numerical approach. A possible way to tackle this kind of optimization problems has been inspired by the Darwinian evolution of species and individuals and coded in the GA BIANCA. A brief description of the general structure of this code is given in the following section. BIANCA is a code that has been developed in the Ph.D. thesis of A. Vincenti first (Vincenti 2002) and completed in that of M. Montemurro then (Montemurro 2012). The interested reader is addressed to these works and to Vincenti et al. (2010), Montemurro et al. (2012b, c) for a deeper insight in the matter.

6.8.1 *The Code BIANCA*

Like any other GA, the code BIANCA works on a *population of individuals* that undergo the genetic operations of *crossover* and *mutation* to obtain a new population. The individuals that are crossed are chosen in some way, but according to a simple criterion: the best the individual, the highest the chance of reproduction. This mimics

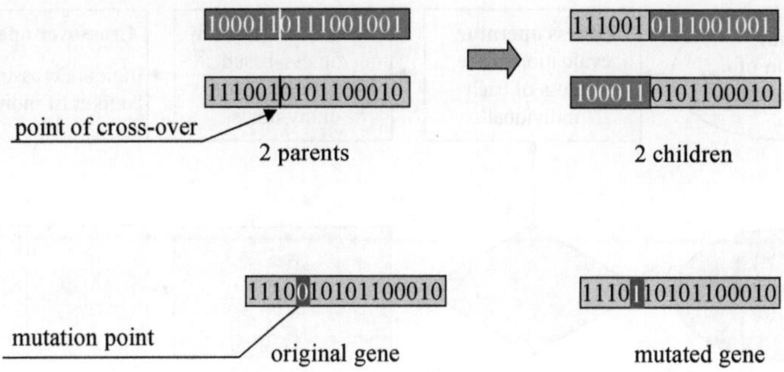

Fig. 6.44 General scheme of standard crossover and mutation operators

the *Darwinian selection in Nature*: better individuals are more fitted to environmental conditions and likely to survive longer than less adapted individuals, so they have a greater chance to reproduce.

In optimization, an individual is simply a *candidate* to be the solution of a given problem; hence, generally speaking, it corresponds to an array collecting the design variables. The individuals are then chosen according to their *fitness*, i.e. according to a measure of the value of the objective function: better individuals have a greater value of their fitness.[18]

The individuals chosen for reproduction undergo *crossover*; this is an algebraic operation inspired by genetic recombination: mixing the genome of two well fitted individuals, a better individual can, hopefully, be obtained. Mathematically speaking, there are different ways to operate crossover; the most well known, is that of splitting the binary string of two individuals at the same point and then recombine the separated chunks of the two different individuals, Fig. 6.44.

The operation of *mutation* consists in changing, randomly, a bit of the binary string of a variable, see again Fig. 6.44; this is done to introduce *biodiversity*, i.e. to better explore the search domain and hence to increase the chance to escape from possible local minima. Other methods exist for crossover and mutation, namely operating directly on the array of real numbers (Michalewicz 1992).

Once obtained the new population, a check of convergence is done; if this is satisfied, then the best individual of the last population is taken as optimal solution, otherwise a new loop is done with the new population until convergence. Another operation, often used because accelerating convergence, is *elitism*: the best individual of each generation is automatically conserved, which allows also to have a non-decreasing fitness of the best individual through generations. The general scheme of a standard GA is shown in Fig. 6.45.

BIANCA was constructed on this same general scheme: binary coding of variables, elitism, selection by roulette wheel or tournament (Michalewicz 1992), and

[18]Often, in GAs the fitness is simply the objective function; however, it is a good practice to traduce in some way the objective function into another function, the fitness, scaled from 0, the worst individual, to 1, the best one. This allows to have a precise, normalized ordering of individuals.

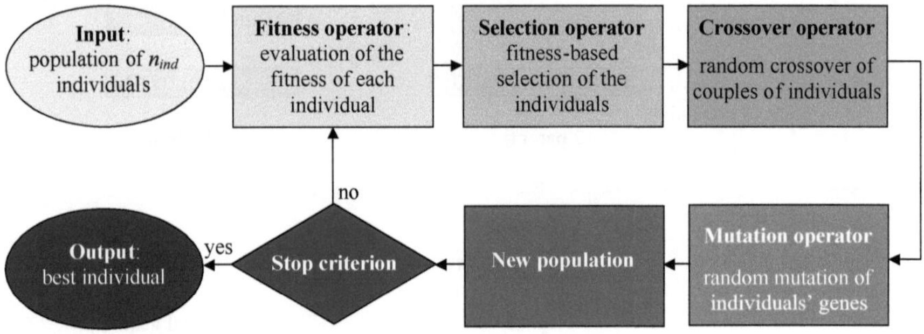

Fig. 6.45 General scheme of a standard genetic algorithm

also multipopulation: for improving the exploration of the search domain, it is possible to let evolve in parallel different populations.

However, the true peculiarity of BIANCA is the organization of the information, that lets treat complex problems and namely cases with varying number of design variables, so in particular modular optimization. Unlike the most part of GAs, that have a mono-chromosome algebraic structure, in BIANCA the information is organized in a *genome* composed of *chromosomes* which in turn are composed by *genes*. An individual, i.e. the candidate to be a solution to the given problem, is characterized by a genome composed by a number n_{chrom} of chromosomes; such number, determines the *species* which the individual belongs to. So, individuals of different species are characterized by a different number of chromosomes. Each chromosome, in turn, is composed by n_{gene} genes, corresponding to the number of design variables needed to determine the individual, i.e. the solution: each gene is a binary coding of a design variable.

For a modular problem, each module is represented by a chromosome and n_{chrom} determines hence the number of modules composing the system, i.e. the species, in the biological paradigm, of the individual. Because in a modular system μ, the number of modules, is a design variable, n_{chrom} is a variable and in the population there are individuals with different n_{chrom}, i.e. belonging to different species.

In other words, the selection is done not only on the individuals, but *also* on species; the idea is that of an extended Darwinian selection, letting emerge new better fitted species and inside them the best individual. Computationally speaking, this needs the introduction of *species-oriented genetic operators*, similar to those used for individuals, crossover and mutation, but acting on species, i.e. on the number of chromosomes.

In BIANCA, such species-oriented operators have been implemented and actually the code makes in parallel and without distinction the optimization of the species and of the individuals. In particular, the typical reproduction operators of crossover and mutation have been specially conceived for crossing species and individuals during the same iterations. A deep mixing of the individual genotype is made in BIANCA: crossover and mutation act on every single gene of the individuals, so allowing for a true independent evolution of each design variable.

Practically, when dealing with modular problems, the user must specify the lower and upper bounds for n_{chrom}, besides the other usual parameters driving the optimization, e.g. the probability of crossover and of mutation, the number of populations, the type of selection etc. Then, the calculation proceeds as usual, according to the scheme of Fig. 6.45, but including also the selection of the species. We have observed, for all the cases treated, that while the selection of species and individuals is made simultaneously, it is the convergence of the species that is achieved first and once the best species obtained, then the code goes on in improving the individuals of that species, which seems quite in agreement with natural selection.

We give below two examples of application of these strategy in optimization of anisotropic laminated structures: the first one is a general problem: how to determine the minimal number of layers for a laminated designed to have some given properties, while the second is an application concerning the design of a stiffened composite wing box-girder for a civil aircraft. We just recall that another application of the selection of species with BIANCA has been already presented in Sect. 6.6.5, *Example 3*.

6.8.2 Designing Laminates with Minimal Number of Layers

We have seen in Sect. 6.4.1 how to formulate the design of a laminate with given general properties in a unique problem: the minimization of a positive semi-definite quadratic function $\mathcal{I}(\mathcal{P}_j)$ of the polar parameters $\mathcal{P}_j$ of the laminate. Such quadratic function is actually the sum of different terms, each one representing a property concerning tensors $\mathbb{A}$, $\mathbb{B}$ and $\mathbb{D}$. Then, we have seen in Sect. 6.6.3 how to modify such problem in order to include some requirements on the optimal elastic response of the structure.

The question is hence rather natural: is it *always* possible to find a laminate meeting all the imposed requirements? Of course, the answer is *non*. A simple example is sufficient to show it: it is not possible to obtain in-plane isotropy with only two layers, three is the minimum number. This suggests us that the question should be changed into: *which is the minimum number of given layers for a laminate will satisfy all the asked requirements*?

Apart from few, elementary cases, the result to such a question is unknown. Actually, finding the minimum number of plies for which a given optimum laminate design problem can be solved is a very difficult task. In fact, the solution, i.e. the number of plies, varies with the number and type of requirements to be met. In other words, the result is strictly problem-dependent and unfortunately the optimum solutions are unknown, so it does not exist an analytical model which could describe their evolution with the number of layers. Therefore, a numerical investigation seems to be the only possibility for this investigation.

It is worth noting that the optimum design of a laminate in terms of the number and properties (orientation, materials, and thickness) of its layers is a combinatorial optimization problem, which is harder to solve for small numbers of layers. In fact, the smaller the number of plies, the smaller becomes the design space, and the number of available solutions decreases. Also, solutions with the least number of plies are interesting for lightweight applications.

We describe here an approach (Montemurro et al. 2012e), to the resolution of this problem in the form of the search for the minimum of a positive semidefinite form, including the number of layers n among the variables in a penalization term able to strongly drive the search for an optimum solution towards laminates with the lowest number of layers.

The new minimum problem, including n among the variables, is

$$\min_{n,\delta_k,h_k} n^s\, \mathscr{I}(\mathscr{P}_j(\delta_k, h_k)), \tag{6.135}$$

with s an exponent to be fixed in order to improve the convergence towards a minimum n. The large number of numerical tests conducted suggests that the best results are obtained when $s \in [1, 4]$. We have included the thicknesses of the layers, h_k, among the design variables for a more general formulation.

The above formulation naturally drives the solution towards low values of n, because the factor n^s is strongly penalizing the positive-semidefinite form $\mathscr{I}(\mathscr{P}_j)$ and the solutions correspond to $\mathscr{I}(\mathscr{P}_j) = 0$.

A problem formulated in this way is clearly a modular problem: the plies are the modules, each one defined at most by two variables, the orientation δ_k and the thickness h_k.[19] So, it can be tackled using the code BIANCA with different species: the number of chromosomes of a species is $n_{chrom} = n$ and each chromosome is composed by two genes, coding δ_k and h_k. The number of variables of each individual is hence $2n$ and it varies from species to species, i.e. with n. Different examples have been treated in Montemurro et al. (2012e), we give in the next section an account of two of them. The material of the layers is, for both the examples, carbon-epoxy T300/5208, cf. Table 4.9; moreover, each problem is treated in the two cases of layers of equal or different thickness and the orientations can take only integer values, while the thicknesses can vary continuously in the interval $[0.1, 0.2]$ mm (for the case of identical layers, $h = 0.125$ mm). Finally, n can take values in the interval $[4, 16]$, the value chosen for s is 2 and each run of BIANCA is made with one population of 500 individuals and for 500 generations.

6.8.2.1 Numerical Examples

Example 1

The first example is that of a laminate designed to be uncoupled and fully orthotropic with $K^A = K^D = 1$ and coincidence of the orthotropy axes of $\mathbb{A}$ and $\mathbb{D}$. The objective function for this problem is[20]

[19] Actually, a more general problem could be considered exactly in the same manner, the one where also the layers are composed by different materials.

[20] For the case of identical layers, the terms $\mathscr{P}_7$ and $\mathscr{P}_8$ can be omitted because they are automatically null, see Eq. (5.52).

$$n^s \mathscr{I}(\mathscr{P}(\delta_k, h_k)) = n^s \left[\mathscr{P}_5^2 + \mathscr{P}_7^2 + \mathscr{P}_8^2 + \mathscr{P}_9^2 + \mathscr{P}_{10}^2 + \mathscr{P}_{17}^2 + (\mathscr{P}_6 - \mathscr{P}_{18})^2\right], \tag{6.136}$$

with $K^A = K^D = 1$. The best solutions found by BIANCA are[21]

- constant thicknesses: $n = 6$, [21.5°, −27.5°, −29.5°, 29.5°, 27.5°, −21.5°], with a residual 1.75×10^{-2};
- variable thicknesses: $n = 4$, [21.5°, −27.5°, 27.5°, −21.5°] and [0.1, 0.2, 0.2, 0.1] mm, with a residual 7.33×10^{-3}.

It can be observed that if the thicknesses are design variables, then the number of layers decreases and the residual too. This is quite obvious, as the introduction of new variables broadens the search space and allows the algorithm to reach more easily lower values of the objective function. Also, both the solutions are antisymmetric stacks.

We show in Fig. 6.46 the directional diagrams of A_{1111}, B_{1111} and D_{1111} for the two solutions above; once more, it is apparent that the requirements are met ($\mathbb{B} \sim \mathbb{O}$, as the almost invisible diagrams of B_{1111} show).

Example 2

The second example concerns the search for quasi-homogeneous laminates; in this case the objective function is[22]

$$\begin{aligned} n^s \mathscr{I}(\mathscr{P}(\delta_k, h_k)) =& n^s \left[\mathscr{P}_7^2 + \mathscr{P}_8^2 + \mathscr{P}_9^2 + \mathscr{P}_{10}^2 + (\mathscr{P}_1 - \mathscr{P}_{13})^2 + \right. \\ & (\mathscr{P}_2 - \mathscr{P}_{14})^2 + (\mathscr{P}_3 - \mathscr{P}_{15})^2 + (\mathscr{P}_4 - \mathscr{P}_{16})^2 + \\ & \left. (\mathscr{P}_5 - \mathscr{P}_{17})^2 + (\mathscr{P}_6 - \mathscr{P}_{18})^2\right]. \end{aligned} \tag{6.137}$$

The best solutions found by BIANCA are

- constant thicknesses: $n = 7$, [−3°, 4°, 7°, −2°, −1°, −4°, 3°], with a residual 1.75×10^{-2};
- variable thicknesses: $n = 6$, [−5°, 8°, −2°, 0°, −11°, 4°] and [0.156, 0.188, 0.158, 0.151, 0.111, 0.182] mm, with a residual 2.99×10^{-5}.

Also for this case, we show in Fig. 6.47 the directional diagrams of A_{1111}, B_{1111} and D_{1111} for the two solutions above; the same commentaries can be done. In addition, it is apparent the identical behavior in extension and bending, for the diagrams of A_{1111} and D_{1111} are perfectly superposed and for this case there is practically no difference in the elastic behavior between the two solutions.

---

[21]Only apparently these sequences do not meet the requirement of integer orientations; actually, they have given in this form to have the orthotropy axes aligned with the coordinate axes and to put in light that the sequences are antisymmetric. Actually, all the orientations differ for an integer value.

[22]For the case of identical layers, the terms $\mathscr{P}_7$, $\mathscr{P}_8$, $(\mathscr{P}_1 - \mathscr{P}_{13})$ and $(\mathscr{P}_2 - \mathscr{P}_{14})$ can be omitted because they are automatically null, see Eqs. (5.52) and (5.53).

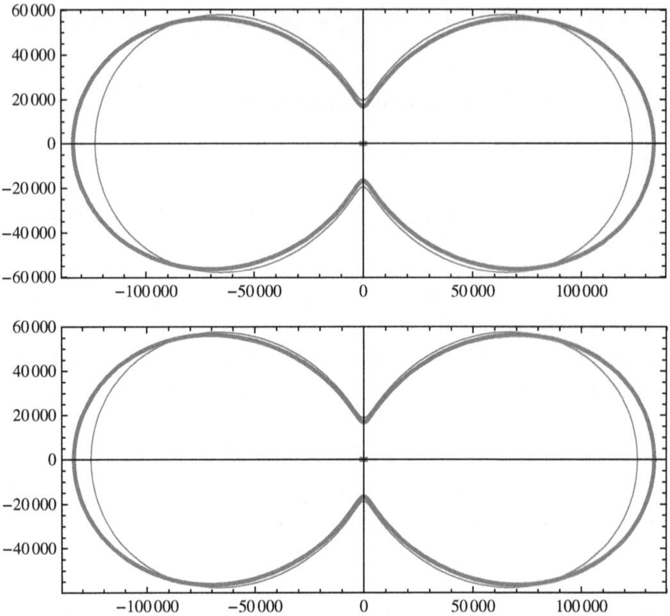

Fig. 6.46 Example 1: directional diagrams of A_{1111}, *thin line*, B_{1111}, in the *center*, and D_{1111}, *thick line* (units: MPa); *top* identical thicknesses, *bottom* different thicknesses

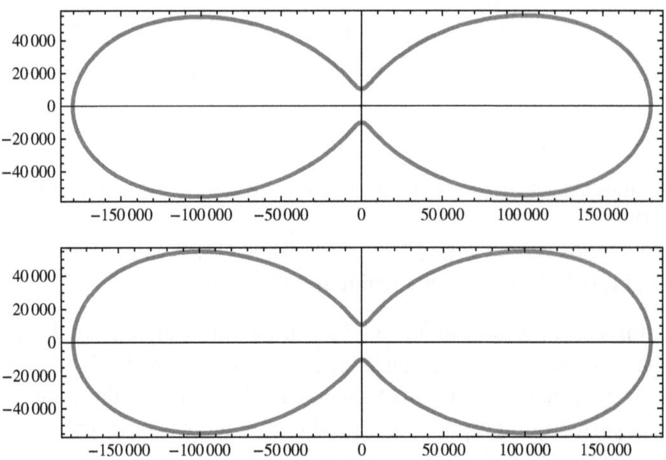

Fig. 6.47 Example 2: directional diagrams of A_{1111} and D_{1111}, *thick line*, and of B_{1111}, in the *center* (units: MPa); *top* identical thicknesses, *bottom* different thicknesses

6.8.3 An Application: The Design of an Aircraft Wing Box-Girder

We give in this section an application of the optimization techniques for anisotropic laminates to a real world case: the design of a stiffened wing box-girder for a civil aircraft. This example has been treated in Montemurro et al. (2012c, d) and the topic of this work was to show that it is actually possible to design without a-priori restrictions a complete anisotropic laminated structure and that, in doing so, the gain in lightness can be really interesting. The objective, in fact, is the reduction of the weight, under a mechanical constraint on the buckling load.

In Fig. 6.48 we show a general scheme of the structure and the steps leading to the construction of the approximate model of the box-girder section. We have considered the wing section located at the 60% of the wing span, whose typical dimensions are

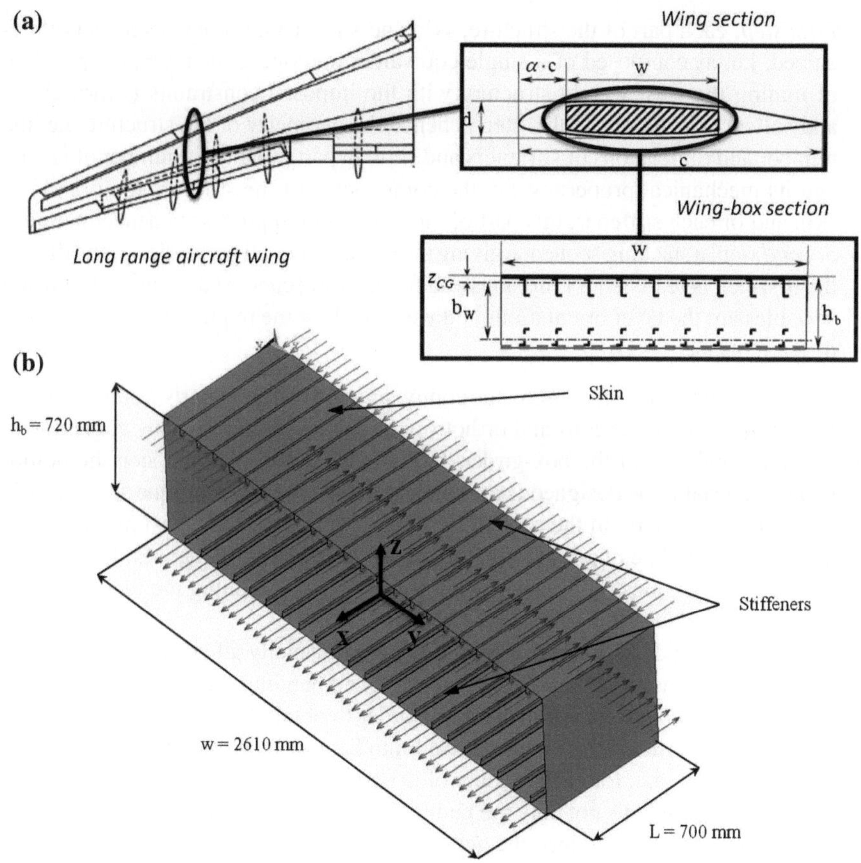

Fig. 6.48 Scheme of the problem: **a** conceptual phases for the construction of the box-girder model; **b** model of the wing box-girder section and applied loads

also shown in Fig. 6.48. These values represent specific dimensions for a civil aircraft with a design range of about 9300 km, 350 passengers, two engines, cruising altitude between ~7600 and 10,700 m and Mach number of about 0.82, cf. Raymer (2006).

The structure has a width $w = 2610$ mm, height $h_b = 720$ mm and length $L = 700$ mm and represents a portion of the wing between two consecutive ribs. We consider this portion of the wing simply-supported on these ribs. Figure 6.48 shows also the loads acting on the structure in normal flight conditions, when the upper panel is subject to compression and hence to buckling phenomena. The whole wing-box is a laminated structure made of carbon-epoxy T300/5280 plies, cf. Table 4.9. Both upper and lower panels have Z-shaped stiffeners with equal flanges. The core and the flanges of each stiffener have the same thickness, i.e. the same number of plies.

The number of plies has to be determined; hence, this is a problem with modular variables, the stiffeners, but also with other non-modular variables, those representing the skin, i.e. the plates composing the box. The whole problem is tackled with the two-step strategy:

- *First step*: each part of the structure, skin and stiffeners, is considered as homogenized, i.e. as composed of a single equivalent homogeneous layer; the problem of finding the least weight structure with the imposed constraints is formulated and solved. The output of this step is hence the geometry of the structure, i.e. the number and dimensions of stiffeners and skin, in particular their number of layers, and its mechanical properties, i.e. the components of the stiffness tensors of the skin and of each stiffener; this part of the optimization process is hence modular;
- *Second step*: a stacking sequence giving the optimal overall properties found during the first step is searched for the skins and for each stiffener. At this stage, the design variables are the layer orientations and we can add some requirements concerning different aspects.

The requirements on the laminates composing the different parts of the structure are to be quasi-homogeneous and orthotropic, with an orthotropy axis aligned with the longitudinal axis of the box-girder. In this way, during the first step the homogeneous material to be designed is completely determined by a unique set of elastic orthotropic parameters; in the second step, the requirements of quasi-homogeneity and orthotropy will be imposed to the search of the stacking sequences. In addition, the assumption of quasi-homogeneity allows for exactly defining the feasible set of polar parameters describing the elastic behavior.

In fact, cf. Sect. 5.2.13, the choice of quasi-homogeneity allows for an exact definition of the geometrical bounds for extension and bending. Such bounds are to be considered *for each part of the structure*, in order to obtain feasible laminates for the second step. In particular, the geometrical bounds must be written for the skins and for each stiffeners, i.e. for each module of the structure. Because these are variable, for this kind of problems not only the number of design parameters vary, but also the number of imposed constraints, the geometrical bounds.

6.8.3.1 The First Step Problem

The parameters to be determined in this phase are:

- the number of the stiffeners;
- the thickness, and hence the number of layers, of the skin and of each stiffener;
- the geometrical dimensions of each stiffener;
- the mechanical properties of the skin and of each stiffener.

Considering the assumptions specified above, the only mechanical variables needed in this phase are the anisotropy polar parameters in extension, i.e. $\mathscr{P}_3$, $\mathscr{P}_4$ and $\mathscr{P}_5$, for the skin and for each stiffener. However, because the choice is done of an orthotropic behavior, $\mathscr{P}_5$ can be incorporated into $\mathscr{P}_3$, simply posing $\mathscr{P}_3 = (-1)^{K^A} R_0^A$, so we are left with just two mechanical variables for each part of the structure.

About the geometrical design variables, these are shown in Fig. 6.49 and are:

- the number of stiffeners N;
- the thickness of each stiffener t_i, $i = 1, \ldots, N$;
- the height of each stiffener h_i $(i = 1, \ldots, N)$;
- the thickness of the skin t.

All the geometrical variables are discrete valued; the ranges of their variation, along with their steps, are shown in Table 6.2. To remark that the thicknesses are chosen as multiples of the thickness of the basic layer, $t^L = 0.125$ mm, so that the number of layers, needed for the second step, is automatically determined.

For technological reasons, the width of the flange of each stiffener, d_i, $i = 1, \ldots, N$, is not a design variable and depends on the height of the stiffener:

$$d_i = 0.3\, h_i, \quad i = 1, \ldots, N. \tag{6.138}$$

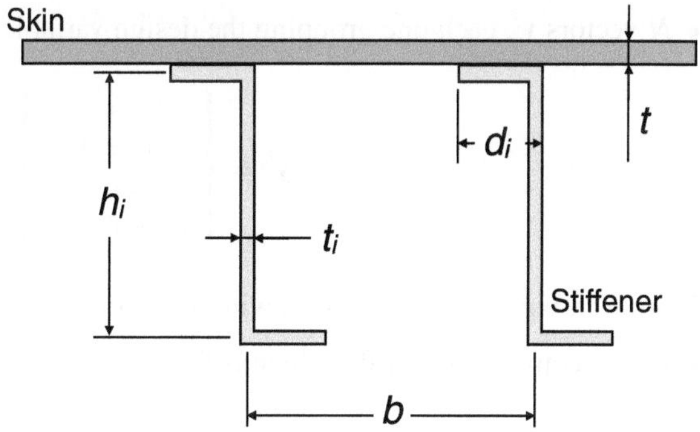

Fig. 6.49 Geometrical design variables

Table 6.2 Design variables of the first step problem

Part	Design variable	Type	Lower bound	Upper bound	Step
Stiffeners	N	Discrete	18	23	1
	t_i (mm)	Discrete	2.0	5.0	0.125
	h_i (mm)	Discrete	40.0	90.0	0.5
	$\mathscr{P}_3^i$ (GPa)	Continuous	−19.710	19.710	–
	$\mathscr{P}_4^i$ (GPa)	Continuous	0.0	21.433	–
Skin	t (mm)	Discrete	2.0	5.0	0.125
	$\mathscr{P}_3$ (GPa)	Continuous	−19.710	19.710	–
	$\mathscr{P}_4$ (GPa)	Continuous	0.0	21.433	–

The stiffeners are equally spaced of b which depends upon the number of stiffeners:

$$b = \frac{w}{N+1}, \tag{6.139}$$

where w is the width of the whole box-girder.

The objective is to minimize the weight of the wing-box; we have already specified that this must be done satisfying on one side the constraint on the buckling load, and on the other side the geometrical bounds for the elastic moduli.

To mathematically formulate the first step problem, let us introduce the following symbols:

- the vector $\mathbf{x}$ collecting all the design variables concerning the overall structure and the skin:

$$\mathbf{x} = \begin{Bmatrix} x_1 = N \\ x_2 = t \\ x_3 = \mathscr{P}_3 \\ x_4 = \mathscr{P}_4 \end{Bmatrix}; \tag{6.140}$$

- N vectors $\mathbf{y}^i$, each one grouping the design variables of the ith stiffener:

$$\mathbf{y}^i = \begin{Bmatrix} y_1 = h_i \\ y_2 = t_i \\ y_3 = (\mathscr{P}_3)_i \\ y_4 = (\mathscr{P}_4)_i \end{Bmatrix}. \tag{6.141}$$

Moreover, we introduce the following functions:

- the objective function W, which is the weight of the structure:

$$W = W(\mathbf{x}, \mathbf{y}^i); \tag{6.142}$$

- the function for the constraint on the critical buckling load:

$$f(\mathbf{x}, \mathbf{y}^i) = p_{ref} - p_{cr}(\mathbf{x}, \mathbf{y}^i); \tag{6.143}$$

- the functions for the four geometric constraints on the polar parameters of the skin, cf. Eq. (5.58):

$$\begin{aligned}
g_1(x_3) &= -\frac{\mathscr{M} x_3}{R_0^L} - 1,\\
g_2(x_3) &= \frac{\mathscr{M} x_3}{R_0^L} - 1,\\
g_3(x_4) &= -x_4,\\
g_4(x_3, x_4) &= 2\left(\frac{\mathscr{M} x_4}{R_1^L}\right)^2 - 1 - (-1)^{K^L}\frac{\mathscr{M} x_3}{R_0^L};
\end{aligned} \tag{6.144}$$

- the functions for the four geometric constraints on the polar parameters of the ith stiffener, $i = 1, \ldots, N$:

$$\begin{aligned}
h_1^i(y_3^i) &= -\frac{\mathscr{M} y_3^i}{R_0^L} - 1,\\
h_2^i(y_3^i) &= \frac{\mathscr{M} y_3^i}{R_0^L} - 1,\\
h_3^i(y_4^i) &= -y_4^i,\\
h_4^i(y_3^i, y_4^i) &= 2\left(\frac{\mathscr{M} y_4^i}{R_1^L}\right)^2 - 1 - (-1)^{K^L}\frac{\mathscr{M} y_3^i}{R_0^L}.
\end{aligned} \tag{6.145}$$

In the above equations, $\mathscr{M}$ is the norm used to normalize the polar parameters, Eq. (6.75), which for the T300/5208 carbon-epoxy is $\mathscr{M} = 64.592$ GPa, see Footnote 10, while p_{ref} is a fixed bound for the critical buckling load p_{cr}. To remark also that the geometrical bounds have been written in a slightly different form; this is because we have put in this case $\mathscr{P}_3 = (-1)^{K^A} R_0^A$: in practice, we need to replace the bounds $0 \leq R_0^A \leq R_0^L$ by $-R_0^L \leq R_0^A \leq R_0^L$, which in terms of variables x_3 and y_3^i corresponds to Eqs. $(6.144)_{1,2}$ and $(6.145)_{1,2}$.

Finally, the first step minimum problem can be stated in the standard form:

$$\begin{aligned}
&\min_{\mathbf{x}, \mathbf{y}^i} W(\mathbf{x}, \mathbf{y}^1, \ldots, \mathbf{y}^N),\\
&\text{with}\\
&f(\mathbf{x}, \mathbf{y}^i, \ldots, \mathbf{y}^N) \leq 0,\\
&g_j(\mathbf{x}) \leq 0, \quad j = 1, \ldots, 4,\\
&g_l^i(\mathbf{y}^i) \leq 0, \quad i = 1, \ldots, N, \quad l = 1, \ldots, 4.
\end{aligned} \tag{6.146}$$

Problem (6.146) is non-linear, in terms of both the geometrical and mechanical variables. Its non-linearity is given not only by the objective function and the geometrical constraints but, in a stronger way, by the constraint on the value of the buckling load, $p_{cr} \geq p_{ref}$. In this particular case, moreover, all the functions but $f(\mathbf{x}, \mathbf{y}^i)$ can be computed analytically: there is no way to calculate p_{cr} but a numerical approach like the finite element method.

The dimension of the design space, i.e. the number of design variables, and the number of constraint equations depend on the number N of the stiffeners: the total number of design variables is $4N + 4$ (there are in fact 4 variables for each stiffener, 3 variables for the skin and the number of stiffeners, N), while the total number of constraint equations is $4N + 5$: the buckling constraint, 4 constraints for the skin and finally 4 constraints for each stiffener. However, though the number of constraints is variable, each one of the four constraints added by a module i depends only on the unknowns concerning that module, i.e. upon $\mathbf{y}^i$.

Finally, the first step problem has been solved using the GA BIANCA, for handling the variable dimension of the problem, coupled to a finite element code for the calculation of p_{cr}.

6.8.3.2 The Second Step Problem

Once solved the first step problem, we know for each stiffener and for the skin the optimal parameters $\mathscr{P}_3^{opt}$ and $\mathscr{P}_4^{opt}$. In addition, we know the number of layers for the skin and for the stiffeners:

$$n = \frac{t}{t^L}, \quad n_i = \frac{t_i}{t^L} \; i = 1, \ldots, N. \tag{6.147}$$

Hence, we need to solve, for the skin and for each one of the stiffeners, a classical second step problem, that accounting for all the requirements specified in Sect. 6.8.3, is

$$\begin{aligned} \min_{\delta_k} \; & (\mathscr{P}_3 - \mathscr{P}_3^{opt})^2 + (\mathscr{P}_4 - \mathscr{P}_4^{opt})^2 + \mathscr{P}_6^2 + \mathscr{P}_9^2 + \mathscr{P}_{10}^2 + \\ & (\mathscr{P}_3 - \mathscr{P}_{15})^2 + (\mathscr{P}_4 - \mathscr{P}_{16})^2 + \mathscr{P}_{18}^2. \end{aligned} \tag{6.148}$$

Also the second step problem has been solved using the code BIANCA, with a population of 500 individuals and for 500 generations. Moreover, the orientations have to belong to the integer set.

6.8.3.3 Results

We show in Fig. 6.50 the geometry, mesh, loads and boundary conditions of the finite element model used. As already said, the box-girder is considered simply supported at its end sections on two wing ribs. The upper panel is subjected to a uniform compression and the lower one to a uniform tension, both of unitary value. These

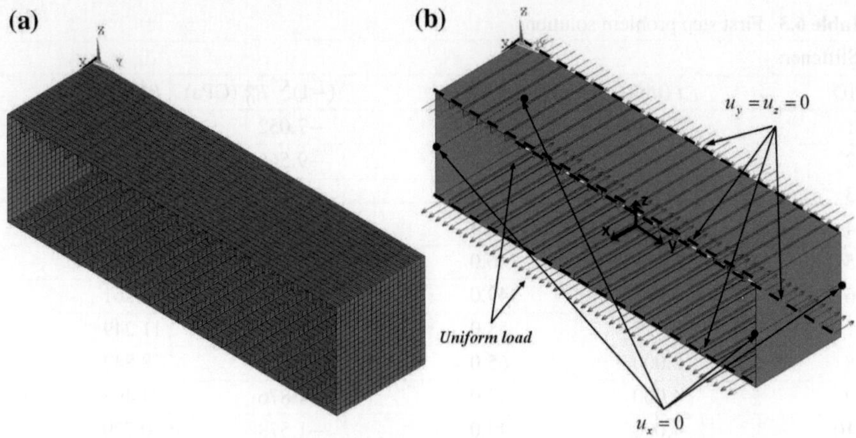

Fig. 6.50 Scheme of the finite element model: **a** mesh; **b** loads and boundary conditions

loads are representative of the condition of the structure in normal flight conditions: only the upper panel and the corresponding stiffeners can buckle.

According to a preliminary mesh sensitivity analysis, the average dimensions of the shell elements have been chosen equal to 14 × 14 mm^2. The number of shell elements in the whole wing-box structure can vary from 14,080 to 17,680, depending on the number of stiffeners N, whilst the number of degrees of freedom of the whole model can vary from 270,744 to 340,164. For each individual, a linear buckling analysis is performed for assessing the first buckling load of the structure.

The reference values $p_{ref} = 1928$ N/mm of the first buckling load and $W_{ref} = 1222.62$ N of the box-girder weight have been evaluated in a preliminary simulation made on a standard structure made by Al-7075-T6 alloy (Young's modulus $E = 72395$ MPa, Poisson's ratio $\nu = 0.33$, density $\rho = 2760$ kg/m^3), stiffened with 20 identical stiffeners with thickness 2.96 mm and height 62.33 mm, while the skin thickness is 4.93 mm. This metallic case is interesting to compare what can be done using an optimized anisotropic structure with what is usually done in practical applications.

In Montemurro et al. (2012d) different cases have been considered: all the stiffeners identical, a symmetric distribution of stiffeners with respect to the axis of the wing, all the stiffeners different. We just present here the last case; in this problem, considering the bounds specified in Table 6.2, the number of design variables can vary between 76 and 96 while that of constraints between 77 and 97. In BIANCA, the number of individuals has been fixed to 70 and that of generations to 80, which means 5600 finite element evaluations of the first buckling load of the structure. The best solution found by BIANCA is given in Table 6.3: the optimal number of stiffeners is 23, while the buckling load and the wing-box weight are $p_{cr} = 1931$ N/mm and $W = 620.19$ N respectively. From Table 6.3, it can be noticed that the orthotropy type can be different for the stiffener laminates: despite every laminate is quasi-homogeneous and orthotropic, for some of them $K^A = 0$ while for others

Table 6.3 First step problem solution

Stiffeners				
ID	t (mm)	h (mm)	$(-1)^{K^A} R_0^A$ (GPa)	R_1^A (GPa)
1	2.750	86.5	−7.052	11.899
2	4.750	55.5	−9.565	1.970
3	2.625	55.0	−1.129	14.037
4	2.125	73.5	18.888	14.574
5	3.625	46.0	−5.404	2.750
6	4.625	49.0	5.701	13.261
7	2.125	58.0	5.924	11.249
8	2.000	65.0	−8.450	8.847
9	4.000	48.0	14.876	4.495
10	4.000	43.0	−1.578	0.739
11	3.000	43.5	1.801	7.574
12	3.750	41.5	8.042	4.290
13	3.000	59.0	−1.095	11.495
14	4.250	52.0	17.811	1.149
15	4.375	54.0	10.865	2.832
16	4.000	84.0	12.536	13.178
17	2.125	48.5	3.993	10.633
18	3.125	48.5	12.276	14.349
19	3.000	56.0	12.610	11.536
20	2.125	56.5	−6.333	7.615
21	4.375	43.0	15.322	8.950
22	3.375	56.0	17.551	5.994
23	3.625	41.0	13.242	7.020
Skin				
t (mm)			$(-1)^{K^A} R_0^A$ (GPa)	R_1^A (GPa)
4.0			12.945	0.882

$K^A = 1$. This gives a reduction of the weight of about 49% with respect to the reference solution and it is very close to the boundary of the feasible domain: $p_{cr} \simeq p_{ref}$. The minimum has been found after 57 generations, see Fig. 6.51 left, where the evolution of the best individual along generations is shown, while on the right we show the convergence of the species, i.e. of the number of stiffeners. As previously said, also in this case we observe that species converge faster than individuals: the best number of stiffeners is found after 16 generations. Since the individuals are randomly sorted in the first generation, also unfeasible individuals are possible, i.e. solutions non fulfilling all the constraints. Nevertheless, the ADP programmed in BIANCA can handle also such cases and it is apparent in Fig. 6.51 that the code is able to drive the search towards feasible individuals.

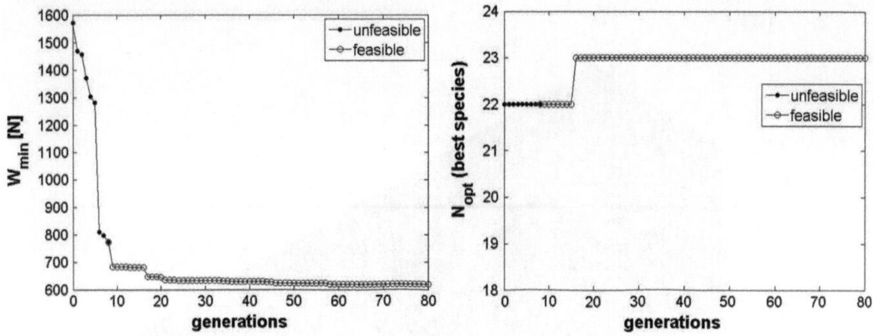

Fig. 6.51 convergence diagrams; *left* best individual, *right* best species

We show in Fig. 6.52 the deformed shape of the structure when the applied load is equal to p_{cr}. In Table 6.4 we report the best stacking sequences found by BIANCA for the skin and the stiffeners. In Fig. 6.53 we show as usually the directional diagrams of A_{1111}, B_{1111} and D_{1111} for the skin and for the first stiffener in Table 6.3. We can remark once more that the imposed requirements are met: the perfect superposition of the diagrams of A_{1111} and D_{1111} along with the practically null diagram of B_{1111} reflect the condition of quasi-homogeneity, while orthotropy with aligned axes is clearly visible.

6.8.3.4 Comparison with Standard Laminates Design

The results obtained in the case presented above deserve some commentary. The complete optimization problem is rather complicate, but it is possible to formulate and solve it in the most general case, i.e. without assuming special simplifying hypotheses, like for instance the symmetry of the stack. This, together with the use of any possible direction for the ply orientations, has given a very interesting result: the weight of the structure has been reduced to the half of a classical metallic structure. This is an example of how, eliminating all the limitations given by simplifying, technical assumptions, the design of anisotropy can give its best results.

However, the reader familiar with laminates design has perhaps noticed that in this, like in all the previous examples in this chapter, we have not followed the usual rule of choosing the possible orientations in the restricted set of *standard orientations*: $0°, \pm 45°, 90°$. This is *not* because the general approach to the laminates design introduced in this chapter and used for the resolution of almost all the examples in not able to handle exclusively standard ply angles. In particular, both the codes ALE-PSO and BIANCA can perfectly work with a limited set of admissible values for the design variables.

The true reason of solving the examples using non standard orientations was actually to give the reader the idea that a design problem concerning anisotropic laminates can *really be formulated and solved in the most general way*. Solutions

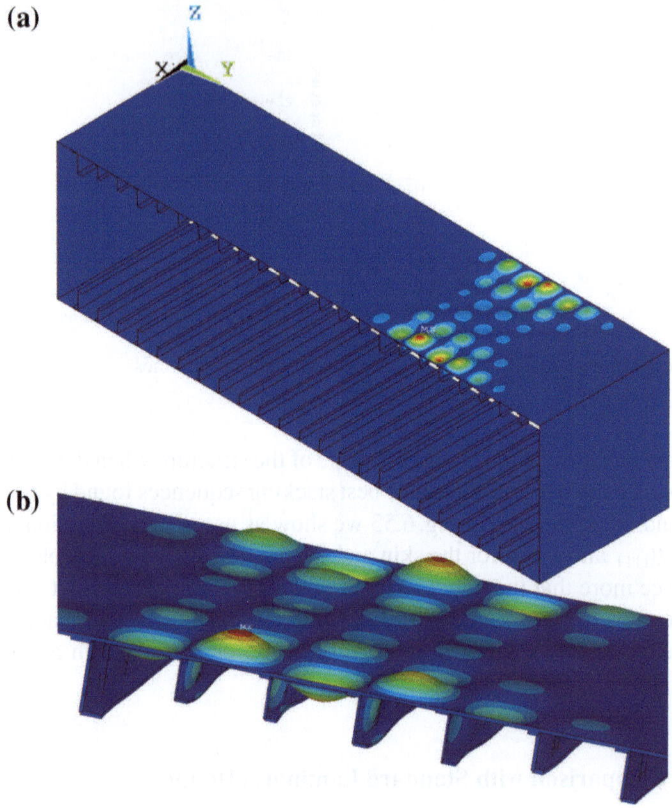

Fig. 6.52 Buckled shape; **a** the whole structure, **b** detail of the upper panel and stiffeners

using standard orientations are just a very particular case of a much larger set and, actually, they are rarely the best solution, especially when lightness is the main goal.

This fact can be seen exactly on the example of the wing box-girder. In Montemurro et al. (2012d), the first case of the wing box-girder, the one with all the stiffeners identical, has been solved once in the general case, and once with the classical rule of symmetric stacks and standard orientations. The best solutions for the two cases are shown in Table 6.5.

To obtain a standard sequence solution with a sufficiently small residual, 30 layers are needed for the stiffeners and 40 for the skin, i.e. more than with the solution of the general case, that has anyway a lower residual. In other words, the solution found using symmetric stacking sequences and standard orientations is *not a global minimum*: the weight of the wing box-girder section passes from 587.28 N, for the solution in the general case, to 710.23 N, with an increase of about 20%, for the symmetric stacks with standard ply angles. It is also to be noticed that the symmetric solutions shown in Table 6.5 are not *completely standard*. In fact, they are not balanced, as often used to obtain in-plane orthotropy, but not bending orthotropy. Nevertheless,

Table 6.4 Solution of the second step problem

Stiffeners			
ID	n	Stacking sequence (°)	Residual
01	22	[28, −29, −30, 28, 25, −29, −29, 28, −26, 42, −26, 26, 24, 27, −29, −26, −27, 25, 30, −31, −22, 29]	1.5×10^{-3}
02	38	[40, −36, −72, −38, 59, −45, 24, 42, 85, −20, 12, 56, 43, 13, −41, −35, −41, 45, −61, −39, 56, 39, −65, 7, 51, −70, −50, 45, −28, −20, −24, 45, −32, 45, 41, −74, 32, −31]	3.3×10^{-4}
03	21	[20_2, −17, −35, −16, 38, −23, −24, −10, 17, 29, 22, 19, −23, 26, −27, 24, −42, −18, −18, 30]	1.9×10^{-3}
04	17	[0, −7, 0, 86, 5, 2_2, 89, 2, 0, −24, 1, 0, 2, −87, 6, −3]	2.1×10^{-3}
05	29	[37, −57, −3, 40, 58, −44, −11, −87, 50, −22, −65, −38, −41, 23, −33, −70, 18, 29, 63, 41, 42, 2, 61, −14, −50, 87, −19, 29, −48]	2.9×10^{-4}
06	37	[6, 8, 1, 61, −45, 5, −24, 22, −50, 6, −7, 8, 7, 51, −21, −34, 12, −21, −11, 6, 33, −6, 16, 78, 5, 7, −6, −51, 25, −4, 6, 18, −42, 54, −12, −20, 9]	2.4×10^{-5}
07	17	[5, 38, −27, −56, −9, 7, 60, 16, −8, 7, −29, 27, 9, −40, 81, −8, 6]	1.5×10^{-3}
08	16	[−25, −47, 28, 45, −16, 23, 35, 1, −42, 48, −41, −36, 0, −33, 45, 24]	5.6×10^{-4}
09	32	[6, −1, 3, 6, 77, −74, −85, −21, 71, −17, −86, −1, 90, −1, 67, −85, 83, 1_2, 3_2, 1, 3, −66, −6, 23, 8, −77, −11, −12, 85, 83]	1.4×10^{-4}
10	32	[−37, 79, 36, −77, −1, −43, 46, 52, −35, −13, 90, 30, −31, −45, −3, 36, 69, −76, 87, 89, −28, −57, 51, 21, −34, 35, 14, 17, 9, −73, −37, 71]	1.3×10^{-3}
11	24	[7, 65, 15, −66, −10, −9, −49, −45, −22, 50, 13, 5, 70, −22, 49, 17, 19, 89, 6, −33, -5_2, −45, 43]	6.8×10^{-4}
12	30	[2, −89, 81, −15, −83, −50, −19, 29, 6, 34, 10, 4, −10, 19, 75, −10, 86, 3, 90, −69, −39, 3, −32, 87, 84, 51, 0, −33, 4, 21]	5.0×10^{-4}
13	24	[34, −45, 5_2, −30, 40, 28, 3, −29, −9, −25, 25, 9, −21, −24, −56, 29, 20, 55, 48, 4, −39, −5, −15]	1.0×10^{-3}
14	34	[85, −4, −75, 4, −2, 84, −2, 83, 88, −16, 6, 2, −83, 0, 87, 4, 3, 84, −7, 84, −85, −74, 86, −9, 15, −86, 10, 1, −6, 88, −3, −89, 88, 5]	1.9×10^{-4}

(continued)

Table 6.4 (continued)

Stiffeners			
ID	*n*	Stacking sequence (°)	Residual
15	35	[4, −79, 82, 37, −2, −10, −38, −70, 76, 90, 10, −9, 2, −2, 18, −77, −5, 90, −3, 78, 81, −21, −87, 28, 9, −66, −13, 45, 8, −14, 82, −71, 78, -2_2]	1.6×10^{-4}
16	32	[12, 11, −7, 5, −85, −9, −14, −66, 50, -5_2, −16, 2, 87, 4, 3, 3, −8, −6, 19, 17, −15, 17, −9, 17, 14, −67, −9, 67, −12, 4, −6]	2.8×10^{-4}
17	17	[−19, 20, −29, 20, −71, −4, 35, 62, −7, −8, 3, 7, −45, −35, 7, 70, 12]	9.2×10^{-4}
18	25	[13, 0, 4, −23, −31, 82, 18, 22, −1, 2, -3_2, −30, 2, 3, 14, −7, 18, 3, 70, −77, −2, −14, −6, 12]	4.4×10^{-4}
19	24	[2, −79, −17, 27, −5, −2, −5, −4, 68, −5, 24, −2, −83, −1, 4, −69, −9, 8, −18, −10, 25, 81, −11, 7]	3.7×10^{-5}
20	17	[31, 46, −14, −61, −38, −22, −17, 25, −50, 30, 47, 6, 62, 28, −29, 3, −44]	7.4×10^{-4}
21	35	[89, −10, 9, 5_2, −6, −75, 58, −4, -3_2, −2, 73, −74, −2, −86, −2, −3, −1, −3, 22_2, −5, −11, −6, −86, −82, −6, 71, −9, 16, 4, −82, 5, −10]	2.2×10^{-4}
22	27	[−5, 7, −83, 5, 10, −15, −85, 75, 90, −4, −4, 83, 0, 3_3, 2, -85_3, −6, 1, 3, 83, 7, −8, 89]	9.3×10^{-4}
23	29	[89, 0, 12, −25, 62, 10, −6, −10, −1, −74, −3, −1, −82, 73, −86, 22, −1, 26, −2, 86, 10, −23, −8, −10, −18, −81, 13, −89, 9]	3.3×10^{-4}
Skin			
	n	Stacking sequence (°)	Residual
32		[3, −79, 9, −12, 85, 47, 78, −5, −8, −78, 81, −73, 85, −37, −86, −4, −2, −90, 6_2, −90, 5, −5, 25, 86, 40, −8, −5, −69, 90, −10, 88]	9.6×10^{-5}

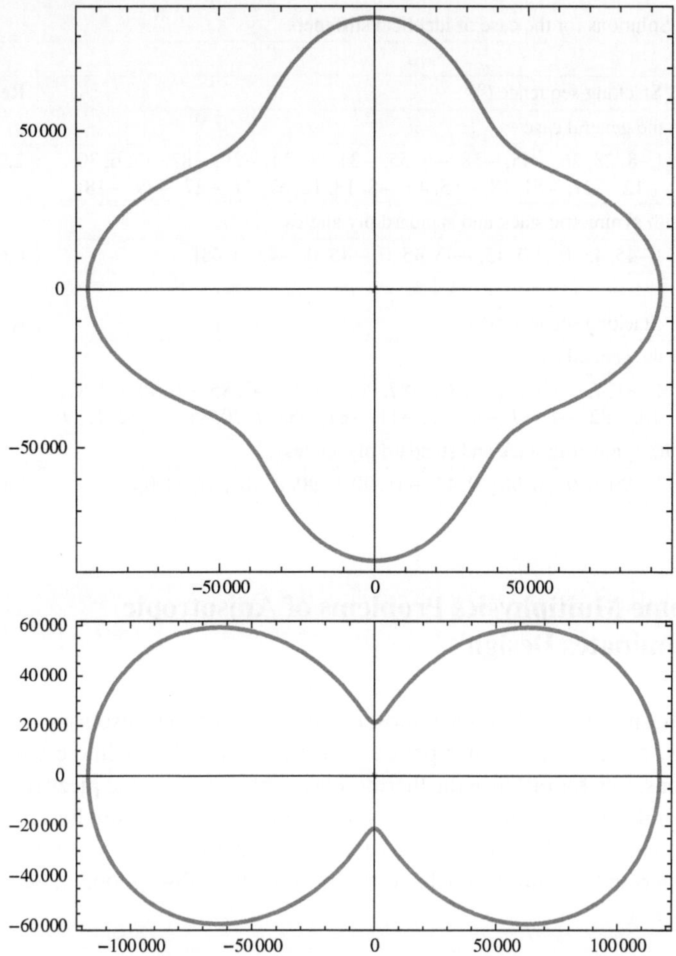

Fig. 6.53 Directional diagrams of A_{1111} and D_{1111}, *thick line*, B_{1111}, in the *center* (units: MPa): *top*, skin laminate; *bottom*, stiffener 1 in Table 6.3

they are fully orthotropic, thanks to the requirement of quasi-homogeneity. The use of balanced stacks will obviously lead to a further increase of the final weight of the structure.

This example shows, on one side, that with the design procedure used in this chapter, it is possible to design standard laminates, but at the same time that this is not, almost never, the best thing to do: disregarding standard rules in the design of anisotropic laminates can be really interesting.

Table 6.5 Solutions for the case of identical stiffeners

Stiffeners		
n	Stacking sequence (°)	Residual
Solution in the general case		
29	$[-8, 28, 26, -45, -58, -3, 55, -31, 76, 34, -39, -87, -7, 6, 30, -12, -21, -51, 18, -55, 49, -8, 18, 12, 57, 44, -27, -79, -18]$	2.9×10^{-4}
Solution with symmetric stack and standard ply angles		
30	$[-45, 45, 0_3, 90, 45, -45, 45, 0, -45, 0, -45, 0, 45]_s$	6.6×10^{-4}
Skin		
n	Stacking sequence (°)	Residual
Solution in the general case		
32	$[-81, 7, -3, -12, 82, 86, -87, 20, -6, 76, -7, 85, -6, 90, -7, 87, 10, -82, -4, -7, -82, 18, -11, -84, -83, 7, 70, 85, 1, -12, 1, 89]$	8.4×10^{-5}
Solution with symmetric stack and standard ply angles		
40	$[0, 90, 0, 90, 0, 90_2, 0, 45, -45, 90, 0, 90, 0, 90_2, 0_2, 90, 0]_s$	9.6×10^{-5}

6.9 Some Multiphysics Problems of Anisotropic Laminates Design

We have been concerned, so far, almost exclusively with the anisotropy of the elastic properties. Nevertheless, other physical properties can be of interest for modern applications, like for instance the thermo-elastic properties or the piezoelectric ones. We give in this section some examples of design with such multiphysics problems. We just remark here that the problems concerning thermal properties outlined below can be rephrased *verbatim* for what concerns moisture absorption, cf. Sect. 5.4.

6.9.1 Tailoring the Thermo-Elastic Properties of an Anisotropic Laminate

Just like the elastic properties, also the thermal ones can be tailored for an anisotropic laminate, i.e. the angular variation of the coefficients of thermal expansion or curvature can be designed to meet some given requirements. In fact, unlike classical materials, anisotropic laminates can be designed in such a way to reduce the thermal coefficients along some given directions to a prescribed value, particularly to zero, or to have thermal isotropy also for an anisotropic elastic behavior. Such a kind of problems has hence attracted the attention of some researchers in the past, see e.g. Ishikawa and Fukunaga (1986), Le Riche and Gaudin (1998), Vannucci and Vincenti (2007). Following this last work, we give below a polar approach to the design of the thermal properties of a laminate.

Let us consider a laminate composed of identical plies and uncoupled; by Eqs. (5.132) and (5.133) we know that it is also $\mathbf{V} = \mathbf{v}_1 = \mathbf{v}_2 = \mathbf{O}$ and that the tensors of thermal expansion and curvature coefficients are given by, respectively,

$$\mathbf{u} = \mathbb{A}^{-1}\mathbf{U}, \quad \mathbf{w} = \mathbb{D}^{-1}\mathbf{W}. \tag{6.149}$$

Like any other second-rank symmetric tensor, $\mathbf{u}$ and $\mathbf{w}$ can be expressed by their polar parameters as in Eq. (4.63). Be hence ξ, η and ζ the polar parameters of $\mathbf{u}$ or $\mathbf{w}$: the problem being the same for the in- and out of-plane thermal properties, we will refer from now on to an *expansion tensor* $\mathbf{z}$, representing indifferently $\mathbf{u}$ or $\mathbf{w}$, and we will speak indifferently of expansion coefficients:

$$\mathbf{z} \rightarrow \begin{cases} z_{11}(\theta) = \xi + \eta\cos 2(\zeta - \theta), \\ z_{12}(\theta) = \eta\sin 2(\zeta - \theta), \\ z_{22}(\theta) = \xi - \eta\cos 2(\zeta - \theta). \end{cases} \tag{6.150}$$

Observing the above equations, it is immediate to remark that (we make the assumption that $\xi > 0$, i.e. that the average thermal response of the laminate to an increase of the temperature is an expansion):

- if $\xi = \eta$, i.e. if the isotropic part of $\mathbf{z}$ is equal to the anisotropic one, then the expansion coefficient z_{11} is null for $\theta = \zeta \pm \frac{\pi}{2}$;
- if $\xi < \eta$ the plane is divided into four opposite sectors, with different signs for z_{11}: if

$$\theta^* = \frac{1}{2}\arccos\left(-\frac{\xi}{\eta}\right), \tag{6.151}$$

  then

$$\begin{aligned} z_{11} \geq 0 \ \text{for} \ & -\pi \leq \theta \leq \pi + \zeta + \theta^*, \ \ \zeta - \theta^* \leq \theta \leq \zeta + \theta^*, \\ & \pi + \zeta - \theta^* \leq \theta \leq \pi; \\ z_{11} < 0 \ \text{for} \ & -\pi + \zeta + \theta^* \leq \theta \leq \zeta - \theta^*, \ \ \zeta + \theta^* \leq \theta \leq \pi + \zeta - \theta^*; \end{aligned} \tag{6.152}$$

  each one of the two sectors with $z_{11} \geq 0$ has amplitude $2\theta^*$, while that of the sectors with $z_{11} < 0$ is $\pi - 2\theta^*$. In the limit case of $\xi = 0$, $\theta^* = \pi/4$ and each sector has an amplitude of $\pi/2$. In this case, a laminate subject to a gradient of temperature warps like in Fig. 6.54, where the four sectors are clearly visible;
- the tensor $\mathbf{z}$ is isotropic, i.e. the thermal expansion coefficients are independent from the direction, if and only if $\eta = 0$.

In principle, if tensors $\mathbb{A}$, $\mathbb{D}$, $\mathbf{U}$ and $\mathbf{W}$ are expressed as functions of the polar parameters of $\mathbb{Q}$ and $\boldsymbol{\alpha}$, cf. Sect. 5.4, and of the ply orientations δ_k, one can obtain the polar components of $\mathbf{u}$ and $\mathbf{w}$ as functions of these last quantities. However, though possible in practice, this gives rises to so long and complicate expressions that it is impossible to write them here.

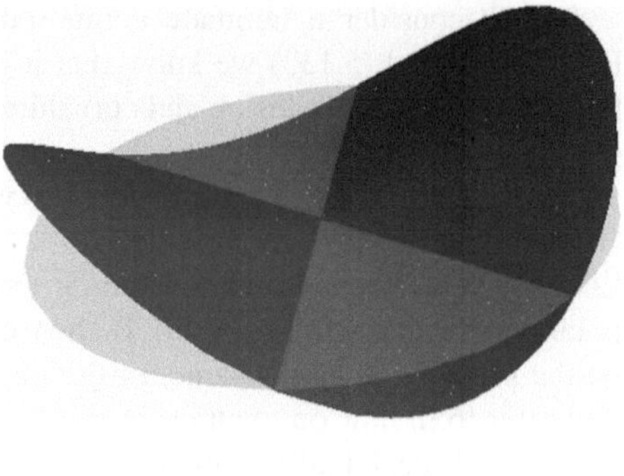

Fig. 6.54 Warping of a plate subject to a temperature gradient when $\xi < \eta$

Hence, normally the search for laminates having required thermal properties can be done only numerically. Like in the case of the elastic properties, a possible approach is the one described in Sect. 6.4.1, modified to take into account for also the thermal requirements:

$$\min_{\delta_k}\left\{\mathscr{I}(\delta_k)+\left[\frac{c_1\xi_u(\delta_k)-c_2\eta_u(\delta_k)}{|\boldsymbol{\alpha}|}\right]^2+\left[\frac{c_3\xi_w(\delta_k)-c_4\eta_w(\delta_k)}{|\boldsymbol{\alpha}|}\right]^2\right\}. \tag{6.153}$$

The quadratic form above is still positive semi-definite, i.e. the minima are its roots; the subscripts u and w denotes quantities relative to tensors $\mathbf{u}$ and $\mathbf{w}$ respectively, while $|\boldsymbol{\alpha}|$ is the norm of the tensor of the thermal expansion coefficients of the basic layer, $|\boldsymbol{\alpha}| = \sqrt{\alpha_{11}^2+\alpha_{22}^2}$. Coefficients c_1 to c_4 determine the problem to be treated:

- to have $\xi_u = \eta_u$, one must put $c_1 = c_2 = 1$;
- to have $\xi_u < \eta_u$, it has to be $c_1 = 1,\ 0 < c_2 < 1$;
- to have $\xi_u = 0$, one has to chose $c_1 = 0$.

The same properties for the bending behavior, i.e. for ξ_w and η_w, can be obtained replacing c_1 and c_2 by c_3 and c_4 respectively; of course, combinations are possible too.

We give here just an example (several others can be found in Vannucci and Vincenti 2007): a 12-ply laminate has to be isotropic in extension, orthotropic in bending with $K^D = 0$ and uncoupled. In addition, it must be thermally isotropic in extension, while in bending it must have a direction of null thermal curvature. The material composing the identical plies is carbon-epoxy T300/5208, whose thermal properties are $\alpha_{11} = 2\times10^{-8}\,^\circ\mathrm{C}^{-1}, \alpha_{22} = 2.25\times10^{-5}\,^\circ\mathrm{C}^{-1}, \alpha_{12} = 0$. According to Eq. (6.153), the problem can be formulated as

$$\min_{\delta_k}\left[\mathscr{P}_3^2+\mathscr{P}_4^2+\mathscr{P}_9^2+\mathscr{P}_{10}^2+\mathscr{P}_{17}^2+\left(\frac{\eta_u}{|\boldsymbol{\alpha}|}\right)^2+\left(\frac{\xi_w-\eta_w}{|\boldsymbol{\alpha}|}\right)^2\right]. \tag{6.154}$$

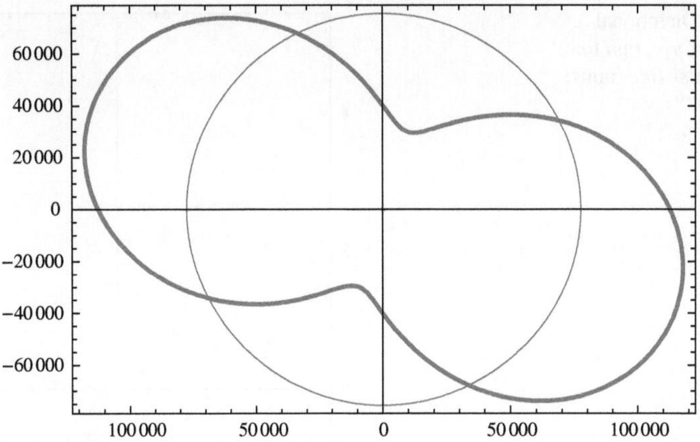

Fig. 6.55 Directional diagrams of A_{1111}, *thin line*, B_{1111}, in the *center*, and D_{1111}, *thick line* (units: MPa)

The problem has been solved using the GA BIANCA and the best solution so found is the sequence

$$[0°, -30°, 44.3°, -61.8°, 89.3°, 61.8°, 31.6°,\\ -89.1°, 33.4°, -71.7°, -11.6°, -28.1°],$$

with a residual $\mathscr{I} = 8.24 \times 10^{-4}$. The directional diagrams of A_{1111}, B_{1111} and D_{1111} are plotted in Fig. 6.55 and clearly show the fulfillment of the prescribed requirement for the elastic behavior. The thermal coefficients u_{11} and w_{11} are plotted in Fig. 6.56 and also in this case we can observe that the prescriptions are met. To remark that the direction of least D_{1111} is that of highest w_{11}: where the elastic bending stiffness is lower, the thermal curvature coefficient is higher.

6.9.2 Thermally Stable Laminates

We have seen in Sect. 5.4.4 that for laminates composed by identical plies, elastic uncoupling implies thermal uncoupling: $\mathbb{B} = \mathbb{O} \Rightarrow \mathbf{V} = \mathbf{v}_1 = \mathbf{v}_2 = \mathbf{O}$ but the converse is not true: $\mathbf{V} = \mathbf{O} \nRightarrow \mathbb{B} = \mathbb{O}$ and $\mathbf{v}_1 = \mathbf{v}_2 = \mathbf{O}$. In other words, when the laminate is elastically coupled, also if $\mathbf{V} = \mathbf{O}$ the thermal coupling tensors $\mathbf{v}_1$ and $\mathbf{v}_2$ do not vanish: a change of temperature t makes the plate warp and a gradient of the temperature ∇t stretches the plate. The use of laminates with $\mathbb{B} \neq \mathbb{O}$ can be interesting in different applications concerning smart structures, but it can present a problem linked to the technology of laminates fabrication. In fact, laminates are often realized by heating, e.g. pre-preg epoxy laminates are polymerized heating them to $\sim$180 °C, so that at room temperature they are actually in a sort of *frozen state*, so

Fig. 6.56 Directional diagrams of u_{11}, *thin line*, and w_{11}, *thick line* (units: $°C^{-1} \times 10^{-6}$)

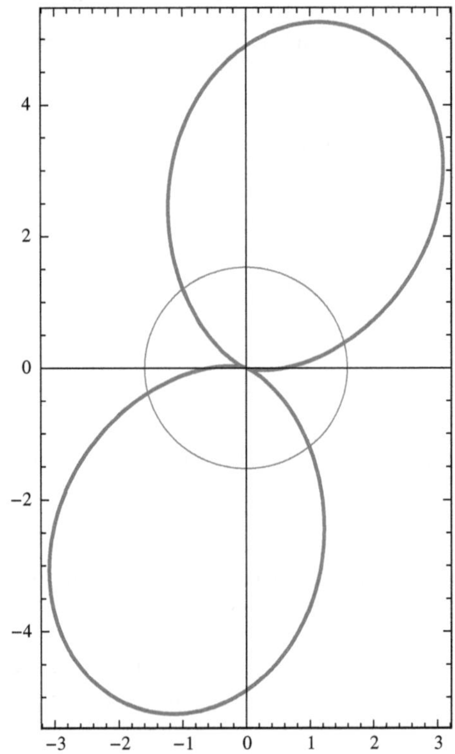

to speak. If a coupled laminate is manufactured with this technology, once cooled down from its manufacturing temperature, it will necessarily warp, i.e. it will loose its design shape. Of course, this can considerably affect the use of coupled laminates, unless the laminate is *thermally stable*: a coupled laminate is said to be thermally stable when temperature changes do not produce geometrical variations of the plate.

The first studies on the thermally stable laminates, actually papers on the phenomenon of thermal warping, date back to 1981 (Hyer 1981a, b; Hahn 1981). Winckler, in his Ph.D. thesis and in some related papers (Winckler 1985, 1986), has proposed the first some sufficient rules for the design of warp-free coupled laminates. Chen has then given some other sufficient rules on the subject (Chen 2003), as well as Cross et al. (2008), in a paper presenting some families of warp-free laminates, work extended in a subsequent paper by Haynes and Armanios (2010). In these last two papers, the authors also look for stacking sequences optimizing the thermal warping. The polar formalism has been used for this problem in Vannucci (2013a), where a general theory of thermally stable laminates is given.

Following this last work, we define as thermally stable a laminate when

$$\mathbf{v}_1 = \mathbf{v}_2 = \mathbf{O}. \tag{6.155}$$

The effect of $\mathbf{v}_1$ is by far more important than that of $\mathbf{v}_2$. In fact, when $\mathbf{v}_2 \neq \mathbf{O}$ the plate warps for a change of temperature and because normally laminates are thin plates, warping can be very important, sometimes even spectacular, with large displacements. That is why, usually thermally stable laminates are analyzed only with respect to $\mathbf{v}_1$, disregarding the stretching effects produced by a temperature gradient described by $\mathbf{v}_2$. In Vannucci (2013a), the following possible cases are considered:

- *warp-free* coupled laminates: they do not warp when a uniform change of temperature t is applied;
- *expansion-free* coupled laminates: they are not stretched by a temperature gradient ∇t;
- *warp- and expansion-free* coupled laminates: they have both the properties above;
- *completely warp-free or expansion-free* coupled laminates: they are completely insensitive to any temperature change.

Let us briefly consider these four cases.

6.9.2.1 Warp-Free Coupled Laminates

Upon definition, a coupled laminate is warp-free when, Eq. (5.124)$_2$,

$$\mathbf{v}_1 = \frac{2}{h}(\mathscr{B}^\top\mathbf{U} + 3\mathscr{D}\mathbf{V}) = \frac{6}{h}(\mathbb{D} - 3\mathbb{B}\mathbb{A}^{-1}\mathbb{B})^{-1}(\mathbf{V} - \mathbb{B}\mathbb{A}^{-1}\mathbf{U}) = \mathbf{O}. \tag{6.156}$$

This equation gives two different types of warp-free coupled laminates.

First type

The first type of warp-free coupled laminates is composed by all the plates that fulfill the following sufficient conditions for Eq. (6.156) to be satisfied:

$$\begin{cases} \mathbb{B} \neq \mathbb{O}, \\ \mathbf{V} = \mathbf{O}, \\ \mathscr{B}^\top\mathbf{U} = \mathbf{O}. \end{cases} \tag{6.157}$$

Some remarks can be done:

- It is possible to have laminates with $\mathbf{V} = \mathbf{O}$ but $\mathbf{v}_1 \neq \mathbf{O}$; it is sufficient that Eq. (6.157)$_3$ is not satisfied.
- Because $\mathbf{V} = \mathbf{O}$ and $R^\gamma \neq 0, R_1^B = 0$, cf. Eqs. (5.130)$_2$ and (5.52)$_3$: $\mathbb{B}$ is necessarily square symmetric and defined by only one invariant, R_0^B:

$$\mathbb{B} = R_0^B \begin{bmatrix} \cos 4\Phi_0^B & -\cos 4\Phi_0^B & \sin 4\Phi_0^B \\ -\cos 4\Phi_0^B & \cos 4\Phi_0^B & -\sin 4\Phi_0^B \\ \sin 4\Phi_0^B & -\sin 4\Phi_0^B & -\cos 4\Phi_0^B \end{bmatrix}. \tag{6.158}$$

- $\mathbb{B}$ is singular: $\det \mathbb{B} = 0$, as can be easily checked from Eq. (6.158): whenever $\mathbf{V} = \mathbf{O}$, $\mathbb{B}$ is singular.

- Because $\mathbf{U} \neq \mathbf{O}$, condition $(6.157)_3$ can be satisfied only if

$$\det \mathscr{B} = 0. \tag{6.159}$$

Using the Binet's theorem in Eq. $(5.33)_2$ gives

$$\det\left(-\frac{\mathscr{B}}{3}\right) = \det(\mathscr{A}\,\mathbb{B}\mathbb{D}^{-1}) = \frac{\det \mathscr{A}\ \det \mathbb{B}}{\det \mathbb{D}}, \tag{6.160}$$

and because for this case $\det \mathbb{B} = 0$, condition (6.159) is automatically satisfied: the special orthotropy of $\mathbb{B}$ is a sufficient condition for satisfying Eq. (6.159) too.
- Condition $(6.157)_3$ can be interpreted as an eigenvector problem: $\mathbf{U}$ has to be the eigentensor of $\mathscr{B}^\top$ corresponding to the null eigenvalue. Nevertheless, one should remember that in a design problem all the tensors are related, because changing the orientations of the layers changes all the tensors describing the elastic and thermo-elastic behavior of the plate. Though it is too cumbersome to explicit all the links between the different tensors, these links exist: in Eq. $(6.157)_3$, $\mathscr{B}$ and $\mathbf{U}$ are not independent. So, the design problem of this first type of warp-free coupled laminates corresponds to look for laminates respecting conditions (6.157): what said above shows that these conditions are compatible, so these laminates can exist and resuming they have

$$\mathbb{B} \neq \mathbb{O},\ \ \mathscr{B} \neq \mathbb{O},\ \ \mathbf{V} = \mathbf{v}_1 = \mathbf{O},\ \ \det \mathbb{B} = \det \mathscr{B} = 0,\ \ R_1^B = 0. \tag{6.161}$$

- It is not possible to give general solutions for the design of warp-free laminates of this first type. Nevertheless, it is possible to give some sufficient conditions for obtaining such laminates. The first one is rather simple: *laminates composed of square symmetric layers automatically satisfy conditions* (6.157). In fact, through Eqs. (5.51)–(5.54) and (5.129)–(5.131),

$$R^\gamma = R_1^L = 0\ \Rightarrow\ R_1^A = R_1^B = R_1^D = R^U = R^V = R^W = 0. \tag{6.162}$$

Then, conditions $(6.157)_2$ and (6.159) are satisfied; it can be verified that also condition $(6.157)_3$ is always satisfied in this case. This can be done using a code for formal mathematics; unfortunately, the analytical developments are too long to show them here in detail. So, *laminates composed by identical layers reinforced by balanced fabrics are always warp free*.
- Another, less restrictive, sufficient condition is the following one:

$$R_1^A = R_1^B = 0. \tag{6.163}$$

This is the case of laminates $\mathbb{A}$ and $\mathbb{B}$ square symmetric. Condition (6.163) implies that also $R^U = 0$, Eqs. $(5.51)_4$ and $(5.129)_2$; moreover, we have already seen that $R_1^B = 0 \Rightarrow \mathbf{V} = \mathbf{O}$, hence condition $(6.157)_2$ is satisfied. Also in this case it can

be proved that condition $(6.157)_3$ is satisfied too. So, a coupled stacking sequence that satisfies Eqs. (6.163) is warp-free.
The solutions proposed by Chen (2003)

$$\xi_3 + i\xi_4 = 0, \;\; \xi_7 + i\xi_8 = 0, \tag{6.164}$$

$$\left[\delta_1, \left(\delta_1 + \frac{\pi}{2}\right), \delta_2, \left(\delta_2 + \frac{\pi}{2}\right), \ldots, \delta_m, \left(\delta_m + \frac{\pi}{2}\right)\right] \text{ such that } R_1^B = 0, \tag{6.165}$$

$$\left[\left[\delta_1, \left(\delta_1 + \frac{\pi}{2}\right), \ldots, \delta_m, \left(\delta_m + \frac{\pi}{2}\right)\right]_{sym}\right]_{-sym}, \tag{6.166}$$

are all of this type (the subscript *sym* stands for symmetric and *-sym* for antisymmetric). In fact, conditions (6.164), the most general case among those proposed by Chen, simply correspond to conditions (6.163), cf. Eqs. $(5.51)_4$ and $(5.52)_4$. The sequences in Eq. (6.165) are sufficient to ensure $R_1^A = 0$, while the angles of the sequences are to be searched using a numerical technique in order to satisfy also $R_1^B = 0$. The third case, sequences of the type (6.166), satisfies automatically Eqs. (6.163), while $R_0^B \neq 0$, as it is easy to check using Eqs. $(5.51)_4$, $(5.52)_4$ and $(5.45)_2$ or $(5.62)_1$ and $(5.63)_1$. So, sequences (6.166) are sufficient to obtain a warp-free laminate, regardless of the values of the orientation angles δ_j. The solutions proposed by Winckler (1985), are of the type

$$\left[\delta, \left(\delta + \frac{\pi}{2}\right)_2, \delta, -\delta, -\left(\delta + \frac{\pi}{2}\right)_2, -\delta\right], \tag{6.167}$$

and it is evident that they are a particular case of the third one proposed by Chen, sequence (6.166). It can be also remarked, as already done in Cross et al. (2008), that condition (6.164) implies also that $R^U = 0$, Eq. $(5.129)_2$, which means that $\mathbf{U}$ is isotropic. Finally, the solutions proposed by Winckler and Chen are warp-free laminates of the first type. Anyway, other solutions can exist, but they need a numerical approach to be found.

Second type

The second type of warp-free coupled laminates is composed by all the plates that fulfill the other sufficient conditions for Eq. (6.156) to be satisfied:

$$\begin{cases} \mathbb{B} \neq \mathbb{O}, \\ \mathbf{V} = \mathbb{B}\mathbb{A}^{-1}\mathbf{U}. \end{cases} \tag{6.168}$$

In this case too some remarks can be done:

- Being $\mathscr{D} \neq \mathbb{O}$, conditions (6.157) and (6.168) are the only possible conditions giving a warp-free coupled laminate: no other types of warp-free coupled laminates can exist. Hence, a coupled laminate is warp-free if and only if it satisfies Eq. (6.157) or (6.168). These are the necessary and sufficient conditions for a coupled laminate to be warp-free, contrarily to what affirmed in Cross et al. (2008), where only solutions of the type (6.164) are considered.

- Considering Eq. (6.168) it is clear that in this second case it is

$$\mathbb{B} \neq \mathbb{O}, \ \mathbf{V} \neq \mathbf{O}, \ \mathbf{v}_1 = \mathbf{O}. \tag{6.169}$$

- Comparing Eq. (6.169) with (6.161) shows that the set of the first type laminates can be considered, in some sense, as a subset of the second type laminates set. In fact, Eqs. (5.33)$_4$ and (5.124)$_2$, in the second type it is

$$\begin{aligned}&\frac{6}{h}(\mathbb{D}-3\mathbb{B}\mathbb{A}^{-1}\mathbb{B})^{-1}(\mathbf{V}-\mathbb{B}\mathbb{A}^{-1}\mathbf{U}) = \frac{6}{h}\mathscr{D}(\mathbf{V}-\mathbb{B}\mathbb{A}^{-1}\mathbf{U}) = \\ &\frac{6}{h}(\mathscr{D}\mathbf{V}-\mathscr{D}\mathbb{B}\mathbb{A}^{-1}\mathbf{U}) = \frac{6}{h}\left(\mathscr{D}\mathbf{V}+\frac{1}{3}\mathscr{B}^{\top}\mathbf{U}\right) \rightarrow \\ &\mathbf{V} = \mathbb{B}\mathbb{A}^{-1}\mathbf{U} \Rightarrow \mathscr{D}\mathbf{V} = -\frac{1}{3}\mathscr{B}^{\top}\mathbf{U},\end{aligned} \tag{6.170}$$

while in the first type it is

$$\mathscr{D}\mathbf{V} = -\frac{1}{3}\mathscr{B}^{\top}\mathbf{U} = 0, \tag{6.171}$$

which is a particular case of the second one. Namely, laminates composed of square symmetric layers belong to the first type, as seen above, because in such a case

$$\mathbf{V} = \mathbb{B}\mathbb{A}^{-1}\mathbf{U} = \mathbf{O}. \tag{6.172}$$

Equation (6.172) shows that in this peculiar case, in which $\mathbf{U}$ is isotropic, $\mathbf{U}$ is the eigentensor of tensor $\mathbb{B}\mathbb{A}^{-1}$ relative to the null eigenvalue; this is possible because in this particular case $\mathbb{B}\mathbb{A}^{-1}$ is singular: using again Binet's theorem, one gets

$$\det\left(\mathbb{B}\mathbb{A}^{-1}\right) = \frac{\det\,\mathbb{B}}{\det\,\mathbb{A}} = 0, \tag{6.173}$$

the last equality being ensured because $\det\mathbb{B} = 0$, as shown above. The same happens when the layers are R_0-orthotropic, though in such a case $\mathbf{U}$ is no more isotropic.

- Looking at Eq. (6.169) it is evident that in this second case, coupled warp-free laminates have nonetheless the thermo-elastic coupling stiffness tensor $\mathbf{V}$ which is not null. Actually, thermal stability is a condition depending essentially upon the compliance coupling tensors and it can exist also when stiffness coupling tensors are not null, as in this case. This, along with a remark made above, confirms that thermo-elastic coupling tensors can be null separately. Physically, Eq. (6.169) implies that non mechanical moments, i.e. moments produced by thermal effects, can be nonzero, also without warping, which contradicts what said in Cross et al.

(2008). This point is rather important, because it shows that the distinction made in this paper between first and second type of coupled warp-free laminates is not a merely taxonomic one, but really a physical one.

- Equation (6.168$_2$) is a link between tensors that are not independent. It can be used to find some sufficient conditions for obtaining warp-free laminates of the second type. In fact, the polar components of the tensor **V** computed using the result provided by Eq. (6.168)$_2$ are to be the same that can be computed using Eq. (5.130), in particular the isotropic part must be null. If V_{11}, V_{22} and V_{12} denote the Cartesian components of **V** computed as in (6.168)$_2$, the three conditions are

$$\begin{cases} V_{11}+V_{22}=0, \\ \dfrac{V_{11}-V_{22}}{2}+i\,V_{12}=\dfrac{h^2}{2}R^{\gamma}\,e^{2i\Phi^{\gamma}}\,(\xi_7+i\xi_8)\,. \end{cases} \tag{6.174}$$

The above conditions can be directly given as functions of the polar parameters of $\mathbb{B}$, $\mathbb{A}$ and **U**, simply introducing in (6.168) the polar expressions of $\mathbb{A}$, $\mathbb{B}$ and **U**; the expressions are very long and cumbersome, so they are omitted here. Nevertheless, it can be checked by some standard, though lengthy passages, that a sufficient condition for Eq. (6.174) to be satisfied is that

$$\frac{T^{\gamma}}{R^{\gamma}}=\frac{T_1^L}{R_1^L}. \tag{6.175}$$

Equation (6.175) is a relation for the polar parameters of $\mathbb{Q}$ and $\boldsymbol{\gamma}$, i.e. for the physical properties of the basic layer. This means that there is a class of warp-free laminates whose existence depends upon the physical properties of the material, not only upon the geometry of the stacking sequence. This circumstance confirms what has already been observed on other problems of optimal elastic properties of a laminate (Vannucci 2012).

The mathematical structure of Eq. (6.175) is rather elegant: it is based upon the ratio of two homologous invariant quantities, one ratio concerning tensor $\boldsymbol{\gamma}$, the other one tensor $\mathbb{Q}$. Nevertheless, it is worth to rewrite Eq. (6.175) with the Cartesian components of $\mathbb{Q}$ and $\boldsymbol{\gamma}$; some simple manipulations, making use of Eqs. (4.48), (4.50) and (4.51) written for $\boldsymbol{\gamma}$ and (4.58) applied to tensor $\mathbb{Q}$, give

$$(Q_{2222}+Q_{1122})\,\gamma_{11}=(Q_{1111}+Q_{1122})\,\gamma_{22}, \tag{6.176}$$

which can still be transformed, using Eq. (5.117), to let appear the components of the thermal expansion coefficients tensor, $\boldsymbol{\alpha}$:

$$\left(Q_{1111}\,Q_{2222}-Q_{1122}^2\right)(\alpha_{22}-\alpha_{11})=0. \tag{6.177}$$

This last condition shows that there are two possible cases: the first one is trivial, it is the case of isotropic thermal expansion coefficients, i.e. of $\alpha_{11}=\alpha_{22}$, a rare but possible case also for anisotropic laminae. The other one is given by the condition

$$Q_{1111}\,Q_{2222} = Q_{1122}^2, \tag{6.178}$$

which written in terms of technical moduli reads like

$$E_1 = 2\nu_{12}^2 E_2. \tag{6.179}$$

What is rather surprising, is that Eqs. (6.175) and (6.176) concern both the thermal and elastic properties of the material, but when they are transformed into Eq. (6.177), then one can immediately see that solutions can depend exclusively either on the thermal or the elastic properties of the material.
It is worth writing Eq. (6.178) with the polar formalism; starting from the polar expression of $\mathbb{Q}$, some simple passages lead to

$$T_1^L\left[T_0^L + (-1)^{K^L} R_0^L\right] = 2\left(R_1^L\right)^2. \tag{6.180}$$

Condition (6.180) corresponds to a point on the boundary of the existence domain of the polar invariants for a planar material, see Eq. $(4.149)_2$. Hence, material dependent solutions satisfying condition (6.178) are only a limit case, while those satisfying $\alpha_{11} = \alpha_{22}$ are possible, though rare.
It is to be noticed that condition (6.175), and the other ones that follow from it, does not correspond to that given by Cross et al. in (2008) as a material dependent solution. In fact, it can be easily checked, using the polar expression of $\boldsymbol{\gamma}$, that the solution proposed by these authors corresponds to the case of a material with $R^\gamma = 0$. This is a situation that always occurs for a square symmetric layer, that is when $R_1^L = 0$, because it has an isotropic tensor of thermal expansion coefficients, i.e. with $R^\alpha = 0$, for obvious material symmetry reasons, and these two facts give always $R^\gamma = 0$, as it can be proved putting $R_1^L = 0$ in the polar expression of $\mathbb{Q}$, and $R^\alpha = 0$ in that of $\boldsymbol{\alpha}$ and then inserting the Cartesian components so found into Eq. (5.117): the result gives $\gamma_{11} = \gamma_{22} \Rightarrow R^\gamma = 0$. Hence, the material-dependent solution proposed in Cross et al. (2008) is actually a solution of the type in Eq. (6.162) which is, contrarily to what said in Cross et al. (2008), not material dependent (because, unlike the cases given by Eq. (6.175), it depends only upon the existence of a special type of orthotropy of the basic layer and not upon the numerical value of its mechanical properties), really possible and even the easiest case to be obtained.
To remark that when $R_1^A = 0$, $\mathbf{V}$ is not null, but when $R_1^B = 0$, then $\mathbf{V}=\mathbf{O}$ and Eq. $(6.157)_3$ is satisfied too: the solution is of the first type.

6.9.2.2 Expansion-Free Coupled Laminates

According to the definition given above, a coupled laminate is expansion-free when, Eq. $(5.124)_3$,

$$\mathbf{v}_2 = \frac{h}{6}(3\mathscr{A}\mathbf{V} + \mathscr{B}\mathbf{W}) = \frac{h}{2}(\mathbb{A} - 3\mathbb{B}\mathbb{D}^{-1}\mathbb{B})^{-1}(\mathbf{V} - \mathbb{B}\mathbb{D}^{-1}\mathbf{W}) = \mathbf{O}. \tag{6.181}$$

This gives two different possibilities for having $\mathbf{v}_2 = \mathbf{O}$ when $\mathbb{B} \neq \mathbb{O}$:

$$\begin{cases} \mathbb{B} \neq \mathbb{O}, \\ \mathbf{V} = \mathbf{O}, \\ \mathscr{B}\mathbf{W} = \mathbf{O}. \end{cases} \tag{6.182}$$

and

$$\begin{cases} \mathbb{B} \neq \mathbb{O}, \\ \mathbf{V} = \mathbb{B}\mathbb{D}^{-1}\mathbf{W}. \end{cases} \tag{6.183}$$

Comparing conditions (6.182) and (6.183) respectively with (6.157) and (6.168), it is easy to recognize that the case of expansion-free laminates is a sort of dual case of the warp-free laminates one. In fact, it is sufficient to replace, in (6.157) and (6.168), $\mathbf{U}$ by $\mathbf{W}$ and $\mathbb{A}$ by $\mathbb{D}$ to obtain, apart the transpose of $\mathscr{B}$, the same conditions of (6.182) and (6.183).

This means that all the remarks done for warp-free laminates can be repeated almost *verbatim* for expansion-free laminates. Namely, it is possible to obtain laminates having $\mathbf{V} = \mathbf{O}$ and $\mathbf{v}_2 \neq \mathbf{O}$, or with $\mathbf{v}_2 = \mathbf{O}$ and $\mathbf{V} \neq \mathbf{O}$, still confirming what already said above. In addition, Eq. (6.162) gives also sufficient conditions to obtain expansion-free laminates of the first type, i.e. satisfying conditions (6.182): laminates composed by identical layers reinforced by balanced fabrics are also expansion-free. Another sufficient condition for obtaining expansion-free laminates of the first type can be obtained by Eq. (6.163) replacing tensor $\mathbb{A}$ with tensor $\mathbb{D}$:

$$R_1^D = R_1^B = 0. \tag{6.184}$$

Conditions (6.174) are valid also in this case for the solutions of the second type, Eq. (6.183): condition (6.175) is a sufficient condition also for obtaining material dependent expansion-free laminates; of course, all the comments made for warp-free laminates are valid also in this case.

To be remarked that sequences (6.164)–(6.167) do not have, generally speaking, $R_1^D = 0$, so they are not expansion-free laminates.

6.9.2.3 Warp- and Expansion-Free Coupled Laminates

We consider now thermally stable laminates, i.e. laminates being at the same time warp-free and expansion-free. To be thermally stable, a laminate must satisfy both conditions (6.155). Considering hence conditions (6.157), (6.168), (6.182) and (6.183), one can see immediately that there are still two types of solution to this problem.

First type

This is the set composed by the laminates that satisfy at both conditions (6.157) and (6.182):

$$\begin{cases} \mathbb{B} \neq \mathbb{O}, \\ \mathbf{V} = \mathbf{O}, \\ \mathscr{B}^\top \mathbf{U} = \mathbf{O}, \\ \mathscr{B}\mathbf{W} = \mathbf{O}. \end{cases} \tag{6.185}$$

It is self evident that conditions (6.162) are again sufficient conditions for having expansion-free and warp-free laminates. Another sufficient condition is obtained by the union of conditions (6.163) and (6.184):

$$R_1^A = R_1^B = R_1^D = 0; \tag{6.186}$$

as a consequence, cf. Eqs. (5.51)–(5.54) and (5.129)–(5.131),

$$R^U = R^V = R^W = 0, \tag{6.187}$$

i.e. these laminates have a perfectly uncoupled isotropic thermal behavior: the coefficients of thermal expansion or curvature do not depend upon the direction, they can be computed using Eq. (5.124)$_{1,4}$ and their common value ε is

$$\varepsilon = \xi^u = \xi^w = \frac{T^\gamma}{4T_1^L}. \tag{6.188}$$

In the equation above, ξ^u and ξ^w indicate the isotropic polar component of tensors $\mathbf{u}$ and $\mathbf{w}$ respectively. To be remarked that such a value of ε cannot be modified, and by consequence it cannot be optimized by a proper choice of the layer orientations, because it depends only on the isotropic parts of $\boldsymbol{\alpha}$ and $\mathbb{Q}$, properties of the basic layer. The value of ε can be given as function of the polar components of only $\mathbb{Q}$ and $\boldsymbol{\alpha}$: using Eq. (5.117) and indicating with t^α, r^α and φ^α the polar components of the tensor $\boldsymbol{\alpha}$ of thermal expansion coefficients of the layer, one gets, after some standard manipulations,

$$\varepsilon = t^\alpha + (-1)^\beta r^\alpha \frac{R_1^L}{T_1^L}, \tag{6.189}$$

which shows that this isotropic thermal expansion coefficient is nonetheless affected by the anisotropic coefficients r^α, of thermal expansion of the layer, and R_1^L. Exponent β is an integer that can take two values:

$$\beta = \begin{cases} 0, & \text{if } \varphi^\alpha = \Phi_1^L, \quad \text{i.e. if } \alpha_{11} > \alpha_{22}, \\ 1, & \text{if } \varphi^\alpha = \Phi_1^L + \frac{\pi}{2}, \quad \text{i.e. if } \alpha_{11} < \alpha_{22}. \end{cases} \tag{6.190}$$

Second type

The second set of warp-free and expansion-free laminates, is composed by the laminates that satisfy at the same time both conditions (6.168) and (6.183):

$$\begin{cases} \mathbb{B} \neq \mathbb{O}, \\ \mathbf{V} = \mathbb{B}\mathbb{A}^{-1}\mathbf{U}, \\ \mathbf{V} = \mathbb{B}\mathbb{D}^{-1}\mathbf{W}. \end{cases} \tag{6.191}$$

Solutions to Eq. (6.191) are necessarily material dependent. Of course, conditions $(6.191)_2$ and $(6.191)_3$ give the following one, that can be obtained also through Eqs. (5.125) and $(6.168)_2$:

$$\mathbb{A}^{-1}\mathbf{U} = \mathbb{D}^{-1}\mathbf{W}. \tag{6.192}$$

It can be checked easily that coupled quasi-homogeneous laminates automatically satisfy Eq. (6.192), it is sufficient to take a look at Eqs. (5.26) and (5.136) and remember that we are dealing with identical layers laminates, for which, cf. Sect. 5.4.5, $\mathbb{C} = \mathbb{O} \Rightarrow \mathbf{Y} = \mathbf{O}$. So, whenever quasi-homogeneous laminates are used, the second type of thermally stable laminates can be found simply satisfying conditions $(6.191)_1$ and $(6.191)_2$ or alternatively $(6.191)_1$ and $(6.191)_3$. In particular, if a quasi-trivial sequence is used,[23] cf. Sect. 5.3.3, the only sufficient condition to be satisfied to obtain thermally stable laminates of the second type is $(6.191)_2$, or alternatively $(6.191)_3$.

Finally, mixed cases, i.e. laminates satisfying conditions (6.157) and (6.183), or (6.168) and (6.182) cannot exist, because in both cases equation $\mathbf{V} = \mathbf{O}$ enter the set of conditions, forcing the solution to be one of the first type.

6.9.2.4 Completely Warp-Free or Expansion-Free Coupled Laminates

The case of coupled laminates is interesting, in principle, for another reason. In fact, let us consider again a laminate which is not acted upon by internal actions, i.e. $\mathbf{N} = \mathbf{M} = \mathbf{O}$. Then, from Eq. (5.123) we see that it is possible to obtain a laminate which is completely warp-free, *for any possible distribution of the temperature*, i.e. $\forall t$ *and* $\forall \nabla t$, when

$$\begin{cases} \mathbf{v}_1 = \mathbf{O}, \\ \mathbf{w} = \mathbf{O}. \end{cases} \tag{6.193}$$

Condition $(6.193)_2$ is never satisfied for uncoupled laminates, because in such a case $\mathbf{v}_1$ and $\mathscr{B}$ are null while $\mathbf{w}$ reduces to, cf. $(5.124)_4$,

$$\mathbf{w} = \mathscr{D}\mathbf{W}, \tag{6.194}$$

[23] A *coupled quasi-homogeneous laminate* is a laminate satisfying condition $\mathbb{C} = \mathbb{O}$ but not $\mathbb{B} = \mathbb{O}$. Coupled quasi-trivial sequences automatically ensure conditions $(6.191)_1$ and (6.192) for any possible choice of orientation angle of the layers.

which is never null. Unlike this case, when a laminate is coupled, $\mathbf{w}$ is null when, see again $(5.124)_4$,

$$\mathbf{V} = \frac{1}{3}\mathbb{A}\mathbb{B}^{-1}\mathbf{W}. \tag{6.195}$$

If this condition is added to (6.168), then the laminate will have

$$\mathbb{B} \neq \mathbb{O}, \ \mathbf{V} \neq \mathbf{O}, \ \mathscr{B} \neq \mathbb{O}, \ \mathbf{v}_1 = \mathbf{O}, \ \mathbf{w} = \mathbf{O}, \tag{6.196}$$

i.e. it will never warp for a thermal field.

In a similar way, it is also possible to find laminates that are expansion-free for any possible thermal field, when

$$\begin{cases} \mathbf{v}_2 = \mathbf{O}, \\ \mathbf{u} = \mathbf{O}. \end{cases} \tag{6.197}$$

As before, condition $(6.197)_2$ is never satisfied by an uncoupled laminate, for which, see Eq. $(5.124)_1$,

$$\mathbf{u} = \mathscr{A}\mathbf{U}, \tag{6.198}$$

which is never null. Nevertheless, if the laminate is coupled, $\mathbf{u}$ is null if, once more cf. $(5.124)_1$,

$$\mathbf{V} = \frac{1}{3}\mathbb{D}\mathbb{B}^{-1}\mathbf{U}, \tag{6.199}$$

condition that added to (6.183) gives

$$\mathbb{B} \neq \mathbb{O}, \ \mathbf{V} \neq \mathbf{O}, \ \mathscr{B} \neq \mathbb{O}, \ \mathbf{v}_2 = \mathbf{O}, \ \mathbf{u} = \mathbf{O}. \tag{6.200}$$

These conditions show that such a laminate will never have a deformation of the mid-plane produced by a thermal field. To remark that $\mathbb{A}^{-1} \neq \mathscr{A}$ and $\mathbb{D}^{-1} \neq \mathscr{D}$ because $\mathbb{B} \neq \mathbb{O}$.

Finally, if conditions (6.193) and (6.197) are imposed at the same time, then such a laminate will never freely deform or warp for the only action of a thermal field. These four conditions are not independent: using Eqs. $(6.168)_2$, $(6.183)_2$, (6.195) and (6.199) gives the following three independent conditions:

$$\begin{cases} \mathbf{V} = \mathbb{B}\mathbb{A}^{-1}\,\mathbf{U}, \\ \mathbf{W} = 3(\mathbb{B}\mathbb{A}^{-1})^2\,\mathbf{U}, \\ \mathbb{D} = 3\mathbb{B}\mathbb{A}^{-1}\,\mathbb{B}. \end{cases} \tag{6.201}$$

This result implies that it is not possible to have 3 tensors simultaneously null among $\mathbf{u}$, $\mathbf{v}_1$, $\mathbf{v}_2$ and $\mathbf{w}$: the only possible situations are one, two or all the four tensors null. Being conditions $(6.168)_2$ and $(6.183)_2$ dependent upon the material, solutions to Eq. (6.201) are material dependent too.

Using the same above conditions, it can be proved, by the polar method, that laminates with $\mathbb{C} = \mathbb{O}$ cannot satisfy at the same time conditions (6.193) and (6.197).

To end this section, these rather theoretical results show, on one side, how wide is the set of all the possible combinations for anisotropic laminates, and on the other side how much it is profitable to work with the invariants of the polar formalism in such a kind of problems.

6.9.3 *Tailoring the Piezo-Electric Properties*

A problem similar to that considered in Sect. 6.9.1 concerns piezoelectricity: how to find laminates that, under the action of piezoelectricity actuators, have null expansion coefficients in one direction or that, though being anisotropic, have an isotropic response?

This problem has been tackled in Vannucci and Pouget (2006) and rather surprisingly, unlike the thermal case, analytical solutions are possible for the SPL case.[24]

In the framework of linear piezoelectricity,[25] the mid-plane extension and curvature tensors are given by,

$$\begin{aligned} \boldsymbol{\varepsilon}^0 &= -(\mathscr{A}\mathbf{F} + \mathscr{B}\mathbf{G})\mathbf{V}_e, \\ \boldsymbol{\kappa} &= -(\mathscr{B}\mathbf{F} + \mathscr{D}\mathbf{G})\mathbf{V}_e, \end{aligned} \tag{6.202}$$

where $\mathbf{F}$ and $\mathbf{G}$ are $3 \times n$ matrices[26] whose columns are the *piezoelectric coupling coefficients*, while $\mathbf{V}_e$ is the n-dimensional column vector of the electric potential differences in the n plies of the laminate (including the piezoelectric actuators).

In the case of a SPL, $\mathbb{B} = \mathbb{O}$, so the above equations reduce to

$$\begin{aligned} \boldsymbol{\varepsilon}^0 &= -\mathscr{A}\mathbf{F}\mathbf{V}_e = -\mathbb{A}^{-1}\mathbf{F}\mathbf{V}_e, \\ \boldsymbol{\kappa} &= -\mathscr{D}\mathbf{G}\mathbf{V}_e = -\mathbb{D}^{-1}\mathbf{G}\mathbf{V}_e. \end{aligned} \tag{6.203}$$

Being $\mathbf{V}_e$ the external action applied to the SPL, we can act only on the two $3 \times n$ matrices

$$\mathfrak{A} = \mathscr{A}\mathbf{F}, \quad \mathfrak{D} = \mathscr{D}\mathbf{G}, \tag{6.204}$$

that define the in-plane deformation and curvature response of the SPL. We develop hereafter only the analysis for $\mathfrak{A}$, that for $\mathfrak{D}$ can be recovered in a similar way.

---

[24]SPL stands for *Standard Piezoelectric Laminate*, a particular piezoelectric laminate defined in *Example 2* of Sect. 6.6.5.

[25]The reader can refer to Tiersten (1969), Ikeda (1996), Rao and Sunar (1994), Saravanos and Heyliger (1999) or Fernandes and Pouget (2002) for a survey of the mechanical models used in piezoelectricity.

[26]For the sake of simplicity, with a slight abuse of notation, we indicate by bold-faced capital letters, normally reserved to second-rank tensors, these matrices.

Because in a SPL the piezoelectric actuators are isotropic, **F** depends upon a unique piezoelectric coupling coefficient, that we indicate by e_{33}, and which is different from zero only for the two outer layers, the piezoelectric actuators. The matrix **F** is hence of the type

$$\mathbf{F} = \begin{bmatrix} e_{33}\ 0 \cdots 0\ e_{33} \\ e_{33}\ 0 \cdots 0\ e_{33} \\ 0\ \ 0 \cdots 0\ \ 0 \end{bmatrix}, \tag{6.205}$$

which gives

$$\mathfrak{A} = e_{33} \left\{ \begin{matrix} \mathscr{A}_{1111} + \mathscr{A}_{1122} \\ \mathscr{A}_{1122} + \mathscr{A}_{2222} \\ \mathscr{A}_{1112} + \mathscr{A}_{1222} \end{matrix} \right\}, \tag{6.206}$$

and finally the in-plane deformation

$$\boldsymbol{\varepsilon}^0 = -e_{33}(V_{bot} + V_{top}) \left\{ \begin{matrix} \mathscr{A}_{1111} + \mathscr{A}_{1122} \\ \mathscr{A}_{1122} + \mathscr{A}_{2222} \\ \mathscr{A}_{1112} + \mathscr{A}_{1222} \end{matrix} \right\}, \tag{6.207}$$

where V_{bot} is the electric potential difference applied to the lower piezoelectric actuator and V_{top} that in the upper one (V is null for the non-piezoelectric layers of the laminate).

The last result shows, on one hand, that e_{33} plays just the role of a scalar multiplier for $\boldsymbol{\varepsilon}^0$: only the mechanical properties influence the variation of $\mathfrak{A}$. On the other hand, we can see that if $V_{bot} = -V_{top}$, then $\boldsymbol{\varepsilon}^0 = \mathbf{O}$ regardless of the compliance $\mathscr{A}$. This is a very well known case, which however bends the plate: $\boldsymbol{\kappa} \neq \mathbf{O}$. Nevertheless, we ponder if it is possible, at least in some directions, to have *always* such a result, i.e. $\forall\ V_{bot}$ and V_{top}. Equation (6.207) shows that this depends exclusively upon $\mathscr{A}$.

If now we introduce the polar formalism, Eq. (4.67), indicating by t_1^A, t_0^A, r_0^A, r_1^A, φ_0^A and φ_1^A the polar components of $\mathscr{A}$, Eq. (6.207) becomes

$$\mathfrak{A} = -4\, e_{33}(V_{bot} + V_{top}) \left\{ \begin{matrix} t_1^A + r_1^A \cos 2(\varphi_1^A - \theta) \\ t_1^A - r_1^A \cos 2(\varphi_1^A - \theta) \\ r_1^A \sin 2(\varphi_1^A - \theta) \end{matrix} \right\}. \tag{6.208}$$

This result is formally identical to Eq. (6.150), concerning the thermal problems considered in Sect. 6.9.1. Hence, the results can be rephrased *verbatim*:

- if $t_1^A = r_1^A$, then $\varepsilon_{11} = 0$ for $\theta = \varphi_1^A \pm \frac{\pi}{2}$;
- if $t_1^A < r_1^A$ the plane is divided into four opposite sectors, with different signs for ε_{11}: if

$$\theta^* = \frac{1}{2} \arccos\left(-\frac{t_1^A}{r_1^A}\right), \tag{6.209}$$

then

$$
\begin{aligned}
\varepsilon_{11} \geq 0 \ \text{for} \ & -\pi \leq \theta \leq \pi + \varphi_1^A + \theta^*, \ \varphi_1^A - \theta^* \leq \theta \leq \varphi_1^A + \theta^*, \\
& \pi + \varphi_1^A - \theta^* \leq \theta \leq \pi; \\
\varepsilon_{11} < 0 \ \text{for} \ & -\pi + \varphi_1^A + \theta^* \leq \theta \leq \varphi_1^A - \theta^*, \\
& \varphi_1^A + \theta^* \leq \theta \leq \pi + \varphi_1^A - \theta^*;
\end{aligned} \tag{6.210}
$$

each one of the two sectors with $\varepsilon_{11} \geq 0$ has amplitude $2\theta^*$, while that of the sectors with $\varepsilon_{11} < 0$ is $\pi - 2\theta^*$. In the limit case of $t_1^A = 0$, $\theta^* = \pi/4$ and each sector has an amplitude of $\pi/2$;
- $\boldsymbol{\varepsilon}^0$ is spherical, i.e. the in-plane deformation is the same for each direction, if and only if $r_1^A = 0$.

Let us now introduce the polar parameters of $\mathbb{A}$ for the SPL: applying Eq. (5.46) gives

$$
\begin{aligned}
T_0^A &= T_0^L + \eta T_0^P, \\
T_1^A &= T_1^L + \eta T_1^P, \\
R_0^A e^{4i\Phi_0^A} &= (-1)^{k^L} R_0^L(\xi_1 + i\xi_2), \\
R_1^A e^{4i\Phi_1^A} &= (-1)^{k^L} R_1^L(\xi_3 + i\xi_4),
\end{aligned} \tag{6.211}
$$

where the superscript L stands for *layer* and P for *piezoelectric*, i.e. the two actuators, and

$$
\eta = \frac{2\, h_P}{h}, \tag{6.212}
$$

see Fig. 6.19. To remark that, because the actuators are isotropic, they do not affect the anisotropic part of $\mathbb{A}$. The design variables, once the materials of the layers and of the actuators chosen, are n: η and $n-1$ orientations (one of the ply angles can always be fixed arbitrarily to fix the principal frame of the laminate and the other ply orientations are actually the differences with the one that has been fixed).

Using Eqs. (4.68) and (6.211) we can explicit the conditions for the different cases of $\boldsymbol{\varepsilon}^0$ listed above as functions of the polar components of $\mathbb{A}$. In particular:

- the condition $t_1^A \leq r_1^A$ corresponds to

$$
(T_0^A)^2 - (R_0^A)^2 \leq 2R_1^A\sqrt{(T_0^A)^2 + (R_0^A)^2 - 2T_0^A R_0^A \cos 4(\Phi_0^A - \Phi_1^A)}, \tag{6.213}
$$

which for $\mathbb{A}$ orthotropic, i.e. with $\cos 4(\Phi_0^A - \Phi_1^A) = (-1)^{K^A}$, becomes

$$
R_1^A \geq \frac{1}{2}(T_0^A + (-1)^{K^A} R_0^A); \tag{6.214}
$$

- the limit case $t_1^A = 0$, giving a null spherical part of $\boldsymbol{\varepsilon}^0$, i.e. an isochoric deformation, becomes

$$T_0^A = R_0^A; \tag{6.215}$$

- the case of $r_1^A = 0$, giving a spherical deformation, corresponds to

$$R_1^A \sqrt{(T_0^A)^2 + (R_0^A)^2 - 2T_0^A R_0^A \cos 4(\Phi_0^A - \Phi_1^A)} = 0, \tag{6.216}$$

which for $\mathbb{A}$ orthotropic becomes

$$R_1^A (T_0^A - (-1)^{K^A} R_0^A) = 0; \tag{6.217}$$

a sufficient condition for the above equations be satisfied is hence $R_1^A = 0$; for that, it is sufficient to use square-symmetric layers, i.e. layers with $R_1^L = 0$, cf. Sect. 5.2.11.

It is easy to check that Eqs. (6.213) to (6.217) fulfill the bounds (4.149) for orthotropic laminates. However, Eq. (6.215) corresponds to a limit case, practically impossible to be obtained, so we discard it.

Using Eq. (6.211), conditions (6.213) and (6.217) can be rewritten as

$$\begin{aligned} &R_1^L \sqrt{\xi_3^2 + \xi_4^2} \geq \frac{1}{2}\left[T_0^L + \eta T_0^P + (-1)^{K^A} R_0^L \sqrt{\xi_1^2 + \xi_2^2}\right], \\ &R_1^L \sqrt{\xi_3^2 + \xi_4^2}\left[T_0^L + \eta T_0^P - (-1)^{K^A} R_0^L \sqrt{\xi_1^2 + \xi_2^2}\right] = 0. \end{aligned} \tag{6.218}$$

Let us consider the case of a cross-ply laminate; then, Eq. (6.218) becomes

$$\begin{aligned} &R_1^L |2\mu - 1| \geq \frac{1}{2}\left[T_0^L + \eta T_0^P + R_0^L\right], \\ &R_1^L |2\mu - 1| \left[T_0^L + \eta T_0^P - R_0^L\right] = 0 \end{aligned} \tag{6.219}$$

with

$$\mu = \frac{n_0}{n} \tag{6.220}$$

the volume fraction of layers at 0°. Relations (6.219) are linear in η and μ; in particular, Eq. $(6.219)_1$ can be given in the explicit form:

$$\eta \leq \frac{2R_1^L |2\mu - 1| - T_0^L - R_0^L}{T_0^P}, \tag{6.221}$$

while the solutions to Eq. $(6.219)_2$ are

$$R_1^L = 0, \quad \mu = \frac{1}{2}, \quad \eta = \frac{R_0^L - T_0^L}{T_0^P}. \tag{6.222}$$

The last solution above is not acceptable, because of Eq. (4.149)$_1$ and the fact that η is a strictly positive quantity. Finally, for a cross-ply laminate, the deformation can be purely spherical only if $R_1^A = 0$, i.e. if $\mathbb{A}$ is square symmetric.

Let us now consider an angle-ply laminate; in that case, Eq. (6.218) becomes

$$\begin{aligned} &R_1^L|\cos 2\delta| \geq \frac{1}{2}\left[T_0^L + \eta T_0^P + (-1)^{K^A} R_0^L|\cos 4\delta|\right],\\ &R_1^L|\cos 2\delta|\left[T_0^L + \eta T_0^P - (-1)^{K^A} R_0^L|\cos 4\delta|\right] = 0, \end{aligned} \tag{6.223}$$

with $K^A = 0$ if $\delta < \pi/8$ while $K^A = 1$ if $\pi/8 < \delta < \pi/4$, $\pm\delta$ being the two possible orientations. Once more, Eq. (6.223)$_1$ can be put in the explicit form

$$\eta \leq \frac{2R_1^L|\cos 2\delta| - T_0^L - (-1)^{K^A} R_0^L|\cos 4\delta|}{T_0^P}, \tag{6.224}$$

while the solutions of Eq. (6.223)$_2$ are

$$R_1^L = 0, \quad \delta = \frac{\pi}{4}, \quad \eta = \frac{(-1)^{K^A} R_0^L|\cos 4\delta| - T_0^L}{T_0^P}, \tag{6.225}$$

and the third one is again non acceptable, while the first two ones correspond, of course, to the cross-ply case.

The case of a laminate having null deformation in some directions, i.e. the solution to Eq. (6.221) for the cross-ply laminates, or (6.224) for the angle-ply ones, can be solved only numerically. So, as an example, let us consider the case of a laminate composed of carbon-epoxy T300/5208 plies, cf. Table 4.9, with two PZT-4 actuators, whose polar parameters are $T_0^P = 30.6$ GPa and $T_1^P = 30.2$ GPa. In such a case, Eqs. (6.221) and (6.224) become

$$\begin{aligned} &\eta \leq 1.372|2\mu - 1| - 1.522,\\ &\eta \leq 1.4|\cos 2\delta| - (-1)^{K^A} 0.644|\cos 4\delta| - 0.878. \end{aligned} \tag{6.226}$$

It is evident that Eq. (6.226)$_1$ has no solutions, because η should be negative, while for the angle ply case the function $\eta(\delta)$ is plotted in Fig. 6.57. Solutions are possible in the range $\delta_1 = 0.247 \simeq 14.14°$ and $\delta_2 = 0.681 \simeq 39.05°$. For $\delta = \delta_{max} = 0.498 \simeq 28.54°$, $\eta = \eta_{max} = 0.146$. If $\delta = \delta_{max}$, the piezoelectric actuators have the highest thickness, equal to the 14.6% of the thickness h of the carbon-epoxy layers; in that case, the piezoelectric effect is the highest one.

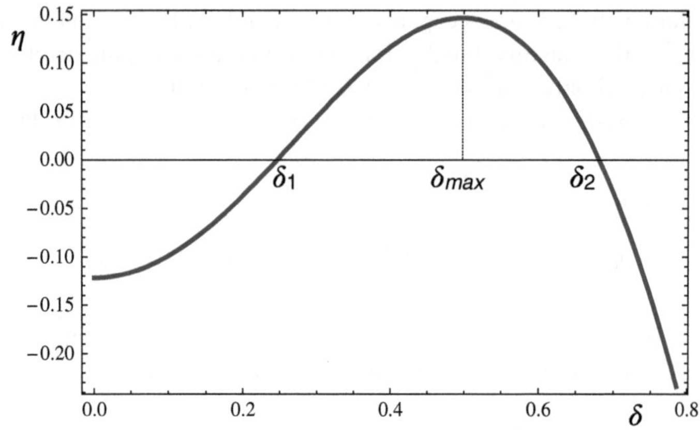

Fig. 6.57 Function $\eta(\delta)$ for a carbon-epoxy T300/5208 laminate with PZT-4 actuators

References

M.M. Abdalla, S. Setoodeh, Z. Gürdal, Design of variable stiffness composite panels for maximum fundamental frequency using lamination parameters. Compos. Struct. **81**, 283–291 (2007)

S. Abrate, Optimal design of laminated plates and shells. Compos. Struct. **29**, 269–286 (1994)

M.R. Ahmadian, A. Vincenti, P. Vannucci, A general strategy for the optimal design of composite laminates by the polar-genetic method. Mater. Des. **32**, 2317–2327 (2011)

G. Allaire, R. Kohn, Optimal design for minimum weight and compliance in plane stress using extremal micro structures. Eur. J. Mech. A/Solids **12**, 839–878 (1993)

G. Allaire, R. Kohn, Shape optimization by the homogenization method. Numer. Math. **76**, 27–68 (1997)

C.A.C. Antonio, A multilevel genetic algorithm for optimization of geometrically nonlinear stiffened composite structures. Struct. Multidiscip. Optim. **24**, 372–386 (2002)

C.A.C. Antonio, Optimal dynamic response of composite structures using a hierarchical genetic algorithm, in *Proceedings of 7th International Conference on Computational Structures Technology*, ed. by B.H.V. Topping, C.A. Mota Soares (Civil-Comp Press, Lisbon, Portugal, 2004)

M. Autio, Determining the real lay-up of a laminate corresponding to optimal lamination parameters by genetic search. Struct. Multidiscip. Optim. **20**, 301–310 (2000)

M. Autio, Optimization of coupled thermal-structural problems of laminated plates with lamination parameters. Struct. Multidiscip. Optim. **21**, 40–51 (2001)

F. Aymerich, M. Serra, Optimization of laminate stacking sequence for maximum buckling load using the ant colony optimization (aco) metaheuristic. Compos. A **39**, 262–272 (2008)

N.V. Banichuk, Optimization problems for elastic anisotropic bodies. Arch. Mech. **33**, 347–363 (1981)

N.V. Banichuk, *Problems and Methods of Optimal Structural Design* (Plenum Press, New York, 1983)

N.V. Banichuk, Optimization of anisotropic properties for continuum bodies and structural elements using spectral methods of tensor analysis. Mech. Struct. Mach. **24**, 71–87 (1996)

H.G. Bargh, M.H. Sadr, Stacking sequence optimization of composite plates for maximum fundamental frequency using particle swarm optimization algorithm. Meccanica **47**, 719–730 (2012)

A.W. Blom, M.M. Abdalla, Z. Gürdal, Optimization of course locations in fiber-placed panels for general fiber angle distributions. Compos. Sci. Technol. **70**, 564–570 (2010)

J.A. Blume, Elastic materials with coincident principal stress and strain axes. J. Elast. **35**, 275–280 (1994)

C.W. Carroll, The created response surface technique for optimizing nonlinear restrained systems. Oper. Res. Int. J. **9**, 169–184 (1961)

A. Catapano, *Stiffness and Strength Optimisation of the Anisotropy distribution for Laminated Structures*. Ph.D. thesis, University Pierre et Marie Curie, Paris, France (2013), https://tel.archives-ouvertes.fr/tel-00952372

A. Catapano, B. Desmorat, P. Vannucci, Invariant formulation of phenomenological failure criteria for orthotropic sheets and optimisation of their strength. Math. Methods Appl. Sci. **35**, 1842–1858 (2012)

A. Catapano, B. Desmorat, P. Vannucci, Stiffness and strength optimization of the anisotropy distribution for laminated structures. J. Optim. Theory Appl. **167**, 118–146 (2015)

N. Chang, W. Wang, W. Yang, J. Wang, Ply stacking sequence optimization of composite laminate by permutation discrete particle swarm optimization. Struct. Multidiscip. Optim. **41**, 179–187 (2010)

H.P. Chen, Study of hygrothermal isotropic layup and hygrothermal curvature-stable coupling composite laminates, in *Proceedings of 44th AIAA/ASME/ASCE/AHS/ASC Structures, Structural Dynamics and Materials Conference*, volume Paper No. AIAA-2003-1506 (2003)

C.A. Coello-Coello, Theoretical and numerical constraint-handling techniques used with evolutionary algorithms: a survey of the state of the art. Comput. Methods Appl. Mech. Eng. **191**, 1245–1287 (2002)

A. Colorni, M. Dorigo, F. Maffioli, V. Maniezzo, G. Righini, M. Trubian, Heuristics from nature for hard combinatorial optimization problems. Int. Trans. Oper. Res. **3**, 1–21 (1996)

R. Courant, Variational methods for the solution of problems of equilibrium and vibrations. Bull. Am. Math. Soc. **49**, 1–23 (1943)

S.C. Cowin, Optimization of the strain energy density in linear anisotropic elasticity. J. Elast. **34**, 45–68 (1994)

S.C. Cowin, Remarks on coaxiality of strain and stress in anisotropic elasticity. J. Elast. **47**, 83–84 (1997)

S.C. Cowin, G. Yang, Material symmetry optimization by Kelvin modes. J. Eng. Math. **37**, 27–43 (2000)

R.J. Cross, R.A. Haynes, E.A. Armanios, Families of hygrothermally stable asymmetric laminated composites. J. Compos. Mater. **42**, 697–716 (2008)

D. Dasgupta, Z. Michalewicz, *Evolutionary Algorithms in Engineering Applications* (Springer, Berlin, 1997)

D.M. De Leon, C.E. de Souza, J.S.O. Fonseca, R.G.A. Da Silva, Aeroelastic tailoring using fiber orientation and topology optimization. Struct. Multidiscip. Optim. **46**, 663–677 (2012)

C.G. Diaconu, M. Sato, H. Sekine, Layup optimization of symmetrically laminated thick plates for fundamental frequencies using lamination parameters. Struct. Multidiscip. Optim. **24**, 302–311 (2002)

M. Dorigo, *Optimization, Learning and Natural Algorithms*. Ph.D. thesis, Politecnico di Milano, Milan, Italy (1992)

R.C. Eberhart, J. Kennedy, Particle swarm optimization, in *Proceedings of IEEE International Conference on Neural Networks* (Piscataway, NJ, 1995), pp. 1942–1948

O. Erdal, F.O. Sonmez, Optimum design of composite laminates for maximum buckling load capacity using simulated annealing. Compos. Struct. **71**, 45–52 (2005)

A.H. Ertas, Optimization of fiber reinforced laminates for maximum fatigue life using particle swarm optimization. Procedia Eng. **38**, 473–478 (2012)

A. Fernandes, J. Pouget, An accurate modelling of piezoelectric multilayer plates. Eur. J. Mech. A/Solids **21**, 629–651 (2002)

A.V. Fiacco, G.P. McCormick, Extensions of SUMT for nonlinear programming: equality constraints and extrapolation. Manage. Sci. **12**, 816–828 (1968)

J.P. Foldager, J.S. Hansen, N. Olhoff, A general approach forcing convexity of ply angle optimization in composite laminates. Struct. Optim. **16**, 201–211 (1998)

H. Fukunaga, H. Sekine, Stiffness design method of symmetric laminates using lamination parameters. AIAA J. **30**, 2791–2793 (1992)

H. Fukunaga, G.N. Vanderplaats, Stiffness optimization of orthotropic laminated composites using lamination parameters. AIAA J. **29**, 641–646 (1991)

H. Ghiasi, D. Pasini, L. Lessard, Optimum stacking sequence design of composite materials part 1: constant stiffness design. Compos. Struct. **90**, 1–11 (2009)

H. Ghiasi, K. Fayazbakhsh, D. Pasini, L. Lessard, Optimum stacking sequence design of composite materials part 2: variable stiffness design. Compos. Struct. **93**, 1–13 (2010)

D.E. Goldberg, *Genetic Algorithms* (Addison-Wesley, Reading, 1991)

M. Grédiac, A procedure for designing laminated plates with required stiffness properties. application to thin quasi-isotropic quasi-homogeneous uncoupled laminates. J. Compos. Mater. **33**, 1939–1956 (1999)

Z. Gürdal, R.T. Haftka, P. Hajela, *Design and Optimization of Laminated Composite Materials* (Wiley, New York, 1999)

Z. Gürdal, B.F. Tatting, K.C. Wu, Variable stiffness panels: effects of stiffness variation on the in-plane and buckling responses. Compos. A **39**, 11–22 (2008)

H.T. Hahn, Warping of unsymmetric cross-ply graphite/epoxy laminates. Compos. Technol. Rev. **3**, 114–117 (1981)

V.B. Hammer, M.P. Bendsoe, R. Lipton, P. Pedersen, Parametrization in laminate design for optimal compliance. Int. J. Solids Struct. **34**, 415–434 (1997)

R.A. Haynes, E.A. Armanios, New families of hygrothermally stable composite laminates with optimal extension-twist coupling. AIAA J. **48**, 2954–2961 (2010)

J.H. Holland, *Adaptation of Natural and Artificial Systems* (University of Michigan Press, Ann Arbor, 1975)

M.W. Hyer, Some observations on the cured shape of unsymmetric laminates. J. Compos. Mater. **15**, 175–194 (1981a)

M.W. Hyer, Calculations of room-temperature shapes of unsymmetric laminates. J. Compos. Mater. **15**, 296–310 (1981b)

M.W. Hyer, H.H. Lee, The use of curvilinear fiber format to improve buckling resistance of composite plates with central circular holes. Compos. Struct. **18**, 239–261 (1991)

T. Ikeda, *Fundamentals of Piezoelectricity* (Oxford University Press, Oxford, 1996)

T. Ishikawa, H. Fukunaga, Taylored laminates with null or arbitrary coefficient of thermal expansion, in *Proceedings of Composites86: Recent Advances in Japan and the USA*, ed. by K. Kawata, S. Umekawa, A. Kobayashi. Japan Society for Composite Materials (Tokyo, Japan, 1986), pp. 701–708

F. Javidrad, R. Nouri, A simulated annealing method for design of laminates with required stiffness properties. Compos. Struct. **93**, 1127–1135 (2011)

A. Jibawy, C. Julien, B. Desmorat, A. Vincenti, F. Léné, Hierarchical structural optimization of laminated plates using polar representation. Int. J. Solids Struct. **48**, 2576–2584 (2011)

R.M. Jones, *Mechanics of Composite Materials*, 2nd edn. (Taylor & Francis, Philadelphia, 1999)

N. Kandil, G. Verchery, New methods of design for stacking sequences of laminates, in *Proceedings of CADCOMP88 - Computer Aided Design in Composite Materials 88* (Southampton, UK, 1988), pp. 243–257

V.M. Kartvelishvili, V.V. Kobelev, Rational schemes for reinforcing laminar plates from composite materials. Prikl. Matem. Mekhan. (English translation in J. Appl. Math. Mech. **48**, 40–49) **48**, 68–80 (1984)

W. Karush, Minima of functions of several variables with inequalities as side constraints. Master's thesis, Department of Mathematics, University of Chicago, Chicago, IL (1939)

J.-S. Kim, C.-G. Kim, C.-S. Hong, Optimum design of composite structures with ply drop using genetic algorithm and expert system shell. Compos. Struct. **46**, 171–187 (1999)

S. Kirkpatrick, C.D. Gelatt, M.P. Vecchi, Optimization by simulated annealing. Science **220**, 671–680 (1983)

N. Kogiso, L.T. Watson, Z. Gurdal, R.T. Haftka, Genetic algorithms with local improvement for composite laminate design. Struct. Optim. **7**, 207–218 (1994)

H.W. Kuhn, A.W. Tucker, Nonlinear programming, in *Proceedings of 2nd Berkeley Symposium* (University of California Press, Berkeley, CA, 1951), pp. 481–492

R. Le Riche, J. Gaudin, Design of dimensionally stable composites by evolutionary optimization. Compos. Struct. **41**, 97–111 (1998)

R. Le Riche, R.T. Haftka, Optimization of laminate stacking sequence for buckling load maximization by genetic algorithm. AIAA J. **31**, 951–956 (1993)

R. Le Riche, R.T. Haftka, Improved genetic algorithm for minimum thickness composite laminate design. Compos. Part B - Eng. **5**, 143–161 (1995)

B. Liu, R.T. Haftka, Single-level composite wing optimization based on flexural lamination parameters. Struct. Multidiscip. Optim. **26**, 111–120 (2004)

B. Liu, R.T. Haftka, M.A. Akgun, A. Todoroki, Permutation genetic algorithm for stacking sequence design of composite laminates. Comput. Methods Appl. Mech. Eng. **186**, 357–372 (2000)

B. Liu, R.T. Haftka, P. Trompette, Maximization of buckling loads of composite panels using flexural lamination parameters. Struct. Multidiscip. Optim. **26**, 28–36 (2004)

J. Majak, Orientational design of nonorthotropic two-dimensional materials. Mech. Compos. Mater. **37**, 27–38 (2001)

J. Majak, S. Hannus, Orientational design of anisotropic materials using the hill and tsai-wu strength criteria. Mech. Compos. Mater. **39**, 509–520 (2003)

J. Majak, M. Pohlak, Decomposition method for solving optimal material orientation problems. Compos. Struct. **92**, 1839–1845 (2010a)

J. Majak, M. Pohlak, Optimal material orientation of linear and non-linear elastic 3d anisotropic materials. Meccanica **45**, 671–680 (2010b)

Z. Michalewicz, *Genetic Algorithms + Data Structures = Evolution Programs*, Artificial Intelligence (Springer, Berlin, 1992)

Z. Michalewicz, M. Schoenauer, Evolutionary algorithms for constrained parameter optimization problems. Evol. Comput. **4**, 1–32 (1996)

M. Miki, Material design of composite laminates with required in-plane elastic properties, in *Proceedings of ICCM 4 - Fourth International Conference on Composite Materials* (Tokio, Japan, 1982), pp. 1725–1731

M. Miki, A graphical method for designing fibrous laminated composites with required in-plane stiffness. Trans. Jpn. Soc. Compos. Mater. **9**, 51–55 (1983)

M. Miki, Design of laminated fibrous composite plates with required flexural stiffness, in *Recent advances in composites in the USA and Japan - ASTM STP 864*, ed. by J.R. Vinson, M. Taya (American Society for Testing and Materials, Philadelphia, PA, 1985), pp. 387–400

M. Miki, Y. Sugiyama, Optimum design of laminated composite plates using lamination parameters, in *Proceedings of AIAA 32nd Structures, Structural Dynamics and Materials Conference* (Baltimore, MD, 1991), pp. 275–283

M. Montemurro, *Optimal Design of Advanced Engineering Modular Systems through a New Genetic Approach*. Ph.D. thesis, University Pierre et Marie Curie, Paris, France (2012), https://tel.archives-ouvertes.fr/tel-00955533

M. Montemurro, A. Catapano, A new paradigm for the optimum design of variable angle tow laminates, in *Variational Analysis and Aerospace Engineering - Mathematical Challenges for the Aerospace of the Future*, vol. 116, Springer Optimization and Its Applications (SOIA), ed. by A. Frediani, B. Mohammadi, O. Pironneau, V. Cipolla (Springer, Berlin, 2016), pp. 375–400

M. Montemurro, A. Vincenti, P. Vannucci, A. Makradi, Optimisation en poids de structures composites stratifiées, in *Proceedings of JNC17 - Journées Nationales sur les Composites* (Poitiers, France, 2011)

M. Montemurro, Y. Koutsawa, S. Belouettar, A. Vincenti, P. Vannucci, Design of damping properties of hybrid laminates through a global optimization strategy. Compos. Struct. **94**, 3309–3320 (2012a)

M. Montemurro, P. Vannucci, A. Vincenti, BIANCA, A Genetic Algorithm for Engineering Optimisation. Technical report, University Pierre et Marie Curie (2012b), https://hal.archives-ouvertes.fr/hal-00767468

M. Montemurro, A. Vincenti, P. Vannucci, A two-level procedure for the global optimum design of composite modular structures - application to the design of an aircraft wing. Part 1: theoretical formulation. J. Optim. Theory Appl. **155**, 1–23 (2012c)

M. Montemurro, A. Vincenti, P. Vannucci, A two-level procedure for the global optimum design of composite modular structures - application to the design of an aircraft wing. Part 2. J. Optim. Theory Appl. **155**, 24–53 (2012d)

M. Montemurro, A. Vincenti, P. Vannucci, Design of elastic properties of laminates with minimum number of plies. Mech. Compos. Mater. **48**, 369–390 (2012e)

M. Montemurro, A. Vincenti, P. Vannucci, The automatic dynamic penalisation method (ADP) for handling constraints with genetic algorithms. Comput. Methods Appl. Mech. Eng. **256**, 70–87 (2013)

M. Montemurro, A. Vincenti, Y. Koutsawa, P. Vannucci, A two-level procedure for the global optimisation of the damping behaviour of composite laminated plates with elastomer patches. J. Vib. Control **21**, 1778–1800 (2015)

A. Muc, W. Gurba, Genetic algorithms and finite element analysis in optimization of composite structures. Compos. Struct. **54**, 275–281 (2001)

S. Nagendra, D. Jestin, Z. Gurdal, R.T. Haftka, L.T. Watson, Improved genetic algorithm for the design of stiffened composite panels. Comput. Struct. **58**, 543–555 (1996)

S.N. Omkar, D. Mudigere, G. Narayana Naik, S. Gopalakrishnan, Vector evaluated particle swarm optimization (VEPSO) for multi objective design optimization of composite structures. Comput. Struct. **86**, 1–14 (2008)

S.N. Omkar, R. Khandelwal, T.V.S. Ananth, G. Narayana Naik, S. Gopalakrishnan, Quantum behaved Particle Swarm Optimization (QPSO) for multi-objective design optimization of composite structures. Expert Syst. Appl. **36**, 11312–11322 (2009)

L. Parnas, S. Oral, U. Ceyhan, Optimum design of composite structures with curved fiber courses. Compos. Sci. Technol. **63**, 1071–1082 (2003)

P. Pedersen, On optimal orientation of orthotropic materials. Struct. Optim. **1**, 101–106 (1989)

P. Pedersen, Bounds on elastic energy in solids of orthotropic materials. Struct. Optim. **2**, 55–62 (1990)

P. Pedersen, A note on design of fiber-nets for maximum stiffness. J. Elast. **73**, 127–145 (2003)

S.S. Rao, *Engineering Optimization - Theory and Practice* (Wiley, Hoboken, 2009)

S.S. Rao, M. Sunar, Piezoelectricity and its use in disturbance sensing and control of flexible structures: a survey. Appl. Mech. Rev. **47**, 113–123 (1994)

D.P. Raymer, *Aircraft Design: A Conceptual Approach* (AIAA Education Series, Reston, 2006)

J. Rechenberg, Cybernetic solution path of an experimental problem. Technical Report Liberty Translation 1122, Royal Aircraft Establishment, Farnborough, UK (1965)

M. Rovati, A. Taliercio, Optimal orientation of the symmetry axes of orthotropic materials, in *Engineering Optimization in Design Processes*, ed. by N. Olhoff H.A. Eschenauer, C. Mattheck, (Springer, Heidelberg, 1991), pp. 127–154

M. Rovati, A. Taliercio, Stationarity of the strain energy density for some classes of anisotropic solids. Int. J. Solids Struct. **40**, 6043–6075 (2003)

G. Royer-Carfagni, Optimal fiber-mesh layout for a composite anisotropic elastic wedge. J. Elast. **60**, 103–117 (2000)

G. Sacchi-Landriani, M. Rovati, Optimal design of two-dimensional structures made of composite materials. ASME J. Eng. Mater. Technol. **113**, 88–92 (1991)

G. Sacchi-Landriani, M. Rovati, A. Taliercio, Optimal orientation of anisotropic properties in continuum bodies, in *Proceedings of the workshop Optimal structural design: analytical approach and computer aided design*, ed. by C. Cinquini, (GIMC - Italian Group for Computational Mechanics, Pavia - Italy, 1993)

D. Sadagopan, R. Pitchumani, Application of genetic algorithms to optimal tailoring of composite materials. Compos. Sci. Technol. **58**, 571–589 (1998)

D.A. Saravanos, P.R. Heyliger, Mechanics and computational models for laminated piezoelectric beams, plates and shells. Appl. Mech. Rev. **52**, 305–320 (1999)

T.A. Sebaey, C.S. Lopes, N. Blanco, J. Costa, Ant colony optimization for dispersed laminated composite panels under biaxial loading. Compos. Struct. **94**, 31–36 (2011)

T.A. Sebaey, C.S. Lopes, N. Blanco, J.A. Mayugo, J. Costa, Two-pheromone ant colony optimization to design dispersed laminates for aeronautical structural applications. Adv. Eng. Softw. **66**, 10–18 (2013)

G.A. Seregin, V.A. Troitskii, On the best position of elastic symmetry planes in an orthotropic body. Prikl. Matem. Mekhan. (English translation in J. Appl. Math. Mech. **45**, 139–142) **45**, 185–189 (1981)

S. Setoodeh, M.M. Abdalla, Z. Gürdal, Design of variable stiffness laminates using lamination parameters. Compos. B **37**, 301–309 (2006a)

S. Setoodeh, M.M. Abdalla, S.T. IJsselmuiden, Z. Gürdal, Design of variable stiffness composite panels for maximum buckling load. Compos. Struct. **87**, 109–117 (2006b)

C. Sgarra, M. Vianello, Directions of coaxiality between pure strain and stress in linear elasticity. J. Elast. **46**, 263–265 (1997a)

C. Sgarra, M. Vianello, Rotations which make strain and stress coaxial. J. Elast. **47**, 217–224 (1997b)

G. Soremekun, Z. Gurdal, R.T. Haftka, L.T. Watson, Composite laminate design optimization by genetic algorithm with generalized elitist selection. Comput. Struct. **79**, 131–143 (2001a)

G. Soremekun, Z. Gurdal, R.T. Haftka, L.T. Watson, Composite laminate design optimization by genetic algorithm with generalized elitist selection. Comput. Struct. **79**, 131–143 (2001b)

G. Soremekun, Z. Gürdal, C. Kassapoglou, D. Toni, Stacking sequence blending of multiple composite laminates using genetic algorithms. Compos. Struct. **56**, 53–62 (2002)

P.Y. Tabakov, Multi-dimensional design optimisation of laminated structures using an improved genetic algorithm. Compos. Struct. **54**, 349–354 (2001)

H.F. Tiersten, *Linear Piezoelectric Plate Vibrations* (Plenum Press, New York, 1969)

A. Todoroki, R.T. Haftka, Stacking sequence optimization by a genetic algorithm with a new recessive gene like repair strategy. Compos. B **29B**, 277–285 (1998)

I.C. Trelea, The particle swarm optimization algorithm: convergence analysis and parameter selection. Inf. Process. Lett. **85**, 317–325 (2003)

J. Van Campen, C. Kassapoglou, Z. Gürdal, Generating realistic laminate fiber angle distributions for optimal variable stiffness laminates. Compos. B **43**, 354–360 (2011)

P. Vannucci, *Thèse d'Habilitation à Diriger les Recherches*. Ph.D. thesis, ISAT - University of Burgundy, Nevers, France (2002), http://tel.archives-ouvertes.fr/docs/00/62/59/58/PDF/HDR.pdf

P. Vannucci, Designing the elastic properties of laminates as an optimisation problem: a unified approach based on polar tensor invariants. Struct. Multidiscip. Optim. **31**, 378–387 (2006)

P. Vannucci, ALE-PSO: an adaptive swarm algorithm to solve design problems of laminates. Algorithms **2**, 710–734 (2009a)

P. Vannucci, Influence of invariant material parameters on the flexural optimal design of thin anisotropic laminates. Int. J. Mech. Sci. **51**, 192–203 (2009b)

P. Vannucci, On special orthotropy of paper. J. Elast. **99**, 75–83 (2010)

P. Vannucci, A new general approach for optimizing the performances of smart laminates. Mech. Adv. Mater. Struct. **18**, 558–568 (2011)

P. Vannucci, Strange laminates. Math. Methods Appl. Sci. **35**, 1532–1546 (2012)

P. Vannucci, General theory of coupled thermally stable anisotropic laminates. J. Elast. **113**, 147–166 (2013a)

P. Vannucci, The design of laminates as a global optimization problem. J. Optim. Theory Appl. **157**, 299–323 (2013b)

P. Vannucci, J. Pouget, Laminates with given piezoelectric expansion coefficients. Mech. Adv. Mater. Struct. **13**, 419–427 (2006)

P. Vannucci, A. Vincenti, The design of laminates with given thermal/hygral expansion coefficients: a general approach based upon the polar-genetic method. Compos. Struct. **79**, 454–466 (2007)

P. Vannucci, B. Desmorat, A. Vincenti, Design problems of anisotropic structures: some recent results, in *Variational Analysis and Aerospace Engineering: Mathematical Challenges for Aerospace Design*, vol. 66, Springer Optimization and Its Applications (SOIA), ed. by G. Buttazzo, A. Frediani, chap. 17 (Springer, Boston, MA, 2012), pp. 395–426

M. Vianello, Coaxiality of strain and stress in anisotropic linear elasticity. J. Elast. **42**, 283–289 (1996a)

M. Vianello, Optimization of the stored energy and coaxiality of strain and stress in finite elasticity. J. Elast. **44**, 193–202 (1996b)

A. Vincenti, *Conception et optimisation des composites stratifiés par méthode polaire et algorithme génétique*. Ph.D. thesis, ISAT - University of Burgundy, Nevers, France (2002)

A. Vincenti, B. Desmorat, Optimal orthotropy for minimum elastic energy by the polar method. J. Elast. **102**, 55–78 (2011)

A. Vincenti, P. Vannucci, G. Verchery, Design of composite laminates as an optimization problem: a new genetic algorithm approach based upon tensor polar invariants, in *Proceedings of WCSMO 5 - Fifth World Congress on Structural and Multidisciplinary Optimization* (Lido di Jesolo, Italy, 2003)

A. Vincenti, M.R. Ahmadian, P. Vannucci, BIANCA: a genetic algorithm to solve hard combinatorial optimisation problems in engineering. J. Global Optim. **48**, 399–421 (2010)

A. Vincenti, P. Vannucci, M.R. Ahmadian, Optimization of laminated composites by using genetic algorithm and the polar description of plane anisotropy. Mech. Adv. Mater. Struct. **20**, 242–255 (2013)

M. Walker, R.E. Smith, A technique for the multiobjective optimisation of laminated composite structures using genetic algorithms and finite element analysis. Compos. Struct. **62**, 123–128 (2003)

W. Wang, S. Guo, N. Chang, F. Zhao, W. Yang, A modified ant colony algorithm for the stacking sequence optimisation of a rectangular laminate. Struct. Multidiscip. Optim. **41**, 711–720 (2010)

S.J. Winckler, Hygrothermally curvature stable laminates with tension-torsion coupling. J. Am. Helicopter Soc. **31**, 56–58 (1985)

S.J. Winckler, *A theory of warp-free unsymmetric composite plates and beams with extension-bending coupling*. Ph.D. thesis, Rensselaer Polytechnique Institute, Troy, NY (1986)

MIX
Papier aus verantwortungsvollen Quellen
Paper from responsible sources
FSC® C105338

If you have any concerns about our products,
you can contact us on
ProductSafety@springernature.com

In case Publisher is established outside the EU,
the EU authorized representative is:
Springer Nature Customer Service Center GmbH
Europaplatz 3, 69115 Heidelberg, Germany

Printed by Libri Plureos GmbH
in Hamburg, Germany